High-Temperature Polymeric Materials

High-Temperature Polymeric Materials

Preparation, Properties, Modeling and Engineering Applications

Edited by

Emmanuel Rotimi Sadiku

Department of Chemical, Metallurgical and Materials Engineering, Tshwane University of Technology, Pretoria, Gauteng, South Africa; Institute of NanoEngineering Research (INER), Tshwane University of Technology, Pretoria, Gauteng, South Africa

Idowu David Ibrahim

Department of Industrial Engineering, Operations Management and Mechanical Engineering, Vaal University of Technology, Vanderbijlpark, South Africa

Williams Kehinde Kupolati

Department of Civil Engineering, Tshwane University of Technology, Pretoria, Gauteng, South Africa

Stephen C. Agwuncha

Department of Industrial Chemistry, Federal University Lokoja, Lokoja, Kogi State, Nigeria

ELSEVIER

Elsevier
Radarweg 29, PO Box 211, 1000 AE Amsterdam, Netherlands
125 London Wall, London EC2Y 5AS, United Kingdom
50 Hampshire Street, 5th Floor, Cambridge, MA 02139, United States

ISBN: 978-0-443-32792-6

For Information on all Elsevier publications
visit our website at https://www.elsevier.com/books-and-journals

Publisher: Matthew Deans
Acquisitions Editor: Jinia Dasgupta
Editorial Project Manager: Ashi Jain
Production Project Manager: Bharatwaj Varatharajan
Cover Designer: Mark Rogers

Typeset by MPS Limited, Chennai, India

Contents

List of contributors

Chibueze Godwin Achi Department of Civil Engineering, Tshwane University of Technology, Pretoria, Gauteng, South Africa

Oluwaseye Samson Adedoja Department of Chemical, Metallurgical and Materials Engineering, Tshwane University of Technology, Pretoria, Gauteng, South Africa; Institute of NanoEngineering Research (INER), Tshwane University of Technology, Pretoria, Gauteng, South Africa

Stephen C. Agwuncha Department of Industrial Chemistry, Federal University Lokoja, Lokoja, Kogi State, Nigeria

Mutairu Ajiboye Department of Materials Science and Engineering, Obafemi Awolowo University, Ile Ife, Nigeria

Musa Aliyu Department of Chemistry, Federal University of Technology, Minna, Niger State, Nigeria

Chioma Grace Anusionwu Department of Chemistry, University of Johannesburg, Johannesburg, South Africa

Esther Aro Department of Chemistry, Federal University of Technology, Minna, Niger State, Nigeria

Thomas Adedayo Ayorinde Department of Agricultural and Environmental Engineering, Obafemi Awolowo University, Ile Ife, Nigeria; Biofuel Pilot Plant and Laboratory, National Biotechnology Research and Development Agency, Abuja, Nigeria; Department of Mechanical and Mechatronics Engineering, Tshwane University of Technology, Pretoria, Gauteng, South Africa

Yakubu Azeh Department of Chemistry, Ibrahim Badamasi Babangida University, Lapai, Nigeria

Umar Muhammad Badeggi Department of Chemistry, Ibrahim Badamasi Babangida University, Lapai, Nigeria

Mokae Bambo Department of Natural Sciences, Vaal University of Technology, Vanderbijlpark, South Africa

Everardt Andre Burger Department of Civil Engineering, Tshwane University of Technology, Pretoria, Gauteng, South Africa

Yejun Cao College of Civil Engineering, Hunan University, Changsha, Hunan, P.R. China

Mohammed Zago Dagaci Department of Chemistry, Ibrahim Badamasi Babangida University, Lapai, Nigeria

Ayodele Abeeb Daniyan Department of Industrial Engineering, University of South Africa, Pretoria, South Africa; Department of Materials Science and Engineering, Obafemi Awolowo University, Ile Ife, Nigeria

Dawood Desai Department of Mechanical and Mechatronics Engineering, Tshwane University of Technology, Pretoria, Gauteng, South Africa

Azunna Agwo Eze Department of Industrial Engineering, Operations Management and Mechanical Engineering, Vaal University of Technology, Vanderbijlpark, South Africa

Emmanuel Ezeani Department of Polymer Engineering, Faculty of Engineering, Nnamdi Azikiwe University, Awka, Nigeria

Anthony Chidi Ezika Chemical, Metallurgical, and Material Engineering Department, Faculty of Engineering and Built Environment, Tshwane University of Technology, Pretoria, Gauteng, South Africa

Festus Oluwadare Fameso Centre for Climate Change, Water Security and Disaster Management, Department of Civil Engineering, Tshwane University of Technology, Pretoria, Gauteng, South Africa

Yskandar Hamam Ecole Superieure d'Ingenieurs en Electrotechnique et Electronique, 2 Boulevard Blaise Pascal, Noisy-Le-Grand, France

Martin Ibenta Department of Polymer Engineering, Faculty of Engineering, Nnamdi Azikiwe University, Awka, Nigeria

Idowu David Ibrahim Department of Industrial Engineering, Operations Management and Mechanical Engineering, Vaal University of Technology, Vanderbijlpark, South Africa

Abel Inobeme Department of Chemistry, Edo State University, Uzairue, Edo, Nigeria

Tamba Jamiru Department of Mechanical and Mechatronics Engineering, Tshwane University of Technology, Pretoria, Gauteng, South Africa

Rosisky M. Kalombe Department of Civil and Mechanical Engineering, University of Missouri-Kansas City, Kansas City, Missouri, United States

Chewe Kambole Department of Civil Engineering and Construction, School of Engineering, The Copperbelt University, Kitwe, Zambia

Vinny N. Katambwe Department of Chemical Engineering, Cape Peninsula University of Technology, Cape Town, Western Cape, South Africa; Department of Environmental Studies for Advanced Society, Graduate School of Environmental Studies, Tohoku University, Sendai, Japan

Williams Kehinde Kupolati Department of Civil Engineering, Tshwane University of Technology, Pretoria, Gauteng, South Africa

Makungu M. Madirisha Department of Chemistry, University of Dar es Salaam, Dar es Salaam, Tanzania

Izuchukwu Madu Department of Polymer Engineering, Faculty of Engineering, Nnamdi Azikiwe University, Awka, Nigeria

Sivuyisiwe Mapukata Mintek, Advanced Materials Division, Randburg, South Africa

John Tsado Mathew Department of Chemistry, Ibrahim Badamasi Babangida University, Lapai, Nigeria

Charles Mbohwa Department of Industrial Engineering, University of South Africa, Pretoria, South Africa

Nonkululeko Miya DSTI/Mintek-Nanotechnology Innovation Centre, Randburg, South Africa; Mintek, Advanced Materials Division, Randburg, South Africa

Jonas Mochane Department of Chemistry, University of South Africa, Johannesburg, South Africa

Kobe Samuel Mojapelo Department of Civil Engineering, Tshwane University of Technology, Pretoria, Gauteng, South Africa

Teboho Clement Mokhena DSTI/Mintek-Nanotechnology Innovation Centre, Randburg, South Africa; School of Chemical and Physical Sciences, University of Mpumalanga, Mbombela, South Africa

Teboho Patrick Mokoena Department of Physics, University of the Free State, Phuthaditjhaba, South Africa

William Moloto DSTI/Mintek-Nanotechnology Innovation Centre, Randburg, South Africa; Mintek, Advanced Materials Division, Randburg, South Africa

Musah Monday Department of Chemistry, Ibrahim Badamasi Babangida University, Lapai, Nigeria

Mathato Motaung DSTI/Mintek-Nanotechnology Innovation Centre, Randburg, South Africa; Mintek, Advanced Materials Division, Randburg, South Africa

Mpho Muloiwa Department of Civil Engineering, Tshwane University of Technology, Pretoria, Gauteng, South Africa

Julius Musyoka Ndambuki Department of Civil Engineering, Tshwane University of Technology, Pretoria, Gauteng, South Africa

Amos Nhlapo Department of Physics, Medunsa, Sefako Makgatho Health Sciences University, South Africa

Evral Ntsa Department of Civil and Mechanical Engineering, University of Missouri-Kansas City, Kansas City, Missouri, United States

Bulelwa Ntsendwana DSTI/Mintek-Nanotechnology Innovation Centre, Randburg, South Africa; Mintek, Advanced Materials Division, Randburg, South Africa

Themba Ntuli DSTI/Mintek-Nanotechnology Innovation Centre, Randburg, South Africa; Mintek, Advanced Materials Division, Randburg, South Africa

Stephen Shaibu Ochigbo Department of Chemistry, Federal University of Technology, Minna, Niger State, Nigeria

Tunde Victor Ojumu Department of Chemical Engineering, Cape Peninsula University of Technology, Cape Town, Western Cape, South Africa

Tochukwu Okonkwo Department of Chemistry, Ibrahim Badamasi Babangida University, Lapai, Niger State, Nigeria

Victor Okpechi Department of Polymer Engineering, Faculty of Engineering, Nnamdi Azikiwe University, Awka, Nigeria

Rita Okpuwhara Department of Chemical Engineering, University of Johannesburg, Johannesburg, South Africa

Babatunde Oladipo Department of Chemical Engineering, Cape Peninsula University of Technology, Cape Town, Western Cape, South Africa

Yemi Olayinka Olasebikan Department of Agricultural Technology, Ekiti State Polytechnic, Isan Ekiti, Nigeria

Peter Apata Olubambi Centre for Nanoengineering and Advanced Materials, University of Johannesburg, Johannesburg, South Africa

Henry Oyeoka Department of Polymer Engineering, Faculty of Engineering, Nnamdi Azikiwe University, Awka, Nigeria

Leslie F. Petrik Department of Chemistry, University of the Western Cape, Cape Town, Western Cape, South Africa

Md. Zillur Rahman Department of Mechanical Engineering, Ahsanullah University of Science and Technology, Dhaka, Bangladesh

Peet Roux Department of Electrical Engineering, Tshwane University of Technology, Pretoria, Gauteng, South Africa

Emmanuel Rotimi Sadiku Institute of NanoEngineering Research (INER), Tshwane University of Technology, Pretoria, Gauteng, South Africa; Department of Chemical, Metallurgical and Materials Engineering, Tshwane University of Technology, Pretoria, Gauteng, South Africa

Smith Adamu Salifu Department of Chemical, Metallurgical and Materials Engineering, University of Johannesburg, Johannesburg, South Africa

Hussaina Salihu Department of Chemistry, Federal University of Technology, Minna, Niger State, Nigeria

Tumelo Seadira Department of Chemical and Materials Engineering, University of South Africa, Johannesburg, South Africa

Mantsopa Koena Selatile Zamisa Department of Chemical and Materials Engineering, University of South Africa, Johannesburg, South Africa

Elijah Yanda Shaba Department of Chemistry, Federal University of Technology Minna, Minna, Nigeria

Katekani Shingange School of Physics, University of the Witwatersrand, Johannesburg, South Africa

Jacques Snyman Department of Civil Engineering, Tshwane University of Technology, Pretoria, Gauteng, South Africa

Abiola Ezekiel Taiwo Faculty of Engineering, Mangosuthu University of Technology, Durban, South Africa

Shadrack Ugwu Department of Polymer Engineering, Faculty of Engineering, Nnamdi Azikiwe University, Awka, Nigeria

Deju Zhu College of Civil Engineering, Hunan University, Changsha, Hunan, P.R. China

About the editors

Prof. Emmanuel Rotimi Sadiku, currently a professor at the Department of Chemical, Metallurgical, and Materials Engineering at Tshwane University of Technology (TUT), South Africa, earned his BSc [Hons] and a PhD from the University of Strathclyde, Glasgow, United Kindgom, specializing in polymer physics with a focus on polymer processing, X-ray diffraction, thermal characteristics, rheology, and microscopy of polymer techniques. His academic journey includes a role as a lecturer II at the Federal University of Technology in Nigeria, followed by a postdoctoral research fellowship at the University of Genova/CNR, Italy. Upon returning to Nigeria, he continued his career by joining the Federal University of Technology, where he served as a senior lecturer. He also spent a year as a research fellow at the KTH Institute of Technology in Stockholm, Sweden. In January 2004, he assumed his current position as a research professor at TUT, South Africa.

Dr. Idowu David Ibrahim holds dual doctoral degrees in mechanical engineering from Tshwane University of Technology, South Africa, and Automation from Université Paris-Saclay, France. He is a C2-rated researcher with South Africa's NRF. His research expertise encompasses advanced composite materials, nanotechnology, sustainable manufacturing, renewable energy, additive manufacturing, material recycling, climate change, and engineering innovation, with the aim of crafting a sustainable environment. Furthermore, Dr. Ibrahim has presented his research findings at several international conferences and is an esteemed member of several professional organizations. He is a registered and practicing mechanical engineer in both South Africa and Nigeria. Dr. Ibrahim is dedicated to advancing globally competitive engineering education through interdisciplinary research, industry collaboration, postgraduate mentorship, and technology-driven solutions that address sustainable development challenges.

Prof. Williams Kehinde Kupolati holds a PhD in civil engineering from the University of Ibadan, Nigeria, following a BSc (Honors) in civil engineering and an MSc in construction management from the University of Lagos, Nigeria. He is a C2-rated researcher with South Africa's National Research Foundation (NRF). After 13 years of professional practice across key areas of civil engineering, he joined the Department of Civil Engineering at the University of Ibadan as a lecturer. He subsequently served as a full professor in the Department of Civil Engineering at Tshwane University of Technology, Pretoria, South Africa, where he remains actively involved in supervising master's and doctoral students. His research interests span civil engineering materials, nanomaterials, the beneficial utilization of waste, transportation engineering, and modeling. His multidisciplinary and translational research focuses on advancing the application of waste-derived materials, pozzolans, nanomaterials, and smart materials in civil engineering, with particular emphasis on promoting green infrastructure and a circular economy.

Dr. Stephen C. Agwuncha is a Fulbright research scholar currently with the Department of Forest Biomaterials, North Carolina State University, United States. He is a member of notable professional bodies, including the Royal Society of Chemistry, the Polymer Institute of Nigeria, the Chemical Society of Nigeria, and the South African Chemical Institute. His areas of research include fiber isolation, characterization and recycling, polymers, nanotechnology, composites, and renewable energy.

PART 1

Preparation and properties of high temperature polymers

1

High-temperature polymers: an introduction in the context of engineering

Idowu David Ibrahim[1], Kobe Samuel Mojapelo[2], Stephen C. Agwuncha[3], Emmanuel Rotimi Sadiku[4], Williams Kehinde Kupolati[2]

[1]DEPARTMENT OF INDUSTRIAL ENGINEERING, OPERATIONS MANAGEMENT AND MECHANICAL ENGINEERING, VAAL UNIVERSITY OF TECHNOLOGY, VANDERBIJLPARK, SOUTH AFRICA [2]DEPARTMENT OF CIVIL ENGINEERING, TSHWANE UNIVERSITY OF TECHNOLOGY, PRETORIA, GAUTENG, SOUTH AFRICA [3]DEPARTMENT OF INDUSTRIAL CHEMISTRY, FEDERAL UNIVERSITY LOKOJA, LOKOJA, KOGI STATE, NIGERIA [4]DEPARTMENT OF CHEMICAL, METALLURGICAL AND MATERIALS ENGINEERING, TSHWANE UNIVERSITY OF TECHNOLOGY, PRETORIA, GAUTENG, SOUTH AFRICA

1.1 Introduction to high-temperature polymeric materials

High-temperature polymer materials represent a special category of polymers designed to maintain their structural integrity, thermal stability, and functional properties in elevated temperature environments. These materials feature high glass transition temperatures (T_g), exceptional resistance to thermal degradation, and enhanced mechanical performance, making them essential for engineering applications where traditional polymers do not meet operational requirements. Notable examples of high-temperature polymers include polyimides (PIs), polyetherimides (PEI), and fluoropolymers, each tailored to specific thermal and mechanical requirements (Ren et al., 2022; Cho et al., 2022).

The importance of high-temperature polymer materials in technology lies in their ability to reliably function under rough conditions, including extreme thermal fluctuations, mechanical stress, and corrosive environments. This reliability is of crucial importance for progress in industries such as aerospace, electronics, automotive, and energy, in which materials have to provide excellent services in demanding applications, including high-performance dielectric films, structural components, and energy storage systems (Lulu Ren et al., 2021; Zhao et al., 2024). The dielectric polymers with high energy density and thermal stability for modern electronics and renewable energy technologies are of essential importance and ensure efficiency and durability for increased operating temperatures (Lei et al., 2022; Deshmukh et al., 2023). Recent progress in polymer science has led to innovative strategies aimed at improving the performance of high-temperature polymers. These strategies include the inclusion of nanofillers such as boron nitride (BN) and hafnium oxide (HFO_2) nanoparticles to improve the dielectric properties, as well as the development of new combustion structures

High-Temperature Polymeric Materials. DOI: https://doi.org/10.1016/B978-0-443-32792-6.00002-6

and molecular falling level to strengthen thermal and electrical stability (Ai et al., 2020). In addition, bottle-breast-polymers and other unique architectures have shown adjustable properties that react to external stimuli and provide varied solutions for temperature-sensitive technical applications (Basak, Chatterjee and Bandyopadhyay, 2024).

In addition to their intrinsic properties, the environmental and economic effects of high-temperature polymers further highlight their importance. Lightweight and durable polymer composites have the potential to reduce energy consumption in the transportation and construction sectors, while their scalability and compatibility with additive manufacturing (AM) processes facilitate cost-effective production and customized design (Bjørken et al., 2022). Moreover, innovations in environment-friendly synthesis and recyclability are paving the way for sustainable practices in polymer engineering, aligning with global initiatives aimed at minimizing environmental impact (Zhang et al., 2023). The high-temperature polymer materials are fundamental to modern engineering and meet the urgent need for materials that can withstand extreme conditions. Their diverse applications, coupled with continued advances in materials science, highlight their essential role in driving technological progress across numerous sectors. Continued research into high-temperature polymers is expected to significantly impact the future of engineering, providing solutions to challenges related to thermal stability, mechanical performance, and environmental sustainability.

Fig. 1-1 categorizes polymers into three principal groups: natural polymers, synthetic elastomers, and synthetic polymers, as outlined by Bala et al. (2014). Natural polymers, which include cellulose, gelatin, and chitosan, are derived from biological sources and are often used due to their inherent biocompatibility and biological degradation in biomedical and environmental applications. Synthetic elastomers such as hydride rubber and neoprene have an exceptional elasticity and durability, making them indispensable in industries such as the

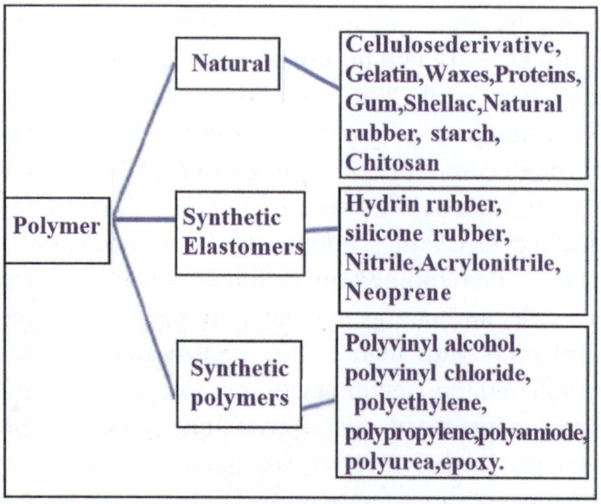

FIGURE 1–1 A flowchart classifying polymers into natural, synthetic elastomers and synthetic polymers (Bala et al., 2014). *Bala, P., Jathar, S., Kale, S., Pal, K., 2014. Transdermal drug delivery system (TDDS)-a multifaceted approach for drug delivery. Journal of Pharmaceutical Research 8 (12), 1805–1835.*

automobile and aerospace. Synthetic polymers, including polyethylene, polyamide, and epoxy, are developed for high-performance applications and appreciated for their thermal stability, mechanical strength, and versatility in various technical areas. This classification underlines the adaptability of polymers and their specific functions in combating the different requirements for contemporary industrial and environmental applications, especially those that demand high temperature resistance.

1.2 Historical overview and milestones of high-temperature polymers in engineering

The development of high-temperature polymers is characterized by significant advances designed to meet the rigorous requirements of extreme thermal and mechanical environments. Early milestones in the 20th century laid the foundation for this field, with the introduction of PIs in the 1960s representing a key turning point. PIs, known for their exceptional thermal stability and mechanical strength, were originally used in aerospace applications and electrical insulation, setting a benchmark for materials that can withstand high temperatures (Jeong et al., 2023; Ren et al., 2022). Despite their innovative properties, these early materials faced challenges including high coefficients of thermal expansion (CTE) and limited scalability, which required further research and development (Jeong et al., 2023). The 1970s and 1980s witnessed significant progress in polymer chemistry, resulting in the emergence of polyetheretherketone (PEEK) and polyethersulfone (PES), two thermoplastics that expanded the scope of high-temperature polymers. PEEK, distinguished by its superior thermal and chemical resistance, has become a preferred choice for structural components within the aerospace and automotive industries (Hanemann et al., 2023). As the demand for improved performance increased, challenges such as dimensional stability under thermal stress were addressed through the incorporation of nanofillers, including carbon nanotubes and silicon dioxide (Ai et al., 2020). Likewise, PES was developed as an engineering thermoplastic that exhibited improved mechanical and thermal properties, enabled by innovative techniques that integrated biodegradable nanofillers such as lignin, which effectively reduced its high CTE and improved environmental compatibility (Jeong et al., 2023).

The 1990s and early 2000s marked a transition to customer-specific adaptation of high-temperature polymers for special applications. For example, PIs were modified using nanotechnology to improve their dielectric properties and thus facilitate their application in energy storage and microelectronics. Studies by Feng et al. (2022) and Zhang et al. (2020) characterize this time as a time of quick progress in dielectric polymer technology, in which the installation of nanofiller fabrics with a low volume in half-crystalline dipolar polymers significantly improved the energy density and the charging and discharging efficiency at increased temperatures. These innovations removed the restrictions of conventional polymers and paved the way for applications in capacitors, batteries, and renewable energy systems. In recent decades, the focus on high-temperature polymers has expanded to include considerations of sustainability and scalability. Advances in crosslinked fluoropolymers and hybrid polymer systems have significantly improved energy efficiency and operational stability, rendering these materials essential for the electrification of transportation and the development of renewable energy technologies (Li et al., 2020; Zhao et al., 2022) introduced poly(cyclosiloxane-carborane) that can withstand temperatures of up to 1000°C, marking a substantial breakthrough that extends the operational range of high-temperature polymers for applications in extreme environments, such as aerospace and defence.

In the 21st century, there was also an expansion of high-temperature polymers beyond traditional industrial applications into emerging areas such as biomedical technology and sustainability. Basak, Chatterjee and Bandyopadhyay (2024) report on the development of bottle brush polymers, which show remarkable adjustability and reaction ability to thermal stimuli and make them ideal for temperature-regulated systems in advanced technical applications. In addition, heat-sensitive polymer blends have pushed innovations in tissue engineering by providing biocompatible and thermally stable solutions for medical applications (Nahra et al., 2022). This historical progression highlights the development of high-temperature polymers in response to the increasing demands of technical applications. From their initial use in aerospace insulation to their contemporary roles in energy storage, microelectronics, and biomedical engineering, these materials have demonstrated significant versatility and innovation. The integration of nanotechnology, molecular design, and AM has enhanced their capabilities and established a foundation for future breakthroughs in sustainability and performance. These advances highlight the enduring importance of high-temperature polymers in shaping modern engineering and overcoming the challenges of extreme environments.

Table 1–1 presents a comprehensive overview of the principal challenges encountered in the development and use of high-temperature polymers, along with innovative solutions that maximize their engineering potential and anticipate future progress. Thermal degradation, a

Table 1–1 Challenges and solutions for high-temperature polymers.

Challenge	Description	Proposed solutions	Sources
Thermal degradation	Polymers experience chain scission and oxidation at high temperatures	Incorporate nanofillers like carbon nanotubes to improve crosslinking density	Yang et al. (2023), Gao et al. (2024), Ayornu, Anwar and Thakur (2022), Ren et al. (2021)
Environmental concerns	Petrochemical-based synthesis contributes to the carbon footprint	Adopt biobased polymers, implement RCP	Jones et al. (2023), Beaumont et al. (2022),Hosseini et al. (2022), Modh (2023)
Scalability	High production costs and energy-intensive processes	Use additive manufacturing and develop cost-effective fillers	Sun et al. (2020), Kade and Dalton (2021), Zhao et al. (2022), Ma et al. (2024)
Mechanical limitations	Reduced performance in extreme environments	Optimize molecular design, surface modifications with boron nitride	Wahlen et al. (2020), Fotoohi et al. (2023), Dong et al. (2021), Mcguinness Leng and Losego (2020)

RCP, Reverse controlled polymerization.

prevalent issue resulting from chain scission and oxidation at elevated temperatures, can be effectively mitigated through the incorporation of nanofillers, such as carbon nanotubes, and by enhancing crosslinking density. These enhancements not only improve thermal stability but also maintain the structural integrity of polymers under extreme conditions. To address environmental concerns associated with the petrochemical-based production of these materials, it is imperative to adopt sustainable alternatives, including biobased polymers and advanced methodologies such as reverse controlled polymerization (RCP), which significantly reduce their carbon footprint and align with global sustainability objectives. The issue of scalability, particularly regarding high production costs and energy-intensive manufacturing processes, can be alleviated through the integration of AM techniques and the development of cost-effective, scalable fillers, thereby enhancing the feasibility of industrial adoption. In addition, mechanical limitations, such as reduced tensile strength and wear resistance in challenging environments, can be addressed by optimizing molecular designs and implementing surface modifications with materials such as boron nitride. These solutions not only enhance durability and adaptability but also facilitate the creation of multifunctional materials capable of meeting the stringent requirements of industries such as aerospace, energy storage, electronics, and advanced manufacturing. By addressing these challenges through innovative strategies, high-temperature polymers can continue to evolve as vital materials for contemporary engineering applications.

1.2.1 Milestones in high-temperature polymer discovery

Early explorations of engineering polymers, particularly PIs and poly(ether ketone), laid the groundwork for polymer systems with high-temperature capabilities. In the initial decades, the emphasis was on investigating the inherent thermal stability and high glass transition temperatures characteristic of these aromatic polymers (Ai et al., 2020). A logical continuity in polymer engineering led from the earliest high-temperature materials to more sophisticated systems involving designed molecular architectures and nanocomposite approaches. The incorporation of cross-linkable groups in melt-processable fluoropolymers provided enhanced performance metrics, such as superior energy density and charge-discharge efficiency, specifically engineered to meet extreme operational conditions (Li et al., 2020). More recent developments have also broadened the field, with the introduction of anisotropic semicrystalline homopolymers and local particle-induced structural modifications of the polymer matrix, which further extend the dielectric constant maximization and breakdown strength (Xu et al., 2024). These incremental developments demonstrate an evolution from traditional polymer systems to designing materials that can withstand rigorous high-temperature regimes.

1.2.2 Prominent contributors and major milestones

Previous studies have significantly advanced the development of high-temperature polymers, where the in situ formation of PI nanocomposites represented a critical step in understanding and exploiting the interplay between the polymer matrix and nanofillers for improved capacitive energy storage. A pioneering crosslinking strategy addresses the fundamental limitations of typical dielectric polymers, resulting in enhanced performance under high-temperature conditions (Li et al., 2020). The ongoing structures confirmed the reductions and emphasized the steady incremental improvements over nearly 5 decades, effectively placing the progress within the context of electronic energy (Yang et al., 2022). Innovations in the processing and structuring of polymers, such as the curly-packed structure approach and the recent anisotropic alignment strategy, further confirm the importance of structural control and material design strategies that have emerged from previous research contributions (Xu et al., 2024; Zhou et al., 2022).

1.3 Fundamental properties and characteristics of high-temperature polymers

Liu et al. (2022) and Cheng et al. (2021) underline the decisive importance of high-temperature polymers for engineering, especially because of their ability to maintain thermal stability, mechanical strength, and functional properties under extreme conditions. High-temperature polymers such as PIs and PEEK have exceptional resistance to thermal degradation, making them essential in progressive technical applications, including electric vehicles, aerospace, and underground exploration. These materials are specially designed in such a way that they withstand elevated temperatures and, at the same time, maintain structural integrity and operational efficiency, thus positioning them at the top of modern technology.

Y. Ren et al. (2021) and Zhao et al. (2020) further investigate the thermal and mechanical properties of high-temperature polymers and emphasize their performance in extreme environments. For example, PIs possess high thermal resistance, with decomposition temperatures exceeding 500°C, and exhibit excellent mechanical properties, making them ideal for energy storage devices and high-performance energy storage dielectric applications. Similarly, PEEK's semicrystalline structure provides remarkable chemical and wear resistance, which is critical for applications requiring long-term durability under thermal and mechanical stress (Mohammed, Oleiwi and Al-Hassani, 2022). Furthermore, these polymers have high energy density, facilitating their use in capacitors and batteries (Liu et al., 2022). The versatility of high-temperature polymers has been significantly improved through material modifications and nanocomposite technology. Zhang et al. (2023) and Ai et al. (2020) discuss how incorporating nanofillers such as amine-functionalized reduced graphene oxide and HFO_2 nanoparticles into polymer matrices can significantly improve thermal conductivity, dielectric properties, and mechanical performance. These modifications help mitigate issues such as thermal expansion and brittleness, thereby improving the adaptability of these polymers for advanced aerospace and energy storage applications. For example, nanocomposite PI films have demonstrated superior degradation strength and thermal resistance, extending the operational limits of dielectric materials in high-temperature environments (Yang et al., 2022).

Pan et al. (2023) and Fotoohi et al. (2023) investigate the distinct properties of branched polymers and polymer blends and highlight their superior mechanical strength, shear resistance, and thermal stability compared to conventional linear polymers. The branched structures contribute to increased molecular rigidity, thereby improving performance in high-temperature and high-salinity environments, such as those encountered in oil and gas exploration. Additionally, surface-modified nanocomposites, including exfoliated graphite-filled polypropylene blends, demonstrate improved elongation at break and fracture toughness, highlighting the importance of material modifications in optimizing high-temperature polymers for demanding applications (Fotoohi et al., 2023).

Hong and Cho (2023) and Gu et al. (2020) emphasize the adaptability of high-temperature polymers in structural applications. Polyphenylene sulfide and PI base show excellent dimensional stability and chemical resistance, making them suitable for use in capacitors, aerospace components, and light insulation materials. In addition, the hybrid organic structures of organosilicon polymers provide small surface tension and flame inhibition that are advantageous for high-frequency electronic applications (Hou, Sun and Fang, 2023). These innovations illustrate the continuous development of high-temperature polymers, which are driven by progress in molecular design, nanotechnology, and manufacturing techniques. The high-temperature polymers are characterized by their ability to maintain performance under extreme thermal, mechanical, and environmental conditions. Their inherent properties, such as thermal stability, mechanical strength, and dielectric performance, have been further enhanced through nanotechnology and material modifications, thereby expanding their applications in critical engineering sectors. From aerospace and electronics to energy storage and oil exploration, high-temperature polymers are essential for developing cutting-edge technologies that meet the strict requirements of modern engineering. These advancements underline the critical role of high-temperature polymers in shaping the future of materials science and engineering.

Fig. 1–2 illustrates the influence of structural modifications on the thermal properties of PEEK. The introduction of functional groups such as carboxylic acid, sulfone (SO_2), and nitrile results in variations in the glass transition temperature (T_g), which serves as a critical indicator of thermal stability. Among the modifications analysed, PEEK-COOH demonstrates the highest T_g at 213°C, exceeding that of PEEK-SO_2 at 156°C and PEEK-CN at 155°C. The improved thermal stability of PEEK-COOH can be attributed to hydrogen bonding interactions that restrict molecular mobility, thereby enhancing resistance to thermal degradation. This characteristic renders PEEK-COOH particularly suitable for demanding applications in high-temperature environments, including aerospace and energy storage.

Differential scanning calorimetry data further underlines the influence of these modifications on the heat flow behavior of PEEK. The different T_g values associated with each modification illustrate how the incorporation of functional groups can tune the thermal performance of high-temperature polymers. This discussion, within the broader context of high-temperature polymers in engineering, emphasizes the importance of molecular design in optimizing material properties for advanced applications. By modifying the PEEK backbone with specific functional groups, it becomes possible to develop special polymer systems with

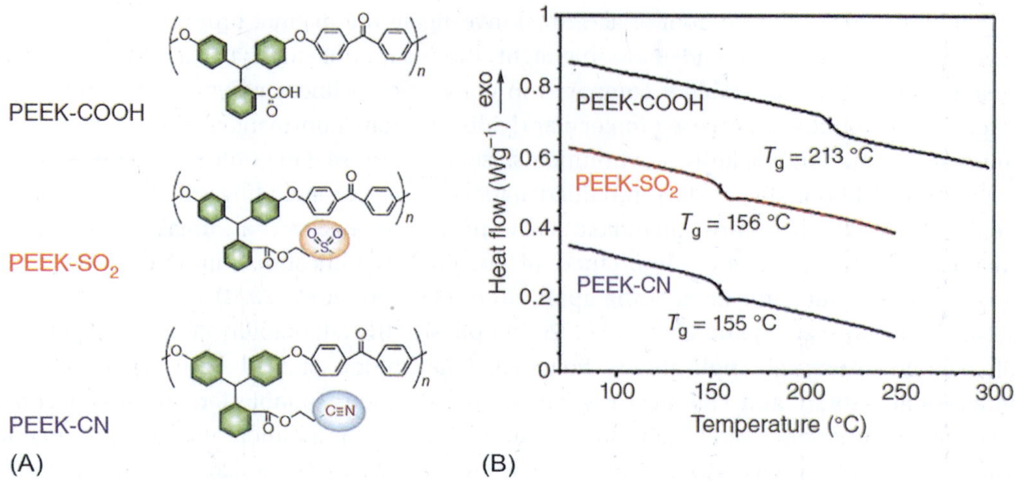

FIGURE 1–2 Pictorial representation of (A) the structural formula and (B) the DSC of polyetheretherketone (PEEK) polymers (Zha and Dang, 2023). *Zha, J.W., Dang, Z.M., 2023. High Temperature Polymer Dielectrics: Fundamentals and Applications in Power Equipment. Wiley, China, pp. 1–382. https://doi.org/10.1002/9783527841059.*

Table 1–2 Properties of high-temperature polymers and applications.

Polymer type	Properties	Applications	Sources
Polyimides	Thermal stability (>500°C), excellent mechanical strength	Aerospace (thermal insulation), microelectronics, dielectric films	Ren et al. (2022), Zhao et al. (2024), Basak, Chatterjee and Bandyopadhyay (2024), Jeong et al. (2023)
Polyetheretherketone	High chemical resistance, durability under thermal stress	Automotive (engine components), medical implants	Hanemann et al. (2023), Mohammed, Oleiwi and Al-Hassani (2022), Kade and Dalton (2021), Yang et al. (2023)
Polyetherimides	High dielectric strength, thermal stability	Energy storage, flexible electronics	Deshmukh et al. (2023), Ai et al. (2020), Pan et al. (2023), Zhao et al. (2020)
Polybenzimidazoles	Exceptional thermal resistance, chemical stability	Insulation materials, fuel cells	Li et al. (2020), Kade and Dalton (2021), Wang et al. (2021), Xu et al. (2022)

increased heat resistance for critical industries such as automotive, electronics, and renewable energy technology. The figure impressively shows how advances in polymer chemistry enhance the versatility and adaptability of high-temperature polymer materials.

Table 1-2 presents the exceptional properties of high-temperature polymers and their extensive engineering applications, particularly in extreme operating environments. PIs are characterized by remarkable thermal stability, exceeding 500°C, along with high mechanical strength. These materials are widely utilized in the aerospace sector for thermal insulation and in microelectronics for dielectric films. Their ability to perform under harsh conditions renders them indispensable in demanding fields such as aviation and electronics. Similarly, PEEK is esteemed for its high chemical resistance and durability under thermal stress, making it a preferred choice in the automotive industry for engine components and in the medical sector for implants. These attributes ensure PEEK's reliability in applications that require longevity and resilience in extreme conditions.

Additionally, the table highlights the advanced dielectric properties and thermal stability of PEI, which position them as essential materials in energy storage devices and flexible electronics. The high dielectric strength of PEI enhances its effectiveness in capacitors and other electronic applications operating at elevated temperatures, thereby addressing the growing demand for durable and efficient energy storage solutions. Polybenzimidazoles (PBI) are recognized for their exceptional thermal resistance and chemical stability, making them ideal candidates for insulation materials and fuel cells. The superior performance of PBI in extreme environments ensures its utility in critical applications, such as high-performance fuel cells, where both efficiency and durability are paramount.

1.4 Manufacturing processes and synthesis methods for high-temperature polymers

The further development of high-temperature polymers depends significantly on sophisticated manufacturing processes and innovative synthesis methods that enable these materials to endure extreme thermal and mechanical conditions. Xu et al. (2022) and Pyronneau et al. (2022) identify thermal treatment and hybridization as crucial techniques in this area. Xu et al. (2022) show that thermal annealing of polyamic acid in PI improves both thermal stability and mechanical strength in graphene oxide composite films, thereby optimizing the effectiveness of electromagnetic interference. In a parallel study, Xu et al. (2022) highlight the utility of vapor phase infiltration (VPI) in the production of organic–inorganic hybrid materials, with processing temperatures significantly affecting durability and structural integrity. These methods highlight the need to adjust material composition and thermal processing parameters to meet technical specifications.

Li et al. (2020) and Yuan et al. (2020) emphasize scalable synthesis techniques for high-temperature polymers suitable for energy storage and electronic applications. Li et al. (2020) present networked fluoropolymers as significant progress in dielectric high-temperature materials that offer improved energy density and charging-enthusiastic efficiency and, at the same time, ensure compatibility with established industrial processing practices. At the same time, Yuan et al. (2020) present molecular semiconductor dielectric composite materials with superior energy storage efficiency under extreem temperature conditions. These scalable methods reveal the potential of high-temperature polymers for applications in renewable energy systems and the electrification of the transport system.

In addition, advanced surface technologies and precision production techniques are crucial for improving the performance of high-temperature polymers. Kafle et al. (2020) describe a coating technology supported by dynamic templates to produce high-age polymer films with optimized electronic properties that address the prevailing challenges in optoelectronic applications. In a related study, Wang et al. (2021) explore surface modification strategies using boron nitride nanoleaves to improve punch strength and electrical insulation in a polymer-based dielectric. These methods illustrate how surface technology significantly improves the thermal and electrical properties of high-temperature polymers and thus promotes their applications in flexible electronics and energy storage devices. The integration of composite manufacturing and functional layering has significantly improved the capabilities of high-temperature polymers. Durmaz, Erdil and Avci (2021) investigate the reinforcement of high-density polyethylene with glass and carbon fibers, resulting in remarkable improvements in its mechanical and thermal properties for structural applications. Dong et al. (2021) focus on the in situ surface functionalization of PI films, leading to improved dielectric energy storage efficiency at temperatures up to 200°C. These studies highlight the importance of designing composite materials and laminated structures to meet the rigorous requirements of modern engineering.

Technological advancements such as AM and selective laser sintering (SLS) have revolutionized the production of high-temperature polymers. Sun et al. (2020) identify SLS as an effective method for fabricating porous scaffolds with controlled geometries, suitable for tissue engineering and biomedical applications. This technique enables processing synthetic polymers at elevated melting points without compromising bioactive substances, thereby expanding the applicability of high-temperature polymers in specialized domains. Similarly, Kade and Dalton (2021) emphasize melt electro writing (MEW), a three-dimensional printing technique that generates high-resolution polymer scaffolds exhibiting excellent thermal and mechanical properties. These methodologies illustrate how AM techniques enable the individual adjustment and scalability of high-temperature polymer components. Innovative chemical synthesis methods have also improved the thermal, electrical, and mechanical properties of high-temperature polymers. Saito et al. (2022) describe the use of polymer-coated nanocrystals to fabricate proton-conductive membranes for fuel cells, resulting in improvements in both thermal stability and proton conductivity. In a related investigation, Ma et al. (2024) examine the synthesis of polymer-derived ceramics (SiAlBCN) via precise pyrolysis processes, resulting in materials characterized by exceptional temperature resistance and oxidation stability. These approaches highlight the critical role of advanced chemistry in developing high-performance polymers that can work in harsh environments.

The adoption of sustainable and environment-friendly manufacturing processes is increasingly important in the production of high-temperature polymers. Beaumont et al. (2022) discuss green fabrication techniques, such as inkjet printing and water-based processes, for producing self-doped conducting polymers suitable for flexible electronics. These methods minimize environmental impact while maintaining high thermal and mechanical performance. Additionally, Mcguinness, Leng and Losego (2020) emphasize the scalability of VPI for creating hybrid polymer materials with enhanced chemical stability and thermal resistance, thereby contributing to global initiatives focused on sustainable materials engineering. The manufacturing processes and synthesis methods for high-temperature polymers have developed considerably to meet the requirements of contemporary technical applications. Innovations, such as thermal treatments, nanocomposite materials, AM, and sustainable manufacturing techniques, have positioned high-temperature polymers at the forefront of material innovation. The various methods reported in the literature underscore the versatility of high-temperature polymers and their potential for ongoing development in aerospace, energy, electronics, and related fields.

1.5 Challenges and limitations in the application of high-temperature polymers

High-temperature polymers play a crucial role in meeting the requirements of contemporary technical applications, especially in extreme environments where conventional materials are limited. However, several significant challenges hinder their widespread introduction, especially with regard to thermal performance, the breakdown, scalability, and cost-effectiveness (Yang et al., 2023; Hou, Sun and Fang, 2023). Coping with these challenges requires continuous innovation in polymer synthesis, processing techniques, and application design to ensure reliable and efficient performance under demanding conditions. A major limitation of high-temperature polymers is their reduced performance under extreme thermal and mechanical stress. For example, even advanced materials such as PI and PEEK show degradation in mechanical properties, including tensile strength and impact resistance, after prolonged exposure to elevated temperatures (Deshmukh et al., 2023; Li and Liu, 2023; Zhao et al., 2022). This emphasizes that polymers often undergo chain scission, oxidation, and void formation at elevated temperatures, compromising their structural integrity and shortening their lifespan in applications such as aerospace and energy storage.

Material compatibility and performance trade-offs also pose challenges in multifunctional applications. While the incorporation of fillers such as boron nitride or carbon nanotubes can enhance thermal conductivity and dielectric properties, these additives often lead to poor dispersion and interfacial adhesion, ultimately resulting in a decrease in overall mechanical strength (Gao et al., 2024; Yang et al., 2023). Furthermore, although fluorinated polymers offer exceptional chemical resistance and dielectric stability, their complex and costly synthesis processes limit their scalability for industrial applications (Ren et al., 2021; Zhou et al., 2023). Processing limitations further inhibit the broader application of high-temperature polymers. Manufacturing techniques such as SLS and MEW facilitate precise control of the polymer architecture but are often associated with high production costs, low throughput, and stringent feedstock material requirements (Kade and Dalton, 2021; Sun et al., 2020). Additionally, pyrolysis methods employed for polymer-derived ceramics require extreme precision and energy-intensive processes, complicating scalability while maintaining consistent material properties (Murdock and Benicewicz, 2024).

Environmental and mechanical degradations present a significant concern, particularly under harsh operational conditions. Modh (2023) indicates that while polymers containing S-triazine rings exhibit resistance to chemical degradation, their performance is compromised in humid environments due to moisture absorption. Furthermore, prolonged exposure to ultraviolet radiation and oxidative conditions accelerates material aging, resulting in surface cracking and a decline in mechanical integrity (Ayornu, Anwar and Thakur, 2022; Ren et al., 2021). These limitations impede their application in environments requiring long-term stability, such as outdoor structural components and high-temperature electrical systems. In the case of energy storage applications, high-temperature polymers face challenges in achieving an optimal balance between thermal stability and electrical performance. Conventional dielectric materials such as biaxially oriented polypropylene can only withstand temperatures up to 105°C due to low energy densities and insufficient breakdown strength (Dong et al., 2023). Although fluorinated polyoxanorbornene polymers and hybrid composites exhibit improved dielectric properties at elevated temperatures, they still encounter obstacles related to scalability and processing costs (Deshmukh et al., 2023). Dielectric losses and conduction currents increase exponentially with temperature, thereby restricting the practical use of these materials in advanced capacitors and energy storage devices (Zhou et al., 2023).

Sustainability is a major challenge in the development of high-temperature polymers. Conventional synthesis methods often depend on forecasters on a petrochemical-based and contribute to considerable environmental waste (Beaumont et al., 2022; Hosseini et al., 2022). Although initiatives for the development of biobased or recyclable high-temperature polymers are underway, these materials often have defects in mechanical strength and thermal stability (Modh, 2023; Hosseini et al., 2022). If industries are looking for greener manufacturing solutions, the intention of environmental compatibility with the performance requirements of high-temperature polymers remains an urgent challenge. Another critical restriction is the effort associated with the raw material and processing technologies for high-temperature polymers. Adding additives such as carbon nanostructures or rare elements significantly improves the properties of polymers but escalates production costs at the same time (Ayornu, Anwar and Thakur, 2022). This economic challenge limits its introduction to cost-sensitive sectors such as automotive and consumer electronics, where alternative materials are preferred for their affordability, though less efficient. The high-temperature polymers show significant potential for advanced technical applications. However, their widespread acceptance is hindered by challenges such as thermal deterioration, material compromises, scalability problems, and sustainability problems. With regard to these restrictions, a multidisciplinary approach requires progressive material synthesis, innovative processing technologies, and the integration of environment-friendly sustainable practices. The overcoming of these challenges is crucial to fully identify the potential of high-temperature polymers in various industrial sectors, including aerospace and renewable energies.

1.6 Engineering applications and future perspectives of high-temperature polymeric materials

High-temperature polymer materials are becoming increasingly important due to their extraordinary thermal stability, mechanical strength, and versatility in progressive technical applications. These materials play a central role in various industries, including aerospace, automobile, electronics, energy storage, and biomedical technology, where extreme environmental conditions require high-performance solutions (Wang et al., 2020). Their ability to withstand temperatures above 150°C without any significant deterioration made them indispensable for applications in which conventional materials do not meet performance standards (Gao et al., 2024; Recio-Colmenares et al., 2022). One of the most promising applications of high-temperature polymers lies in the aerospace sector, where the demand for light and thermally stable materials for improving fuel efficiency and reducing carbon emissions is of crucial importance. Due to their superior thermal stability and mechanical properties under extreme conditions, polymides such as PIs and PEEK are often used in aircraft components and space (Mollaoğlu and Kaptan, 2022). In addition, progress in polymer nanocomposites that integrate materials such as carbon nanotubes and graphene has significantly improved the thermal conductivity and mechanical strength of these polymers, thereby expanding their applicability in progressive air and space systems (Wahlen et al., 2020).

In the electronics industry, high-temperature polymer materials are critical to the development of dielectric films, flexible circuits, and insulation for high-performance devices. For example, PEI and PBI exhibit excellent dielectric properties, making them ideal for applications in capacitors and other electronic components that operate at elevated temperatures (Deacy et al., 2021; Yang et al., 2023). Furthermore, innovations in polymer-based energy storage materials, including high-temperature polymer dielectrics, hold the potential to revolutionize the performance of advanced capacitors and enable efficient energy storage even in demanding environments (Shao and Wang, 2020). Energy storage applications benefit significantly from high-temperature polymer materials. Polymers, such as fluorinated polyoxanorbornenes and hybrid composites, exhibit excellent dielectric properties and thermal stability, making them suitable for advanced capacitors and batteries (Malucelli and Mariani, 2024). However, challenges such as dielectric loss, thermal degradation, and limited scalability present significant obstacles to their widespread adoption. Addressing these challenges requires innovative molecular engineering strategies and the incorporation of new fillers to improve performance while maintaining processability (Bao, 2021; Nazrin et al., 2020).

Sustainability is another crucial point of view for the future of high-temperature polymer materials. Conventional synthesis processes are mainly dependent on petrochemical raw materials and contribute to significant environmental impacts (Nicolau, Mutch and Thickett, 2024; Singha et al., 2024). To alleviate these effects, researchers examine biobased polymers and recycling methods such as RCP to establish circular material life cycles (Jones et al., 2023; Vijayan, Prabhu and Jineesh, 2024). This progress can enable the more sustainable use of high-temperature polymers in industries that want to reconcile performance and environmental responsibility. New technologies, including AM and frontal polymerization (FP), are also influencing the future of high-temperature polymer materials. AM enables the creation of complex geometries with precise control of material properties and is, therefore, particularly suitable for prototyping and customized technical solutions (Mollaoğlu and Kaptan, 2022; Wahlen et al., 2020). Conversely, FP offers a rapid and energy-efficient approach to producing thermally stable polymer composites, particularly in contexts where time and cost efficiency are paramount (Malucelli and Mariani, 2024). With a view to the future, the integration of nanotechnology, intelligent material systems, and advanced manufacturing techniques will drive the development of high-temperature materials of the next generation. These innovations are ready to enable the creation of multifunctional materials with tailor-made properties, including self-healing functions, adaptive thermal conductivity, and improved recyclability to ensure their persistent relevance for state-of-the-art technical applications (Recio-Colmenares et al., 2022; Shao and Wang, 2020).

It can be said that high-temperature polymer materials have considerable potential to change technical applications in various industries. Due to their extraordinary thermal stability, mechanical strength, and versatility, they are well-suited for coping with the challenges of extreme operating conditions. Continuous innovations in synthesis, processing, and sustainability practices are of crucial importance to overcome current restrictions and to fully exploit their skills. These materials will play a crucial role in the further development of technologies and the fulfillment of the requirements of an increasingly dynamic technical landscape.

Table 1–3 illustrates the significant progress in manufacturing processes for high-temperature polymers and their influence on enhancing performance and broadening applicability. AM techniques, such as SLS and MEW, facilitate the creation of intricate polymer structures characterized by high levels of customization and minimal material waste. These attributes render them particularly suitable for prototyping and complex designs in the aerospace and biomedical fields (Sun et al., 2020; Wahlen et al., 2020). VPI is noteworthy due to the ability to produce hybrid polymer materials that have improved chemical stability and thermal resistance, thereby addressing critical challenges in electronics and energy storage systems (Mcguinness, Leng and Losego, 2020; Xu et al., 2022).

Table 1–3 Advances in manufacturing processes.

Process	Description	Advantages	Sources
Additive manufacturing	3D printing technologies like SLS and MEW for precise polymer architecture	High customization, minimal material waste	Sun et al. (2020), Wahlen et al. (2020), Kade and Dalton (2021), Mollaoğlu and Kaptan (2022)
Vapor phase infiltration	Technique for creating hybrid polymer materials	Enhanced chemical stability, thermal resistance	Mcguinness, Leng and Losego (2020), Wang et al. (2021), Zhao et al. (2022), Xu et al. (2022)
Frontal polymerization	Rapid polymerization using thermal or light ignition	Energy efficiency, faster production	Malucelli and Mariani (2024) Recio-Colmenares et al. (2022), Deshmukh et al. (2023), Modh (2023)
Surface functionalization	Modifications to improve thermal and mechanical properties	Improved energy density, durability	Dong et al. (2021), Kafle et al. (2020), Fotoohi et al. (2023), Ren Mujib and Singh (2021)

3D, Three-dimensional; *MEW*, melt electro writing; *SLS*, selective laser sintering.

FP offers a rapid and energy-efficient method of polymerization via thermal or light ignition, significantly reducing production time while maintaining material integrity, which is especially advantageous for applications requiring cost-effective scalability (Malucelli and Mariani, 2024; Recio-Colmenares et al., 2022). Surface functionalization methods, including nanocoating and the incorporation of fillers, enhance thermal conductivity, mechanical durability, and energy density, thereby ensuring that these polymers meet the demanding requirements of contemporary engineering sectors such as energy and electronics (Kafle et al., 2020; Dong et al., 2021). Collectively, these advanced processes underscore the evolving strategies for optimizing high-temperature polymers for sustainable, efficient, and high-performance applications.

1.7 Conclusion

High-temperature polymer materials have become indispensable in today's technology and meet the increasing demands for performance, durability, and sustainability in various industries. Their exceptional thermal stability, mechanical strength, and versatility make them essential for applications in aerospace, automotive, electronics, energy storage, and biomedical engineering. The historical development of PIs in the 1960s and subsequent advances in nanocomposites, AM, and environment-friendly synthesis methods reflect the development of high-temperature polymers that must be driven by the need to overcome extreme operational challenges. Despite considerable progress, there are some critical challenges, including thermal breakdown, scalability, environmental impact, and cost efficiency. Innovations such as molecular technology, hybrid polymer systems, and advanced manufacturing techniques continuously expand the possibilities of these materials. In addition, the increasing emphasis on sustainability underlines the importance of biobased polymers, recycling technologies, and energy-efficient manufacturing processes, thus paving the way for a more environmentally conscious future. With regard to the future, the integration of newly emerging technologies such as nanotechnology, intelligent material systems, and adaptive polymers is a significant promise for the development of multifunctional materials that are specially tailored to the technical requirements. This progress is ready to improve both the performance and the durability of high-temperature polymers and, at the same time, expand their applicability to new domains, including climate-resilient infrastructure, systems for renewable energies, and advanced medical technologies.

High-temperature polymer materials are of fundamental importance for technological innovation and technical excellence. Continuous development will be of crucial importance to master the challenges of extreme environments, promote sustainability, and shape the future of engineering in several sectors. Through the promotion of multidisciplinary research and cooperation between science and industry, we can fully exploit the potential of high-temperature polymers and ensure their decisive contribution to technological progress and social progress.

AI disclosure

During the preparation of this work the author(s) used Quillbot to improve readability and language. After using this tool/service, the author(s) reviewed and edited the content as needed and take(s) full responsibility for the content of the publication.

References

Ai, D., Li, H., Zhou, Y., Ren, L., Han, Z., Yao, B., et al., 2020. Tuning nanofillers in in situ prepared polyimide nanocomposites for high-temperature capacitive energy storage. Advanced Energy Materials 10 (16). Available at https://doi.org/10.1002/aenm.201903881.

Ayornu, H., Anwar, S., Thakur, Y., 2022. Fabrication and characterization of semi-crystalline and amorphous dielectric polymer films for energy storage. Materials Sciences and Applications 13 (11), 559–568. Available at https://doi.org/10.4236/msa.2022.1311034.

Bala, P., Jathar, S., Kale, S., Pal, K., 2014. Transdermal drug delivery system (TDDS)-a multifaceted approach for drug delivery. Journal of Pharmaceutical Research 8 (12), 1805–1835.

Bao, Y., 2021. Controlling molecular aggregation-induced emission by controlled polymerization. Molecules 26 (20), 6267. Available at https://doi.org/10.3390/molecules26206267.

Basak, S., Chatterjee, R., Bandyopadhyay, A., 2024. Beyond traditional stimuli: exploring salt-responsive bottlebrush polymers—trends, applications, and perspectives. ACS Omega 9 (31), 33365–33385. Available at https://doi.org/10.1021/acsomega.4c06137.

Beaumont, C., Lapointe, R., Beaupré, P., Trudeau, C., Le Bouch, N., Leclerc, M., 2022. Printed temperature sensor based on self-doped conducting polymers. Flexible and Printed Electronics 7 (4), 044006. Available at https://doi.org/10.1088/2058-8585/ac9e65.

Bjørken, O.U., Andresen, B., Eikevåg, S.W., Steinert, M., Elverum, C.W., 2022. Thermal layer design in fused filament fabrication. Applied Sciences 12 (14), 7056. Available at https://doi.org/10.3390/app12147056.

Cheng, Y., Zhang, X., Qin, Y., Dong, P., Yao, W., Matz, J., et al., 2021. Super-elasticity at 4 K of covalently crosslinked polyimide aerogels with negative Poisson's ratio. Nature Research, China Nature Communications 12 (1). Available at https://doi.org/10.1038/s41467-021-24388-y.

Cho, C., Nam, S.L., de la Mata, A.P., Harynuk, J.J., Elias, A.L., Chung, H.J., et al., 2022. Investigation of the accelerated thermal aging behavior of polyetherimide and lifetime prediction at elevated temperature. Journal of Applied Polymer Science 139 (15). Available at https://doi.org/10.1002/app.51955.

Deacy, A.C., Gregory, G.L., Sulley, G.S., Chen, T.T.D., Williams, C.K., 2021. Sequence control from mixtures: switchable polymerization catalysis and future materials applications. Journal of the American Chemical Society 143 (27), 10021–10040. Available at https://doi.org/10.1021/jacs.1c03250.

Deshmukh, A.A., Wu, C., Yassin, O., Chen, L., Shukla, S., Zhou, J., et al., 2023. Effect of fluorine in redesigning energy-storage properties of high-temperature dielectric polymers. ACS Applied Materials and Interfaces 15 (40), 46840–46848. Available at https://doi.org/10.1021/acsami.3c08858.

Dong, J., Li, L., Qiu, P., Pan, Y., Niu, Y., Sun, L., et al., 2023. Scalable polyimide-organosilicate hybrid films for high-temperature capacitive energy storage. Advanced Materials 35 (20). Available at https://doi.org/10.1002/adma.202211487.

Dong, J., Hu, R., Xu, X., Chen, J., Niu, Y., Wang, F., et al., 2021. A facile in situ surface-functionalization approach to scalable laminated high-temperature polymer dielectrics with ultrahigh capacitive performance. Advanced Functional Materials 31 (32). Available at https://doi.org/10.1002/adfm.202102644.

Durmaz, S., Erdil, Y.Z., Avci, E., 2021. Improvement of technological properties of wood plastic composites reinforced with glass and carbon fibre fabric. Polymers and Polymer Composites 29 (9_suppl), S1457. Available at https://doi.org/10.1177/09673911211054266.

Feng, Q.K., Zhong, S.L., Pei, J.Y., Zhao, Y., Zhang, D.L., Liu, D.F., et al., 2022. Recent progress and future prospects on all-organic polymer dielectrics for energy storage capacitors. Chemical Reviews 122 (3), 3820–3878. Available at https://doi.org/10.1021/acs.chemrev.1c00793.

Fotoohi, A., Ashenai Ghasemi, F., Fasihi, M., Rajaee, P., 2023. Mechanical behavior and essential work of fracture of surface-modified exfoliated graphite filled polypropylene/ethylene-vinyl acetate blend nanocomposites. Journal of Applied Polymer Science 140 (20). Available at https://doi.org/10.1002/app.53841.

Gao, W., Duan, K., Jin, Q., Wang, H., Zhou, X., 2024. 3D printing high-performance photopolymers based on high-temperature photopolymerization technology. ACS Applied Polymer Materials 6 (19), 12123–12130. Available at https://doi.org/10.1021/acsapm.4c02240.

Gu, W., Wang, G., Zhou, M., Zhang, T., Ji, G., 2020. Polyimide-based foams: fabrication and multifunctional applications. ACS Applied Materials & Interfaces 12 (43), 48246–48258. Available at https://doi.org/10.1021/acsami.0c15771.

Hanemann, T., Klein, A., Baumgärtner, S., Jung, J., Wilhelm, D., Antusch, S., 2023. Material extrusion 3D printing of PEEK-based composites. Polymers 15 (16), 3412. Available at https://doi.org/10.3390/polym15163412.

Hong, S.H., Cho, B.G., 2023. Effect of cooling temperature on crystalline behavior of polyphenylene sulfide/glass fiber composites. Polymers 15 (15). Available at https://doi.org/10.3390/polym15153179.

Hosseini, M.J.M., Yang, Y., Prendergast, A.J., Donati, E., Faezipour, M., Indiveri, G., et al., 2022. An organic synaptic circuit: toward flexible and biocompatible organic neuromorphic processing. Neuromorphic Computing and Engineering 2 (3). Available at https://doi.org/10.1088/2634-4386/ac830c.

Hou, J., Sun, J., Fang, Q., 2023. Recent advance in low-dielectric-constant organosilicon polymers. Chinese Journal of Chemistry 41 (18), 2371–2381. Available at https://doi.org/10.1002/cjoc.202300125.

Jeong, J., Kim, S., Yun, S., Yang, X., Kim, Y.J., 2023. Preparation and characterization of low CTE poly (ethersulfone) using lignin nano composites as flexible substrates. Polymers 15 (14). Available at https://doi.org/10.3390/polym15143113.

Jones, G.R., Wang, H.S., Parkatzidis, K., Whitfield, R., Truong, N.P., Anastasaki, A., 2023. Reversed controlled polymerization (RCP): depolymerization from well-defined polymers to monomers. Journal of the American Chemical Society 145 (18). Available at https://doi.org/10.1021/jacs.3c00589.

Kade, J.C., Dalton, P.D., 2021. Polymers for melt electrowriting. Advanced Healthcare Materials 10 (1). Available at https://doi.org/10.1002/adhm.202001232.

Kafle, P., Zhang, F., Schorr, N.B., Huang, K.Y., Rodríguez-López, J., Diao, Y., 2020. Printing 2D conjugated polymer monolayers and their distinct electronic properties. Advanced Functional Materials 30 (12). Available at https://doi.org/10.1002/adfm.201909787.

Lei, S., Sun, J., Lv, K., Zhang, Q., Yang, J., 2022. Types and performances of polymer gels for oil-gas drilling and production: a review. Gels 8 (6), 386. Available at https://doi.org/10.3390/gels8060386.

Li, F., Liu, W., 2023. Progress in the catalyst for ethylene/α-olefin copolymerization at high temperature. The Canadian Journal of Chemical Engineering 101 (9), 4992–5019. Available at https://doi.org/10.1002/cjce.24854.

Li, H., Gadinski, M.R., Huang, Y., Ren, L., Zhou, Y., Ai, D., et al., 2020. Crosslinked fluoropolymers exhibiting superior high-temperature energy density and charge-discharge efficiency. Energy and Environmental Science 13 (4), 1279–1286. Available at https://doi.org/10.1039/c9ee03603b.

Liu, X.J., Zheng, M.S., Chen, G., Dang, Z.M., Zha, J.W., 2022. High-temperature polyimide dielectric materials for energy storage: theory, design, preparation and properties. Energy and Environmental Science 15 (1), 56–81. Available at https://doi.org/10.1039/d1ee03186d.

Ma, C., Liu, K., Shao, P., Han, D., Wang, K., Yang, M., et al., 2024. Structural evolution and high-temperature sensing performance of polymer-derived SiAlBCN ceramics. Journal of Advanced Ceramics 13 (4), 478–485. Available at https://doi.org/10.26599/jac.2024.9220870.

Malucelli, G., Mariani, A., 2024. Frontal polymerization for the rapid obtainment of polymer composites. ACS Omega 9 (24), 25448–25456. Available at https://doi.org/10.1021/acsomega.4c03017.

Mcguinness, E.K., Leng, C.Z., Losego, M.D., 2020. Increased chemical stability of vapor-phase infiltrated AlOx-poly(methyl methacrylate) hybrid materials. AACS Applied Polymer Materials 2 (3), 1335–1344. Available at https://doi.org/10.1021/acsapm.9b01207.

Modh, J.V., 2023. Synthetic protocols and significance of heterocyclic fluorescent - reinforcing polymers: a comprehensive review. Journal of Advanced Scientific Research 14 (02), 9–39. Available at https://doi.org/10.55218/jasr.202314202.

Mohammed, A.A., Oleiwi, J.K., Al-Hassani, E.S., 2022. Heat treatment effect on biological behavior of polyetheretherketone composites. Journal of Biomimetics, Biomaterials and Biomedical Engineering 54, 119–128. Available at https://doi.org/10.4028/www.scientific.net/JBBBE.54.119.

Mollaoğlu, E.Y., Kaptan, A., 2022. Evaluation of temperature changes in the pulp chamber during bulk-fill composite polymerization: an in-vitro study. Journal of Advanced Oral Research 13 (2), 207–215. Available at https://doi.org/10.1177/23202068221103971.

Murdock, L.A., Benicewicz, B.C., 2024. Teaching an old dog new tricks: synthesis, processing, and application of polybenzimidazole (PBI) membranes. ACS Applied Energy Materials 7 (1), 239–249. Available at https://doi.org/10.1021/acsaem.3c02560.

Nahra, S.R., Oliveira, M.P., de Macedo, E.F., Hurtado, C.R., Tada, D.B., Guerrini, L.M., et al., 2022. Development of scaffolds based in blends of poly(N-vinylcaprolactam) and poly(N-vinylcaprolactam-co-butylacrylate) with poly (3-hexylthiophene) for tissue engineering: Synthesis, processing, characterization, and biological assay. Journal of Applied Polymer Science 139 (41). Available at https://doi.org/10.1002/app.53006.

Nazrin, A., Sapuan, S.M., Zuhri, M.Y.M., Ilyas, R.A., Syafiq, R., Sherwani, S.F.K., 2020. Nanocellulose reinforced thermoplastic starch (TPS), polylactic acid (PLA), and polybutylene succinate (PBS) for food packaging applications. Frontiers in Chemistry 8. Available at https://doi.org/10.3389/fchem.2020.00213.

Nicolau, A., Mutch, A.L., Thickett, S.C., 2024. Applications of functional polymeric eutectogels. Macromolecular Rapid Communications 45 (21). Available at https://doi.org/10.1002/marc.202400405.

Pan, Y., Bai, J., Yang, G., Li, Z., 2023. Progress in the research on branched polymers with emphasis on the Chinese petrochemical industry. Molecules 28 (23), 7934. Available at https://doi.org/10.3390/molecules28237934.

Pyronneau, K., McGuinness, E.K., Gonzalez, A.J., Jean, B.C., Manno, H.V., McClelland, N.R., et al., 2022. Wash fastness of hybrid AlOx-PET fabrics created via vapor-phase infiltration. ACS Applied Polymer Materials 4 (5), 3304–3314. Available at https://doi.org/10.1021/acsapm.1c01851.

Recio-Colmenares, C.L., Ortíz-Rios, D., Pelayo-Vázquez, J.B., Moreno-Medrano, E.D., Arratia-Quijada, J., Torres-Lubian, J.R., et al., 2022. Polystyrene macroporous magnetic nanocomposites synthesized through deep eutectic solvent-in-oil high internal phase emulsions and Fe_3O_4 nanoparticles for oil sorption. ACS Omega 7 (25), 21763–21774. Available at https://doi.org/10.1021/acsomega.2c01836.

Ren, L., Li, H., Xie, Z., Ai, D., Zhou, Y., Liu, Y., et al., 2021. High-temperature high-energy-density dielectric polymer nanocomposites utilizing inorganic core–shell nanostructured nanofillers. Advanced Energy Materials 11 (28). Available at https://doi.org/10.1002/aenm.202101297.

Ren, L., Yang, L., Zhang, S., Li, H., Zhou, Y., Ai, D., et al., 2021. Largely enhanced dielectric properties of polymer composites with HfO2 nanoparticles for high-temperature film capacitors. Composites Science and Technology 201, 108528. Available at https://doi.org/10.1016/j.compscitech.2020.108528.

Ren, W., Yang, M., Zhou, L., Fan, Y., He, S., Pan, J., et al., 2022. Scalable ultrathin all-organic polymer dielectric films for high-temperature capacitive energy storage. Advanced Materials 34 (47). Available at https://doi.org/10.1002/adma.202207421.

Ren, Y., Zhang, L., Xie, G., Li, Z., Chen, H., Gong, H., et al., 2021. A review on tribology of polymer composite coatings. Friction 9 (3), 429–470. Available at https://doi.org/10.1007/s40544-020-0446-4.

Ren, Z., Mujib, S.B., Singh, G., 2021. High-temperature properties and applications of Si-based polymer-derived ceramics: a review. Materials 14 (3), 614. Available at https://doi.org/10.3390/ma14030614.

Saito, T., Nohara, T., Tabata, K., Nakazaki, H., Makino, T., Matsuo, Y., et al., 2022. Relationship between the proton conductive performance and water uptake ratio on a filler-filled polymer electrolyte membrane. Energy & Fuels 36 (22), 13924–13929. Available at https://doi.org/10.1021/acs.energyfuels.2c02527.

Shao, S., Wang, L., 2020. Through-space charge transfer polymers for solution-processed organic light-emitting diodes. Aggregate 1 (1), 45–56. Available at https://doi.org/10.1002/agt2.4.

Singha, S., Pan, S., Tallury, S.S., Nguyen, G., Tripathy, R., De, P., 2024. Recent developments on cationic polymerization of vinyl ethers. ACS Polymers Au 4 (3), 189–207. Available at https://doi.org/10.1021/acspolymersau.3c00055.

Sun, Z., Wu, F., Gao, H., Cui, K., Xian, M., Zhong, J., et al., 2020. A dexamethasone-eluting porous scaffold for bone regeneration fabricated by selective laser sintering. ACS Applied Bio Materials 3 (12), 8739–8747. Available at https://doi.org/10.1021/acsabm.0c01126.

Vijayan, J.G., Prabhu, T.N., Jineesh, A.G., 2024. Nanocellulose as an ecofriendly source in nanoscience: synthesis, characterization, properties, and applications—a review. Macromolecular Symposia 413 (2). Available at https://doi.org/10.1002/masy.202300113.

Wahlen, C., Blankenburg, J., Von Tiedemann, P., Ewald, J., Sajkiewicz, P., Müller, A.H.E., et al., 2020. Tapered multiblock copolymers based on farnesene and styrene: impact of biobased polydiene architectures on material properties. Macromolecules 53 (23), 10397–10408. Available at https://doi.org/10.1021/acs.macromol.0c02118.

Wang, H., Liu, H., Cao, Z., Li, W., Huang, X., Zhu, Y., et al., 2020. Room-temperature autonomous self-healing glassy polymers with hyperbranched structure. Proceedings of the National Academy of Sciences 117 (21), 11299–11305. Available at https://doi.org/10.1073/pnas.2000001117.

Wang, P., Yao, L., Pan, Z., Shi, S., Yu, J., Zhou, Y., et al., 2021. Ultrahigh Energy Storage Performance of Layered Polymer Nanocomposites over a Broad Temperature Range. Advanced Materials 33 (42). Available at https://doi.org/10.1002/adma.202103338.

Xu, L., Wang, L., Zhang, W., Xue, J., Hou, S., 2022. The reinforced electromagnetic interference shielding performance of thermal reduced graphene oxide films via polyimide pyrolysis. ACS Omega 7 (13), 10955–10962. Available at https://doi.org/10.1021/acsomega.1c06767.

Xu, W., Zhou, C., Ji, W., Zhang, Y., Jiang, Z., Bertram, F., et al., 2024. Anisotropic semicrystalline homopolymer dielectrics for high temperature capacitive energy storage. Angewandte Chemie International Edition 63 (24). Available at https://doi.org/10.1002/anie.202319766.

Yang, M., Yuan, F., Shi, W., Ren, W., Guo, M., Ouyang, C., et al., 2023. Sub-nanowires boost superior capacitive energy storage performance of polymer composites at high temperatures. Advanced Functional Materials 33 (12). Available at https://doi.org/10.1002/adfm.202214100.

Yang, M., Ren, W., Guo, M., Shen, Yang, 2022. High-energy-density and high efficiency polymer dielectrics for high temperature electrostatic energy storage: a review. Small (Weinheim an der Bergstrasse, Germany) 18 (50). Available at https://doi.org/10.1002/smll.202205247.

Yuan, C., Zhou, Y., Zhu, Y., Liang, J., Wang, S., Peng, S., et al., 2020. Polymer/molecular semiconductor all-organic composites for high-temperature dielectric energy storage. Nature Communications 11 (1). Available at https://doi.org/10.1038/s41467-020-17760-x.

Zha, J.W., Dang, Z.M., 2023. High Temperature Polymer Dielectrics: Fundamentals and Applications in Power Equipment. Wiley, pp. 1–382. Available at 10.1002/9783527841059.

Zhang, T., Chen, X., Thakur, Y., Lu, B., Zhang, Q., Runt, J., et al., 2020. A highly scalable dielectric metamaterial with superior capacitor performance over a broad temperature. American Science Advances 6 (4). Available at https://doi.org/10.1126/sciadv.aax6622.

Zhang, Y., Fu, L., Martinez, M.R., Sun, H., Nava, V., Yan, J., et al., 2023. Temperature-responsive bottlebrush polymers deliver a stress-regulating agent in vivo for prolonged plant heat stress mitigation. ACS Sustainable Chemistry and Engineering 11 (8), 3346–3358. Available at https://doi.org/10.1021/acssuschemeng.2c06461.

Zhao, L., Li, T., Wang, B., Chen, K., Hu, X., Liu, M., et al., 2022. Poly(cyclosiloxane–carborane)s for harsh environments. Polymer Chemistry 13 (10), 1328–1334. Available at https://doi.org/10.1039/d1py01323h.

Zhao, T., Peng, J., Zhang, Y., Chen, J., Chen, Y., Sun, W., et al., 2020. Synthesis of ultra-high concentration of salt-resistant polyacrylamide. Polymers for Advanced Technologies 31 (12), 2980–2989. Available at https://doi.org/10.1002/pat.5021.

Zhao, W., Qiu, J., Sakai, E., Wu, H., Zhao, Y., Feng, H., et al., 2024. Ultra-high bonding strength of carbon fiber–doped polyamide–aluminum alloy composite structure achieved by changing crystallinity. Polymer Composites 45 (14), 13089–13098. Available at https://doi.org/10.1002/pc.28688.

Zhou, C., Xu, W., Zhang, B., Zhang, Y., Shen, C., Xu, Q., et al., 2022. Curly-packed structure polymers for high-temperature capacitive energy storage. Chemistry of Materials 34 (5), 2333–2341. Available at https://doi.org/10.1021/acs.chemmater.1c04220.

Zhou, Y., Zhu, Y., Xu, W., Wang, Q., 2023. Molecular trap engineering enables superior high-temperature capacitive energy storage performance in all-organic composite at 200 °C. Advanced Energy Materials 13 (11). Available at https://doi.org/10.1002/aenm.202203961.

2

High-temperature polymers: Methodology and factors affecting development

Babatunde Oladipo[1], Abiola Ezekiel Taiwo[2], Tunde Victor Ojumu[1]

[1]DEPARTMENT OF CHEMICAL ENGINEERING, CAPE PENINSULA UNIVERSITY OF TECHNOLOGY, CAPE TOWN, WESTERN CAPE, SOUTH AFRICA [2]FACULTY OF ENGINEERING, MANGOSUTHU UNIVERSITY OF TECHNOLOGY, DURBAN, SOUTH AFRICA

2.1 Introduction

High-temperature polymers, also referred to as high-performance polymers, belong to an advanced material category known for their remarkable thermal stability and mechanical characteristics. These polymers are engineered to endure elevated temperatures and the often challenging environmental circumstances (Odian, 2004). A practical definition that is gaining recognition is that a high-performance/high-temperature polymer maintains its functional properties at temperatures exceeding 177 °C, even when subjected to various stressors, such as electrical, mechanical, chemical, airflow (static or flowing), radiation, and contact with fluids over both short and long durations (Hergenrother, 2003). The thermal resistance of high-temperature polymers is typically achieved through the incorporation of rigid backbone structures, such as aromatic rings, heterocyclic units, or strong intermolecular forces, which limit segmental motion at elevated temperatures. The presence of these structural motifs not only raises the glass transition and melting temperatures but also improves the dimensional stability, creep resistance, and thermal oxidative stability (Liu et al., 2024). Consequently, these polymers find extensive usage across various sectors and industries, such as in aerospace, automotive, electronics, power generation, oil and gas, and defense (Al Christopher et al., 2021; Luise, 2018), where conventional polymers deteriorate or fail when subjected to high-temperature operational conditions.

The methodology for the development of high-temperature polymers encompasses a range of polymerization techniques, such as the step-growth and chain-growth mechanisms, which employ monomers and conditions specifically designed to yield thermally robust architectures. Advanced polymerization strategies, such as emulsion, suspension, solution, and bulk techniques, are optimized to control the molecular weight, dispersity, and morphology of the resulting polymer, which in turn influences the thermal and mechanical behavior of the final polymer (Azapagic et al., 2003; Odian, 2004).

High-Temperature Polymeric Materials. DOI: https://doi.org/10.1016/B978-0-443-32792-6.00018-X

Numerous factors affect the development and ultimate performance of high-temperature polymers. These include appropriate monomer selection, polymer backbone rigidity, degree of crystallinity, crosslinking density, and processing conditions, such as the curing temperature and solvent removal rate (McKeen, 2014). Moreover, additives, such as fillers, plasticizers, and stabilizers, can be incorporated to enhance thermal performance, improve processability, or impart multifunctionality (Kosmalska and Masek, 2022). An understanding of how these variables interact is critical for the design of polymers for specific high-temperature applications.

This chapter, therefore, offers a comprehensive overview of the methodologies and the variables that influence the advancement of high-temperature polymers. It delves into the crucial elements of the polymerization techniques that are utilized to engineer polymers with outstanding thermal stability and mechanical attributes. Additionally, the chapter examines the pivotal roles that are essential and played by these factors, hence influencing the development and applicability of these advanced materials.

2.2 Classes of polymerization

In a simple definition, polymers are made of very large molecules, also called macromolecules, resulting from the combination of numerous small molecules, known as monomers. The chemical process for the synthesis of polymers from monomers is referred to as polymerization, which encompasses a range of chemical mechanisms in various polymerization reactions. Since the functionality of monomers directly influences the properties of the resulting polymers, the methods for the development of high-temperature polymers typically commence with the selection of the appropriate monomers and the desirable polymerization techniques (Su, 2013). Polymers can be categorized based on the various acceptable standards necessitated by their multifunctional nature. Based on the structure desired, polymerization is classified as addition and condensation polymerizations, whereas it can further be classified into step-growth polymerization and chain polymerization, based on its kinetic mechanism (Saldivar-Guerra and Vivaldo-Lima, 2013). The subsequent sections detail the classification of polymerization based on the types of reactions employed in the synthesis.

2.2.1 Classification based on mechanism

An understanding of the kinetic mechanism of polymerization is critically essential for the control of the properties of the resulting polymers, hence leading to optimized processes and the development of new materials for specific applications. The major classification of polymerization, based on the mechanism of polymerization, is explained as follows:

2.2.1.1 Step-growth polymerization

Step-growth polymerization is a type of polymerization mechanism where bifunctional (containing two reactive sites) or multifunctional (with more than two reactive sites) monomers undergo condensation or addition reactions to form oligomers and, subsequently, long-chain polymers. In this mechanism, the polymer chains grow gradually in a stepwise process, where the functional groups of the monomers react through covalent bonds, leading to the gradual formation of larger molecular structures (Yokozawa and Ohta, 2016). For example, monomers react to form dimers, which then further react with monomers to form trimers. The rate of polymerization is independent of the molecular size of the reacting species, and the reaction continues until most of the functional groups have reacted. Step-growth polymerization employs various chemical reactions, depending on the nature of the functional groups involved. The most common reactions include esterification, amidation, urethanization, and etherification, which are reversible (Asua, 2008; Rogers et al., 2003). The reversibility of these reactions is influenced by factors such as temperature, catalyst presence, and reaction conditions. For instance, esterification and amidation can be driven to completion by continuously removing the by-product, whereas urethanization and etherification may involve mechanisms that reduce reversibility. Step-growth polymerization is characterized by a progressive increase in molecular weight over time (Gupta and Kumar, 2012). To produce a polymer with an adequate molecular weight for real-world usage, highly significant conversions of the reactive groups are necessary. Thus, the step-growth polymerization technique is subject to strict conditions for the development of the desired polymers, which include the absence of side reactions and the need for a favorable equilibrium (Yilmaz and Yagci, 2020).

Notable examples of high-temperature polymers, synthesized by using the step-growth polymerization route, include polyimides (PIs), polybenzimidazoles (PBIs), polyphenylene sulfides (PPSs), polyether-ether-ketones, and polyquinoxalines (Choi and McAuley, 2007). A representative reaction for a step-growth polymerization is the synthesis of a polyester through the esterification reaction between a diol (a compound with two alcohol groups) and a dicarboxylic acid (a compound with two carboxyl groups). The reaction occurs in a stepwise manner as follows:

Step 1: Formation of a dimer

$$\mathrm{HO-R-OH+HOOC-R'-COOH\rightarrow HO-R-OCO-R'-COOH+H_2O}$$

Step 2: Growth into an oligomer

$$\mathrm{HO-R-OCO-R'-COOH+HO-R-OH\rightarrow HO-R-OCO-R'-COO-R-OH+H_2O}$$

Step 3: Further polymerization

$$\mathrm{HO-R-OCO-R'-COO-R-OH+HOOC-R'-COOH\rightarrow HO-R-OCO-R'-COO-R-OCO-R'-COOH+H_2O}$$

Step 4: Formation of a long-chain polymer

$$n\mathrm{(HO-R-OH)}+n\mathrm{(HOOC-R'-COOH)}\rightarrow \mathrm{[O-R-OCO-R'-CO]}_n+n\mathrm{H_2O}$$

The reaction continues until most of the functional groups have reacted, thereby ensuring the development of long polymer chains that are essential for high-performance material applications.

2.2.1.2 Chain-growth polymerization

Chain-growth polymerization is a form of polymerization mechanism where polymer chains grow by the successive addition of monomers to an active center (also known as the initiation sites). During this process, monomers can bond exclusively with active chains, induced by either an initiator or a catalyst. Initiators are compounds that produce reactive species (such as free radicals, cations, anions, or coordination complexes) that are necessary to initiate/start polymerization, while initiation sites are the specific locations on the monomers, where these reactive species interact to form the initial active centers of the polymer chains (Gody et al., 2016). The polymerization process occurs in distinct stages: initiation, propagation, termination, and, in some cases, chain transfer. The polymer chain grows rapidly until termination occurs, either by coupling, disproportionation, or chain transfer, hence halting polymer growth. Unlike step-growth polymerization, chain polymerization often results in polymers with narrower molecular weight distributions. This mechanism of polymerization can be categorized based on the active center involved. These categories include coordination polymerization, free-radical polymerization, anionic polymerization, and cationic polymerization, which have catalyst, radical, anion, and cation as the active centers, respectively (Asua, 2008). These polymerization mechanisms are further explained in the following subsections.

2.2.1.2.1 Coordination polymerization

Coordination polymerization is a specialized type of chain-growth polymerization that involves the use of transition metal catalysts to initiate and control the polymerization of monomers. The mechanism of coordination polymerization begins with the coordination of the monomer to an active metal center within the catalyst. The metal complex, often a transition metal, such as titanium, zirconium, or nickel, forms a coordination bond with the monomer, activating it for insertion into the growing polymer chain. Propagation occurs through the repeated monomer insertions, guided by the steric and electronic effects of the metal catalyst (Chen, 2009; Chien, 2012). The catalyst efficiently regulates the insertion process, thereby enabling precise control over the microstructure of the polymer. Coordination polymerization prevents unwanted chain-transfer and termination reactions, hence leading to polymers with uniform structures and superior properties (Asua, 2008; Chen, 2009). Various catalytic systems are employed in coordination polymerization, with Ziegler–Natta catalysts and metallocene catalysts being the most widely used. Ziegler–Natta catalysts comprise transition metal chlorides (e.g., $TiCl_4$) combined with organoaluminum compounds (e.g., $AlEt_3$), which enables the polymerization of α-olefins, such as ethylene and propylene (Soares and Pérez, 2013). Metallocene catalysts, based on single-site organometallic complexes, offer enhanced control over polymer stereochemistry, resulting in isotactic, syndiotactic, or atactic polymers with tailored physical properties. Postmetallocene catalysts and single-site catalysts have further expanded the capabilities of coordination polymerization, thereby allowing for the precise synthesis of specialty polymers (Resconi et al., 2000).

Examples of high-temperature polymers, synthesized through the coordination polymerization route, are ultrahigh molecular weight polyethylene, isotactic polypropylene, poly(1-butene), poly(4-methylpentene-1), and certain fluorinated polyolefins.

The polymerization of ethylene ($CH_2{=}CH_2$) by using a Ziegler–Natta catalyst is a common example of a coordination polymerization reaction. The catalyst system consisted of a transition metal compound (titanium tetrachloride: $TiCl_4$) and a cocatalyst (triethylaluminium: $Al(C_2H_5)_3$). The reactions are as follows:

Step 1: Activation of catalyst

The Ziegler–Natta catalyst is prepared by reacting $TiCl_4$ with $Al(C_2H_5)_3$. The alkyl aluminum reduces Ti(IV) to an active Ti(III) species and simultaneously forms a Ti–C bond that serves as the initiation site.

$$TiCl_4 + Al(C_2H_5)_3 \rightarrow [Ti - C_2H_5]^* + \text{byproducts}$$

Step 2: Monomer coordination

Ethylene coordinates to the vacant site on the titanium metal center.

$$CH_2 = CH_2 + [Ti - C_2H_5]^* \rightarrow [Ti] - (C_2H_5)(CH_2 = CH_2)$$

Step 3: Insertion into the growing chain (propagation)

The coordinated ethylene moiety undergoes migratory insertion into the Ti–C bond, thereby extending the polymer chain.

$$[Ti] - CH_2CH_3 + CH_2 = CH_2 \rightarrow [Ti] - CH_2CH_2CH_2CH_3$$

This step is repeated, leading to chain propagation:

$$[Ti] - (CH_2CH_2)_n - CH_2CH_3 + CH_2 = CH_2 \rightarrow [Ti] - (CH_2CH_2)_{n+1} - CH_2CH_3$$

Step 4: Termination (optional)

Termination can occur through various mechanisms, such as the β-hydride elimination or reaction with water or alcohols (e.g., during catalyst deactivation).

$$[Ti] - (CH_2CH_2)_n - CH_2CH_3 + H_2O \rightarrow (CH_2CH_2)_n - CH_2CH_3(\text{polyethylene}) + Ti - OH$$

2.2.1.2.2 Free-radical polymerization

Free-radical polymerization is a widely utilized polymerization process in which the polymer chains grow via the successive addition of monomers to active free-radical sites. This polymerization mechanism is particularly useful for the synthesis of a variety of thermoplastic and elastomeric materials. The process is initiated by free radicals, which are highly reactive species, generated through the thermal decomposition of organic peroxides, photochemical generation of radicals by using ultraviolet light, and redox initiation involving oxidizing and reducing agents (Asua, 2008). The initiation step involves the decomposition of an initiator, such as benzoyl peroxide (BPO) or azobisisobutyronitrile (AIBN), to generate free radicals. These radicals react with monomer molecules, forming a reactive monomer radical that serves as the starting point for the polymer chain growth. During propagation, additional monomer units are successively added to the growing polymer radical, leading to chain elongation. The reaction continues until termination occurs, which can happen through combination (two radicals react to form a single polymer molecule) or disproportionation (hydrogen transfer between radicals, hence, leading to distinct polymer chains) (Matyjaszewski and Davis, 2002). Free-radical copolymerization is appealing due to its ability to polymerize a vast array of monomers. Its compatibility with various media (such as organic and aqueous) and its comparatively high resistance to impurities are desirable attributes (Hutchinson and Penlidis, 2007; Saldivar-Guerra and Vivaldo-Lima, 2013). The choice of reaction conditions and initiator systems directly influences the molecular weight, polydispersity, and structural properties of the synthesized polymers. The control of the temperature of the reactor is critical for ensuring the quality of the polymer and the safety of the operation, since free-radical polymerization is an extremely exothermic process (Furuncuoglu et al., 2010). Most radical polymerizations are irreversible at temperatures that are widely employed in the industry. Nevertheless, a complete conversion is unattainable due to the reversibility of the reaction at elevated temperatures, which occurs as a result of its exothermic nature (Meira and Kiparissides, 2007; Saldivar-Guerra and Vivaldo-Lima, 2013).

Some notable examples of the high-performance polymers that are often developed through free radical polymerization routes include poly(methyl methacrylate), polyacrylonitrile, polytetrafluoroethylene (PTFE), and polyvinylidene fluoride (PVDF).

An example of a free-radical polymerization reaction is the polymerization of styrene ($C_6H_5CH{=}CH_2$) by using BPO as an initiator:

Step 1: Initiation

The process begins with the thermal decomposition of BPO, which generates two benzoyloxy radicals:

$$C_6H_5COO - OOCPh \rightarrow 2C_6H_5COO \bullet$$

where Ph represents the C_6H_5 group

Each benzoyloxy radical ($C_6H_5COO\bullet$) rapidly loses carbon dioxide to form a phenyl radical:

$$C_6H_5COO\bullet \rightarrow C_6H_5 \bullet + CO_2$$

The phenyl radical initiates the polymerization by attacking a styrene monomer:

$$C_6H_5 \bullet + CH_2 = CHPh \rightarrow C_6H_5 - CH_2 - CH \bullet Ph$$

This forms the first activated styrene radical.

Step 2: Propagation
The active radical reacts with additional styrene monomers, hence propagating the chain:

$$C_6H_5 - CH_2 - CH \bullet Ph + nCH_2 = CHPh \rightarrow C_6H_5 - [CH_2 - CH(Ph)]_n \bullet$$

This step repeats multiple times, leading to the growth of the polystyrene (PS) chain.
Step 3: Termination

Termination occurs either by combination or disproportionation. One common route is the radical coupling, where two growing radical chains combine:

$$[CH_2 - CH(Ph)]_n \bullet + [CH_2 - CH(Ph)]_m \bullet \rightarrow [CH_2 - CH(Ph)]_{n+m}$$

This results in the formation of the final polymer, PS, with a stable, nonradical end.

2.2.1.2.3 Anionic polymerization

Anionic polymerization occurs when negatively charged species, typically an anion, initiate the polymerization of the monomers that contain electron-withdrawing groups (Morton, 2012). In the initiation step, a strong nucleophile, such as an alkali metal alkoxide, organolithium compound, or sodium naphthalene radical anion, reacts with a monomer, generating reactive anionic species. The commonly used monomers in anionic polymerization include styrene, butadiene, isoprene, and methacrylate derivatives, as they contain electron-withdrawing groups that stabilize the anionic propagating species. Propagation occurs through the continuous addition of monomers to the growing polymer chain, thereby maintaining the anionic active site (Asua, 2008; Hadjichristidis et al., 2001). Anionic polymerization does not involve chain transfer or spontaneous termination. The polymerization can be terminated intentionally by using electrophiles, such as water, alcohols, or carbon dioxide, to cap the chain end with a functional group (Quirk, 2013). In living anionic polymerization, the active species remain intact, hence allowing for a block copolymer formation by the sequential monomer addition route (Hirao et al., 2014).

The most common reaction types employed in anionic polymerization include the nucleophilic addition to double bonds, ring-opening polymerization of cyclic ethers, and the controlled deprotonation of acidic monomers (Morton, 2012). Examples of high-temperature polymers, synthesized by using the anionic polymerization procedure, include poly(styrene-butadiene-styrene), polyisoprene, and polydienes, which are commonly used in high-performance elastomers and thermoplastic materials.

An example of anionic polymerization is the polymerization of styrene ($C_6H_5CH{=}CH_2$) by using butyllithium (C_4H_9Li) as an initiator:

Step 1: Initiation
In this step, the butyllithium donates an electron pair to the styrene monomer, forming a carbanion structure at the β-position of the vinyl group ($-CH{=}CH_2$), stabilized by the phenyl ring, along with lithium as a counterion.

$$C_4H_9Li + CH_2 = CH - C_6H_5 \rightarrow C_4H_9 - CH_2 - CH^{-}C_6H_5 + Li^{+}$$

Step 2: Propagation
The anionic species reacts with additional styrene monomers in a stepwise fashion, thereby extending the polymer chain with each new addition, while retaining the anionic active center.

$$C_4H_9 - CH_2 - CH^{-}C_6H_5 + nCH_2 = CH - C_6H_5 \rightarrow C_4H_9 - [CH_2 - CH(C_6H_5)]_n^{-} + Li^{+}$$

Step 3: Termination

Termination occurs when the active chain end reacts with a proton source, e.g., water, hence, yielding the final neutral polymer product and a hydroxide ion.

$$C_4H_9 - [CH_2 - CH(C_6H_5)]_n^{-} + H_2O \rightarrow C_4H_9 - [CH_2 - CH(C_6H_5)]_n - H + OH^{-}$$

2.2.1.2.4 Cationic polymerization

Cationic polymerization is initiated by an electrophilic species, typically a strong acid or Lewis acid, which generates a positively charged active center. This polymerization technique is highly effective for monomers that contain electron-donating groups, such as alkenes with electron-rich substituents (Asua, 2008; Plesch, 2016). In the initiation step, a strong electrophile, such as boron trifluoride (BF_3) or aluminum chloride ($AlCl_3$), interacts with a coinitiator, e.g., water or an alcohol, to form a reactive protonated species. This electrophilic species then reacts with a monomer, thereby generating a carbocation that serves as the active center for polymerization. Propagation occurs as additional monomer molecules successively add to the growing carbocation, thus leading to chain elongation. This step continues rapidly, as the highly reactive carbocation intermediate facilitates a rapid monomer addition. Termination can occur *via* several mechanisms, such as proton transfer to a nucleophile, counterion recombination, or chain transfer to a monomer or solvent (Aoshima and Kanaoka, 2009). Additionally, the presence of adventitious moisture or impurities can lead to premature termination, making stringent reaction conditions necessary for the control of the polymer growth.

Most common cationic polymerization reactions include electrophilic addition to alkenes, ring-opening polymerization of cyclic ethers, and controlled cationic rearrangements (Tardy et al., 2017). High-temperature polymers, synthesized by using cationic polymerization, include polyisobutylene, poly(vinyl ether), and PPS.

A typical example of cationic polymerization is the polymerization of isobutylene ($CH_2{=}C(CH_3)_2$) by using boron trifluoride (BF_3) as an initiator:

Step 1: Initiation
Formation of a protonic acid species:

$$BF_3 + H_2O \rightarrow H^{+} + BF_3OH^{-}$$

The proton (H^+) is generated and then reacts with isobutylene to form a stable tertiary carbocation:

$$H^+ + CH_2 = C(CH_3)_2 \rightarrow CH_3 - C^+(CH_3) - CH_3$$

This tertiary carbocation is stabilized by the inductive and hyper-conjugative effects of the methyl groups, hence making it an ideal propagating center for cationic polymerization.

Step 2: Propagation

The reactive carbocation attacks another isobutylene monomer, hence continuing the chain growth:

$$CH_3 - C^+(CH_3) - CH_3 + CH_2 = C(CH_3)_2 \rightarrow CH_3 - C(CH_3)_2 - CH_2 - C^+(CH_3) - CH_3$$

This step repeats, with the growing polymer chain, thereby retaining a carbocation at the chain end:

$$\rightarrow [CH_2 - C(CH_3)_2]_n - C^+(CH_3) - CH_3$$

Step 3: Termination

Termination typically occurs by reaction with water or other proton donors, thereby quenching the active carbocation:

$$[CH_2 - C(CH_3)_2]_n - C^+(CH_3) - CH_3 + H_2O \rightarrow [CH_2 - C(CH_3)_2]_n - CH(CH_3)2 + H_3O^+$$

2.2.2 Classification based on structure

The classification of the different polymerization schemes, based on the resulting polymer structure, provides a framework for the understanding of the different types of polymers that can be synthesized. This is crucial for the prediction of their properties, processing capabilities, and behavior for potential applications.

2.2.2.1 Addition polymerization

Addition polymerization involves the process in which the unsaturated monomers bond together to form a polymer without the loss of any small molecules, for example, water or alcohol. Typically, addition polymerization is carried out with the use of unsaturated monomers, such as olefins, acetylenes, and aldehydes. This process is alternatively referred to as chain-growth polymerization because the reactions are recognized to progress step-by-step via the reactive intermediates. The reaction produces, very rapidly, a polymer with an exceptionally high molecular weight. Addition polymerization reactions are the most prevalent and are the thermodynamically preferred chemical transformations of olefins (Ivin, 2000; Kariduraganavar et al., 2014). In general, these polymers can be synthesized via emulsion, bulk, solution, and suspension polymerization methods.

2.2.2.2 Condensation polymerization

Condensation polymerization is a process whereby monomers containing two or more functional groups are combined, during which time, small molecules, such as water, alcohol, or ammonia, are released as by-products. The nature of the final product obtained by condensation polymerization is determined by the quantity of the reactive functional end groups that are capable of reacting together to establish covalent bonds that are present in the monomer (Yokozawa and Yokoyama, 2009). If the reacting functional groups/species are within the same molecule, it is called intramolecular condensation. Conversely, when the functional groups are in different molecules, the process is termed intermolecular condensation (Saldivar-Guerra and Vivaldo-Lima, 2013). The monomers that participate in condensation polymerization differ from those involved in addition polymerization. The monomers possess two primary attributes, namely, they include functional groups, such as -OH, $-NH_2$, or -COOH, instead of double bonds, and each monomer has a minimum of two reactive sites (Kariduraganavar et al., 2014). In this method, the achievement of a high molecular weight product is possible only at high levels of conversion, and most of the reactions often have significant activation energies and, therefore, typically necessitate the use of heat.

2.3 Techniques for developing high-temperature polymers

The polymerization techniques can essentially be categorized as homogeneous or heterogeneous according to the various methods of preparation (Asua, 2008; Kariduraganavar et al., 2014). In the context of homogeneous processes, the reaction takes place in a medium that is generated by the blending of the reactants when diluted or pure monomers are directly introduced to one another, whereas the phase boundaries serve as the interphases in heterogeneous processes, where the reactions take place (Kariduraganavar et al., 2014). The classes of polymerization mentioned previously can be carried out through different methods, such as bulk polymerization, solution polymerization, slurry polymerization, suspension polymerization, and emulsion polymerization, some of which are briefly highlighted in the subsequent sections below.

2.3.1 Bulk polymerization

Bulk polymerization is a fundamental polymerization technique in which monomers are polymerized in their pure form without the use of solvents or dispersing agents. The absence of a solvent results in high monomer concentrations, leading to rapid polymerization rates, high polymer yields, and low levels of residual impurities. Bulk polymerization takes place internally, that is, within the monomer. Under the influence of light or heat, additives, including initiators and transfer agents, catalyze the reaction. The controllability of this polymerization process is compromised due to its high exothermicity; consequently, the resulting polymer typically exhibits an uneven molecular mass distribution (Demir et al., 2007). Nevertheless, a chain transfer agent can be used to efficiently alter the molecular-weight distribution. The properties of the polymer may also be altered by manipulating the

Table 2–1 Examples of polymers synthesized via bulk polymerization.

Polymer	Polymerization process	Structure	References
Polyether–ether–ketone	Step-growth polymerization of 4,4'-difluorobenzophenone and hydroquinone	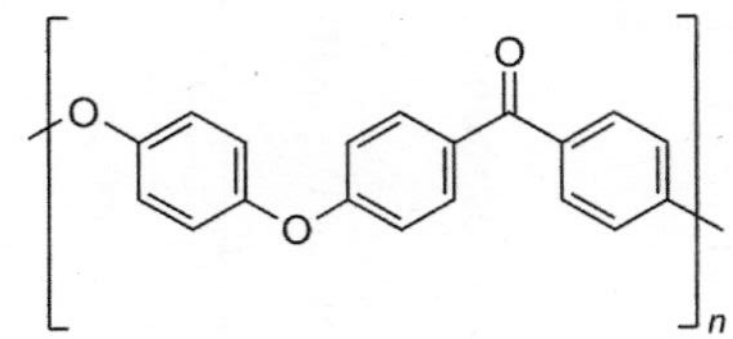	Garcia-Gonzalez et al. (2015), Haleem and Javaid (2019), Rinaldi et al. (2020)
Polyimide	Condensation polymerization of diamines and dianhydrides	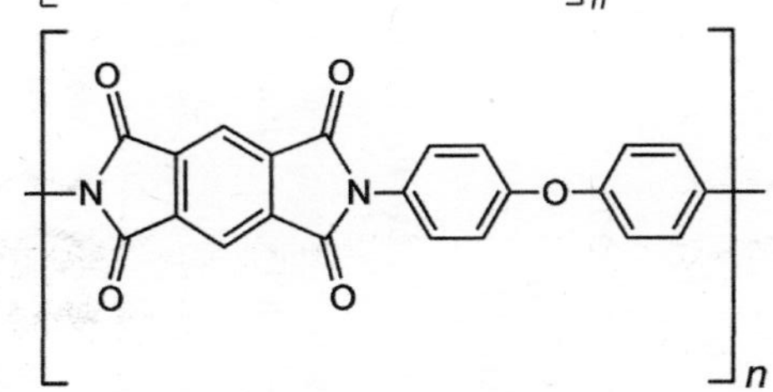	Gouzman et al. (2019), Liaw et al. (2012), Wilson et al. (2013)
Polybenzoxazines	Cationic ring-opening polymerization of phenolic compounds and primary amines	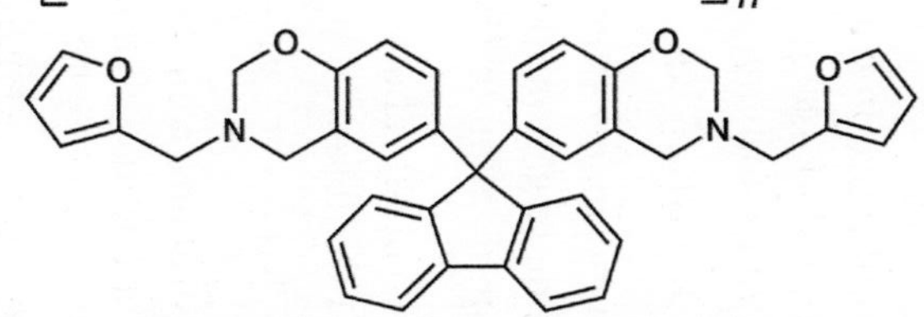	Ghosh et al. (2007), Takeichi and Agag (2006)

temperature and pressure conditions. The polymer is acquired in the form of a porous or granular solid if its monomer is not soluble in the medium. As the formulation predominantly comprises the monomers, the resulting polymer is usually of high purity (Kariduraganavar et al., 2014; Saldivar-Guerra and Vivaldo-Lima, 2013). Examples of the polymers formed through this polymerization process are given in Table 2–1.

A representative of bulk polymerization reaction is the condensation polymerization of diamines (e.g., 4,4'-oxydianiline [ODA]: $H_2N\text{-}C_6H_4\text{-}O\text{-}C_6H_4\text{-}NH_2$) and dianhydrides (e.g., pyromellitic dianhydride [PMDA]: $C_{10}H_2O_6$)

Step 1: Formation of poly(amic acid) (PAA)
In the initial stage, PMDA reacts with ODA via the nucleophilic addition of the amine groups to the anhydride rings, forming a PAA intermediate.

$$nH_2N - R - NH_2 + nO = C - Ar(CO)_2O \rightarrow -[NH - R - NH - CO - Ar - COOH]_n$$

where R=aromatic spacer from the diamine (e.g., $-C_6H_4\text{-}O\text{-}C_6H_4-$) and Ar=aromatic ring from dianhydride (e.g., from PMDA).
Step 2: Thermal imidization to form PI

Upon heating (typically above 200°C), the PAA undergoes intramolecular cyclodehydration to form the PI via the loss of water:

$$-[NH - R - NH - CO - Ar - COOH]_n \rightarrow -[CO - Ar - CO - NH - R - NH]_n + 2nH_2O$$

The resulting structure contains imide rings, 5-membered rings, consisting of two carbonyl groups bound to a nitrogen atom.

2.3.2 Solution polymerization

Solution polymerization is an industrial type of polymerization process, in which a monomer is dissolved in a catalyst-containing nonreactive solvent, which facilitates the reaction and controls the viscosity of the growing polymer chains (Owolabi et al., 2018). Both the monomer and the polymer, produced by this process, are soluble in the solvent. As a result of the solvent absorbing the heat produced during the reaction, the rate of the reaction is decreased. Thus, the solvent acts as a heat sink, thereby reducing the risk of localized overheating, which is particularly important in exothermic polymerization reaction scenarios. After the desired or utmost conversion has been achieved, any remaining solvent must be eliminated to achieve a pure polymer (Saldivar-Guerra and Vivaldo-Lima, 2013). The polymers obtained via this technique have low molecular weights due to the potential for chain transfer. This method is well-suited for the synthesis of wet polymers due to the difficulty of solvent removal. Examples of polymers formed through this polymerization process are given in Table 2–2. Consequently, this method of polymerization is utilized in industrial processes that require solutions of the polymers for ready use, such as in adhesives and surface coatings.

A representation of a typical solution polymerization is the solution radical polymerization of vinylidene fluoride (VDF: $CH_2{=}CF_2$) in polar solvents, for example, dimethylacetamide (DMAc) or *N*-methyl-2-pyrrolidone (NMP) to form PVDF:

Table 2–2 Examples of polymers synthesized via solution polymerization.

Polymer	Polymerization process	Structure	References
Polyacrylonitrile	Free-radical polymerization or anionic polymerization of acrylonitrile monomers in polar solvents such as dimethyl sulfoxide or dimethylformamide	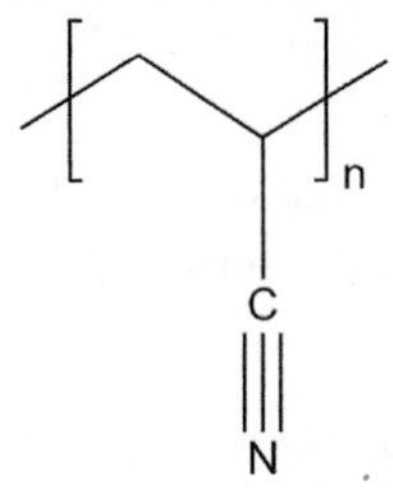	Nataraj et al. (2012), Scharnagl and Buschatz (2001)
Polyvinylidene fluoride	Radical polymerization of vinylidene fluoride monomers in polar solvents like dimethylacetamide or *N*-methyl-2-pyrrolidone	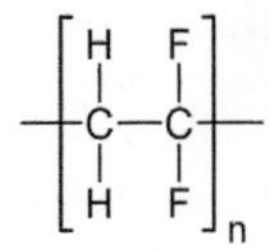	Kang and Cao (2014), Saxena and Shukla (2021)
Polysulfones	Polycondensation of bisphenol-based monomers with chlorinated solvents like dichloromethane or chloroform	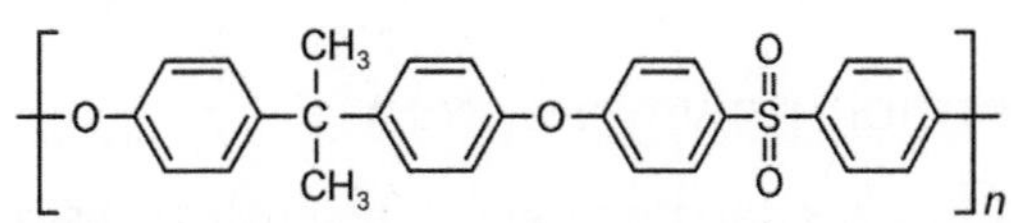	Abdelrasoul et al. (2015), Barth et al. (2000)

Step 1: Initiator (e.g., AIBN) decomposition
Upon heating, AIBN generates two free radicals:

$$AIBN \rightarrow 2\,R\bullet$$

where $R\bullet = \bullet C(CH_3)_2 - CN$ (a 2-cyano-2-propyl radical)
This decomposition typically occurs between 60°C and 80°C in polar aprotic solvents, for example, DMAc or NMP, which help to stabilize the growing radical and dissolve both the initiator and monomer.

Step 2: Initiation
The generated radical (R•) attacks the double bond of VDF ($CH_2{=}CF_2$), thereby forming a new carbon-centered radical on the fluorinated carbon:

$$R\bullet + CH_2 = CF_2 \rightarrow R - CH_2 - C\bullet F_2$$

This marks the start of the polymer chain propagation/growth, with the unpaired electron on the fluorinated carbon being highly reactive for further monomer addition.

Step 3: Propagation

The active radical site continues to react with additional VDF monomers, thereby extending the polymer chain:

$$R - [CH_2 - CF_2]_n - CH_2 - C \bullet F_2 + CH_2 = CF_2 \rightarrow R - [CH_2 - CF_2]_{n+1} - CH_2 - C \bullet F_2$$

Each propagation step adds one repeat unit of the $-CH_2-CF_2-$, hence increasing molecular weight and maintaining the radical on the terminal CF_2 carbon.

Step 4: Termination

Termination typically occurs by radical–radical recombination, where two growing chains couple together to form a final polymer molecule:

$$R - [CH_2 - CF_2]_n - CH_2 - C \bullet F_2 + R' - [CH_2 - CF_2]_m - CH_2 - C \bullet F_2 \rightarrow R - [CH_2 - CF_2]_n - CH_2 - CF_2 - CH_2 - CF_2 - [CH_2 - CF_2]_m - R'$$

This results in the formation of high molecular weight PVDF, a semicrystalline, high-temperature fluoropolymer.

2.3.3 Suspension polymerization

Suspension polymerization represents a heterogeneous radical polymerization technique. In this process, the monomer droplets, containing various components, for example, initiators and modifiers, are dispersed in a solvent, usually water, through a vigorous agitation, hence allowing the polymerization to occur within these droplets (Hedayati et al., 2022; Tian et al., 2022). Due to their insolubility in the liquid phase, the monomer and initiator congregate to form beads or granules within the liquid medium. The monomer droplets act as individual reaction vessels, where polymerization occurs, thereby allowing efficient heat dissipation and minimizing issues, for example, high viscosity, which are common in bulk polymerization (Brooks, 2010). To maintain stability and prevent the agglomeration of the monomer droplets, suspension of reagents, such as polyvinyl alcohol (PVA) or methyl cellulose, is often incorporated. The resulting polymer particles can be easily separated and processed into various forms, therefore making suspension polymerization a valuable method for industrial-scale polymer synthesis (Chaudhary and Sharma, 2019). An inherent benefit of this process lies in the efficient heat transfer during the polymerization process, thereby ensuring the direct control of the reaction conditions. Examples of polymers developed through the suspension polymerization procedure are given in Table 2–3.

Representative reaction for the suspension polymerization of vinyl chloride to produce polyvinyl chloride (PVC), include

Step 1: Preparation of the suspension

Vinyl chloride monomer (VCM: CH_2=CHCl) is dispersed in water with a protective colloid (e.g., PVA or gelatin) to stabilize the droplets. An oil-soluble initiator, for example, BPO (BPO: $C_{14}H_{10}O_4$), is added to the monomer phase.

Table 2–3 Examples of polymers synthesized via suspension polymerization.

Polymer	Polymerization process	Structure	References
Poly(*p*-phenylene sulfide)	Suspension polymerization of *p*-dichlorobenzene and sodium sulfide	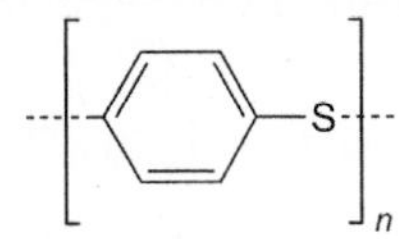	Rahate et al. (2013), Yang et al. (2009)
Polyvinyl chloride	Suspension polymerization of vinyl chloride monomer in water, aided by dispersing agent, and polymerized using initiators	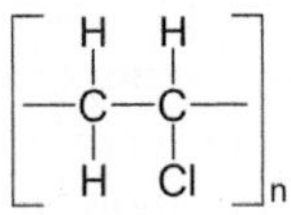	Patrick (2005)
Poly(methyl methacrylate)	Suspension polymerization of methyl methacrylate monomers in water with a dispersing agent	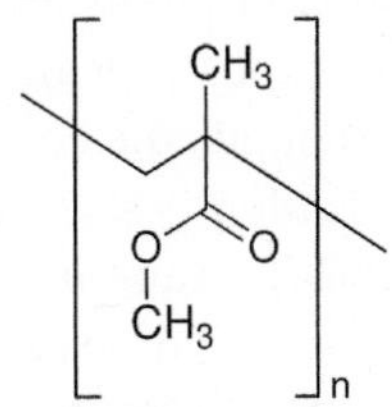	Ali et al. (2015)

Step 2: Initiation

The initiator decomposes upon heating (typically between 50°C and 70°C) to form free radicals:

$$BPO \rightarrow 2\,PhCOO\bullet$$

where Ph stands for the phenyl group, which is the aromatic ring C_6H_5, and COO• is the benzoyloxy radical

$$PhCOO\bullet \rightarrow Ph\bullet + CO_2$$

$$Ph\bullet + CH_2 = CHCl \rightarrow Ph - CH_2 - CH\bullet Cl$$

Step 3: Propagation

The radical reacts with additional VCMs, thereby propagating the chain:

$$Ph - CH_2 - CH\bullet Cl + CH_2 = CHCl \rightarrow Ph - CH_2 - CHCl - CH_2 - CH\bullet Cl$$

This continues:

$$Ph - [CH_2 - CHCl]_n - CH_2 - CH\bullet Cl$$

Step 4: Termination occurs via the combination or disproportionation of two growing chains:

$$[CH_2 - CHCl_2]_n\bullet + [CH_2 - CHCl_2]_m\bullet \rightarrow [CH_2 - CHCl]_{n+m}$$

As the polymer chains grow, the resulting PVC becomes insoluble in the monomer droplets and the precipitates within the suspended beads. These beads are collected, washed, and dried to obtain the raw PVC resin.

2.3.4 Precipitation polymerization

Precipitation polymerization is a form of heterogeneous polymerization that first begins in the continuous phase as a homogeneous system. In this technique, polymer formation occurs in a solvent that dissolves the monomer, but not the resulting polymer (Zhang et al., 2022). As the polymer chains grow and attain a critical molecular weight, they become insoluble in the reaction medium, and they precipitate out as fine particles. By separating the precipitated polymer into particles or gel, the centrifugation or simple filtration procedure may be utilized. The level of polymerization is significantly elevated owing to the absence of heat dissipation issues (Yang et al., 2020). Examples of polymers developed through this means are poly(aryl ether ketone), which is synthesized by polymerizing aromatic dihalides and aromatic bisphenols in high-boiling solvents, and PBI, developed by polymerizing tetra-aminobiphenyl and diphenyl isophthalate in polyphosphoric acid (Chen et al., 2017; Chung, 1997).

Representative reaction is the precipitation polymerization of tetrafluoroethylene (TFE: CF_2=CF_2) to form PTFE:

Step 1: Initiation
In water-insoluble initiators such as ammonium persulfate (APS: $(NH_4)_2S_2O_8$), under pressure and heat, to generate free radicals, with polymerization procedures, thus:

$$\text{APS} \rightarrow 2\text{SO}_4^- \bullet$$

Step 2: Propagation
TEF monomers are introduced, and the radicals will then initiate the polymerization:

$$\text{SO}_4^- \bullet + \text{CF}_2 = \text{CF}_2 \rightarrow \text{SO}_4^- - \text{CF}_2 - \text{CF}_2 \bullet$$

$$\text{SO}_4^- - \text{CF}_2 - \text{CF}_2 \bullet + n(\text{CF}_2 = \text{CF}_2) \rightarrow \text{SO}_4^- - [\text{CF}_2 - \text{CF}_2]_n - \text{CF}_2 - \text{CF}_2 \bullet$$

Step 3: Precipitation
As the chain grows, PTFE becomes insoluble in the aqueous medium and hence precipitates out:

$$\text{SO}_4^- - [\text{CF}_2 - \text{CF}_2]_n - \text{CF}_2 - \text{CF}_2\bullet \rightarrow [\text{CF}_2 - \text{CF}_2]_{n+1}(\text{PTFE precipitate})$$

Step 4: Termination

The termination process occurs via a combination of radicals or disproportionation, thereby forming a stable polymer chain:

$$[CF_2 - CF_2]_n \bullet + [CF_2 - CF_2]_m \bullet \rightarrow [CF_2 - CF_2]_{n+m}$$

2.3.5 Emulsion polymerization

Emulsion polymerization is a type of radical polymerization that occurs in an aqueous medium, where monomer droplets are emulsified by using surfactants, and polymerization takes place within micelles. This polymerization technique relies on a water-immiscible monomer that can be dispersed in water and stabilized by surfactants, with the polymerization initiated by water-soluble initiators, such as potassium persulfate ($K_2S_2O_8$). An oil–water emulsion is the prevailing form of emulsion polymerization (Lovell and Schork, 2020). The mechanism of emulsion polymerization involves three main stages, namely micelle formation, particle nucleation, and particle growth (Eliseeva et al., 2012). Initially, the monomers are dispersed in water with the aid of surfactants, thereby forming micelles. A water-soluble initiator generates free radicals in the aqueous phase, which diffuse into the micelles that contain the monomer molecules. Once a radical enters a micelle, it initiates polymerization, leading to the formation of a growing polymer particle. As polymerization continues, additional monomer diffuses from the large monomer droplets into the micelles, sustaining chain propagation. Termination typically occurs by radical–radical recombination within the micelle. This mechanism ensures the formation of colloidal polymer particles suspended in water, often resulting in latexes with high solids content and stability (Chern, 2006). Certain water-soluble polymers, including hydroxyethyl celluloses and PVAs, can also function as emulsifiers and stabilizers. The spontaneous formation of latex particles, during the initial minutes of the process, facilitates polymerization. As the polymer molecules are encapsulated within the particles, the viscosity of the reaction medium closely resembles that of water and remains unaffected by the molecular weight. An additional advantage of emulsion polymerization is its environmental friendliness, as it uses water as the dispersion medium, reducing the need for organic solvents. Emulsion polymerizations are engineered to achieve high monomer-to-polymer conversion rates, which can lead to substantial chain transfer to polymer (Thickett and Gilbert, 2007). Examples of high-performance polymers that can also be synthesized by using this polymerization technique are PPS, PS, PI, polybenzoxazine, and PBI.

Representative reaction for the formation of a high-temperature polymer via emulsion polymerization, using styrene (CH_2=CH–Ph) as a model monomer to form PS:

Step 1: Initiator decomposition (in aqueous phase)
A water-soluble initiator, such as potassium persulfate ($K_2S_2O_8$), decomposes thermally to produce sulfate radicals:

$$K_2S_2O_8 \rightarrow 2K^+ + 2SO_4^- \bullet$$

Step 2: Radical entry into micelles
The generated sulfate radical enters a micelle (a small droplet of monomer that is stabilized by surfactant) and initiates polymerization of styrene:

$$SO_4^{-}\bullet + CH_2 = CH - Ph \rightarrow SO_4 - CH_2 - CH\bullet - Ph$$

Step 3: Propagation
The activated monomer radical adds additional styrene monomers within the micelle, thereby forming a growing polymer chain:

$$SO_4 - CH_2 - CH\bullet - Ph + nCH_2 = CH - Ph \rightarrow SO_4 - [CH_2 - CH(Ph)]_n - CH\bullet - Ph$$

Step 4: Termination (combine or disproportionation)

$$[CH_2 - CH(Ph)]_n - CH\bullet + \bullet CH - [CH(Ph)]_m \rightarrow [CH_2 - CH(Ph)]_{n+m}\,\text{(polystyrene)}$$

2.4 Factors affecting the development of high-temperature polymers

The development of high-temperature polymers is influenced by a myriad of factors, which dictate their thermal stability, mechanical properties, and their overall functionality in demanding environments (Novikov and Rössler, 2013). By understanding these factors, researchers and scientists/engineers can tailor the properties of high-temperature polymers to meet specific application requirements. The subsections delve into such key factors.

2.4.1 Processing conditions

The processing conditions, such as temperature, pressure, curing time, and environmental factors, play a crucial role in determining the structural and functional characteristics of high-temperature polymers. These variables must be meticulously optimized to achieve the desired thermal, mechanical, and chemical properties (Schawe, 2014). Among these, processing temperature is particularly critical, since it influences the degree of polymerization, molecular orientation, crystallinity, and crosslinking density, all of which govern the final thermal performance of the material. High-temperature polymers often require elevated polymerization and curing temperatures to ensure complete conversion of monomers and the formation of a well-defined network (Liao et al., 2023). Inadequate curing can result in under-crosslinked regions, hence reducing the polymer's thermal stability and compromising its mechanical integrity under heat stress. Curing time and pressure also influence the uniformity of the network structure, thereby affecting void formation, residual stress, and dimensional stability. A key property in the evaluation of thermal performance is the glass transition temperature, defined as the temperature range over which a polymer transitions from a hard, glassy state to a soft, rubbery phase. A high glass transition temperature is desirable in high-temperature applications, since it indicates the fact that the material can retain its structural integrity at elevated temperatures (Tao et al., 2021). Many factors, such as molecular weight, chain rigidity, and chemical structure, have been associated with the glass transition temperatures of polymers (Novikov and Rössler, 2013; Tao et al., 2021); however, it is also highly sensitive to processing conditions, which can modify chain packing, mobility, and free volume within the polymer matrix. In addition, processing additives, such as salts, cosolutes, surfactants, and pH modifiers, can significantly alter the phase behavior of thermos-responsive polymers (Qiao and Wang, 2018). The balance of interactions between polymers and water determines the phase transition temperature. Structural features that boost the polymer–polymer attractive interactions decrease the transition temperature, while those that increase the polymer–water attractive interactions increase the transition temperature (Hoogenboom, 2019; Qiao and Wang, 2018). Therefore, the precise control of the processing variables is essential to the fine-tuning of the thermal transitions, ensures reproducibility, and develops robust high-temperature polymers.

2.4.2 Synthetic strategies

Temperature-responsive polymers have been synthesized by using a variety of synthetic approaches (Das et al., 2020; Liao et al., 2023; Musarurwa and Tavengwa, 2022). The thermal characteristics of polymers can be significantly influenced by the chosen synthesis method, which affects the architecture, molecular weight, and functional group distribution of polymers. Techniques, such as step-growth polymerization and ring-opening polymerization, have been widely used to produce polymers with high thermal resistance attributes, particularly due to their ability to form stable, high-molecular-weight chains with controlled structural features. Several strategies, such as the design of heat-resistant monomers, the selection of appropriate initiators, and the use of crosslinking agents, are often employed during synthesis to enhance high-temperature performance (Balani et al., 2015). The choice of the synthetic method influences the transition temperature of a polymer and its self-assembly behavior. Polymer chains can be crosslinked to reduce molecular mobility at high temperatures, thus enhancing the thermal stability of such crosslinked polymers (Barra et al., 2023). Crosslinking can take various forms, such as chemical or physical cross-linking. Chemical crosslinking, involving agents such as peroxides or multifunctional monomers, leads to permanent networks, while physical crosslinking, such as hydrogen bonding or ionic interactions, offers reversible thermal behavior (Tillet et al., 2011). Both approaches can be tailored, depending on the desired balance between thermal stability and mechanical flexibility. It is crucial to choose suitable monomers when synthesizing polymers. Monomers containing aromatic rings, heterocyclic structures, or other rigid frameworks are often selected to enhance thermal stability. The incorporation of bulky side groups can reduce rotational freedom along the polymer backbone, thereby increasing the glass transition temperature. Thus, the synthetic strategy not only determines the transition temperature of the polymer but also governs its self-assembly behavior, degradation profile, and the overall performance of the polymer under thermal stress.

2.4.3 Molecular weight and distribution

Polymers have different phases, depending on their molecular weights. Different molecular weights can cause different transition temperatures. A higher molecular weight polymer is more thermally stable than its lower molecular weight counterpart. This is because its chains are more entangled and less mobile (Balani et al., 2015), while a lower molecular weight polymer tends to degrade more quickly under heat due to the shorter chains and weaker molecular forces than its higher molecular weight counterpart. The narrow variation in molecular weight of well-defined polymers allows for the exact analysis of their thermal behaviors, since the temperature at which the phase transition that occurs in thermo-responsive polymers is influenced by the molecular weight of these polymers (Qiao and Wang, 2018). Entanglement density, which increases with molecular weight, contributes significantly to thermal resistance, since it impedes segmental motion and chain slippage (Kong et al., 2021). Additionally, chain-end effects, more prominent in low-molecular-weight polymers, can lower thermal resistance due to their increased susceptibility to chain scission and oxidation (Pielichowski et al., 2022). Thus, for high-performance applications, the control of both the absolute molecular weight and its distribution is essential, as it affects not only thermal properties but also processing behavior, viscosity, and mechanical strength.

2.4.4 Compositional factors

The transition temperatures of polymers can also be affected by structural or compositional factors (Qiao and Wang, 2018; Tao et al., 2021). It is essential to understand how monomers interact with each other to comprehend the resulting polymer behavior. The chemical composition of the polymer backbone can significantly influence its thermal stability (Koltzenburg et al., 2023). The thermal stability of aromatic and heterocyclic polymers is higher than that of aliphatic polymers. The incorporation of rigid structural units, such as benzene rings or fused aromatic systems, into the polymer backbone generally leads to an increase in the glass transition temperature and decomposition temperature (T_d) of the resulting polymer due to the restricted chain mobility and enhanced thermal resistance (Pixton and Paul, 2018). The nature and the positioning of substituent groups on the polymer chain also play a critical role. Electron-withdrawing groups, such as $-CF_3$ or $-NO_2$, can enhance thermal stability by delocalizing charge and stabilizing the polymer backbone (Wei et al., 2020), while bulky side groups can hinder chain rotation, thus raising the glass transition temperature (Kunal et al., 2008). However, excessive bulkiness may also disrupt the polymer's crystallinity or the packing efficiency, thereby potentially compromising mechanical integrity. Careful selection and arrangement of the monomer units, functional groups, and comonomer ratios are central to obtaining polymers that can withstand elevated temperatures.

2.4.5 Additives

The thermal stability and mechanical properties of polymers can be improved by the incorporation of thermal stabilizers, plasticizers, flame retardants, antioxidants, UV absorbers, and reinforcing fillers (Shi et al., 2022). These additives have specific functions and must be uniformly dispersed and well-integrated into the polymer mix under optimal temperature conditions. As a result of the actions of these additives, heat resistance is improved, degradation is prevented, and the mechanical properties of the resulting polymers are improved at elevated temperatures (Schawe, 2014). In addition, such additives might influence how the polymers crystallize and how they are formed, structurally. During the decomposition of polymers, excessive heat, hazardous smoke, ash particles, and toxic gases can be released, thereby posing significant fire hazards and health risks, especially in enclosed or populated spaces. To mitigate these dangers, flame/smoke retardants have been widely employed, not only to delay ignition but also to reduce heat release and the evolution of toxic fumes during combustion (Shi et al., 2022). It is important to note that some additives may compromise other properties, such as transparency, flexibility, or long-term aging behavior, thus requiring a balanced formulation strategy.

2.4.6 Crystallinity

Crystallinity pertains to the extent to which a polymer displays a highly organized and repetitive molecular structure, as opposed to being amorphous with a disorganized molecular arrangement. The degree of crystallinity has a significant impact on the thermal, mechanical, and chemical characteristics of high-temperature polymers. Crystallinity enhances thermal stability, since it is generally associated with better heat resistance in high-temperature polymers (Das et al., 2020). Polymers with high degrees of crystallinity tend to exhibit high melting points, improved dimensional stability, and reduced segmental mobility, all of which contribute to enhanced performance under elevated temperature conditions. The densely packed molecular arrangement in the crystalline regions creates strong intermolecular interactions (e.g., the van der Waals forces or hydrogen bonding), which help resist thermal degradation and deformation. The degree of polymer crystallization is influenced by several factors, such as the chemical structure of the polymer, tacticity, molecular weight, and the cooling rate after processing (Schawe, 2014). For instance, slow cooling typically promotes high crystallinity by allowing more time for molecular chains to align and be packed into ordered regions. In contrast, rapid cooling tends to result in a predominantly amorphous structure with reduced thermal resistance. Polymers with a high level of crystallinity often exhibit greater tensile strength and modulus than their low-crystallinity counterparts, owing to the stronger intermolecular bonds and more efficient chain packing within the crystalline phase (Murthy, 2018). This ordered alignment also contributes to better dimensional stability, reduced thermal expansion, and resistance to solvent penetration. However, excessive crystallinity may lead to brittleness or hinder processability. Semicrystalline polymers, which consist of both crystalline and amorphous domains, often offer a desirable compromise by combining structural rigidity with some degree of flexibility and impact resistance.

2.4.7 Immiscibility, stability, and compatibility issues in polymer blends

The development of high-temperature polymers is often complicated by immiscibility, instability, and poor compatibility in polymer blends. These issues can result in phase separation, reduced mechanical integrity, poor thermal stability, compromised dimensional accuracy, and increased susceptibility to environmental degradation. Incompatible blends typically exhibit weak interfacial adhesion, thereby leading to discontinuities in the polymer matrix that negatively affect the overall performance under thermal stress. The compatibility of polymer blends is largely governed by intermolecular interactions, such as hydrogen bonding, dipole–dipole interactions, π–π stacking, and van der Waals forces. When such interactions are weak or absent, the polymer phases tend to segregate, resulting in immiscibility. The thermodynamic miscibility of polymers is also influenced by their molecular structure, solubility parameters, and polarity. Processing conditions, particularly melt temperature and pressure, have a direct effect on chain mobility, phase morphology, and the stability of the resulting blend (DeMeuse, 2014; Liao et al., 2023; Utracki, 2014). For effective melt processing, it is essential to maintain a temperature window between the glass transition temperature and the phase separation temperature. If this window is too narrow or not well controlled, premature phase separation may occur during processing, often just a few degrees above glass transition temperature, resulting in coarse morphologies and weak interfacial regions (DeMeuse, 2014; Hoogenboom, 2019; Liao et al., 2023; Murthy, 2018). A key rheological parameter that governs blend morphology is the viscosity ratio of the individual polymer components. A mismatch in viscosities can lead to the formation of nonuniform morphologies, such as dispersed droplets or elongated fibrils, which directly affect the mechanical and thermal behavior of the final product (Ajitha and Thomas, 2020; Liao et al., 2023; Utracki, 2014). Fine-tuning the molecular weight and melt processing conditions allows for control over blend ratios and phase morphology, even in inherently immiscible systems (Das et al., 2020; DeMeuse, 2014; Liao et al., 2023). To enhance compatibility and prevent phase separation, compatibilizers or reactive stabilizing agents can be added. These substances promote interfacial adhesion by forming copolymers or interphases between the distinct polymer domains. Stabilizers, such as antioxidants, UV absorbers, and heat stabilizers, also help to maintain the integrity of the polymer architecture during processing or service life by preventing chain scission and oxidative degradation (Barra et al., 2023; Utracki, 2014). Moreover, the design of matrix resins in high-temperature polymer composites often involves the incorporation of aromatic or heterocyclic units, linked by flexible bridging groups. These architectural features not only contribute to high glass transition temperature values and excellent thermal stability but also improve miscibility when blended with polymers of similar structural motifs (Barra et al., 2023).

2.5 Challenges and future perspectives

Despite the significant advances in high-temperature polymers made in the last two decades, the development of this class of polymers faces several challenges. These challenges span from the fundamental material science issues to the practical manufacturing and the associated environmental concerns. Ensuring thermal stability and resistance to degradation are major challenges. High-temperature environments can cause polymers to degrade, leading to a loss of structural integrity and functional properties. To address this issue, the development of new polymer chemistry and the incorporation of stabilizing additives that can withstand prolonged exposure to extreme conditions without breaking down are needed. Furthermore, in terms of their mechanical properties, the balancing of strength, toughness, and flexibility of high-temperature polymers is difficult. Polymers often become brittle at elevated temperatures, thereby limiting their usage in dynamic and load-bearing applications. New polymer architectures and crosslinking techniques that enhance their mechanical performance, while maintaining their thermal stability, should be explored. The need for polymers in lightweight dielectrics is growing due to the technological advancements in various industries, such as aerospace and aviation, marine, microelectronics, and construction. These industries operate at temperatures of between 300°C and 400°C (Barra et al., 2023; Shi et al., 2022). However, the synthesis and processing of high-temperature polymers are often very costly. Therefore, the exploration of ways to reduce costs, while maintaining performance, is essential for a broader adoption of these materials in various industries (Barra et al., 2023; Das et al., 2020). Many high-temperature polymers are derived from nonrenewable resources and can release toxic by-products during synthesis and degradation. The development of sustainable alternatives, by using biobased monomers and green chemistry principles, is critical. Additionally, improving the recycling methods and the design of polymers for easier degradation at the end of their life cycles can mitigate environmental impact (Das et al., 2020; Tao et al., 2021). The compatibility of high-temperature polymers with nanomaterials is another area that requires attention. The incorporation of nanomaterials to enhance their properties is a promising route. However, the achievement of a uniform dispersion and strong interfacial bonding within the polymer matrix is technically demanding. Advanced dispersion techniques and surface modification of nanomaterials can improve compatibility and performance. Accurate computational models are, likewise, needed to design polymers with desired properties and predict their behaviors in high-temperature conditions. To effectively screen out the best promising temperature-resistant polymers through a high-throughput screening, it is envisaged that future research endeavors will focus on the development of a model by using the machine learning route that can empirically predict the glass transition temperature of polymers based on the structures of the polymers, especially the high-temperature grades.

2.6 Conclusion

The development of high-temperature polymers is a multifaceted field that requires an integrated approach by combining advanced synthesis techniques, careful consideration of influencing factors, and a commitment to sustainability. The advancement of these materials is key for numerous high-performance applications in sectors such as aerospace, automotive, and electronics, where extreme thermal conditions are routine phenomena. Various techniques employed in the synthesis of high-temperature polymers offer unique advantages and, at the same time, present specific challenges. The choice of techniques profoundly impacts the properties and performance of the resulting polymers, thereby necessitating a thorough understanding of the underlying mechanisms and the processing conditions.

Immiscibility issues between different polymer phases can lead to heterogeneous materials with inferior properties, hence underscoring the importance of compatibility and dispersion techniques. The variance in the working temperatures of different polymer components further complicates the synthesis processes, thereby demanding precise control over the thermal conditions to ensure uniform and predictable performance. Additives and nanomaterials significantly enhance the thermal stability, mechanical strength, and chemical resistance properties of high-temperature polymers. They also introduce certain complexities in the processing and dispersion that must be carefully managed. Therefore, continued research and, hence, collaboration across disciplines will be pivotal to overcoming the existing challenges and unlocking new potentials in the next generation of high-temperature polymers that are needed to meet the demanding requirements of modern industry, while aligning with environmental and societal goals.

References

Abdelrasoul, A., Doan, H., Lohi, A., Cheng, C.H., 2015. Morphology control of polysulfone membranes in filtration processes: a critical review. ChemBioEng Reviews 2 (1), 22–43. Available at https://doi.org/10.1002/cben.201400030.

Ajitha, A., Thomas, S., 2020. Introduction: polymer blends, thermodynamics, miscibility, phase separation, and compatibilization. Compatibilization of Polymer Blends. Elsevier, pp. 1–29.

Al Christopher, C., da Silva, Í.G., Pangilinan, K.D., Chen, Q., Caldona, E.B., Advincula, R.C., 2021. High performance polymers for oil and gas applications. Reactive and Functional Polymers 162, 104878. Available at https://doi.org/10.1016/j.reactfunctpolym.2021.104878.

Ali, U., Karim, K.J.B.A., Buang, N.A., 2015. A review of the properties and applications of poly (methyl methacrylate)(PMMA). Polymer Reviews 55 (4), 678–705. Available at https://doi.org/10.1080/15583724.2015.1031377.

Aoshima, S., Kanaoka, S., 2009. A renaissance in living cationic polymerization. Chemical Reviews 109 (11), 5245–5287. Available at https://doi.org/10.1021/cr900225g.

Asua, J., 2008. Polymer Reaction Engineering. John Wiley & Sons.

Azapagic, A., Emsley, A., Hamerton, I., 2003. Polymers: The Environment and Sustainable Development. John Wiley & Sons.

Balani, K., Verma, V., Agarwal, A., Narayan, R., 2015. Biosurfaces: A Materials Science and Engineering Perspective. John Wiley & Sons.

Barra, G., Guadagno, L., Raimondo, M., Santonicola, M.G., Toto, E., Vecchio Ciprioti, S., 2023. A comprehensive review on the thermal stability assessment of polymers and composites for aeronautics and space applications. Polymers 15 (18), 3786.

Barth, C., Gonçalves, M., Pires, A., Roeder, J., Wolf, B., 2000. Asymmetric polysulfone and polyethersulfone membranes: effects of thermodynamic conditions during formation on their performance. Journal of Membrane Science 169 (2), 287–299. Available at https://doi.org/10.1016/S0376-7388(99)00344-0.

Brooks, B., 2010. Suspension polymerization processes. Chemical Engineering & Technology 33 (11), 1737–1744. Available at https://doi.org/10.1002/ceat.201000210.

Chaudhary, V., Sharma, S., 2019. Suspension polymerization technique: parameters affecting polymer properties and application in oxidation reactions. Journal of Polymer Research 26 (5), 102. Available at https://doi.org/10.1007/s10965-019-1767-8.

Chen, E.Y.-X., 2009. Coordination polymerization of polar vinyl monomers by single-site metal catalysts. Chemical Reviews 109 (11), 5157–5214. Available at https://doi.org/10.1021/cr9000258.

Chen, B., Wang, Y., Berretta, S., Ghita, O., 2017. Poly aryl ether ketones (PAEKs) and carbon-reinforced PAEK powders for laser sintering. Journal of Materials Science 52, 6004–6019. Available at https://doi.org/10.1007/s10853-017-0840-0.

Chern, C., 2006. Emulsion polymerization mechanisms and kinetics. Progress in Polymer Science 31 (5), 443–486. Available at https://doi.org/10.1016/j.progpolymsci.2006.02.001.

Chien, J.C., 2012. Coordination Polymerization: A Memorial to Karl Ziegler. Elsevier.

Choi, K.Y., McAuley, K.B., 2007. Step-growth polymerization. In: Asua, J.M. (Ed.), Polymer Reaction Engineering. Wiley, pp. 273–314.

Chung, T.-S., 1997. A critical review of polybenzimidazoles: historical development and future R&D. Journal of Macromolecular Science, Part C: Polymer Reviews 37 (2), 277–301. Available at https://doi.org/10.1080/15321799708018367.

Das, A., Chatham, C.A., Fallon, J.J., Zawaski, C.E., Gilmer, E.L., Williams, C.B., et al., 2020. Current understanding and challenges in high temperature additive manufacturing of engineering thermoplastic polymers. Additive Manufacturing 34, 101218. Available at https://doi.org/10.1016/j.addma.2020.101218.

DeMeuse, M., 2014. Thermodynamics of high temperature polymer blends. High Temperature Polymer Blends. Elsevier, pp. 130–140.

Demir, M.M., Castignolles, P., Akbey, Ü., Wegner, G., 2007. In-situ bulk polymerization of dilute particle/MMA dispersions. Macromolecules 40 (12), 4190–4198. Available at https://doi.org/10.1021/ma070142e.

Eliseeva, V.I., Ivanchev, S.S., Kuchanov, S.Ii, Lebedev, A., 2012. Emulsion Polymerization and Its Applications in Industry. Springer Science & Business Media.

Furuncuoglu, T., Ugur, I., Degirmenci, I., Aviyente, V., 2010. Role of chain transfer agents in free radical polymerization kinetics. Macromolecules 43 (4), 1823–1835. Available at https://doi.org/10.1021/ma902803p.

Garcia-Gonzalez, D., Rusinek, A., Jankowiak, T., Arias, A., 2015. Mechanical impact behavior of polyether–ether–ketone (PEEK). Composite Structures 124, 88–99. Available at https://doi.org/10.1016/j.compstruct.2014.12.061.

Ghosh, N., Kiskan, B., Yagci, Y., 2007. Polybenzoxazines—new high performance thermosetting resins: synthesis and properties. Progress in Polymer Science 32 (11), 1344–1391. Available at https://doi.org/10.1016/j.progpolymsci.2007.07.002.

Gody, G., Zetterlund, P.B., Perrier, S., Harrisson, S., 2016. The limits of precision monomer placement in chain growth polymerization. Nature Communications 7 (1), 10514. Available at https://doi.org/10.1038/ncomms10514.

Gouzman, I., Grossman, E., Verker, R., Atar, N., Bolker, A., Eliaz, N., 2019. Advances in polyimide-based materials for space applications. Advanced Materials 31 (18), 1807738. Available at https://doi.org/10.1002/adma.201807738.

Gupta, S.K., Kumar, A., 2012. Reaction Engineering of Step Growth Polymerization. Springer Science & Business Media.

Hadjichristidis, N., Pitsikalis, M., Pispas, S., Iatrou, H., 2001. Polymers with complex architecture by living anionic polymerization. Chemical Reviews 101 (12), 3747–3792. Available at https://doi.org/10.1021/cr9901337.

Haleem, A., Javaid, M., 2019. Polyether ether ketone (PEEK) and its 3D printed implants applications in medical field: an overview. Clinical Epidemiology and Global Health 7 (4), 571–577. Available at https://doi.org/10.1016/j.cegh.2019.01.003.

Hedayati, H.R., Khorasani, M., Ahmadi, M., Ballard, N., 2022. Preparation of well-defined poly (vinyl alcohol) by hydrolysis of poly (vinyl acetate) synthesized by RAFT suspension polymerization. Polymer 246, 124674. Available at https://doi.org/10.1016/j.polymer.2022.124674.

Hergenrother, P.M., 2003. The use, design, synthesis, and properties of high performance/high temperature polymers: an overview. High Performance Polymers 15 (1), 3–45. Available at https://doi.org/10.1177/095400830301500101.

Hirao, A., Goseki, R., Ishizone, T., 2014. Advances in living anionic polymerization: from functional monomers, polymerization systems, to macromolecular architectures. Macromolecules 47 (6), 1883–1905. Available at https://doi.org/10.1021/ma401175m.

Hoogenboom, R., 2019. Chapter 2 - Temperature-responsive polymers: properties, synthesis, and applications. In: Aguilar, M.R., San Román, J. (Eds.), Smart Polymers and their Applications, second ed. Woodhead Publishing, pp. 13–44. Available at https://doi.org/10.1016/B978-0-08-102416-4.00002-8.

Hutchinson, R.A., Penlidis, A., 2007. Free-radical polymerization: homogeneous systems. In: Asua, J.M. (Ed.), Polymer Reaction Engineering. Wiley, pp. 118–178. Available at https://doi.org/10.1002/9780470692134.ch3.

Ivin, K., 2000. Thermodynamics of addition polymerization. Journal of Polymer Science Part A: Polymer Chemistry 38 (12), 2137–2146.

Kang, G.-d, Cao, Y.-m, 2014. Application and modification of poly (vinylidene fluoride)(PVDF) membranes–a review. Journal of Membrane Science 463, 145–165.

Kariduraganavar, M.Y., Kittur, A.A., Kamble, R.R., 2014. Polymer synthesis and processing. Natural and Synthetic Biomedical Polymers. Elsevier, pp. 1–31.

Koltzenburg, S., Maskos, M., Nuyken, O., 2023. Polymer Chemistry. Springer Nature.

Kong, D.C., Yang, M.H., Zhang, X.S., Du, Z.C., Fu, Q., Gao, X.Q., et al., 2021. Control of polymer properties by entanglement: a review. Macromolecular Materials Engineering Science & Technology Journal 306 (12), 2100536.

Kosmalska, A., Masek, A., 2022. Thermal analysis methods in characterization of Polymer Additives. In: In: Krzysztof, P., Kinga, P. (Eds.), Thermal Analysis of Polymeric Materials: Methods and Developments Vol. 1. Wiley, pp. 457–483.

Kunal, K., Robertson, C.G., Pawlus, S., Hahn, S.F., Sokolov, A.P., 2008. Role of chemical structure in fragility of polymers: a qualitative picture. Macromolecules 41 (19), 7232–7238.

Liao, E., Zhang, Y., Lu, A., Geng, C., 2023. Effect of processing temperature on crystallization, phase morphology and mechanical property of a linear low density polyethylene/ethylene-octene copolymer blend. Journal of Macromolecular Science, Part B 62 (4), 172–181.

Liaw, D.-J., Wang, K.-L., Huang, Y.-C., Lee, K.-R., Lai, J.-Y., Ha, C.-S., 2012. Advanced polyimide materials: syntheses, physical properties and applications. Progress in Polymer Science 37 (7), 907–974.

Liu, X.J., Xiao, M., Huang, W., Yang, X., Zha, J.W., 2024. Overview of high-temperature polymers. In: Zha, J.-W., Dang, Z.-M. (Eds.), High Temperature Polymer Dielectrics: Fundamentals and Applications in Power Equipment. Wiley, pp. 1–19. Available at https://doi.org/10.1002/9783527841059.ch1.

Lovell, P.A., Schork, F.J., 2020. Fundamentals of emulsion polymerization. Biomacromolecules 21 (11), 4396–4441.

Luise, R.R., 2018. Applications of High Temperature Polymers. CRC press.

Matyjaszewski, K., Davis, T.P., 2002. Handbook of Radical Polymerization. Wiley Online Library.

McKeen, L.W., 2014. The Effect Of Temperature and Other Factors on Plastics and Elastomers. William Andrew.

Meira, G.R., Kiparissides, C., 2007. Free-radical polymerization: heterogeneous systems. In: Asua, J.M. (Ed.), Polymer Reaction Engineering. Wiley, pp. 179–208. Available at https://doi.org/10.1002/9780470692134.ch4.

Morton, M., 2012. Anionic polymerization: principles and practice. Elsevier.

Murthy, N.S., 2018. Experimental techniques for understanding polymer crystallization. Crystallization in Multiphase Polymer Systems. Elsevier, pp. 49–72.

Musarurwa, H., Tavengwa, N.T., 2022. Thermo-responsive polymers and advances in their applications in separation science. Microchemical Journal 179, 107554.

Nataraj, S., Yang, K., Aminabhavi, T., 2012. Polyacrylonitrile-based nanofibers—a state-of-the-art review. Progress in Polymer Science 37 (3), 487–513.

Novikov, V., Rössler, E., 2013. Correlation between glass transition temperature and molecular mass in non-polymeric and polymer glass formers. Polymer 54 (26), 6987–6991.

Odian, G., 2004. Principles of Polymerization. John Wiley & Sons.

Owolabi, R.U., Usman, M.A., Kehinde, A.J., 2018. Modelling and optimization of process variables for the solution polymerization of styrene using response surface methodology. Journal of King Saud University-Engineering Sciences 30 (1), 22–30.

Patrick, S., 2005. Practical Guide to Polyvinyl Chloride. iSmithers Rapra Publishing.

Pielichowski, K., Njuguna, J., Majka, T.M., 2022. Thermal Degradation of Polymeric Materials. Elsevier.

Pixton, M., Paul, D., 2018. Relationships between structure and transport properties for polymers with aromatic backbones. Polymeric Gas Separation Membranes. CRC Press, pp. 83–153.

Plesch, P.H., 2016. The Chemistry of Cationic Polymerization. Elsevier.

Qiao, S., Wang, H., 2018. Temperature-responsive polymers: synthesis, properties, and biomedical applications. Nano Research 11, 5400–5423.

Quirk, R., 2013. Anionic polymerization. In: Enrique, S.-G., Eduardo, V.-L. (Eds.), Handbook of Polymer Synthesis, Characterization, and Processing. Wiley, pp. 127–162.

Rahate, A.S., Nemade, K.R., Waghuley, S.A., 2013. Polyphenylene sulfide (PPS): state of the art and applications. Reviews in Chemical Engineering 29 (6), 471–489.

Resconi, L., Cavallo, L., Fait, A., Piemontesi, F., 2000. Selectivity in propene polymerization with metallocene catalysts. Chemical Reviews 100 (4), 1253–1346.

Rinaldi, M., Cecchini, F., Pigliaru, L., Ghidini, T., Lumaca, F., Nanni, F., 2020. Additive manufacturing of polyether ether ketone (PEEK) for space applications: a nanosat polymeric structure. Polymers 13 (1), 11.

Rogers, M.E., Long, T.E., Turner, S.R., 2003. Introduction to synthetic methods in step-growth polymers. In: Rogers, M.E., Long, T.E. (Eds.), Synthetic Methods in Step-Growth Polymers. John Wiley & Sons, Inc., pp. 1–16. Available at https://catalogimages.wiley.com/images/db/pdf/047138769X.c01.pdf.

Saldivar-Guerra, E., Vivaldo-Lima, E., 2013. Handbook of Polymer Synthesis, Characterization, and Processing. John Wiley & Sons.

Saxena, P., Shukla, P., 2021. A comprehensive review on fundamental properties and applications of poly (vinylidene fluoride)(PVDF). Advanced Composites and Hybrid Materials 4, 8–26.

Scharnagl, N., Buschatz, H., 2001. Polyacrylonitrile (PAN) membranes for ultra-and microfiltration. Desalination 139 (1-3), 191–198.

Schawe, J.E., 2014. Influence of processing conditions on polymer crystallization measured by fast scanning DSC. Journal of Thermal Analysis and Calorimetry 116, 1165–1173.

Shi, X.-H., Li, X.-L., Li, Y.-M., Li, Z., Wang, D.-Y., 2022. Flame-retardant strategy and mechanism of fiber reinforced polymeric composite: a review. Composites Part B: Engineering 233, 109663.

Soares, J.B., Pérez, O., 2013. Coordination polymerization. in: *Handbook of Polymer Synthesis, Characterization, and Processing*, S.-G. Enrique, V.-L. Eduardo, pp. 85-104.

Su, W.-F., 2013. Principles of Polymer Design and Synthesis. Springer.

Takeichi, T., Agag, T., 2006. High performance polybenzoxazines as novel thermosets. High Performance Polymers 18 (5), 777–797.

Tao, L., Chen, G., Li, Y., 2021. Machine learning discovery of high-temperature polymers. Patterns 2 (4), 100225. Available at https://doi.org/10.1016/j.patter.2021.100225.

Tardy, A., Nicolas, J., Gigmes, D., Lefay, C., Guillaneuf, Y., 2017. Radical ring-opening polymerization: scope, limitations, and application to (bio) degradable materials. Chemical Reviews 117 (3), 1319–1406. Available at https://doi.org/10.1021/acs.chemrev.6b00319.

Thickett, S.C., Gilbert, R.G., 2007. Emulsion polymerization: state of the art in kinetics and mechanisms. Polymers 48 (24), 6965–6991. Available at https://doi.org/10.1016/j.polymer.2007.09.031.

Tian, P., Zhao, X., Liu, W., Zhang, J., 2022. Preparation and application of cross-linked PVA microspheres with narrow particle size distribution by suspension polymerization using uniform porous tube. Reactive and Functional Polymers 171, 105153. Available at https://doi.org/10.1016/j.reactfunctpolym.2021.105153.

Tillet, G., Boutevin, B., Ameduri, B., 2011. Chemical reactions of polymer crosslinking and post-crosslinking at room and medium temperature. Progress in Polymer Science 36 (2), 191–217. Available at https://doi.org/10.1016/j.progpolymsci.2010.08.003.

Utracki, L., 2014. Characterization methods for high temperature polymer blends. High Temperature Polymer Blends. Elsevier, pp. 14–69.

Wei, Z., Liu, H., Yu, L., Xiao, S., Hou, Y., Chen, X., 2020. Delocalized aromatic molecules with matched electron-donating and electron-withdrawing groups enhancing insulating performance of polyethylene blends. Journal of Applied Polymer Science 137 (39), 49185. Available at https://doi.org/10.1002/app.49185.

Wilson, D., Stenzenberger, H.D., Hergenrother, P.M., 2013. Polyimides. Springer Science & Business Media.

Yang, J., Xu, T., Lu, A., Zhang, Q., Tan, H., Fu, Q., 2009. Preparation and properties of poly (p-phenylene sulfide)/multiwall carbon nanotube composites obtained by melt compounding. Composites Science and Technology 69 (2), 147–153. Available at https://doi.org/10.1016/j.compscitech.2008.08.030.

Yang, X., Jiang, X., Bashir, M.S., Kong, X.Z., 2020. Preparation of highly uniform polyurethane microspheres by precipitation polymerization and Pd immobilization on their surface and their catalytic activity in 4-nitrophenol reduction and dye degradation. Industrial & Engineering Chemistry Research 59 (7), 2998–3007. Available at https://doi.org/10.1021/acs.iecr.9b06367.

Yilmaz, G., Yagci, Y., 2020. Light-induced step-growth polymerization. Progress in Polymer Science 100, 101178. Available at https://doi.org/10.1016/j.progpolymsci.2019.101178.

Yokozawa, T., Ohta, Y., 2016. Transformation of step-growth polymerization into living chain-growth polymerization. Chemical Reviews 116 (4), 1950–1968. Available at https://doi.org/10.1021/acs.chemrev.5b00393.

Yokozawa, T., Yokoyama, A., 2009. Chain-growth condensation polymerization for the synthesis of well-defined condensation polymers and π-conjugated polymers. Chemical Reviews 109 (11), 5595–5619. Available at https://doi.org/10.1021/cr900041c.

Zhang, R., Gao, R., Gou, Q., Lai, J., Li, X., 2022. Precipitation polymerization: a powerful tool for preparation of uniform polymer particles. Polymers 14 (9), 1851. Available at https://doi.org/10.3390/polym14091851.

3

High-temperature polymers: synthesis and characterization

Idowu David Ibrahim[1], Kobe Samuel Mojapelo[2], Williams Kehinde Kupolati[2], Azunna Agwo Eze[1]

[1]*DEPARTMENT OF INDUSTRIAL ENGINEERING, OPERATIONS MANAGEMENT AND MECHANICAL ENGINEERING, VAAL UNIVERSITY OF TECHNOLOGY, VANDERBIJLPARK, SOUTH AFRICA* [2]*DEPARTMENT OF CIVIL ENGINEERING, TSHWANE UNIVERSITY OF TECHNOLOGY, PRETORIA, GAUTENG, SOUTH AFRICA*

3.1 Introduction

This chapter summarizes the emerging field of high-temperature polymers by discussing the molecular design aspects, synthetic approaches, and the thermal and mechanical properties of these materials. It also highlights their importance in innovative engineering applications and hostile environments, as well as the chapter's overall goals and scope. High-temperature polymers, which are high-performance specialized materials, have been developed to endure elevated temperatures while retaining their structural integrity and functional characteristics (Zha et al., 2023). Typically, their molecular backbones are designed to consist of rigid aromatic structures, high glass-transition temperatures, and high crosslink densities, all of which provide excellent thermal stability and degradation resistance (Deshmukh et al., 2023). New structure-guided synthesis methods based on the reversible breaking of covalent hyphens by breakdown forces include energy-efficient frontal polymerization, which has further increased the design space for these polymers and enabled properties such as enhanced dielectric breakdown strength and high energy storage capacity (Feng et al., 2022). In fact, novel architectures, such as propargyl ether-terminated systems, have shown excellent char residue and thermal resistance at high temperatures, demonstrating the promise of such materials for many applications (Sundari et al., 2023).

These materials are essential for advanced engineering in extreme environments, making them ideal candidates for advanced next-generation energy-storage devices, capacitor dielectrics, and aerospace components. Crosslinked fluoropolymers demonstrate an exceptionally high-temperature energy density and charge-discharge efficiency, making them viable candidates for demanding operating conditions (Li et al., 2023). Polymer capacitor dielectrics are also beneficial for the thermal management of power electronics because they allow for more compact and safer designs. In the field of polymer electrolytes, analogous trends have been detected by augmenting with piperazinium ionic liquids to enable stable performance even at high temperatures (Gallastegui et al., 2024). Moreover, these polymers are not only functional and difficult to endure in highly electric fields but also have fire-

High-Temperature Polymeric Materials. DOI: https://doi.org/10.1016/B978-0-443-32792-6.00006-3

retardant properties that make them more effective in performance under extreme environments (Alam et al., 2019).

This chapter has three objectives and a defined scope. First, we aim to critically review and synthesize a range of chemical strategies that underpin the implementation of high-temperature polymers, including novel catalytic systems and innovative cross-linking methods that facilitate the formation of large-bandgap materials and enhance resistance to thermal degradation (Embarek and Sahli, 2020). The approaches described in this chapter include the strategic use of indirect data to monitor project milestones while cautioning against overinterpreting high-level molecular structures owing to the lack of reliable characterization techniques available for correlating molecular structures with high-temperature performance. These techniques include thermogravimetric analysis (TGA), differential scanning calorimetry (DSC), nuclear magnetic resonance (NMR) spectroscopy, and Fourier transform infrared (FTIR) spectroscopy. By linking polymer synthesis strategies to real-world applications, such as energy storage and aerospace, this chapter highlights the current challenges and future innovations in high-temperature polymers (Embarek and Sahli, 2020). This systematic review provides an overview of the fundamental principles and advanced synthesis methods, emphasizing the importance of these materials for transforming applications in extreme environments.

3.2 Classification and chemical structure

High-temperature polymers are categorized by their molecular architecture and backbone bonding elements, most notably by splitting between aromatic and semiaromatic, as well as thermoplastic and thermoset types. The thermal and chemical resistances of aromatic polymers, such as poly(ether-ether-ketone) (PEEK) and wholly aromatic polyamides, are attributed to their rigid chemical structures composed of benzene rings and π-electron systems, which improve dimensional stability and give rise to stronger intermolecular interactions (Tekin et al., 2018; Qi et al., 2019). In contrast, semiaromatic polymers are synthesized from both types of monomers, which provide rigidity while improving processability and reducing density. Semiaromatic polyamides are purposefully designed with flexible aliphatic spacers to enhance melt-processing properties (Ren et al., 2022; Fu et al., 2022). Thermoplastics consist of linear chains held together by noncovalent bonds, making them reusable and recyclable. PEEK is a prime example of a semicrystalline material from this class with high thermal resistance (Yao et al., 2023).

In contrast, thermosets form permanently crosslinked three-dimensional (3D) networks that exhibit excellent dimensional and mechanical stability at elevated temperatures but limited recyclability (Chen et al., 2020). Therefore, emerging covalent adaptable networks provide reversibility by bridging thermosets with thermoplastics (Lucherelli et al., 2023). The structure–property relationship is key to performance. Aromatic systems are well known for their thermal stability, stiffness, and chemical resistance owing to their strong C–C bonds and electron delocalization (Yang et al., 2021). The melting points, T_g, and processability can be tuned by tailoring the substituents and segment composition. Among high-temperature polymers, PEEK, polyimide (PI), polyphenylene sulphide (PPS), and polytetrafluoroethylene are extensively employed in the aerospace, automotive, medical, and chemical industries owing to their heat-bearing and mechanical properties (Ahmad and Niasar, 2025). Advancing synthetic chemistry has broadened its functions and uses (Farajidizaji et al., 2021). High-temperature polymers are classified based on their aromatic contents and network architectures. Rigidity, flexibility, and bonding types work together to make them suitable for harsh environments.

3.3 Synthesis techniques

The design of the monomers is critical for tuning the polymer structure and function. Dynamic covalent linkages, such as imine bonds in covalent organic frameworks (COFs), promote the formation of stable porous networks with excellent optoelectronic properties, particularly triazine-based cores (Wei et al., 2019). A variety of pendant groups have been introduced into the Hantzsch reaction to enhance the thermal and ultraviolet blocking performance (Chen et al., 2025). Computationally assisted selection of monomers that would participate in molecular imprinting by predicting the binding sites (Rajpal et al., 2023).

3.3.1 Condensation polymerization

Imine-based condensation polymerization is a vital method for creating highly ordered COFs, which are crucial for applications involving thermal and chemical stability in catalysis, gas separation, and energy storage. This strategy consists of the formation of imine linkages through a reversible reaction between aldehydes and amines, allowing error correction during the formation of the polymer network, resulting in highly crystalline, porous materials (Geng et al., 2020). The reversibility of these bonds enhances the structural order and the adjustability of the framework properties, including the pore size, electronic structure, and chemical reactivity (Chu et al., 2023). Their thermal stability and tunable porosity make imine-linked COFs promising candidates for high-temperature applications. To further enhance the mechanical and thermal robustness of these materials, condensation reaction-based crosslinking strategies such as the Hantzsch pathway have been employed in imine-based COFs. However, a π-conjugated polymer was first grown in solution, trapping additional functional groups that reinforce interchain interactions and limit conformational mobility, thereby inhibiting thermally induced mechanical failure or degradation at high temperatures (Liu et al., 2024).

These changes lead to enhanced energy density and stability owing to the increased rigidity of the framework and preservation of the function of the molecules. Notably, Liu et al. (2024) reported that binary crosslinked polymers exhibit better performance under thermal stress owing to the confinement effect, which hinders amorphous transitions and molecular rearrangement. Such improvements are essential for deploying COFs under extreme conditions, particularly in structural or energy-related applications. Moreover, imine-based polymerization is modular and can be used to tailor material properties by selecting appropriate monomers for polymerization as well as through postsynthetic modification. The reactivity of COFs can be tuned by changing the type of aldehyde and amine employed, which can facilitate selective gas adsorption, enhance conductivity, or improve catalytic activity (Liang et al., 2023; Ankit et al., 2024). Furthermore, implementing these design strategies alongside a crosslinking approach significantly broadens the functional scope of COFs, while preserving their crystallinity and chemical integrity. Enhanced and crosslinked COFs have demonstrated their ability to maintain order under thermal cycling, as confirmed through advanced characterization techniques such as spectroscopy and diffraction, thereby establishing their suitability for next-generation technologies as high-performing materials (Chu et al., 2023).

3.3.2 Ring-opening polymerization

Ring-opening polymerization (ROP) serves as a platform for creating high-temperature polymers; it allows for precise molecular weight control and yields a narrow dispersity. The methodology exploits ring strain in cyclic monomers to enable controlled polymerization under extreme thermal conditions. Although less frequently used than other polymerization strategies, ROP is increasingly recognized for its ability to yield thermally stable polymers across a wide variety of architectures (Chen et al., 2020). Recent innovations have utilized external stimuli (light, redox, and mechanical forces) as dynamic tools to modulate catalytic activity during ROP. Stimuli-responsive strategies have emerged that facilitate finer adjustments of reaction rates and molecular weight distributions. This is especially crucial for minimizing side reactions in polymer synthesis at elevated temperatures. Moreover, bifunctional catalysis has proven effective in reducing polymerization inhibition, particularly in copolymerization systems involving epoxides and cyclic anhydrides. Lidston et al. (2020) reported that bifunctional catalysts can simultaneously initiate and control several polymer chains, enabling efficient polymerization under milder conditions. Additional developments include the incorporation of ring-opening metathesis polymerization (ROMP) and cascade polymerization. Park and Bielawski (2021) described the use of chain transfer agents in the ROMP of cyclic allenes to enable architectural control, while Yuan et al. (2020a) highlighted cascade reactions for implementing complex functionality during polymer growth. This has also sparked interest in cyclic polymers that eliminate terminal-group effects and increase thermal stability (Liénard et al., 2020).

3.3.3 Other synthetic strategies

Real-time monitoring, iterative synthesis, and flow chemistry are becoming essential requirements for recent advancements in high-temperature polymer synthesis to achieve reproducibility, scalability, and ecofriendliness. In-line monitoring tools, such as in situ methods applied to the electro polymerization of molecularly imprinted polymers, allow the dynamic modulation of reaction parameters to reduce run-to-run heterogeneity (Moschetta et al., 2024). Similarly, in-line monitoring under high-temperature conditions has been used to optimize the dispensation and uniformity of polymer synthesis (Premadasa et al., 2024). These systems are particularly vital in the industry for maintaining accurate control of the stability and selectivity of polymers under high-temperature conditions. Simultaneously, hierarchical chromatographic purification combined with iterative synthetic strategies and continuous-flow processing enables improved control over the polymer architecture. This facilitates the synthesis of discrete polymer chains with defined connectivity, which is essential for the thermal performance (Ötvös and Kappe, 2020). It can further enhance the environmental footprint and scalability of the process by increasing the concentration, decreasing the residence time and minimizing solvent waste through the development of flow chemistry (Galant et al., 2021). Moreover, dynamic covalent chemistry, such as aminoketone amides and boronic ester vitrimers, introduces reversible bonds into polymer backbones, permitting reprocessing and recyclability without thermal mixing loss (Capaldo et al., 2023; Koza et al., 2022).

Another area where flow chemistry expands the horizons of discovery is the scale-up of complex reactions, such as photochemical macrocyclization, which significantly benefits from enhanced heat and mass transfer control (Du and Siewert, 2020; Morin et al., 2024). Microstructure photoreactors are also conducive to this trend, providing energy-efficient, high-yield processes on an industrial scale (Buglioni et al., 2022). Rheological exploration facilitates the accurate adjustment of thermoplastic processing conditions by gathering valuable information about viscosity and flow characteristics across diverse thermal environments (Morin et al., 2024). This combination of real-time control, flow-based platforms, and processable chemistry provides a robust foundation for engineering performance and sustainability in advanced high-temperature polymers.

3.3.4 Scaling and process optimization challenges

One of the most significant challenges is scaling the synthesis of high-temperature polymers to relevant industrial levels, where controlling the polymer chain architecture, molecular weight distribution, and thermal properties is essential to ensure optimal performance. The effectiveness of polymers in critical applications, such as capacitive energy storage and high-performance composites, can be highly sensitive to minor variations in temperature, shear rate, and mixing efficiency (Zhou et al., 2021; Paszkiewicz et al., 2021). A narrow processing window (230°C–240°C) for thermally stable poly(hexamethylene 2,5-furanate) production via melt polycondensation is essential for achieving optimal thermal and mechanical properties. This sensitivity underscores the necessity of process optimization at scale, where kinetic consistency and reaction stability become increasingly complex. Recent studies have emphasized the implementation of kinetic modeling, novel mixing technologies, and in situ characterization to address the existing challenges. Kinetic models that predict polymer growth under various conditions have proven beneficial for modeling reversible addition-fragmentation chain transfer polymerization (Kandelhard et al., 2023). Resonant acoustic mixing provides a scalable and energy-efficient polymerization platform under ambient conditions while ensuring thermal stability (Fowler et al., 2025). Concurrently, instrumental techniques, such as TGA and X-ray diffraction (XRD), are employed in situ to analyze the reaction dynamics and thermal properties during scale-up (Coessens et al., 2023). As polymer formulations grow increasingly complex, particularly in copolymer and terpolymer systems, multivariable optimization becomes essential. The integration of modeling, monitoring, and innovative processing is vital for overcoming the scale-up limitations and producing high-performance polymers in a reproducible manner.

3.4 Thermal and chemical characterization

3.4.1 Thermogravimetric analysis

TGA was used to examine the weight loss as a function of temperature, offering crucial insights into the thermal degradation and stability of high-temperature polymers. This method is essential for determining the onset temperatures of the degradation and decomposition profiles of materials designed for extreme thermal conditions. Liu et al. (2024) employed TGA to evaluate polymer nanocomposites for geothermal applications that require high-temperature stability. Carotenuto et al. (2024) utilized TGA to assess thermal degradation in hybrid green materials, highlighting the technique's potential in exploring matrix-filler interactions related to degradation mechanisms. Furthermore, Shakir et al. (2024) applied TGA methods to investigate the thermal behavior of thermoplastic composites, confirming their broad applicability for establishing degradation profiles and ensuring that material performance meets high-temperature requirements.

3.4.2 Differential scanning calorimetry

DSC is widely used to characterize the thermal transitions of polymers, including glass transition, crystallization, and melting temperatures. DSC data provided information on the degree of curing, reaction kinetics, and morphological stability. Kang et al. illustrated the thermal transitions that are fundamental to processing and service performance and helped correlate the evolution of thermodynamic properties with resin processing. They demonstrated the use of DSC to unravel complex resin chemistry. Kang et al. (2023) employed DSC to investigate the thermodynamic properties of benzoxazine resins. Goes et al. (2022) further demonstrated the utility of DSC in characterizing the morphological transitions of semi-crystalline polymers by describing the relationship between extensional flow conditions and thermal transitions that define crystallinity. These investigations highlight that DSC is a discriminative technique that links the microstructural evolution of high-temperature polymers to their thermal histories.

3.4.3 Dynamic mechanical analysis

Dynamic mechanical analysis (DMA) is one of the most insightful techniques for probing the viscoelastic behavior of high-temperature polymers. By applying a dynamic load, DMA provides insight into temperature- and frequency-dependent parameters, such as the storage modulus, loss modulus, and damping factor. Xiong et al. (2023) used DMA to monitor the curing process of ultrahigh-temperature resistant resins, demonstrating that this technique can track mechanical transitions during thermal cycling. Furthermore, the significance of DMA in evaluating the interfacial interactions and behaviors in polymer-filled composites, which is also crucial for predicting long-term performance, was examined by Bashir (2021) and Kamaraj et al. (2021) demonstrated that DMA measures viscoelastic behaviors at multiple temperatures, supporting a detailed assessment of polymer stiffness, damping, and service integrity. In high-performance applications, tuning the mechanical properties under dynamic loading conditions is a prime requirement, which is effectively addressed by comprehensive DMA studies.

3.4.4 Profiles of thermal stability and decomposition

The thermal stability and decomposition profiles are critical for ensuring the reliability of high-temperature polymers under extreme conditions, highlighting the importance of an integrated assessment. This approach involves using TGA and DSC data to analyze both the initiation of degradation and the thermal transitions that occur prior to the total breakdown. Cho et al. (2022) studied the accelerated thermal ageing of high-performing polymers and proposed a method for reliability extrapolation at elevated temperatures. Furthermore, Liu et al. (2024) employed TGA to achieve a parallel characterization of the degradation behavior of polymer nanocomposites. Stojanović et al. (2023) discussed the thermal stability of silicone-based coatings, which improved the performance of stable siloxane networks. The diverse range of decomposition profiles presented in these studies, through TGA and further confirmed via DSC, collectively ensures that the polymers can be reliably utilized in environments that demand significant thermal performance.

3.4.5 Solvent resistance and chemical durability

In addition to determining the shelf life of a pop-up event, solvent resistance and chemical durability over the service life are important. These two fundamental parameters are essential for assessing the long-term service performance of high-temperature polymers in harsh chemical environments. Solvent resistance is commonly evaluated by measuring the extent of swelling, mass loss, and modification of mechanical properties upon prolonged contact with certain solvents. Karabıyık et al. (2023) added to this discussion by investigating the influence of distinct solvent preferences on the morphological stability and chemical integrity of epoxy-based polymers. Yalcinkaya and Buzgo (2024) showed that polyamide-imide nanofibres exhibit outstanding chemical resistance and durability because of their polymer chain structures. Li et al. (2020) also provided insights into the chemical stability of composite nanofibre aerogels, which was attributed to the robust chemical resistance of their high-temperature polymer constituents. These results highlight the necessity for thorough chemical durability assessments that, when paired with thermal probes, ensure extended use of the material in the fusion, aerospace, and energy sectors. Briefly, the synthesis and characterization of these high-temperature polymers require a comprehensive suite of analytical techniques. TGA, DSC, and DMA are used to describe the thermal and mechanical behavior of materials under extreme conditions. In addition, the characterization of the thermal stability, thermal decomposition profiles, solvent/chemical resistance, and durability offered a holistic overview of the material's performance. Recent literature shows that the application of all these techniques simultaneously guides polymer synthesis and design and the reliability and robustness of the polymers for demanding applications.

3.5 Chemical structure and morphological characterization

A detailed discussion of the chemical structure and morphological characterization of high-temperature polymers is presented in the following subsection, which was done by synthesizing information from recent literature. This includes FTIR and NMR spectroscopy, XRD, electron microscopy techniques, atomic force microscopy (AFM), and crystallinity and molecular orientation analysis.

3.5.1 Fourier transform infrared spectroscopy and nuclear magnetic resonance spectroscopy

The development of high-temperature polymers designed for extreme environments relies heavily on specialized structural characterization techniques such as FTIR and NMR spectroscopy. Li et al. (2022) confirmed the incorporation of thermally stable moieties into the polymer backbone using FTIR, which identifies key functional groups such as esters, amides, and aromatics. NMR provides further information regarding the local chemical environment of protons and carbon atoms, offering more significant insights into chain architecture and end-group fidelity (Anis et al., 2023). These methods effectively yield robust polymers that can maintain their shapes and performance under high thermal loads. These spectroscopic methods are also crucial for advancing the industrial application of high-temperature polymers. Additionally, establishing structural characteristics using FTIR and NMR assists in optimizing materials for extreme environments, such as dielectric insulation, chemical-resistant coatings, and gas separation membranes. Initiated chemical vapor deposition of polymers based coatings that enhance thermodynamic durability are enabled through molecular design, and hydrocarbon ladder polymers employed for gas separation at high temperatures derive their stability from structural control (Lai et al., 2022; Graur et al., 2022). FTIR and NMR are used for quality assurance and for creating new high-performance polymers.

3.5.2 X-ray diffraction analysis

XRD is a powerful technique for characterizing the structure and morphology of high-temperature polymers, particularly for determining crystalline phases and quantifying crystallinity. Crystalline phases are associated with enhanced thermal stability and mechanical properties, which are essential characteristics of polymers that operate in extreme environments (Fauziyah et al., 2022). Advanced XRD analytical strategies, such as high-resolution or time-dependent methodologies, can be employed to explore the molecular orientation and phase changes upon exposure to elevated temperatures, thereby enhancing our understanding of degradation mechanisms and informing studies on process optimization (Khan et al., 2020). XRD provides a comprehensive picture of the polymer morphology and thermal behavior when used in conjunction with complementary techniques, such as DSC and solid-state NMR.

The control and characterization of crystalline morphology via XRD promises significant potential for the application of high-temperature polymers across a wide range of industries, including aerospace, electronics, and energy storage. Designers guided by XRD have enhanced the structural stability of polymer electrolytes used in solid-state lithium batteries and facilitated the formation of thermostable systems for catalysis (Nematdoust et al., 2020). Furthermore, an accurate structural analysis promotes sustainability through recyclability and material recovery measures (Abubaker et al., 2021). Therefore XRD is a key diagnostic tool vital for designing and utilizing novel innovative polymer systems for advanced industrial applications.

3.5.3 Scanning and transmission electron microscopy

The structural features of the high-temperature polymers were characterized using scanning and electron microscopy (SEM) and transmission electron microscopy (TEM). These techniques enable high-resolution imaging of surface morphology, revealing roughness, defect sites, and nanoparticle dispersion, which have been shown to influence mechanical and thermal performance significantly (Mcmullen, 2021). At high temperatures, even subtle surface defects identified via SEM can considerably shorten the lifespan of a material, making this technique essential for both validating research and ensuring quality control in the industrial sector. We enhanced these nanoscale insights into the internal microstructural features provided through SEM by incorporating TEM, which reveals phase separations, interfacial regions, and additive dispersion in heterogeneous polymer systems (Chen et al., 2020). However, recent advancements, such as cryogenic electron microscopy, which accommodates nonaqueous samples, and in situ SEM experiments under heating and mechanical stress, have expanded the methodology (Matatyaho Ya'Akobi and Talmon, 2021). Together, SEM and TEM offer a comprehensive view of the structure-property relationships essential for developing tailored high-temperature polymers for demanding applications, including aerospace, automotive, and electronics.

3.5.4 Atomic force microscopy

AFM has become an essential method for investigating the nanostructure and nanomechanical properties of high-temperature polymers. It offers insights into the surface topography and heterogeneity with the spatial resolution required to correlate nanoscale features with macroscopic thermal performance. This can be achieved using techniques such as single-molecule force spectroscopy, which quantifies viscoelastic responses at subnanometer resolutions that bulk techniques cannot attain, providing insight into how molecular orientation and interactions influence behavior under thermal stress (Murphy et al., 2022). These features help optimize the synthesis processes, where thermal stability and mechanical integrity are paramount. In industrial applications, advanced AFM modes, such as the tapping mode and photothermal actuation, can be utilized to map rheological interfacial properties that are important for high-temperature composites (Piacenti et al., 2024). This knowledge can enhance the nanofiller dispersion, adhesion, and stress transfer efficacy, which are all key factors in the composite lifetime. Recent innovations include dynamic force spectroscopy and the integration of AFM–IR, expanding the scope of characterizing viscoelasticity under thermal conditions and guiding the design of dielectric-enhanced polymer nanocomposites (Yang et al., 2024). AFM provides critical insights into the structure-property relationships of factors influencing the performance of state-of-the-art polymer systems.

3.5.5 Analysis of crystallinity and molecular orientation

The performance of heat-resistant polymers in industrial settings and high-energy applications is governed mainly by the degree of crystallinity and modes of molecular orientation. Commonly used techniques to confirm these parameters include XRD, FTIR, and DSC. XRD assessments provide information on the size and distribution of crystalline domains, whereas polarized FTIR measurements help observe chain orientation, particularly in the case of strained or processed fibers (Laramée and Pellerin, 2023). DSC is used alongside these approaches to offer a comprehensive breakdown of thermal profiles, including melting and crystallization behaviors, which are directly related to the polymer's organizational structure (Hasan et al., 2024). Together, these tools facilitate the quantification of orientation through measurement metrics, such as Herman's orientation factor, and inform the tuning of synthesis and processing conditions. Low crystallinity and poor orientation present challenges from an industrial perspective, particularly for aerospace, automotive, and electronics applications. Achieving high levels of crystalline structure is crucial, as it enhances the thermal resistance, mechanical strength, and dimensional stability (Teng et al., 2020). Other products, such as aramid fibers, exhibit performance from densely packed and well-aligned chains formed during optimized spinning and processing. Integrated characterization approaches also provide feedback that connects structural features to performance outcomes, allowing the maintenance of material quality as scale-up occurs. This structure–property relationship is essential for guiding the design and synthesis of next-generation high-temperature polymers that can function in extreme environments while still ensuring economic viability (Laramée et al., 2020, 2022).

3.6 Mechanical and rheological properties

Engineered high-temperature polymers and their composites exhibit outstanding mechanical properties, including tensile strength, flexural modulus, and impact resistance. Reinforced epoxy resins and hybrid polymer systems have demonstrated excellent performance, even in structural applications, particularly when microstructures are optimized through controlled synthesis and dynamic covalent networks (Yu et al., 2024). Such design strategies allow for customized mechanical responses, providing flexibility in load-bearing and impact-sensitive locations. Fatigue, creep, and stress relaxation behaviors are crucial for durability. Epoxy vitrimer systems feature attractive stress relaxation based on bond exchange mechanisms but are compromised by a lack of creep resistance unless permanent crosslinks are introduced (Duan et al., 2022).

The use of predictive models and microstructural modifications optimizes these time-dependent behaviors while maintaining long-term performance under cyclic loading. In addition, the rheological behavior of high-temperature polymers during processing is crucial. In injection molding and additive manufacturing applications, materials with high particle loading can display non-Newtonian, shear-thinning flow behavior that is essential for flow control and the quality of the final parts (Bek et al., 2020). Amphoteric polymers have proven effective in maintaining viscosity and minimizing fluid loss at extreme temperatures in thermally demanding fluid systems, such as drilling mud (Hamad et al., 2020). Enhanced analyzes, such as discrete relaxation spectrum techniques and the use of hybrid polymer blends, have provided new insights into viscoelastic responses and permitted accurate simulation of flow characteristics (Han et al., 2022). These advancements in mechanical tuning and rheological control will facilitate broader applications of high-performance polymer systems in extreme environments.

3.6.1 Mechanical properties

Conventional studies have highlighted the advantages of hybrid composites in improving the tensile, flexural, and impact characteristics. Sosiati et al. (2024) showed that abaca/carbon/epoxy hybrid composites exhibit promising mechanical properties under tensile, flexural, and impact loads while also providing a characterization of their properties. According to Sunny and Pickering (2022), tensile strength and flexural modulus of aligned hemp fiber mat-reinforced polypropylene matrix composites can be improved by up to 3.0 and 3.4 times, respectively, compared to those of the neat matrix. Similarly, Chimeni-Yomeni et al. (2022) highlighted that the performance of low-density polyethylene composites reinforced with flax fibers can be significantly enhanced in terms of flexural and impact strengths, demonstrating the effectiveness of fiber reinforcement for strengthening composite performance. The performance of high-temperature polymer composites can be improved by incorporating fibers, fillers, and nanoparticles.

3.6.2 Static mechanical and viscoelastic properties

Static mechanical properties, long-term stability, and viscoelastic behaviors, such as fatigue, creep, and stress relaxation, have garnered significant attention in recent studies. Sakahara et al. (2022) conducted rheological experiments on ABS composites modified with elastomeric phases, demonstrating the viscoelasticity of composite specimens with multiple relaxation modes and long relaxation times of up to 100 seconds, thereby confirming the stress relaxation and viscoelastic creep behavior. Furthermore, Arrigo et al. (2021) investigated the influence of elongational flow on polymer morphology and concluded that the development of a structure under applied stress is crucial for understanding creep behavior and fatigue performance. Collectively, these studies indicate that monitoring the viscoelastic response using techniques such as DMA provides insights into the long-term behavior and durability of polymer systems subjected to cyclic or sustained loading (Sakahara et al., 2022; Arrigo et al., 2021). High-temperature polymer composites are often subjected to this kind of condition, hence the need to ensure relevant properties are improved.

3.6.3 Rheological behavior during processing

The rheological properties of materials and their behavior during processing are fundamental factors that determine the manufacturing quality and performance of a manufactured part, particularly in additive manufacturing or processes such as extrusion. Das et al. (2021) linked polymer rheology with material extrusion in additive manufacturing, highlighting that physical properties, such as zero-shear viscosity and shear modulus, have a direct influence on interlayer welding and, as a result, on the mechanical integrity of the printed parts. Similarly, Marzuki et al. (2024) investigated the rheological behavior of 3D printed hydroxyapatite/Polylactic acid (PLA) composites and highlighted the importance of the viscosity of the composite melt on the flow behavior and shape retention. In addition, Han et al. (2022) examined the shear-thinning behavior and filament formation in fiber-reinforced thermoplastic composites, where rheometry was utilized as a predictive tool to gauge printability and optimize processing considerations. Taken together, these studies emphasize the importance of thoroughly characterizing polymer melt viscoelasticity to maximize the efficiency of processing methods such as fused deposition modeling and direct ink writing (Than et al., 2022).

3.7 Applications and industrial relevance

3.7.1 Transportation industry

High-performance materials have emerged as prized resources for the transportation sector, particularly in the aerospace and automotive industries, owing to their applications in extreme operational environments. High-temperature shape memory alloys (HTSMAs), particularly NiTi-based alloys and novel high-entropy systems, offer excellent mechanical adjustability. These advanced alloys introduce the possibility of superelastic recovery of 4% strain at temperatures up to 285°C, achieving a unique balance between the high-temperature resilience of conventional NiTi alloys and their enhanced microstructural stability (Peltier et al., 2020). They are used in actuators, sensors, and energy devices, where high-temperature durability and spatial precision are essential for device reliability and performance. Friction stir processing (FSP) has been established as a revolutionary method for enhancing the structural integrity of metallic components during transportation. FSP significantly improves fatigue life and corrosion resistance, especially for aluminum-based composites, by refining microstructures and reducing residual stresses (Adiga et al., 2022).

This technique is particularly valuable for components subjected to cyclic loading and harsh environments because it has demonstrated a reduction in brittle intermetallic and improved stress uniformity, thereby supporting the development of lightweight and durable transport systems. Concurrently, shape-memory polymers (SMPs) have emerged as promising candidates owing to their low density and significant deformability. Thermoplastic SMPs with a high glass transition temperature (T_g), such as materials derived from PEEK, can maintain mechanical integrity at elevated temperatures, showcasing large recoverable strains (Zhou et al., 2021). These polymers contribute to reducing vehicle weight, enhancing fuel efficiency, and supporting adaptive structural designs. The combination of HTSMAs and FSP with high-temperature SMPs highlights the importance of multifunctional materials and processing technologies for next-generation energy-efficient and safe transportation systems.

3.7.2 Electronics and microfabrication

Nanomaterials have inevitably been integrated into high-temperature polymer matrices to produce self-healing conductors for soft sensors and wearable electronics. Heat or reversible chemical interactions activate nanofillers, reduced graphene oxide, carbon nanotubes, and metallic nanoclusters, forming a dynamic supramolecular network that autonomously self-heals during traversal through the polymers. Such materials can recover their electrical conductivity even after repeated mechanical deformation, thereby considerably improving the durability and reliability of flexible devices used in complex environments (Jamil et al., 2024; Araya-Hermosilla et al., 2021). The microfabrication of high-temperature polymers offers reasonable thermal stability with adjustable rheology, making them suitable for precision microfluidic devices and dense integrated circuits. The incorporation of nanomaterials into these polymers enhances efficient heat dissipation and reduces the likelihood of overheating during the miniaturization of electronics (Neumann and Dickey, 2020; Lee et al., 2022).

Technologies such as liquid metal direct writing have allowed for the fabrication of elastic, self-healing, high conductivity printed circuit boards under mechanical strain, which has promoted the progress of stretchable electronics (Lin et al., 2024). These polymer nanohybrids are essential for future sensors and electronic devices owing to their synergetic self-healing properties, effective thermal management, and reinforcement with nanomaterials. This ensures sustained sensitivity, trace-level detection, and prolonged functionality of the sensors in environmental and industrial monitoring systems (Ibrahim et al., 2025). Furthermore, the ability of these materials to be easily adapted to current manufacturing protocols enables the creation of high-performance, resilient circuitry that is suitable for demanding environments (Wang et al., 2021).

3.7.3 Oil and gas applications

Owing to their excellent resistance to corrosion and thermal degradation, high-temperature polymer composites have emerged as crucial materials in the oil and gas industry. These materials are particularly important in pipelines and drilling operations where pressure, aggressive fluids, and mechanical wear are prevalent. The mechanical strength, thermal stability, and chemical resistance of polymer matrices can be significantly enhanced by the incorporation of nanoparticles, such as metal oxides or carbon-based fillers, which offer more durable protection in extreme environments (Abdikamalova et al., 2023). Recent advances in polymer synthesis and processing have enabled the engineering of composite properties on the molecular scale. The incorporation of nanofillers in drilling fluids also modifies their viscosity and flow properties, resulting in improved pumping efficiency and reduced energy waste (Kharitonov et al., 2024). In addition, these polymers are used to create high-temperature cement composites for oil wells, mitigating the degradation caused by corrosive gases such as CO_2 and H_2S. This demonstrates the versatility and industrial relevance of high-temperature polymer systems for downhole applications and good infrastructure (Tan et al., 2023). Based on this theory, field studies have also corroborated the benefits of polymer composites in reducing maintenance, enhancing seal performance, and decreasing corrosion-based failures and downtime. Corrosion inhibitors and nanoreinforcements reduce unexpected downtimes and improve the sustainability of oil recovery operations (Wang et al., 2021). These advanced materials are increasing the reliability of pipeline and drilling technologies, providing a future-oriented substitute for conventional materials in demanding oil and gas settings (Wu and Koo, 2022).

3.7.4 Energy generation, transmission, and storage

Owing to their unique thermomechanical stability, high-temperature polymers are essential components of modern energy systems. When integrated into advanced manufacturing techniques, such as additive manufacturing, they enable the creation of lightweight and intricate components with low waste generation, which is essential for high-performance energy generation, transmission, and storage (Han et al., 2022). Their dielectric and structural integrities under extreme environmental conditions offer excellent value to the aerospace and power electronics industries, where operational reliability is paramount. High-performance dielectrics such as polyimides and high-temperature polymers such as PEEK have been engineered into nanocomposites with modified dielectric properties and energy densities. Layered nanocoatings and quantum-sized fillers, including hafnium (IV) oxide (HfO_2) nanoparticles, have significantly improved both the charge retention and capacitive performance at temperatures exceeding 200°C (Yang et al., 2023).

Furthermore, all-organic polymer/molecular semiconductor composites have demonstrated considerable potential for achieving energy storage efficiencies comparable to those of room-temperature systems while functioning under ultrathermal conditions (Yuan et al., 2020b). Their industrial significance is enhanced by their environmentally friendly (biobased) nature and compatibility with large-scale, energy-efficient fabrication methods. This trend is notable because the integration of bioproducts into sustainable additive manufacturing is crucial for reducing the environmental impacts associated with production (Zhou et al., 2023). These materials exhibit improved dielectric strength and durability in energy devices, with applications ranging from renewable energy converters to electric vehicle power systems (Rajpoot and Banik, 2023; Min et al., 2022). While the tunability of polymer-embedded nanostructured composites facilitates processing, their high-temperature stability and biocompatibility enhance the versatility of effective functional crosslinking of glassy domains to potentially improve thermal stability.

3.7.5 Biomedical and other specialized applications

High-temperature polymers with shape-memory properties facilitate the development of innovative biomedical devices that offer superior thermal stability, favorable biocompatibility, and excellent actuation performance. For structural applications, shape memory polyurethanes synthesized with biobased chain extenders such as glycerol have demonstrated good shape fixity and outstanding recovery at high temperatures (Başkır et al., 2025), making them ideal for devices such as stents and occlusion devices. These materials have been shown through cyclic thermomechanical testing to endure repeated thermal and mechanical strains without compromising structural integrity, which is essential for long-term implantable technologies (Swamy MK et al., 2022). Precise and minimally invasive actuation in a dynamic physiological environment is critical.

High-temperature SMPs, such as polymer-braided stents, provide superior adaptability to vascular geometries and better mechanical response to pulsatile blood flow (Hu et al., 2024). Mechanical thrombectomy devices have leveraged this thermally activated recovery behavior, achieving rapid device recovery with minimal force. This property facilitates controlled site-specific drug delivery, thereby improving therapeutic efficacy (O'Brien et al., 2023). Wide-temperature polymers are utilized as substrate materials for multifunctional bioelectronic systems in addition to their mechanical applications. Their thermal resistance and hardness form the foundation for innovative implants such as vascular sensors and flexible stents, which continuously monitor haemodynamic states (Herbert, 2024). All these materials collaborate to bridge the gap between biocompatibility, durability, and functional adaptability, leading to the development of next-generation smart biomedical devices that can intelligently respond to complex physiological conditions.

3.8 Future trends

3.8.1 Nanocomposites and high-performance hybrids

In recent years, the advancement of high-temperature polymers has primarily taken the form of nanocomposite materials, achieved through the incorporation of nanoscale fillers and engineered interactions at surface-filled interfaces that enhance thermal endurance and dielectric metrics. Wang et al. (2020) demonstrated that interfacial 2D montmorillonite nanocoatings can produce sandwiched polymer nanocomposites suitable for ultrahigh capacitive energy storage at elevated temperatures, highlighting that nanostructured hybridization is an effective tool for promoting device operation in harsh environments. Similar studies have shown that the incorporation of carbon nanotubes and nanoclays in polymer matrices significantly influences the properties of dynamic mechanics and thermals, while exhibiting high dielectric properties across a wide temperature range (Mat Yazik et al., 2021). This dual-modality strategy preserves the mechanical integrity beyond 150°C and optimizes the energy storage capacity of the composite, rendering these materials promising candidates for aerospace and electrified transportation applications (Wang et al., 2022).

3.8.2 Sustainable and biobased alternatives

In addition to the ongoing pursuit of performance benefits, there is a growing interest in sustainable and biobased materials within the context of high-temperature polymer design. Recent studies have highlighted the potential for synthesizing polymers from renewable feedstocks, such as CO_2 and lignin, which can benefit the environment and promote recycling (Villa et al., 2025). These sustainable polymers show promise for reducing the carbon footprint while delivering competitive thermal and electrical performance, as required by high-temperature engineering polymers. Furthermore, the use of new green solvents and energy-efficient processing methods has significantly enhanced not only the material properties but also the overall sustainability of the process, marking one of the key trends in the development of polymer science (Wang et al., 2022).

3.8.3 Research challenges and preprocesses

Based on recent advances in organic and inorganic nanocomposites and sustainable polymers, they have great potential. However, several significant challenges remain to be overcome. A key challenge lies in the effective control of interfacial phenomena, such as charge trapping and local field enhancement, which can degrade the dielectric stability under high-temperature regimes. The generation of new, high-T_g polymers that blend essential electrical performance with good thermal stability will require tight control of the network structures and chemical cross-linking. Recent studies on crosslinked fluoropolymers have shed light on this issue. Opportunities lie in the advanced modeling of the physical/chemical kinetics of charge-trapping and de-trapping processes, as well as the development of scalable, efficient, and cost-effective (Meng et al., 2024). Combined with new experimental strategies, these research directions provide a strong platform for circumventing the present limits in material performance.

3.8.4 Toward circular economy and recycling

Encouraging a circular approach to high-temperature polymers is essential because industries demand both high performance and resource efficiency. Comprehensive polymer recycling and reprocessing strategies have been well established, and research has demonstrated that CO_2 depolymerization into valuable feedstock is feasible for reprocessing. The future of high-temperature polymers depends on advances in nanocomposite architectures, sustainable high-throughput or green synthesis, and focused research efforts to overcome interfacial issues and recycling challenges. These comprehensive strategies are crucial for fulfilling the requirements of advanced contemporary power systems and robust and sustainable material technologies. In addition to all the discussed future trends, AI-assisted polymer design, stimuli-responsive systems, and real-time in situ analytics offer exciting opportunities to advance both the scientific and industrial frontiers of high-temperature polymers.

3.9 Conclusion

This chapter extensively covers the synthesis, classification, processing, and characterization of high-temperature polymers and their importance in next-generation applications, including aerospace, energy storage materials, electronics, and biomedical systems. The state-of-the-art integration of nanocomposites, the design of sustainable polymers, and advanced processing techniques have significantly improved thermal, mechanical, and functional properties, even under extreme conditions. At the same time, the increasing use of biobased monomers and the principles of a circular economy, including chemical recycling and upcycling, demonstrate the sector's growing alignment with environmental and economic sustainability. The future of high-temperature polymers must strike a balance between material innovation and ecological responsibility. Although complete closed-loop systems remain challenging, particularly in terms of scalability, recyclability, and process optimization, overcoming these barriers is critical to their realization. Ongoing research on intelligent monomer design, dynamic networks, and hybrid nanostructures will drive the development of future high-performance polymers with exceptional durability and functionality that can be integrated into the industry.

AI disclosure

During the preparation of this work the author(s) used Quillbot to improve readability and language. After using this tool/service, the author(s) reviewed and edited the content as needed and take(s) full responsibility for the content of the publication.

References

Abdikamalova, A., Kalilaev, M., Khojanazarova, S., Khalilov, M., Bukharov, S., Kovalev, I., 2023. Polymer structure-forming agents for control the characteristics of drilling fluids. E3S Web of Conferences 390, 05032. Available at https://doi.org/10.1051/e3sconf/202339005032.

Abubaker, S.A., Chaqmaqchee, F.A., Taha, A.H., 2021. Identification and characterization of different types of plastics wastes using X-ray diffraction and X-ray fluorescence techniques. ARO-The Scientific Journal of Koya University 9 (2), 22–25. Available at https://doi.org/10.14500/aro.10840.

Adiga, K., Herbert, M.A., Rao, S.S., Shettigar, A., 2022. Applications of reinforcement particles in the fabrication of aluminium metal matrix composites by friction stir processing - a review. Manufacturing Review 9. Available at https://doi.org/10.1051/mfreview/2022025.

Ahmad, J., Niasar, M.G., 2025. Aging behavior of PEEK, PTFE, and PI insulation materials under thermal oxidative and humid conditions for aerospace applications. Journal of Applied Polymer Science 142 (19). Available at https://doi.org/10.1002/app.56858.

Alam, M.M., Biswas, B., Nedeltchev, A.K., Han, H., Ranasinghe, A.D., Bhowmik, P.K., et al., 2019. Phosphine oxide containing poly(pyridinium salt)s as fire retardant materials. Polymers 11 (7). Available at https://doi.org/10.3390/polym11071141.

Anis, A., Alam, M., Alhamidi, A., Gupta, R.K., Tariq, M., Al-Zahrani, S.M., 2023. Studies on polybenzimidazole and methanesulfonate protic-ionic-liquids-based composite polymer electrolyte membranes. Polymers 15 (13). Available at https://doi.org/10.3390/polym15132821.

Ankit, Saini, N., Pandey, H., Pandey, K., 2024. A systematic review of MOF, COF, and their hybrid-based composite membranes for gas separation. Macromolecular Symposia 413 (1). Available at https://doi.org/10.1002/masy.202300058.

Araya-Hermosilla, E., Giannetti, A., Lima, G.M.R., Orozco, F., Picchioni, F., Mattoli, V., et al., 2021. Thermally switchable electrically conductive thermoset rgo/pk self-healing composites. Polymers 13 (3), 1–19. Available at https://doi.org/10.3390/polym13030339.

Arrigo, R., Malucelli, G., Mantia, F.P.L., 2021. Effect of the elongational flow on the morphology and properties of polymer systems: a brief review. Polymers 13 (20). Available at https://doi.org/10.3390/polym13203529.

Bashir, M.A., 2021. Use of dynamic mechanical analysis (DMA) for characterizing interfacial interactions in filled polymers. Solids 2 (1), 108–120. Available at https://doi.org/10.3390/solids2010006.

Başkır, S., Cankaya, A.O., Tekay, E., Erenay, B., Garipcan, B., Kocatürk, Ö., 2025. Shape memory polyurethane synthesis using glycerol as chain extender for biomedical applications. Polymer International 74 (4), 336–345. Available at https://doi.org/10.1002/pi.6725.

Bek, M., Gonzalez-Gutierrez, J., Kukla, C., Črešnar, K.P., Maroh, B., Perše, L.S., 2020. Rheological behaviour of highly filled materials for injection moulding and additive manufacturing: effect of particle material and loading. Applied Sciences (Switzerland) 10 (22), 1–23. Available at https://doi.org/10.3390/app10227993.

Buglioni, L., Raymenants, F., Slattery, A., Zondag, S.D.A., Noël, T., 2022. Technological innovations in photochemistry for organic synthesis: flow chemistry, high-throughput experimentation, scale-up, and photoelectrochemistry. Chemical Reviews 122 (2), 2752–2906. Available at https://doi.org/10.1021/acs.chemrev.1c00332.

Capaldo, L., Wen, Z., Noël, T., 2023. A field guide to flow chemistry for synthetic organic chemists. Chemical Science 14 (16), 4230–4247. Available at https://doi.org/10.1039/d3sc00992k.

Carotenuto, M.R., Cavallaro, G., Chinnici, I., Lazzara, G., Milioto, S., 2024. Hybrid green materials obtained by PCL melt blending with diatomaceous earth. Molecules (Basel, Switzerland) 29 (6). Available at https://doi.org/10.3390/molecules29061203.

Chen, D., Tiwari, S.K., Ma, Z., Wen, J., Liu, S., Li, J., et al., 2020. Phase behavior and thermo-mechanical properties of if-ws2 reinforced pp-pet blend-based nanocomposites. Polymers 12 (10), 1–15. Available at https://doi.org/10.3390/polym12102342.

Chen, H., Wang, L., Feng, S., Li, L., 2025. Enaminoester-based unsymmetrical hantzsch reaction for one-step hetero-difunctionalization of polymers. Macromolecules 58 (3), 1642–1651. Available at https://doi.org/10.1021/acs.macromol.4c03104.

Chimeni-Yomeni, D., Fazli, A., Dubois, C., Rodrigue, D., 2022. Low density polyethylene composites based on flax fibers modified by a combination of coupling agent. Polymer Engineering and Science 62 (11), 3553–3565. Available at https://doi.org/10.1002/pen.26126.

Cho, C., Nam, S.L., de la Mata, A.P., Harynuk, J.J., Elias, A.L., Chung, H.J., et al., 2022. Investigation of the accelerated thermal aging behavior of polyetherimide and lifetime prediction at elevated temperature. Journal of Applied Polymer Science 139 (15). Available at https://doi.org/10.1002/app.51955.

Chu, F., Hai, G., Zhao, D., Liu, S., Hu, Y., Zhao, G., et al., 2023. Regulating keto-enol tautomerism of β-ketoenamine covalent-organic frameworks for photocatalytic oxidative coupling of amines. ACS Catalysis 13 (20), 13167–13180. Available at https://doi.org/10.1021/acscatal.3c03857.

Coessens, S., Bahramian, B., Bellemans, I., Crivits, T., Detavernier, C., Verbeken, K., 2023. Material selection for fundamental investigation of high temperature processes in lithium-ion cathode production. ECS Meeting Abstracts MA2023-02 (2), 389. Available at https://doi.org/10.1149/ma2023-022389mtgabs 389.

Das, A., Gilmer, E.L., Biria, S., Bortner, M.J., 2021. Importance of polymer rheology on material extrusion additive manufacturing: correlating process physics to print properties. ACS Applied Polymer Materials 3 (3), 1218–1249. Available at https://doi.org/10.1021/acsapm.0c01228.

Deshmukh, A.A., Wu, C., Yassin, O., Chen, L., Shukla, S., Zhou, J., et al., 2023. Effect of fluorine in redesigning energy-storage properties of high-temperature dielectric polymers. ACS Applied Materials and Interfaces 15 (40), 46840–46848. Available at https://doi.org/10.1021/acsami.3c08858.

Du, J.P., Siewert, I., 2020. An electrochemical and spectroscopic study on Re(CO)3(L)Cl in dimethylformamide (L = 2,2'-bipyridine). Zeitschrift fur Anorganische und Allgemeine Chemie 646 (13), 705–711. Available at https://doi.org/10.1002/zaac.201900320.

Duan, X., Yuan, H., Tang, W., He, J., Guan, X., 2022. An engineering prediction model for stress relaxation of polymer composites at multiple temperatures. Polymers 14 (3), 568. Available at https://doi.org/10.3390/polym14030568.

Embarek, N., Sahli, N., 2020. A novel green synthesis method of poly (3-glycidoxypropyltrimethoxysilane) catalyzed by treated bentonite. Bulletin of Chemical Reaction Engineering & Catalysis 15 (2), 290–303. Available at https://doi.org/10.9767/bcrec.15.2.6568.290-303.

Farajidizaji, B., Borrego, E.I., Athukorale, S., Jazi, M., Donnadieu, B., Pittman, C.U., et al., 2021. Triphenylene-enchained perfluorocyclobutyl aryl ether polymers: a modular synthetic route to processable thermoplastics approaching upper limit tgand photostability. Macromolecules 54 (16), 7666–7672. Available at https://doi.org/10.1021/acs.macromol.1c01043.

Fauziyah, N.A., Perwitasari, D.S., Fadly, T.A., Soontaranon, S., Pratapa, S., 2022. A comparative study of the crystallinity value of polyethylene glycol-based composites using XRD and WAXS. Journal of Physics: Conference Series 2392 (1), 012033. Available at https://doi.org/10.1088/1742-6596/2392/1/012033.

Feng, Q.K., Zhong, S.L., Pei, J.Y., Zhao, Y., Zhang, D.L., Liu, D.F., et al., 2022. Recent progress and future prospects on all-organic polymer dielectrics for energy storage capacitors. Chemical Reviews 122 (3), 3820–3878. Available at https://doi.org/10.1021/acs.chemrev.1c00793.

Fowler, H.R., O'Shea, R., Sefton, J., Howard, S.C., Muir, B.W., Stockman, R.A., et al., 2025. Rapid, highly sustainable ring-opening polymerization via resonant acoustic mixing. ACS Sustainable Chemistry and Engineering 13 (5), 1916–1926. Available at https://doi.org/10.1021/acssuschemeng.4c06330.

Fu, X.B., Zhang, T., Yang, J.C., Zhang, G., Zhang, M.L., Wang, X.J., et al., 2022. Structures and properties of newly synthesized semi-aromatic polyamide thermoplastic elastomers. Polymer Chemistry 13 (34), 4980–4991. Available at https://doi.org/10.1039/d2py00541g.

Galant, O., Donmez, H.B., Barner-Kowollik, C., Diesendruck, C.E., 2021. Flow photochemistry for single-chain polymer nanoparticle synthesis. Angewandte Chemie 133 (4), 2070–2074. Available at https://doi.org/10.1002/ange.202010429.

Gallastegui, A., Lingua, G., Lopez-Larrea, N., Carfora, R., Pasini, D., Mantione, D., et al., 2024. Piperazinium poly(ionic liquid)s as solid electrolytes for lithium batteries. Macromolecular Rapid Communications 45 (16). Available at https://doi.org/10.1002/marc.202400184.

Geng, K., He, T., Liu, R., Dalapati, S., Tan, K.T., Li, Z., et al., 2020. Covalent organic frameworks: design, synthesis, and functions. Chemical Reviews 120 (16), 8814–8933. Available at https://doi.org/10.1021/acs.chemrev.9b00550.

Goes, M.A.de, Santos, J.P.F., Carvalho, B.de M., 2022. Impact of controlled extensional flow during extrusion of PP, PVDF and LDPE. Polímeros 32 (2). Available at https://doi.org/10.1590/0104-1428.20210085.

Graur, V., Mukherjee, A., Sebakhy, K.O., Bose, R.K., 2022. Initiated chemical vapor deposition (iCVD) of bio-based poly(tulipalin A) coatings: structure and material properties. Polymers 14 (19). Available at https://doi.org/10.3390/polym14193993.

Hamad, B.A., He, M., Xu, M., Liu, W., Mpelwa, M., Tang, S., et al., 2020. A novel amphoteric polymer as a rheology enhancer and fluid-loss control agent for water-based drilling muds at elevated temperatures. ACS Omega 5 (15), 8483–8495. Available at https://doi.org/10.1021/acsomega.9b03774 Available from: pubs.acs.org/journal/acsodf.

Han, L., Zheng, M., Guo, H., Wang, T., Zhu, J., Hajikarimi, P., 2022. Mechanism behind time dependent elasticity of crumb rubber-nano-asphalt hybrids using discrete relaxation spectrum. Journal of Renewable Materials 10 (6), 1727–1752. Available at https://doi.org/10.32604/jrm.2022.018467.

Hasan, M.M., Rahman, M., Sadeque, M.S.B., Ordu, M., 2024. Self-poled piezoelectric nanocomposite fiber sensors for wireless monitoring of physiological signals. ACS Applied Materials and Interfaces 16 (27), 34549–34560. Available at https://doi.org/10.1021/acsami.4c04908.

Herbert, R., 2024. Wireless vascular bioelectronic systems with printed soft sensors and flexible electronic stents. Proceedings Volume 12948, Soft Mechatronics and Wearable Systems; 129480C (2024). SPIE. Available at https://doi.org/10.1117/12.3012004.

Hu, X., Zhang, C., Wang, B., Liu, J., Cheng, J., Zhao, G., et al., 2024. Mechanical behavior of polymer braided stent at rapid radial loading. Key Engineering Materials 977, 141–145. Available at https://doi.org/10.4028/p-0xC6xG.

Ibrahim, O., Oluwadunsin, J., Antwi, M., Mekunye, F., Oluwatobi, M., Olorunfemi, A., et al., 2025. The evolution of self-healing electrodes: a critical review of nanomaterial contributions. American Journal of Nanosciences 9 (1), 8–31. Available at https://doi.org/10.11648/j.ajn.20250901.12.

Jamil, H., Faizan, M., Adeel, M., Jesionowski, T., Boczkaj, G., Balčiūnaitė, A., 2024. Recent advances in polymer nanocomposites: unveiling the frontier of shape memory and self-healing properties—a comprehensive review. Molecules (Basel, Switzerland) 29 (6), 1267. Available at https://doi.org/10.3390/molecules29061267.

Kamaraj, M., Dodson, E.A., Datta, S., 2021. Thermal and viscoelastic behaviour of graphene nanoplatelets/flax fibre/epoxy composites. Plastics, Rubber and Composites 50 (5), 219–227. Available at https://doi.org/10.1080/14658011.2020.1871241.

Kandelhard, F., Pashayev, E., Schymura, J., Georgopanos, P., 2023. Kinetic modeling of the synthesis of poly(4-vinylpyridine) macro-reversible addition-fragmentation chain transfer agents for the preparation of block copolymers. Industrial & Engineering Chemistry Research 62 (22), 8696–8708. Available at https://doi.org/10.1021/acs.iecr.3c00607.

Kang, N., Yang, S., Xiong, X., Han, A., Ren, R., Wang, J., 2023. The performance and synthesis of alkynyl–functionalized benzoxazine containing phthalide side groups and cyano groups with different molecular weights. Polymers 15 (16), 3478. Available at https://doi.org/10.3390/polym15163478.

Karabıyık, M., Cihanoğlu, G., Ebil, Ö., 2023. CVD deposited epoxy copolymers as protective coatings for optical surfaces. Polymers 15 (3), 652. Available at https://doi.org/10.3390/polym15030652.

Khan, H., Yerramilli, A.S., D'Oliveira, A., Alford, T.L., Boffito, D.C., Patience, G.S., 2020. Experimental methods in chemical engineering: X-ray diffraction spectroscopy—XRD. Canadian Journal of Chemical Engineering 98 (6), 1255–1266. Available at https://doi.org/10.1002/cjce.23747.

Kharitonov, E.V., Antonova, K.Y.U., Sharifullin, A.V., Baybekova, L.R., Yurchenko, Y.A.A., Minnakhmetov, F.F., et al., 2024. The influence of hyperbranched polymers on viscosity values for pipeline pumping of liquid hydrocarbons. Herald of Technological University 27 (5), 100–104. Available at https://doi.org/10.55421/1998-7072_2024_27_5_100.

Koza, P., Koczorowski, T., Mlynarczyk, D.T., Goslinski, T., 2022. Zinc(II) sulfanyltribenzoporphyrazines with bulky peripheral substituents—synthesis, photophysical characterization, and potential photocytotoxicity. Applied Sciences 12 (13), 6825. Available at https://doi.org/10.3390/app12136825.

Lai, H.W.H., Benedetti, F.M., Ahn, J.M., Robinson, A.M., Wang, Y., Pinnau, I., et al., 2022. Hydrocarbon ladder polymers with ultrahigh permselectivity for membrane gas separations. Science (New York, N.Y.) 375 (6587), 1390–1392. Available at https://doi.org/10.1126/science.abl7163.

Laramée, A.W., Pellerin, C., 2023. Raman analysis of orientation and crystallinity in high tg, low crystallinity electrospun fibers. Applied Spectroscopy 77 (11), 1289–1299. Available at https://doi.org/10.1177/00037028231202791.

Laramée, A.W., Lanthier, C., Pellerin, C., 2020. Electrospinning of highly crystalline polymers for strongly oriented fibers. ACS Applied Polymer Materials 2 (11), 5025–5032. Available at https://doi.org/10.1021/acsapm.0c00873 Available from: pubs.acs.org/journal/aapmcd.

Laramée, A.W., Lanthier, C., Pellerin, C., 2022. Raman investigation of the processing structure relations in individual poly(ethylene terephthalate) electrospun fibers. Applied Spectroscopy 76 (1), 51–60. Available at https://doi.org/10.1177/00037028211049242.

Lee, W., Kim, H., Kang, I., Park, H., Jung, J., Lee, H., et al., 2022. Universal assembly of liquid metal particles in polymers enables elastic printed circuit board. Science (New York, N.Y.) 378 (6620), 637–641. Available at https://doi.org/10.1126/science.abo6631.

Li, B., Hu, C., Pang, X., Chen, X., 2023. Valence-variable catalysts for redox-controlled switchable ring-opening polymerization. Chemistry – An Asian Journal 18 (1). Available at https://doi.org/10.1002/asia.202201031.

Li, D., Liu, H., Shen, Y., Wu, H., Liu, F., Wang, L., et al., 2020. Preparation of Pi/PTFE–PAI composite nanofiber aerogels with hierarchical structure and high-filtration efficiency. Nanomaterials 10 (9), 1–18. Available at https://doi.org/10.3390/nano10091806.

Li, J., Richardson, J.J., Ejima, H., 2022. A simple and feasible synthetic strategy towards poly(4-thiostyrene). Macromolecular Chemistry and Physics 223 (17). Available at https://doi.org/10.1002/macp.202200092.

Liang, S., Li, M.H., Qi, M., Wang, L., Yang, Y.W., 2023. Nanoplatforms based on covalent organic frameworks (COFs) for biomedical applications. Chemistry of Materials 35 (20), 8353–8370. Available at https://doi.org/10.1021/acs.chemmater.3c02095.

Lidston, C.A.L., Abel, B.A., Coates, G.W., 2020. Bifunctional catalysis prevents inhibition in reversible-deactivation ring-opening copolymerizations of epoxides and cyclic anhydrides. Journal of the American Chemical Society 142 (47), 20161–20169. Available at https://doi.org/10.1021/jacs.0c10014.

Liénard, R., De Winter, J., Coulembier, O., 2020. Cyclic polymers: advances in their synthesis, properties, and biomedical applications. Journal of Polymer Science 58 (11), 1481–1502. Available at https://doi.org/10.1002/pol.20200236.

Lin, Y., Wang, H., Qiu, W., Ye, C., Kong, D., 2024. Liquid metal-based self-healing conductors for flexible and stretchable electronics. ACS Applied Materials & Interfaces 16 (33), 43083–43092. Available at https://doi.org/10.1021/acsami.4c10541.

Liu, Y., Xu, Y., Gao, J., Xu, J., Wu, M., Liu, Z., et al., 2024. Superior energy storage performance in crosslinked binary polymers at high temperatures via confinement effect. Energy & Environmental Materials 8 (2), e12847. Available at https://doi.org/10.1002/eem2.12847.

Lucherelli, M.A., Duval, A., Averous, L., 2023. Combining associative and dissociative dynamic linkages in covalent adaptable networks from biobased 2,5-furandicarboxaldehyde. ACS Sustainable Chemistry and Engineering 11 (6), 2334–2344. Available at https://doi.org/10.1021/acssuschemeng.2c05956.

Marzuki, A.P., Nasir, M.A.M., Salleh, F.M., Ismail, M.H., Murat, B.I.S., Ibrahim, M., 2024. Initial study on rheological behaviour of hydroxyapatites/polylactic acid composite for 3D printing filament. International Journal of Automotive and Mechanical Engineering 21 (1), 11064–11073. Available at https://doi.org/10.15282/ijame.21.1.2024.10.0856.

Mat Yazik, M.H., Sultan, M.T.H., Jawaid, M., Abu Talib, A.R., Mazlan, N., Md Shah, A.U., et al., 2021. Effect of nanofiller content on dynamic mechanical and thermal properties of multi-walled carbon nanotube and montmorillonite nanoclay filler hybrid shape memory epoxy composites. Polymers 13 (5), 1–21. Available at https://doi.org/10.3390/polym13050700.

Matatyaho Ya'Akobi, A., Talmon, Y., 2021. Extending cryo-EM to nonaqueous liquid systems. Accounts of Chemical Research 54 (9), 2100–2109. Available at https://doi.org/10.1021/acs.accounts.1c00077.

McMullen, R.L., 2021. Atomic Force Microscopy, Transmission Electron Microscopy, Scanning Electron Microscopy, and Fourier Transform-Infrared Spectroscopic Imaging as Tools to Study Various Phenomena Involving Poly(Vinylpyrrolidone). In: Musa, O. (Ed.), Handbook of Pyrrolidone and Caprolactam Based Materials. Wiley, pp. 1759–1829. Available at https://doi.org/10.1002/9781119468769.hpcbm022.

Meng, Z., Zhang, T., Zhang, C., Shang, Y., Lei, Q., Chi, Q., 2024. Advances in polymer dielectrics with high energy storage performance by designing electric charge trap structures. Advanced Materials 36 (52). Available at https://doi.org/10.1002/adma.202310272.

Min, D., Ji, M., Gao, Z., Cai, Z., Wu, Q., Liu, J., et al., 2022. A unified model for conductivity, electric breakdown, energy storage, and discharge efficiency of linear polymer dielectrics. Journal of Physics D: Applied Physics 55 (28), 285501. Available at https://doi.org/10.1088/1361-6463/ac5d02.

Morin, E., Neiderer, W., Cruché, C., Bleton, O., Cave, C., Collins, S.K., 2024. A flow chemistry platform for photochemical macrocyclization of peptides. ACS Sustainable Chemistry and Engineering 12 (16), 6433–6439. Available at https://doi.org/10.1021/acssuschemeng.4c01441.

Moschetta, E.G., Cook, G.C., Edwards, L.J., Ischay, M.A., Lei, Z., Buono, F., et al., 2024. Photochemistry in pharmaceutical development: a survey of strategies and approaches to industry-wide implementation. Organic Process Research and Development 28 (4), 831–846. Available at https://doi.org/10.1021/acs.oprd.3c00499.

Murphy, J.G., Raybin, J.G., Sibener, S.J., 2022. Correlating polymer structure, dynamics, and function with atomic force microscopy. Journal of Polymer Science 60 (7), 1042–1058. Available at https://doi.org/10.1002/pol.20210321.

Nematdoust, S., Najjar, R., Bresser, D., Passerini, S., 2020. Understanding the role of nanoparticles in PEO-based hybrid polymer electrolytes for solid-state lithium–polymer batteries. The Journal of Physical Chemistry C 124 (51), 27907–27915. Available at https://doi.org/10.1021/acs.jpcc.0c08749.

Neumann, T.V., Dickey, M.D., 2020. Liquid metal direct write and 3D printing: a review. Advanced Materials Technologies 5 (9). Available at https://doi.org/10.1002/admt.202000070.

O'Brien, R., Moritz, V., McDonald, P., Devine, D., Srivastava, R., 2023. Design and Development of Shape Memory Polymer-Based Mechanical Thrombectomy Device. Proceedings of the ASME 2023 International Mechanical Engineering Congress and Exposition. Volume 3: Advanced Manufacturing. New Orleans, Louisiana, USA. October 29–November 2, 2023. V003T03A050. ASME. Available at https://doi.org/10.1115/IMECE2023-113295.

Ötvös, S.B., Kappe, C.O., 2020. Continuous-flow amide and ester reductions using neat borane dimethylsulfide complex. ChemSusChem 13 (7), 1800–1807. Available at https://doi.org/10.1002/cssc.201903459.

Park, G., Bielawski, C.W., 2021. Ring opening metathesis polymerization of cyclic allenes. Macromolecules 54 (13), 6135–6143. Available at https://doi.org/10.1021/acs.macromol.1c00571.

Paszkiewicz, S., Irska, I., Walkowiak, K., Zubkiewicz, A., 2021. Influence of synthesis conditions on molecular weight as well as mechanical and thermal properties of poly(hexamethylene 2,5-furanate). Polimery 66 (10). Available at https://doi.org/10.14314/polimery.2021.10.3.

Peltier, L., Lohmuller, P., Meraghni, F., Berveiller, S., Patoor, E., Laheurte, P., 2020. Investigation and composition characterization of a "NiTi-like" alloy combining high temperature shape memory and high entropy. Shape Memory and Superelasticity 6 (2), 273–283. Available at https://doi.org/10.1007/s40830-020-00290-2.

Piacenti, A.R., Adam, C., Hawkins, N., Wagner, R., Seifert, J., Taniguchi, Y., et al., 2024. Nanoscale rheology: dynamic mechanical analysis over a broad and continuous frequency range using photothermal actuation atomic force microscopy. Macromolecules 57 (3), 1118–1127. Available at https://doi.org/10.1021/acs.macromol.3c02052.

Premadasa, U.I., Doughty, B., Custelcean, R., Ma, Y.-Z., 2024. Towards energy-efficient direct air capture with photochemically-driven CO_2 release and solvent regeneration. ChemPlusChem 89 (10). Available at https://doi.org/10.1002/cplu.202300713.

Qi, X., Wang, W., Zhang, J., Yan, C., Zhu, Y., Zhang, F., 2019. Aqueously soluble semi-aromatic copolyamides with dual-functional groups as sizing agent. Journal of Applied Polymer Science 136 (9). Available at https://doi.org/10.1002/app.47132.

Rajpal, S., Mishra, P., Mizaikoff, B., 2023. Rational in silico design of molecularly imprinted polymers: current challenges and future potential. International Journal of Molecular Sciences 24 (7), 6785. Available at https://doi.org/10.3390/ijms24076785.

Rajpoot, A., Banik, N., 2023. A review on the recent development on polymer nanocomposite for energy storage application. Material Science Research India 20 (SpecialIssue1), 13–26. Available at https://doi.org/10.13005/msri.20.special-issue1.02.

Ren, H., Fan, X., Li, P., Ji, M., Zhang, G., Yan, Y., 2022. Novel semi-aromatic polyamides containing ether bond and pendant methyl group: synthesis and properties. Polymer Bulletin 79 (11), 9629–9642. Available at https://doi.org/10.1007/s00289-021-03951-z.

Sakahara, R.M., da Silva, D.J., Wang, S.H., 2022. Composites of ABS with SEBS-g-MA and copper microparticles modified by mussel-bioinspired polydopamine: a comparative rheological study. Journal of Applied Polymer Science 139 (10), 51768.

Shakir, M.H., Singh, A.K., Siddhartha, 2024. Repercussion of fiber content on the mechanical, thermal, and thermomechanical properties of natural fibers reinforced thermoplastic composites for automotive application. Physica Scripta 99 (9). Available at https://doi.org/10.1088/1402-4896/ad6644.

Sosiati, H., Afif, A.L.A., Hudiatma, A., Adi, R.K., Pamasti, A., 2024. Study of abaca/carbon/epoxy hybrid composite properties as an alternative prosthetic socket material. Key Engineering Materials 977, 95–102. Available at https://doi.org/10.4028/p-8jitl7.

Stojanović, I., Škrlec, B., Kurtela, M., Alar, V., Odeljan, M., 2023. Physical and chemical properties of high-temperature silicone-based polymer coatings applied on different surface roughnesses. Coatings 13 (12), 2100. Available at https://doi.org/10.3390/coatings13122100.

Sundari, S.S.K., Rishwana, S.S., Dhanalakshmi, J., Arunjunai Raj, M., Vijayakumar, C.T., 2023. Fumarate metal organic frameworks as reactive and curing reaction alternant in hydrophobic bispropargyl ether based matrix resin system. Journal of Polymer Research 30 (11), P1.

Sunny, T., Pickering, K.L., 2022. Improving polypropylene matrix composites reinforced with aligned hemp fibre mats using high fibre contents. Materials 15 (16). Available at https://doi.org/10.3390/ma15165587.

Swamy MK, R., Mallikarjun, U.S., Udayakumar, V., 2022. Cyclic thermomechanical analysis of polyethylene glycol based shape memory polymers. Journal of Mines, Metals and Fuels 70 (8), 282–289. Available at https://doi.org/10.18311/jmmf/2022/31988.

Tan, C., Wang, M., Chen, R., You, F., 2023. Study on the oil well cement-based composites to prevent corrosion by carbon dioxide and hydrogen sulfide at high temperature. Coatings 13 (4), 729. Available at https://doi.org/10.3390/coatings13040729.

Tekin, S., Cangül, S., Adıgüzel, Ö., Değer, Y., 2018. Areas for use of PEEK material in dentistry. International Dental Research 8 (2), 84–92. Available at https://doi.org/10.5577/intdentres.2018.vol8.no2.6.

Teng, C., Li, H., Liu, J., Gu, H., Kong, H., Yu, M., 2020. Effect of high molecular weight PPTA on liquid crystalline phase and spinning process of aramid fibers. Polymers 12 (5), 1206. Available at https://doi.org/10.3390/polym12051206.

Than, Y.M., Suriyarak, S., Titapiwatanakun, V., 2022. Rheological investigation of hydroxypropyl cellulose–based filaments for material extrusion 3D printing. Polymers 14 (6). Available at https://doi.org/10.3390/polym14061108.

Villa, R., Salas, R., Maciá, M., Velasco, F., Altava, B., García-Verdugo, E., et al., 2025. How to easily depolymerize polyurethane foam wastes by superbase catalysts in ionic liquids below 100 °C. Angewandte Chemie - International Edition 64 (5). Available at https://doi.org/10.1002/anie.202418034.

Wang, D.P., Zhao, Z.H., Li, C.H., 2021a. Universal self-healing poly(dimethylsiloxane) polymer crosslinked predominantly by physical entanglements. ACS Applied Materials and Interfaces 13 (26), 31129–31139. Available at https://doi.org/10.1021/acsami.1c06521.

Wang, X., Liu,, X., Li, B., Yin,, Q., Han, Z., Li, J., et al., 2021. The application of the multi-component thermal fluid huff and puff technology to daqing heavy oil block. Paper presented at the International Petroleum Technology Conference, Virtual, March 2021. . Available at https://doi.org/10.2523/IPTC-21415-MS.

Wang, Y., Li, Z., Moran, T.J., Ortiz, L.A., Wu, C., Konstantinou, A.C., et al., 2022. Interfacial 2D montmorillonite nanocoatings enable sandwiched polymer nanocomposites to exhibit ultrahigh capacitive energy storage performance at elevated temperatures. Advanced Science 9 (35). Available at https://doi.org/10.1002/advs.202204760.

Wang, Z., Yoon, S., Wang, J., 2022. Breaking the paradox between grafting-through and depolymerization to access recyclable graft polymers. Macromolecules 55 (20), 9249–9256. Available at https://doi.org/10.1021/acs.macromol.2c01609.

Wang, Z., Gnanasekar, P., Sudhakaran Nair, S., Farnood, R., Yi, S., Yan, N., 2020. Biobased epoxy synthesized from a vanillin derivative and its reinforcement using lignin-containing cellulose nanofibrils. ACS Sustainable Chemistry and Engineering 8 (30), 11215–11223. Available at https://doi.org/10.1021/acssuschemeng.0c02559.

Wei, S., Zhang, F., Zhang, W., Qiang, P., Yu, K., Fu, X., et al., 2019. Semiconducting 2D triazine-cored covalent organic frameworks with unsubstituted olefin linkages. Journal of the American Chemical Society 141 (36), 14272–14279. Available at https://doi.org/10.1021/jacs.9b06219.

Wu, H., Koo, J.H., 2022. Chapter 8 - Characterization of high-temperature polymers for extreme environments. Analysis of Flame Retardancy In Polymer Science. Elsevier, United States, pp. 299–331. Available at https://doi.org/10.1016/B978-0-12-824045-8.00008-3.

Xiong, X., Ai, J., Ren, R., Wang, J., Li, G., 2023. Curing behavior and properties of ultra-high temperature resistant RTM bismaleimide resin. High Performance Polymers 35 (5), 469–479. Available at https://doi.org/10.1177/09540083221146926.

Yalcinkaya, B., Buzgo, M., 2024. Optimization of electrospun TORLON® 4000 polyamide-imide (PAI) nanofibers: bridging the gap to industrial-scale production. Polymers 16 (11), 1516. Available at https://doi.org/10.3390/polym16111516.

Yang, L., Nickmilder, P., Verhoogt, H., Hoeks, T., Leclère, P., 2024. Probing viscoelastic properties and interfaces in high-density polyethylene vitrimers at the nanoscale using dynamic mode atomic force microscopy. ACS Applied Materials and Interfaces 16 (29), 38501–38510. Available at https://doi.org/10.1021/acsami.4c06809.

Yang, M., Wang, S., Fu, J., Zhu, Y., Liang, J., Cheng, S., et al., 2023. Quantum size effect to induce colossal high-temperature energy storage density and efficiency in polymer/inorganic cluster composites. Advanced Materials 35 (30). Available at https://doi.org/10.1002/adma.202301936.

Yang, Q., Li, W., Stober, S.T., Burns, A.B., Gopinadhan, M., Martini, A., 2021. Molecular dynamics simulation of the stress-strain behavior of polyamide crystals. Macromolecules 54 (18), 8289–8302. Available at https://doi.org/10.1021/acs.macromol.1c00974.

Yao, C., Huang, M., Huang, G., He, S., Ma, X., Li, Z., et al., 2023. Fully closed-loop recyclable thermosetting polythiourethane microspheres and their application in efficient fabrication of composites with segregated structures. ACS Applied Engineering Materials 1 (8), 2092–2103. Available at https://doi.org/10.1021/acsaenm.3c00250.

Yu, J., Breite, C., Van Loock, F., Pardoen, T., Swolfs, Y., 2024. Experimental investigation of yield and hysteresis behaviour of an epoxy resin under cyclic compression in the large deformation regime. Express Polymer Letters 18 (2), 133–143. Available at https://doi.org/10.3144/expresspolymlett.2024.11.

Yuan, C., Zhou, Y., Zhu, Y., Liang, J., Wang, S., Peng, S., et al., 2020a. Polymer/molecular semiconductor all-organic composites for high-temperature dielectric energy storage. Nature Communications 11 (1). Available at https://doi.org/10.1038/s41467-020-17760-x.

Yuan, J., Wang, W., Zhou, Z., Niu, J., 2020b. Cascade reactions in chain-growth polymerization. Macromolecules 53 (14), 5655–5673. Available at https://doi.org/10.1021/acs.macromol.0c00417.

Zha, J.W., Tian, Y., Zheng, M.S., Wan, B., Yang, X., Chen, G., 2023. High-temperature energy storage polyimide dielectric materials: polymer multiple-structure design. Materials Today Energy 31. Available at https://doi.org/10.1016/j.mtener.2022.101217.

Zhou, T., Zhou, Y., Hua, Z., Yang, Y., Zhou, C., Ren, L., et al., 2021. 4D printing high temperature shape-memory poly(ether–ether–ketone). Smart Materials and Structures 30 (11), 115006. Available at https://doi.org/10.1088/1361-665x/ac24f0.

Zhou, Y., Zhu, Y., Xu, W., Wang, Q., 2023. Molecular trap engineering enables superior high-temperature capacitive energy storage performance in all-organic composite at 200 °C. Advanced Energy Materials 13 (11). Available at https://doi.org/10.1002/aenm.202203961.

4

Nanocomposite-based high temperature polymers

Thomas Adedayo Ayorinde[1,2,3], Ayodele Abeeb Daniyan[4,5], Yemi Olayinka Olasebikan[6], Mutairu Ajiboye[5], Charles Mbohwa[4], Tamba Jamiru[3], Peet Roux[7]

[1]DEPARTMENT OF AGRICULTURAL AND ENVIRONMENTAL ENGINEERING, OBAFEMI AWOLOWO UNIVERSITY, ILE IFE, NIGERIA [2]BIOFUEL PILOT PLANT AND LABORATORY, NATIONAL BIOTECHNOLOGY RESEARCH AND DEVELOPMENT AGENCY, ABUJA, NIGERIA [3]DEPARTMENT OF MECHANICAL AND MECHATRONICS ENGINEERING, TSHWANE UNIVERSITY OF TECHNOLOGY, PRETORIA, GAUTENG, SOUTH AFRICA [4]DEPARTMENT OF INDUSTRIAL ENGINEERING, UNIVERSITY OF SOUTH AFRICA, PRETORIA, SOUTH AFRICA [5]DEPARTMENT OF MATERIALS SCIENCE AND ENGINEERING, OBAFEMI AWOLOWO UNIVERSITY, ILE IFE, NIGERIA [6]DEPARTMENT OF AGRICULTURAL TECHNOLOGY, EKITI STATE POLYTECHNIC, ISAN EKITI, NIGERIA [7]DEPARTMENT OF ELECTRICAL ENGINEERING, TSHWANE UNIVERSITY OF TECHNOLOGY, PRETORIA, GAUTENG, SOUTH AFRICA

4.1 Introduction

Nanocomposite-based high-temperature polymers represent a class of advanced high-performance polymeric materials designed to maintain enhanced mechanical strength, electromagnetic properties, and structural integrity under extremely high temperatures above 200°C (Simões, 2024; Gay, 2022; Bershtein and Yakushev, 2023). These materials are made by mixing tiny fillers like carbon nanotubes (CNTs), graphene, nanoclays, or metal oxides into strong polymer types such as polyimides (PIs), polyetheretherketone (PEEK), and polyphenylene sulfide (PPS). The bonding between the polymer matrix and the nanofillers results in significantly enhanced thermal stability, mechanical strength, chemical resistance, and dimensional stability, making them suitable for applications in aerospace, automotive, electronics, and industrial sectors (Li et al., 2021; Darwish et al., 2022).

Incorporating nanocomposites into high-temperature polymers boosts their structural and functional qualities, making them well-suited for demanding operational situations, and exceptional material reliability is critical in aerospace, automotive, electronics, and energy (Ekeocha et al., 2021).

High-temperature polymers are crucial in applications where materials must work reliably under intense heat, stress, and chemical exposure (Li et al., 2024). In aerospace and aviation, they are employed in components like turbine blades and insulation due to their thermal stability. The automotive industry employs them in engines and fuel systems to save weight and enhance durability. In electronics, they serve as insulators and substrates in high-temperature circuits. Industrial and chemical processing applications benefit from their tolerance to

High-Temperature Polymeric Materials. DOI: https://doi.org/10.1016/B978-0-443-32792-6.00015-4

corrosive chemicals and heat. In medical and biotech, they are employed in implants and sterilizable devices for their biocompatibility. Energy and power generation systems rely on them in fuel cells and reactors, while defense and military use them in protective gear, missile parts, and coatings for their strength and heat resistance (Ekeocha et al., 2021; Li et al., 2024).

4.1.1 Nanocomposite application in high temperature

Nanocomposites combine the unique characteristics of polymers with nanoscale fillers that improve the material's overall performance via regulated interactions between the nanofillers and the polymer matrix. This integration facilitates enhanced control over attributes like thermal stability, surface area, magnetic properties, optical characteristics, and catalytic activity. Nanocomposites significantly enhance the performance, reliability, and versatility of polymers subjected to harsh temperature and mechanical conditions (Njuguna et al., 2021; Marquis et al., 2011). This improvement validates its employment in key areas like aircraft structures, automotive components, electrical gadgets, and industrial machinery, where high-temperature reliability is paramount. Moreover, nanocomposite technology facilitates the creation of lighter, more resilient, and multifunctional materials, in accordance with world-wide initiatives aimed at energy efficiency, sustainability, and advancements in material science (Bokobza, 2023).

4.1.2 Nanocomposites

Nanocomposites are advanced materials that combine a continuous phase in a polymer matrix with a dispersed phase of nanoscale reinforcement, where at least one phase of the nanoscale reinforcement must be reduced to a scale of approximately 1–100 nm (Njuguna et al., 2021). These nanoscale reinforcements are nanofillers made from metals, ceramics, or carbon-based biomass substances (Xu et al., 2018; Chen et al., 2019). The thin size of the reinforcement material enhances the properties of polymers with a moderate loading and is uniformly dispersed within the polymer matrix to get composites like graphene, an allotrope of carbon (Huang et al., 2020; Hu et al., 2021). This integration fosters unique interfacial interactions within the material, between the nanofillers and the polymer matrix, resulting in materials with enhanced properties. The strong interfacial adhesion of the material matrix with the nanofillers at a high surface area-to-volume ratio allows for better control over the thermal stability, surface area, magnetic behavior, and optical properties, and it can be made into desired shapes and sizes (Li et al., 2024; Zafar et al., 2024; Fedullo et al., 2007).

The cohesive bond between the polymer matrix and reinforcing material enhances the material's tensile strength, flexural performance, and resistance to environmental degradation. These improved structural properties are highly effective in solving various real-world challenges, making them crucial for the development of innovative technologies with high-temperature polymeric nanocomposites in high demand. Their multifunctionality, high durability, thermal resistance, and mechanical performance make them ideal for next-generation solutions across various fields. Their exceptional mechanical, structural, and electromagnetic performance at high temperatures makes them a major material in the pursuit of next-generation solutions in various fields (Sun et al., 2021; Chow and Mohd Ishak, 2020; Mohd et al., 2008). These versatile materials have applications in packaging, energy systems, safety technologies, transportation, electromagnetic shielding, defense systems, catalysis, sensors, and information and communication technology. Their versatility also makes them suitable for demanding industries like aerospace, automotive, and energy (Zafar et al., 2024; Kumar et al., 2025).

4.1.3 Objectives and scope

This chapter aims to provide an in-depth overview of high-temperature polymeric nanocomposites, concentrating on their structural composition, functional properties, and importance in extremely high-temperature applications. The major purpose is to examine the scientific basis, fabrication methodologies, performance characteristics, and practical applicability of polymer nanocomposites designed to operate under high temperatures and severe environments. Special emphasis is placed on the integration of nanoscale fillers, such as graphene, CNTs, and metal oxides, into thermally stable polymer matrices, and how this synergy leads to the development of high-performance materials with enhanced mechanical, thermal, electrical, and chemical properties.

It addresses crucial issues like the description and classification of polymeric nanocomposites, the reasons behind their better thermal and mechanical behavior, and the criteria for material selection for harsh environment applications. It also illustrates the essential requirements and challenges connected with creating and deploying these materials in industries such as aerospace, automotive, defense, electronics, and energy. Additionally, the chapter examines surface functionalization tactics, dispersion techniques, and property enhancement procedures to solve current limits in nanocomposite technology.

By synthesizing recent advances and ongoing research, this chapter aims to provide readers with a solid understanding of the state-of-the-art developments, potential applications, and future directions for high-temperature polymeric nanocomposites in addressing complex engineering challenges in extreme environments.

4.2 Fundamentals of high-temperature polymers

High-temperature polymers, in their own right, are designed to resist deterioration and keep functionality in severe settings comprising high temperatures, aggressive chemicals, and mechanical stress. They have strong covalent bonding, stiff molecular backbones, and increased glass transition and decomposition temperatures that exhibit superior performance compared to ordinary polymers (Ekeocha et al., 2021).

4.2.1 Classification of high-temperature polymer

These polymers are widely categorized into three types based on their temperature resistance and molecular architecture: thermosetting polymers (e.g., PIs and bismaleimides [BMIs]), thermoplastic polymers (e.g., PEEK and PPS), and liquid crystal polymers (LCPs) (Liang et al., 2023).

1. Thermosetting high-temperature polymer: These polymers undergo irreversible cross-linking during curing to form a rigid three-dimensional (3D) network that improves their mechanical and thermal stability. Common examples are BMIs, which are frequently used in high-performance composite materials; phenolic resins, which are usually employed in heat shields and ablative systems; and PIs, which are extensively used in aerospace and electronic applications because of their exceptional heat resistance (Liu et al., 2022).
2. Thermoplastic high-temperature polymers: These materials may be melted and remolded again, making them recyclable and efficient for processing. Prominent constituents of this group encompass PEEK, renowned for its exceptional resistance to elevated temperatures and corrosive chemicals; PPS, frequently utilized in automotive and industrial applications; and polybenzimidazole (PBI), distinguished by its remarkable thermal and oxidative stability, even under extreme conditions (Das et al., 2020).
3. LCPs are unique materials that display high anisotropic mechanical strength and exceptional chemical resistance. Their molecular orientation and rigid rod-like structure contribute to their remarkable thermal and mechanical properties, making them increasingly employed in electronic devices, aerospace components, and high-performance fiber manufacturing (Chung et al., 2023; Treccani, 2022).

4.2.2 High-temperature polymer matrices

PIs, PEEK, and PPS are typical examples of high-temperature polymers. They have a perfect matrix structure to accommodate polymer nanocomposites and form high-temperature polymer nanocomposites (HTPNCs) when reinforced with nanoscale fillers, as shown in Table 4–1 (Li et al., 2024).

4.2.2.1 Polyimides

PIs are one of the most notable high-temperature polymers, capable of withstanding temperatures approaching 300°C. They are characterized by their excellent thermal stability, mechanical properties, and electrical insulation capabilities. However, PIs can absorb moisture, leading to dimensional changes, which is a consideration in their application. They are widely used in aerospace applications, flexible printed circuits, and high-temperature insulation materials due to their robust performance under extreme conditions (Ma et al., 2023).

4.2.2.2 Polyetheretherketone

PEEK is another prominent high-temperature polymer known for its exceptional mechanical strength and chemical resistance. PEEK can operate effectively in temperatures ranging from –50°C to +250°C, making it suitable for various applications, including medical devices, automotive components, and aerospace parts. Its inert nature and high thermal stability allow

Table 4–1 Fundamental characteristic of high-temperature polymer matrices (Guerrero et al., 2013; Rahimi-Ahar and Rahimi Ahar, 2024).

Polymer	Maximum temperature (°C)	Tensile strength (MPa)	Thermal conductivity (W (m K)$^{-1}$)	Key features	Nanocomposite application
Polyether ether ketone	~260	90–100	~0.25	High strength chemical resistance flame retardant	Used in aerospace and electronics with CNTs for enhanced thermal and mechanical properties
Polyimides	~300–400	100–180	~0.2–0.3	Excellent thermal stability radiation resistance	Used in microelectronics and aerospace for thermal barrier coatings and structural components
Polyphenylene sulphide	~260	55–75	~0.3	Good chemical resistance, inherent flame retardancy	Used in automotive and electronics with nanoclays and BN for insulation and mechanical durability
Polyetherimide	~215	110–130	~0.22	High modulus, good dielectric properties	Structural composites and EMI shielding materials, when combined with graphene or CNTs

CNTs, Carbon nanotubes; *EMI*, electromagnetic interference.

Source: From Rahimi-Ahar, Z., Rahimi Ahar, L., 2024. Thermal, optical, mechanical, dielectric, and electrical properties of nanocomposites. European Polymer Journal 218, 113337. doi: 10.1016/j.eurpolymj.2024.113337.

it to maintain performance in harsh environments, and it is often used in applications requiring high wear resistance and low friction (Shabaev et al., 2024).

4.2.2.3 Polyphenylene

PPS is also a significant high-temperature polymer, recognized for its outstanding chemical resistance and thermal stability, with many grades achieving ratings of 200°C–240°C. PPS is often used in applications that require flame retardancy and dimensional stability, such as electrical connectors and automotive parts. While it is generally more affordable than PEEK and PIs, it has lower long-term thermal resistance and can be more brittle (Gao et al., 2022).

4.3 Classification of nanocomposite

Nanocomposites are classified based on two phases, like other composite materials, namely, the matrix and filler.

4.3.1 Nano matrix phase

Nanocomposites are characterized according to the type of primary matrix phase in the foundation material, in which nanoscale fillers are embedded as reinforcements. The matrix material serves as the continuous phase that binds the nanoparticles, determining the overall properties of the nanocomposite. Three types of matrix materials are present in nanocomposites: ceramic matrix nanocomposites, metal matrix nanocomposites, and biomatrix nanocomposites (Saha et al., 2025).

4.3.1.1 Organic matrix nanocomposites

Organic matrix nanocomposites are categorized into carbon matrix nanocomposites and polymer matrix nanocomposites. Polymer composites are commonly used due to their high stiffness, low mass, and temperature resistance. They can also be reinforced with fibers, whiskers, platelets, and other particles to improve their strength and modulus. Polymers can also be filled with inorganic chemicals to enhance heat and impact resistance, mechanical strength, and flame retardancy. Organic nanocomposites are used when excellent optical properties are required and are classified as organic–inorganic and inorganic–organic materials, as discussed under hybrid composites (Saha et al., 2025).

4.3.1.2 Metal matrix nanocomposites

Metal matrix nanocomposites represent a class of advanced materials that incorporate a metal or alloy matrix, which is enhanced with the implant of nanosized reinforced fillers. These composites combine the mechanical strength and ductility of metal and its alloys with the enhanced properties of nanofillers, leading to improved wear resistance and overall mechanical performance (Gupta et al., 2025). Composites like Fe-Cr/Al_2O_3, Fe/MgO, Co/Cr, Ni/Al_2O_3, and Al/CNT offer superior strength, stiffness, and thermal stability compared to conventional metals (El Hawary et al., 2019; Kumar et al., 2020).

4.3.1.3 Ceramic matrix nanocomposites

Ceramic matrix nanocomposites are inorganic composite materials like alumina, silica, and zirconia, known for their exceptional high-temperature stability, hardness, and resistance to corrosion (Sharma et al., 2020). When reinforced with nanofillers, they demonstrate higher toughness and improved resilience to wear and thermal stress. These composites are commonly employed in high-temperature environments such as turbine blades, high-performance coatings, and even dental applications, where the combination of strength, biocompatibility, and resistance to wear is crucial (Fedullo et al., 2007; Saha et al., 2025).

4.3.2 Nanofiller phase

Nanocomposites can be classified according to the dimension of the nanofillers, which include nanolayer-, nanoparticle-, and nanofiber-reinforced composites. As shown in Fig. 4–1, each nanofiller type exhibits a distinct structural geometry that influences the resulting composite properties; nanolayers provide large interfacial contact for barrier enhancement, nanoparticles improve isotropic mechanical and thermal performance, while nanofibers or nanotubes offer directional reinforcement and electrical conductivity. These

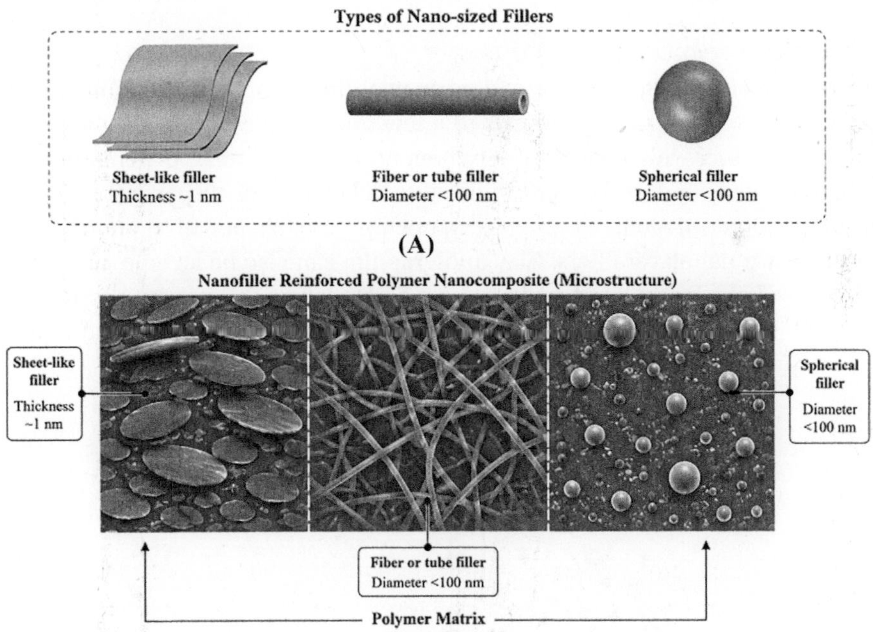

(B)

FIGURE 4–1 Structure of the nanoscale fillers: (A) The dimension of nanofillers; (B) Nanoparticles (1), nanofibers (2), and nanoclays (3) (Al-Mutairi et al., 2022). *From Al-Mutairi, N.H., Mehdi, A.H., Kadhim, B. J, 2022. Nanocomposite materials: definitions, types, and some of their applications: a review. European Journal of Research Development and Sustainability 3, 102–108.*

structural distinctions determine how effectively each nanofiller interacts with the polymer matrix to enhance the high-temperature performance of the resulting nanocomposite.

4.3.2.1 Nanoparticle-reinforced composites

Nanoparticle-reinforced composites are among the most extensively used composites, where microscopic particles of nanofillers are added to enhance the strength and stiffness of a polymer matrix, as shown in Fig. 4–2. By lowering particle size to the nanofillers, new and enhanced material characteristics can be obtained (Thippeswamy et al., 2021). Nanoparticles are particularly preferred as fillers in the formation of high-performance composites because of their simplicity of processing and appropriateness for large-scale manufacture. However, maintaining a uniform dispersion of nanoparticles within the matrix remains a considerable difficulty since they tend to agglomerate due to high interparticle interaction (Saha et al., 2025).

4.3.2.2 Nanofiber-reinforced composite

CNTs and nanofibers are now the most sought-after fillers for their high performance at extremely high temperatures while maintaining structural integrity (Camargo et al., 2009). Carbon nanofibers show high compressive strength when used for reinforcement in various thermoplastic and thermoset polymer matrices (Alshammari et al., 2021).

4.3.2.3 Nanolayer-reinforced composites

Nanolayer reinforcement fillers are formed using two-dimensional (2D) nanomaterials having sheets or layers whose thickness is from 1 to a few nm. As shown in Fig. 4–3, they are often layered over each other with a gap between them at different states: conventional composite, partially intercalated and exfoliated, fully intercalated and dispersed, and exfoliated. Reproduced with permission. Graphene, nanoclay, and sheets of titanium carbide are examples of known nanolayer fillers. Clay and graphite can also be layered as bulk materials. They can be used efficiently by spreading the discrete layers throughout the matrix (Lipton et al., 2020). This enhances the properties of the composite as compared to pure polymers or

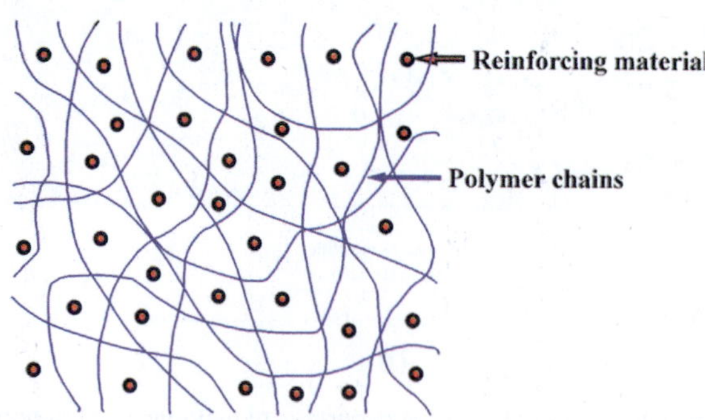

FIGURE 4–2 Structure of nanoparticle-reinforced composites (Saha et al., 2025). *From Saha, R., Munshi, M.H., Akter, M., Shikder, A.A.R., Islam T., 2025. Synthesis techniques, fundamental properties, and emerging applications of nanocomposites: a comprehensive review. SPE Polymers 6 (2). https://doi.org/10.1002/pls2.70004.*

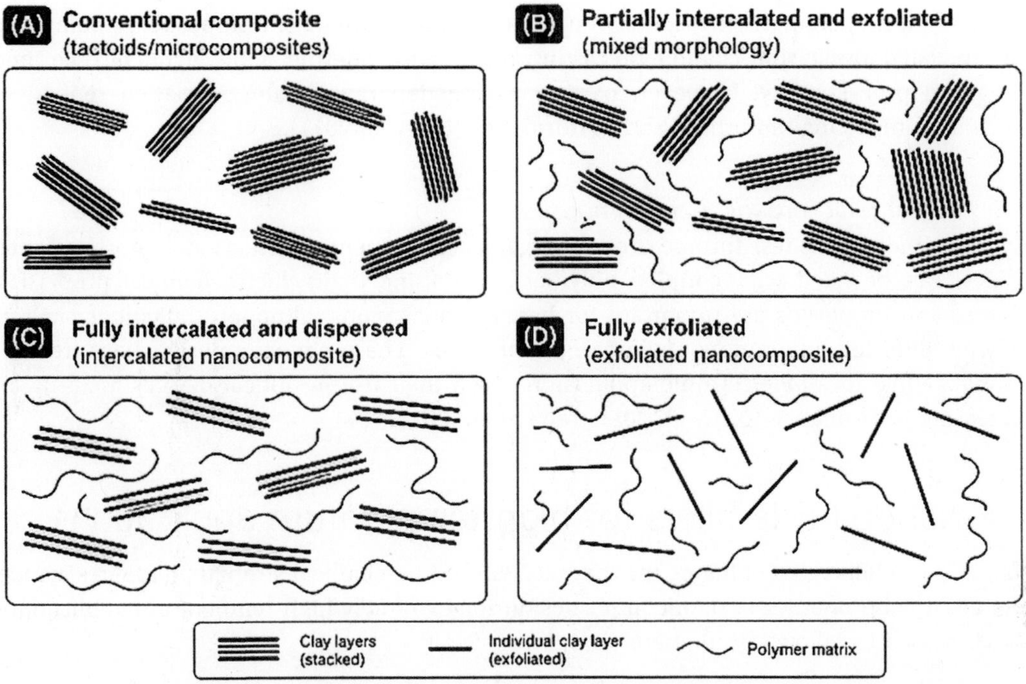

FIGURE 4–3 Nanocomposites (clay/polymer) at different states: (A) conventional composite, (B) partially intercalated and exfoliated, (C) fully intercalated and dispersed, and (D) exfoliated (Saha et al., 2025; Luo and Daniel, 2003). *From Luo, J.J., Daniel, I.M., 2003. Characterization and modeling of mechanical behavior of polymer/clay nanocomposites. Composites Science and Technology 63 (11), 1607–1616. https://doi.org/10.1016/S0266-3538(03)00060-5.*

typical composites with strong strength and heat resistance, low gas permeability and flammability, and high biodegradability (Zhou et al., 2022).

4.3.3 Hybrid nanocomposites

Advanced materials called hybrid nanocomposites combine different nanostructured components, frequently combining inorganic and organic phases (Gerasin et al., 2013). Self-healing, stimuli-responsive behavior, and enhanced corrosion resistance are unique functions resulting from the interactions between various nanomaterials, making them perfect for energy, biomedical, electronics, and aerospace applications. The improved mechanical strength, thermal stability, electrical conductivity, and outstanding barrier qualities provide multipurpose solutions for next-generation technologies by customizing their structure and content (Merhari, 2009).

4.3.3.1 Organic–inorganic nanocomposites

Inorganic nanoparticles are integrated into polymer matrices to form nanocomposites, often using organic ligands as protective layers. These ligands can influence electron transport and ensure uniform nanoparticle dispersion without altering properties. Composite thin films can be fabricated using spin-coating or casting techniques, offering lightweight, flexible, and enhanced processability. However, particle aggregation during film formation remains a challenge, impacting uniformity and performance (Nguyen, 2011).

4.3.3.2 Inorganic–organic nanocomposites

Hybrid nanocomposites, formed from inorganic components such as nanoporous metal oxides, can be filled with coupled polymers, increasing their electrochemical properties. These nanocomposites are important for battery applications, although difficulties such as polymer diffusion into nanopores impair performance. These nanocomposites have tremendous potential for optoelectronic applications, with their use in optical devices focusing on increasing performance (Nguyen, 2011; Saha et al., 2025).

4.4 Nanoparticle fillers for high-temperature applications

Nanoparticle fillers that enhance the thermal stability, mechanical strength, flame resistance, and electrical properties of nanocomposites during extremely high-temperature applications are discussed as follows (Hanemann and Szabó, 2010).

4.4.1 Nanoclay

Nanoclays are layered silicate minerals that are frequently utilized in polymer nanocomposites to improve overall performance. Examples of these materials are saponite, hectorite, and montmorillonite (see Fig. 4–4). Nanoclays are perfect for high-temperature and moisture-resistant applications because of their high aspect ratio, which also helps them to reduce gas permeability, reinforce the polymer matrix to increase mechanical strength and stiffness, and act as heat barriers to greatly improve thermal stability. These properties make nanoclay-reinforced composites valuable in various fields, including aerospace and automotive industries for lightweight, high-strength materials; high-performance coatings with enhanced durability; and advanced packaging solutions that offer superior barrier properties to prolong the shelf life of perishable products (Shunmugasamy et al., 2015).

4.4.2 Carbon nanotubes

CNTs are made entirely of carbon atoms. They have hollow cylindrical structures and exist in two primary forms: single- and multiwalled, as illustrated in Fig. 4–5. They exhibit extremely high thermal conductivity (~3000 W $(m\cdot K)^{-1}$), making them ideal for heat dissipation; possess mechanical strength nearly 100 times greater than steel; and enhance flame resistance in high-temperature polymer matrices, improving safety. These features make CNTs suitable for a wide range of applications, including high-temperature electrical shielding in aerospace and electronics, lightweight yet strong components for aerospace and defense, and structural reinforcement in advanced composites used in the automotive and construction industries.

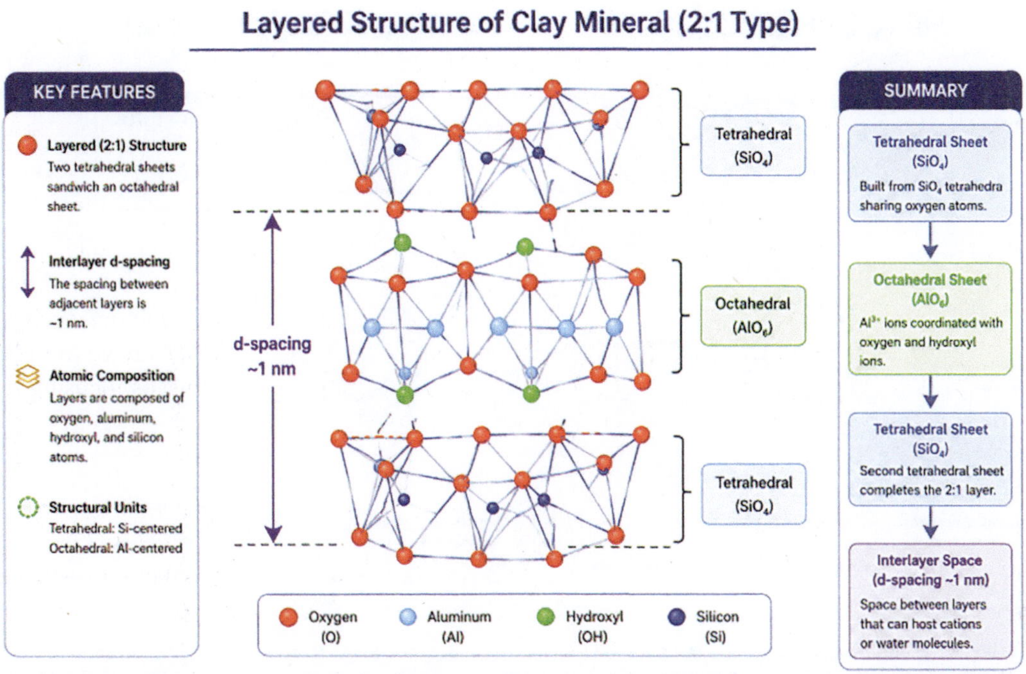

FIGURE 4–4 Nanoclay structure (montmorillonite) (Uddin, 2008). *From Uddin, F., 2008. Clays, nanoclays, and montmorillonite minerals. Metallurgical and Materials Transactions A 39 (12), 2804–2814. https://doi.org/10.1007/s11661-008-9603-5.*

4.4.3 Graphene and graphene oxide

Graphene is a carbon-derived material and has a 2D structure composed of atoms organized in a continuous hexagonal lattice, as illustrated in Fig. 4–6. Graphene and its chemically modified variant, graphene oxide (GO), have enhanced functionalization opportunities, expanding their variety of applications as the strongest known material with remarkable thermal conductivity (~5000 W $(m \cdot K)^{-1}$), effective heat dissipation, outstanding mechanical strength (~130 GPa tensile strength), and improved electrical conductivity when included in polymer matrices. These distinctive characteristics render graphene suitable for various applications, including high-performance aerospace materials that integrate strength with lightweight design, thermally conductive coatings for electronic and industrial applications, and advanced automotive and structural composites designed to enhance fuel efficiency and durability (Deepa et al., 2022).

4.4.4 Nobel transition and precious metal nanoparticles

Nanoparticles from advanced materials like silver (Ag), copper (Cu), gold (Au), and nickel (Ni) are widely valued due to their exceptional conductivity, catalytic activity, and antibacterial qualities. They are noble, transition, and precious metals with enhanced thermal

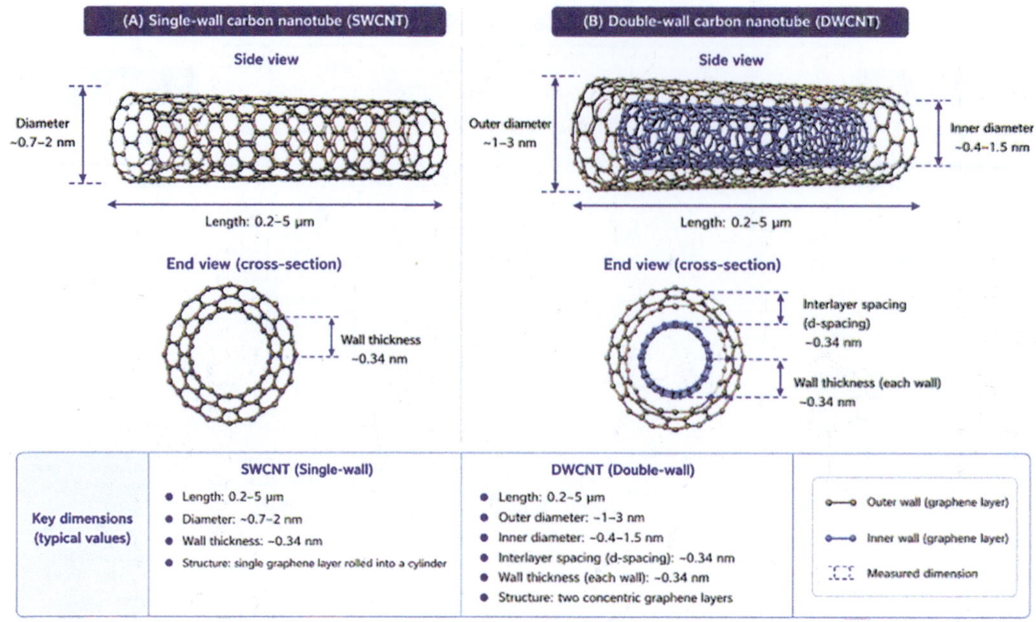

FIGURE 4–5 Single-walled (*left*) and multiwalled (*right*) carbon nanotubes (Al-Mutairi et al., 2022). *From Al-Mutairi, N.H., Mehdi, A.H., Kadhim, B. J, 2022. Nanocomposite materials: definitions, types, and some of their applications: a review. European Journal of Research Development and Sustainability 3, 102–108.*

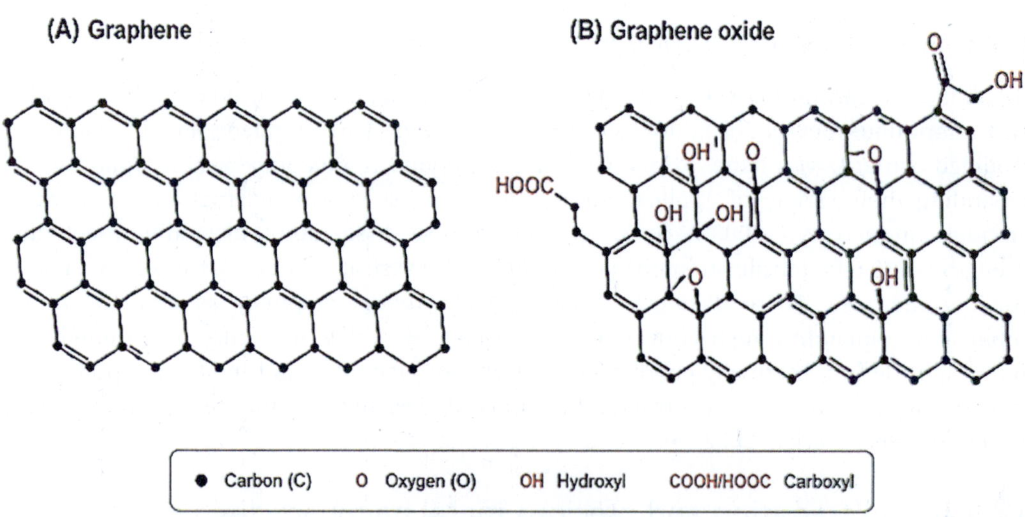

FIGURE 4–6 Honeycomb lattice of graphene and graphene oxide (Jain et al., 2024). *From Jain, P., Verma, C., Raman, A.P.S., Kumari, K., Singh, P., 2024. Biosensors Based on Graphene, Graphene Oxide and Graphynes for Early Detection of Cancer. CRC Press, India, pp. 1–281. https://doi.org/10.1201/9781003491361.*

and electrical conductivity, essential for electronic devices. Their versatile applications include high-performance conductive adhesives for flexible and wearable electronics, thermally conductive components and sensors for improved efficiency in heat-sensitive systems, and heat-resistant coatings for the automotive and aerospace sectors to ensure durability under extreme conditions (Glover et al., 2011).

4.4.5 Ceramic nanoparticles

Ceramic nanoparticles are highly valued for their exceptional thermal and mechanical resilience, making them perfect for strengthening materials. These nanoparticles include silica (SiO_2), alumina (Al_2O_3), titania (TiO_2), and zirconia (ZrO_2), as shown in Fig. 4-7. Their primary advantages are enhanced heat resistance and minimized thermal expansion, which improve material stability at high temperatures; increased resistance to abrasion and wear, making them ideal for high-friction scenarios; and improved fire-retardant properties, which enhance safety in construction and transportation. These ceramic nanoparticles are extensively used in high-temperature insulation coatings for aerospace and industrial applications, engine components in the aerospace and automotive industries to extend lifespan and improve performance, and wear-resistant materials for tribological applications such as bearings and cutting tools (Treccani, 2022).

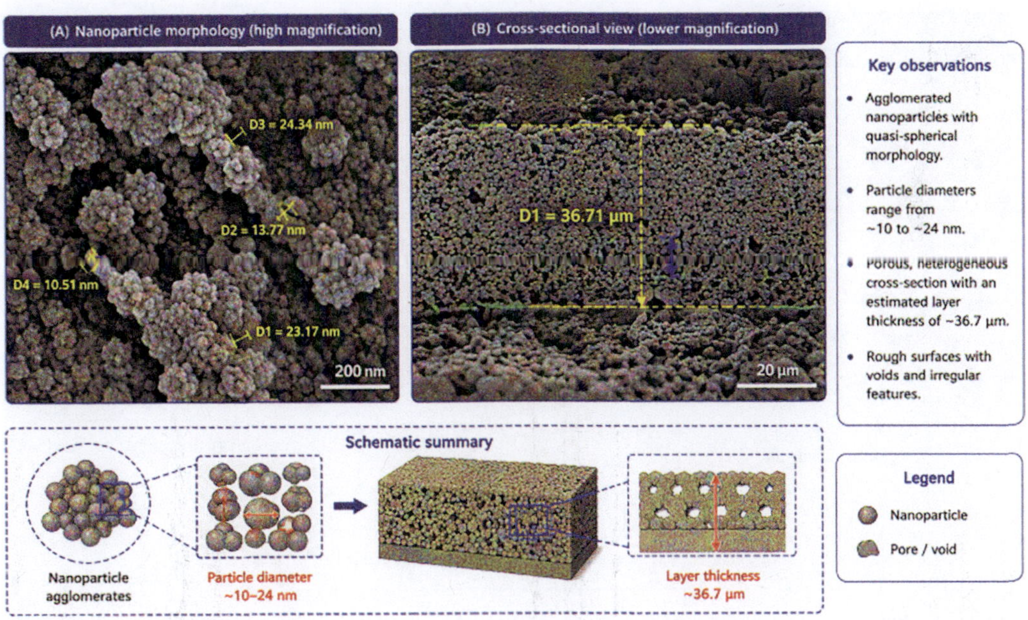

FIGURE 4–7 SEM image of ZrO_2 on ceramic (Chitoria et al., 2023; Treccani, 2022). *From Treccani, L., 2022. Introduction to ceramic materials. In: Surface-Functionalized Ceramics: For Biotechnological and Environmental Applications. Wiley, Italy, pp. 1–46. https://doi.org/10.1002/9783527698042.ch1.*

4.4.6 Boron nitride nanoparticle

Boron nitride (NB) nanoparticle has a hexagonal boron nitride structure, often referred to as "white graphite" due to its structure, as illustrated in Fig. 4–8. This ceramic nanoparticle has attributes comparable to graphite but with greater thermal stability and dielectric characteristics. Its high dielectric strength makes it ideal for electronic insulation; its strong thermal conductivity (~400 W $(m{\cdot}K)^{-1}$) makes it perfect for heat management applications; and its self-lubricating qualities reduce friction in high-temperature situations. Applications for h-BN include heat-resistant coatings for aircraft, which shield materials from high temperatures; lubricating parts in hot conditions, such as industrial machinery and engine parts; and thermally conductive plastics for electronics and packaging, which effectively dissipate heat (Gonzalez-Ortiz et al., 2020).

4.5 Unique high-temperature properties of nanoparticles

Nanoparticles possess unique properties at the nanoscale due to the surface area, quantum effects, and size-dependent characteristics, enhancing the mechanical, thermal, electrical, and chemical performance of polymer matrices, leading to the development of superior nanocomposite materials.

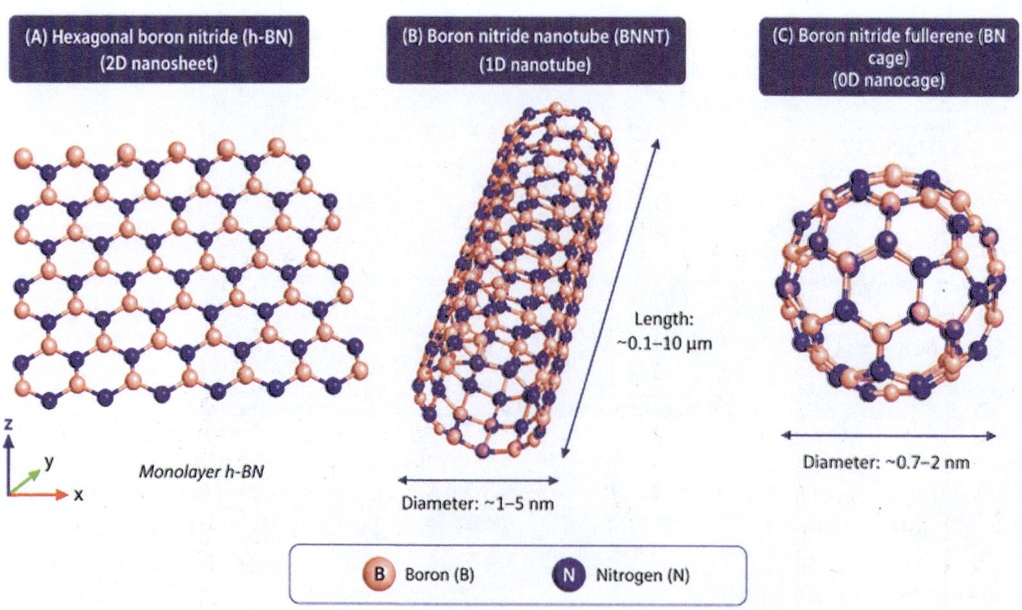

FIGURE 4–8 Nanostructured boron nitride–based materials (A) Hexagonal boron nitride nanosheet; (B) Boron nitride nanotube; (C) boron nitride fullerene nanocage (Gonzalez-Ortiz et al., 2020). *From Gonzalez-Ortiz, D., Salameh, C., Bechelany, M., Miele, P., 2020. Nanostructured boron nitride–based materials: synthesis and applications. Materials Today Advances 8, 100107.*

4.5.1 High surface area to volume ratio

Thenanoparticles possess a significantly large surface area relative to their volume. A 1 nm particle has much more surface area compared to a larger particle of the same material. The increased surface area facilitates stronger bonding between the nanoparticles and the polymer matrix, promotes easier and more uniform dispersion of nanoparticles within the polymer matrix, ensures consistent material properties, and improves interaction between nanoparticles and the polymer matrix, leading to more efficient load transfer and enhancing the overall strength and durability of the composite material. This helps improve the thermal stability of nanocomposites at high temperatures and also ensures mechanical properties like tensile strength, stiffness, and toughness (Sui et al., 2022).

4.5.2 Quantum size effects

The size of nanoparticles affects the unique electrical, optical, and magnetic characteristics of nanofillers (Njuguna et al., 2021). The quantum size effect of the fillers at the nanoscale has a significant effect on the material properties, with an increased reactivity and tunable electronic properties that the bulk materials lack. CNTs and graphene can significantly boost the electrical conductivity of polymers; CNTs and graphene enhance the thermal conductivity of polymers, enabling better heat dissipation in heat-resistant applications, while silver or gold can exhibit localized surface plasmon resonance effects, which improve the optical properties of the polymer, benefiting applications in sensing and imaging (Kambhampati, 2021). The increased effect of quantum size on the engineering properties of these composites at high temperatures makes them suitable for applications in electromagnetic shielding and electrical components, heat dissipation in heat-resistant applications in aerospace and automotive, and sensing and imaging applications (Kong et al., 2021).

4.5.3 Enhanced mechanical properties

Nanoscale particles of fillers have improved the strength and stiffness of composite materials due to their nanoscale dimensions, atomic level homogeneity, and unique structural characteristics. They considerably improve mechanical performance when integrated into polymer matrices (Markandan and Lai, 2023). Strength, durability, and flexibility are enhanced, particularly in high-temperature settings, by reinforcing with metallic nanoparticles or CNTs. These nanoparticles in the fillers aid in stress distribution across the matrix, enhancing performance under dynamic or impact-loading circumstances, which is critical in aerospace and automotive applications. Moreover, ceramic nanoparticles like silica boost wear and abrasion resistance, extending the service life of the composite materials (Sheikh and Behdinan, 2023).

4.5.4 Thermal properties enhancement

Nanoparticles exhibit peculiar thermal characteristics compared to their bulk forms, attributable to their large surface area and size effects. They can be insulators or thermal conductors, depending on their composition (Khan and Hossain, 2022). Nanoparticles like clay minerals and ceramic kinds increase the thermal stability of polymers, allowing them to maintain their thermal integrity at high temperatures. The addition of graphene or nanoclays also improves flame resistance, which makes these composites perfect for fire-safe applications like electrical insulation or car interiors. Lastly, materials with high thermal conductivity, notably graphene and CNTs, improve heat dissipation, which is essential in high-performance systems and electronics (Thakur et al., 2023; Treccani, 2022).

4.5.5 Improved chemical and environmental resistance

The nanoscale dimension and higher surface reactivity of nanoparticles contribute to enhanced chemical interaction with polymer matrices. Nanofillers made with silica or metal nanoparticles lead to greater chemical resistance that shields the polymer from corrosion, oxidation, and degradation caused by chemicals or environmental exposure (Mehta et al., 2022; Kumar and Kanchan, 2024). Additionally, nanoclay fillers generate nanoscale barriers inside the polymer, decreasing the permeability of gases and liquids, which is particularly advantageous in food packaging and protective coatings. Metal nanoparticles can also provide better corrosion resistance, a vital requirement in marine and aerospace applications (Kalita et al., 2025).

4.5.6 Self-healing and antibacterial properties

Silver and copper nanoparticles have sophisticated properties with healing tendencies and conductive fillers, which can autonomously repair microcracks in the polymer matrix after damage, thus extending the material's lifespan and minimizing the need for maintenance (Sabet, 2024). These nanoparticles are notably prized for their antibacterial qualities, making the nanocomposites excellent for biomedical equipment, hospital surfaces, and food packaging, where hygiene and contamination control are critical (Ibrahim et al., 2025).

4.5.7 Enhanced optical properties

The optical behavior of nanoparticles can be modified based on their dimensions and form, adjusting how they interact with light. Nanoparticles of gold and silver create excellent optical effects like plasmonic resonance, luminescence, and color tuning, which help produce polymer composites for optical sensing, medical imaging, and diagnostic applications due to their surface plasmon resonance activity (Kumar and Kanchan, 2024). Furthermore, nanoparticles such as ZnO and TiO_2 improve UV stability and weather resilience, making them excellent for outdoor-use polymers exposed to sunshine and harsh environmental conditions (Sedlmeier and Gorris, 2015).

4.6 Nanoparticle loading and its influence on polymer nanocomposites

Nanoparticle loading represents the quantity of nanoparticles contained within a polymer matrix, playing a crucial role in creating the material's overall features. The level of loading directly influences increases in mechanical strength, thermal resistance, electrical conductivity, and functional performance. While modest loading improves these qualities, excessive inclusion may promote particle agglomeration, delay processing, and elevate production expenses (Khdary et al., 2023; Treccani, 2022). Achieving an optimal loading threshold is critical to reconciling material performance with manufacturing efficiency and cost-effectiveness. This section digs into how variable nanoparticle concentrations influence nanocomposite behavior and offers ways for identifying appropriate loading levels (Castro et al., 2022).

4.6.1 Nanoparticle structure

The structure and morphology of nanoparticles (dimension) play a critical role in regulating the performance of polymer nanocomposites (Njuguna et al., 2021). Different shapes, such as 0D nanodiamonds (NDs), 1D CNTs, and 2D graphene nanoplatelets (GNPs), influence dispersion behavior, interfacial bonding, and mechanical enhancement within the polymer matrix (Arshad et al., 2023). Among these, NDs displayed higher performance due to their spherical geometry and smooth surfaces, which resulted in dispersion, reduced agglomeration, and increased interfacial interactions. These qualities lead to greater load transfer, reduced porosity, and increased mechanical properties, such as tensile strength and wear resistance, in epoxy-based nanocomposites. While form has a modest impact on surface wettability (Xu et al., 2025).

4.6.2 Nanoparticle loading

Nanoparticles in polymer matrices greatly increase the mechanical characteristics of nanocomposites, with concentration and dispersion playing crucial roles. CNTs, including NDs, CNTs, and GNPs, are utilized to reinforce materials like epoxy coatings (Ali et al., 2023). Optimal nanoparticle loading, about 1.0 wt.%, leads to uniform dispersion, reduced viscosity, and lower porosity, resulting in better tensile strength, abrasion resistance, and failure strain compared to pure polymers. This is due to the improved dispersion and bonding of nanoparticles inside the polymer matrix, increasing load transfer and energy dissipation (Arshad et al., 2023; Kausar, 2024).

4.6.3 Effect of nanoparticle loading on composite properties

Nanoparticle loading has a major effect on the mechanical strength, thermal stability, and electrical performance of polymer composites; meticulous nanoparticle concentration optimization is especially advantageous for high-temperature applications (Kutvonen et al., 2012).

4.6.3.1 Mechanical properties

The mechanical performance of polymer nanocomposites is strongly governed by nanoparticle loading and dispersion. The following mechanical properties illustrate these effects:

1. Tensile strength and modulus: Moderate nanoparticle loading improves tensile strength and stiffness due to improved interfacial interactions between nanoparticles and the

polymer matrix. However, excessive loading can promote particle agglomeration, lowering the efficiency of stress transfer and weakening the mechanical characteristics (Yazman et al., 2021).

2. Fracture toughness: Properly dispersed nanoparticles improve fracture toughness by deflecting cracks and absorbing energy. High nanoparticle concentrations, though, may rise to stress concentration sites, which can cause brittle failure under high-stress circumstances (Mousavi et al., 2022).
3. Flexural strength: Optimal dispersion of nanoparticles improves flexural strength by boosting load transmission within the matrix, making the composite more durable under bending loads, which is critical for high-temperature applications (Huang et al., 2023).

4.6.3.2 Thermal properties

Thermal performance is critical in determining how polymer nanocomposites behave in extreme service conditions. Nanoparticle loading directly influences key thermal characteristics by modifying chain mobility, enhancing conductivity, and stabilizing the material against degradation. The effects of nanoparticle incorporation on the thermal performance of composites are outlined as follows:

1. Glass transition temperature (T_g): Moderate nanoparticle loading boosts the glass transition temperature by constraining polymer chain mobility, enhancing the composite's thermal durability at elevated temperatures. However, heavy loading might disrupt uniform dispersion, limiting its efficiency (Sharifzadeh and Cheraghi, 2021).
2. Thermal conductivity: High thermal conductivity nanoparticles, such as CNTs, graphene, and metal oxides, promote heat dissipation. It is particularly important for high-temperature applications, where efficient heat management is critical. Careful control of loading is required to attain the percolation threshold for optimal thermal conductivity (Yasmin et al., 2023).
3. Decomposition temperature: Nanoparticles operate as a protective barrier, lowering polymer chain degradation and delaying heat degradation. This considerably increases the composite's capacity to resist high temperatures without losing integrity (Król-Morkisz and Pielichowska, 2019).

Overall, controlling nanoparticle concentration is essential to maintain optimal heat resistance without compromising structural stability. Well-dispersed nanoparticles act as efficient heat transfer pathways, improving the material's resilience under sustained high-temperature exposure.

4.6.3.3 Electrical properties

The electrical performance of polymer nanocomposites depends on nanoparticle type and loading concentration. The following properties summarize these influences:

1. Electrical conductivity: CNTs like graphene promote electrical conductivity. In high-temperature applications, achieving the percolation threshold is crucial for ensuring

consistent conductivity. Below this threshold, the composite stays insulating, but beyond it, considerable electrical conductivity can be achieved (Lalire et al., 2024).

2. Dielectric properties: Excessive nanoparticle loading might lead to dielectric breakdown due to particle clustering. Therefore maintaining an ideal loading concentration is critical to achieving the correct dielectric characteristics, especially in high-temperature conditions where material performance is more susceptible to variations in particle distribution (Cai et al., 2017).

4.6.4 Optimal nanoparticle loading for high-temperature applications

Determining the optimal nanoparticle loading is essential for balancing enhanced performance with material cost and processability. Proper loading ensures uniform dispersion, strong interfacial bonding, and stable thermal and mechanical behavior under extreme conditions. Excessive nanoparticle addition can increase viscosity, promote agglomeration, and reduce overall composite reliability.

4.6.4.1 Percolation threshold

The concentration at which nanoparticles form a continuous network is critical for high-performance attributes. Optimal loading should be just above this threshold to maximize mechanical, thermal, and electrical qualities without excessive material expenditures (Dlamini et al., 2019).

4.6.4.2 Dispersion quality

Proper dispersion is crucial for maximizing the benefits of nanoparticle loading. Agglomeration can diminish effectiveness; therefore surface modifications or dispersing agents may be needed to increase nanoparticle distribution, especially for high-temperature applications (Huang et al., 2023).

4.6.4.3 Processing challenges

High nanoparticle concentrations might increase viscosity, making manufacturing more problematic. Balancing loading levels is required to provide ease of processing while preserving performance (Apmann et al., 2021).

4.6.4.4 Economic viability

Although higher nanoparticle concentrations can improve performance, they can raise material costs. In high-temperature applications, establishing a balance between performance enhancement and cost efficiency is vital for practical use (Mohajerani et al., 2019).

4.7 Synthesis of high-temperature nanocomposite

The synthesis of nanoparticles into polymers includes diverse processing procedures, determined by parameters like polymer type, nanoparticle properties, and application. The approach greatly affects the dispersion and interaction of nanoparticles inside the polymer matrix, altering the nanocomposite's attributes like mechanical strength, thermal stability, and chemical resistance. The process is discussed as follows:

4.7.1 Solution blending

Solution blending is a process where nanoparticles are initially dispersed in a polymer solution, and then the mixture is cast or evaporated to form a nanocomposite film (see). This approach is often applied to polymers such as PIs, PEEK, and epoxy-based composites. Solution mixing promotes efficient dispersion of nanoparticles, which is critical for producing homogeneous characteristics in the final material. However, thorough clea (Yilmaz et al., 2024) nup of solvents is important to avoid flaws in the nanocomposite. While this approach is effective for high-performance thermosetting polymers, it can be time-consuming, costly, and environmentally difficult due to the use of solvents.

4.7.2 Melt blending

Extrusion, sometimes referred to as melt blending, as shown in Fig. 4–9, is the process of combining the polymer and nanoparticles at high temperatures while they are molten, then molding or extruding. This approach is often used for thermoplastic polymers like PPS and PEEK. Melt blending offers various advantages, including being solvent-free, environmentally benign, and scalable for commercial manufacturing. The high processing temperatures strengthen the interaction between the polymer and nanoparticles, resulting in enhanced composite characteristics. However, the elevated temperatures might lead to the breakdown of certain polymers or nanoparticles, and attaining uniform dispersion may be challenging without optimum processing conditions (Albdiry, 2024).

4.7.3 In situ polymerization

In situ polymerization is a technique where nanoparticles are distributed in a monomer solution (see Fig. 4–10), followed by polymerization to form a nanocomposite. This approach provides robust interfacial interactions between nanoparticles and polymer chains, resulting in enhanced mechanical and thermal properties. It is widely used for thermosetting polymers such as PIs and epoxy resins. In situ polymerization enables exquisite control over the nanocomposite's structure and properties, providing a strong bond between the polymer and nanoparticles. However, it requires careful selection of polymerization conditions to avoid nanoparticle aggregation, and the process can be complex (Xu et al., 2023).

4.7.4 Electrospinning

Electrospinning produces ultrafine polymer nanofibers embedded with nanoparticles by applying a high-voltage electric field to a polymer solution or melt, as illustrated in Fig. 4–11. This method is particularly helpful in aerospace, filtration, and biomedical applications, where high mechanical strength, porosity, and thermal stability are desired. Electrospinning

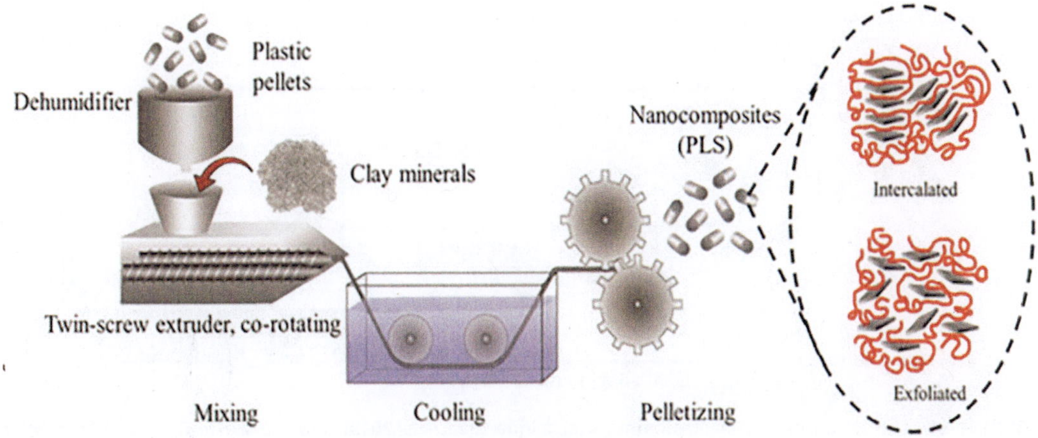

FIGURE 4–9 Melt blending method (Albdiry, 2024; Dossin Zanrosso et al., 2020). *From Dossin Zanrosso, C., Piazza, D., Lansarin, M.A., 2020. PVDF/ZnO composite films for photocatalysis: a comparative study of solution mixing and melt blending methods. Polymer Engineering and Science 60 (6), 1146–1157. https://doi.org/10.1002/pen.25368.*

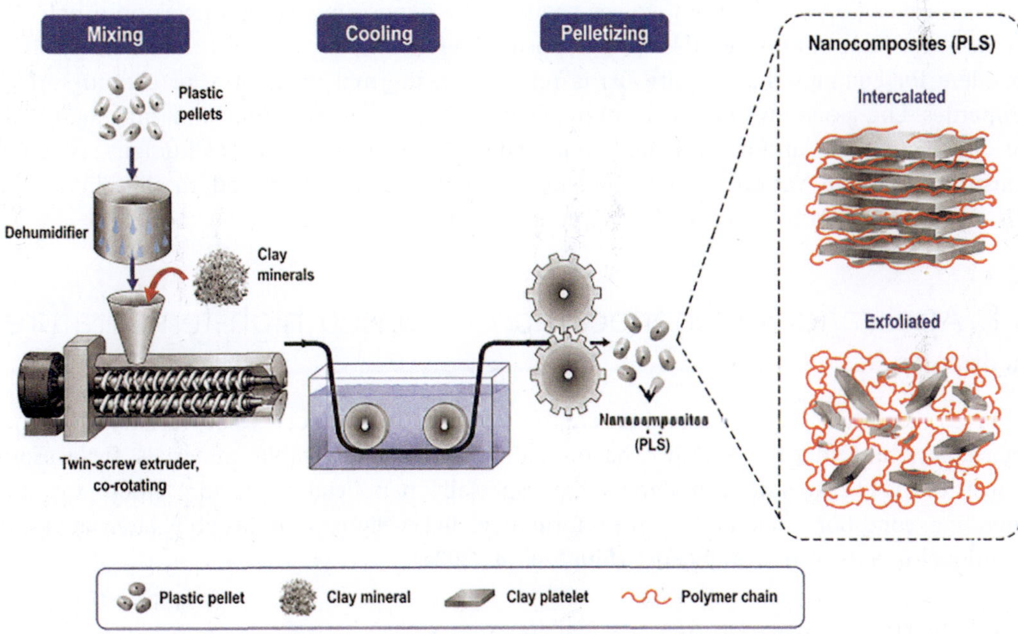

FIGURE 4–10 In situ polymerization method (Beatrice, 2019). *From Beatrice, C.A., 2019. Polymer Nanocomposites with Different Types of Nanocomposites: Recent Evolutions.*

allows precise nanoparticle dispersion throughout the fibers and generates nanofibers with a high surface area. However, it requires specialized equipment and fine process control, limiting its scalability compared to melt blending or extrusion (Ji et al., 2024).

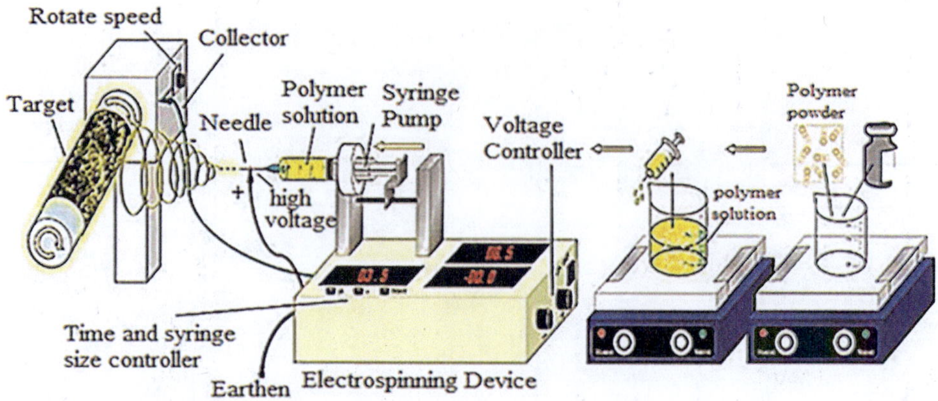

FIGURE 4–11 The set-up of the electrospinning technique (Beatrice, 2019). *From Beatrice, C.A., 2019. Polymer Nanocomposites with Different Types of Nanocomposites: Recent Evolutions.*

4.7.5 Layer-by-layer

Layer-by-layer (LbL) assembly involves sequentially depositing nanoparticles onto polymer surfaces to form multilayered nanocomposites, as shown in Fig. 4–12. This method is excellent for coatings that require flame retardancy, thermal insulation, or certain surface properties. LbL assembly offers control over layer thickness and composition, making it ideal for surface modifications and functional coatings. However, the technology is time-consuming due to various deposition steps and is frequently limited to thin films and coatings rather than bulk solids (Zhang et al., 2019).

4.8 Applications of nanocomposite-based high-temperature polymers

HTPNCs' exceptional mechanical strength, chemical resistance, wear resistance, flame retardancy, and thermal stability have made them indispensable materials for various industries. These advanced materials are especially beneficial in sectors where extreme operating conditions demand high-performance, lightweight, and durable alternatives to traditional metals or thermoplastics (Singh et al., 2024).

4.8.1 Aerospace and aviation applications

In the aerospace sector, components are constantly exposed to elevated temperatures, aggressive environments, and mechanical stress. HTPNCs are favored for being lightweight, heat- and wear-resistant, and capable of maintaining performance under extreme stress. For example, PI/GO composites are employed in turbine blades, jet engine parts, and after-burners, offering superior thermal and oxidative stability while reducing weight (Mat Yazik and Ismail, 2025). Additionally, PEEK/Al_2O_3 and PTFE-based nanocomposites are used in

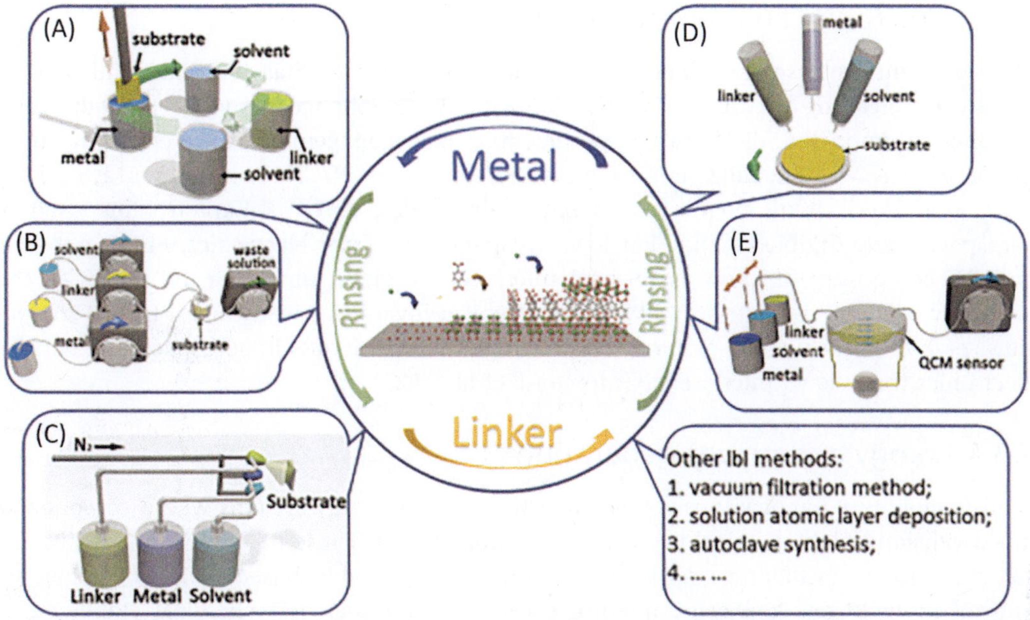

FIGURE 4–12 Schematic depiction of the layer-by-layer assembly techniques for SURMOFs: (A) dipping method, (B) pumping method, (C) spray method, (D) spin coating method, and (E) LbL synthesis using a QCM system (Liang et al., 2023). *LbL*, Layer-by-layer. *From Liang, Y., Sun, W., Li, T., Chen, Y., Cheng, Y., Zhang, L., 2023. All organic polymer dielectrics for high-temperature energy storage from the classification of heat-resistant insulation grades. Journal of Polymer Science 61 (22), 2777–2795. https://doi.org/10.1002/pol.20230334.*

bearings, bushings, seals, and gaskets, where they enhance durability and lubrication at high operating temperatures. In spacecraft, CNT-reinforced PIs are incorporated into thermal protection systems to resist intense reentry heat, with NASA utilizing such composites in thermal shields for lightweight and high heat resistance (Wani et al., 2024).

4.8.2 Automotive applications

In the automobile sector, high-temperature polymers utilizing nanocomposite technology are replacing metal components in vehicles to reduce weight, enhance fuel efficiency, and improve overall durability. These materials are utilized in engine covers, pistons, and fuel system components, with PEEK/CNT and PI/SiC nanocomposites offering stability at temperatures over 300°C (Adem et al., 2024). Phenolic resins combined with CNTs enhance the friction characteristics and diminish wear in braking systems, resulting in more durable brake pads with reduced dust generation. Graphene and boron nitride (BN) composites provide superior thermal insulation in heat shields and under-the-hood components, rendering them optimal for safeguarding critical areas adjacent to exhaust systems. For example, PI/BN nanocomposites are utilized in exhaust heat shields owing to their lightweight nature and exceptional heat resistance (Selvaraj et al., 2021).

4.8.3 Electronics and semiconductor applications

The electronics and semiconductor sectors demand materials that can withstand thermal stress while ensuring electrical reliability. Nanocomposite-enhanced polymers meet this need by offering improved dielectric properties, thermal management, and electromagnetic interference (EMI) shielding. For example, PI/silica and PI/CNT composites are widely used in printed circuit boards and flexible electronics, such as Kapton films used in aerospace-grade flexible circuits. For heat dissipation in power electronics, PEEK/graphene and BN composites function as thermal interface materials, improving the longevity and efficiency of high-power devices like electric vehicle inverters. Meanwhile, PI/CNT composites serve as EMI shielding materials, protecting sensitive circuits in aerospace and defense electronics from signal interference (Mercante et al., 2023).

4.8.4 Energy and power generation

High-temperature nanocomposites play a critical role in energy systems where components must withstand both chemical exposure and extreme heat. These materials are used in fuel cell membranes, insulation, and heat exchangers. Notably, PBI-based nanocomposites are utilized in hydrogen fuel cells, offering exceptional thermal and chemical stability that supports efficient energy conversion in harsh environments. These composites also contribute to the reliability of components in next-generation nuclear and fusion energy systems (Wang et al., 2022).

4.8.5 Applications in medicine and biomedicine

HTPNCs are utilized in the medical industry for drug delivery systems, implants, and surgical tools that need to be both sterilization-resistant and biocompatible. Their ability to withstand repeated autoclaving and exposure to body fluids makes them superior alternatives to conventional metals. PEEK-based nanocomposites, in particular, are widely adopted for orthopedic implants, offering a combination of strength, lightweight properties, and resistance to degradation, which improves patient outcomes and reduces complications (Alqutaibi et al., 2025).

4.9 Future prospects and challenges of high-temperature nanocomposite polymers

High-temperature nanocomposite polymer is evolving rapidly, driven by the demand for materials that exhibit superior thermal stability, mechanical strength, and multifunctional properties. These materials are crucial for advanced applications in aerospace, automotive, electronics, energy, and beyond. However, their widespread adoption still faces several technical and economic hurdles. The following sections highlight both the emerging trends and the challenges shaping their future development (Balasubramanian and Tirumalai, 2013).

4.9.1 Emerging trends and future directions

The development of HTPNCs is evolving rapidly with innovations in nanomaterials, processing techniques, and sustainability considerations. Several emerging trends are shaping the next generation of these advanced materials:

1. Incorporation of advanced nanoparticles: Future progress lies in the integration of novel nanofillers with enhanced thermal, electrical, and mechanical properties (Balasubramanian and Tirumalai, 2013). Materials such as MXenes (e.g., $Ti_3C_2T_x$), boron nitride (BN) nanotubes, and metal-organic frameworks offer superior conductivity, flame retardancy, and gas barrier capabilities. The performance of high-temperature polymers could be greatly improved by these next-generation nanoparticles (Balasubramanian and Tirumalai, 2013; Basha et al., 2024).
2. Development of multifunctional hybrid nanocomposites: Combining various nanofillers (e.g., graphene, CNTs, and ceramics) within a single matrix can produce synergistic effects, improving conductivity, strength, and thermal resistance simultaneously. These hybrid nanocomposites are particularly valuable in aerospace, defense, wearable electronics, and energy systems requiring multifunctional performance (Monteiro and Simões, 2024).
3. Ecofriendly and sustainable composites: Sustainability is a growing priority. Researchers are developing biobased and recyclable polymer matrices, as well as biodegradable or green nanofillers. Efforts are also being made to implement water-based or solvent-free fabrication techniques, contributing to safer, more sustainable material ecosystems (Monteiro and Simões, 2024).
4. Integration into next-generation technologies: Tailored high-temperature nanocomposites are being developed for specialized applications, including hypersonic vehicles, wearable sensors, fusion energy systems, and self-healing aerospace components. It is anticipated that these materials would be essential in harsh settings where conventional materials are ineffective.
5. Advancement in manufacturing and processing: Innovations in processing include plasma-assisted dispersion, ultrasonication, and additive manufacturing (e.g., 3D printing) to ensure better filler distribution and reduced defects. AI and machine learning tools are also being used to design and optimize nanocomposite formulations for targeted properties (Saxena et al., 2023).

4.9.2 Challenges in high-temperature nanocomposite development

Despite impressive progress, several challenges continue to hinder the widespread industrial application of HTPNCs. These challenges stem from technical, economic, and environmental limitations, as outlined below:

1. Nanoparticle dispersion and interfacial compatibility: Achieving uniform dispersion of nanoparticles in the polymer matrix is one of the most enduring issues. Agglomeration, weak bonding, and decreased performance are caused by poor distribution. While surface functionalization and high-shear mixing techniques help improve compatibility, these processes can be technically demanding and increase production costs (Hong and Chen, 2014).

2. Complex processing requirements: Processing high-performance polymers like PIs, PEEK, and PPS sometimes calls for specialized, high-temperature techniques like in situ polymerization or melt blending. Some nanoparticles may degrade during processing, limiting their effectiveness. Lower temperature or additive manufacturing methods are being explored to circumvent these limitations (Li et al., 2024).
3. High cost of materials and processing: The cost of high-performance nanocomposites remains a major barrier due to expensive nanoparticles (e.g., CNTs and graphene), complex processing, and strict quality control requirements. Research is ongoing to develop low-cost synthesis methods, optimize filler loading, and improve manufacturing efficiency without compromising performance (Bhat et al., 2021).
4. Environmental and health concerns: Certain nanoparticles pose toxicity risks during production, use, or disposal. CNTs and some metal oxides have raised concerns about bioaccumulation and long-term exposure effects. Safer nanomaterials, encapsulation strategies, and comprehensive toxicity studies are essential to ensure safe use in real-world applications (Hassan et al., 2021).

4.9.3 Research imperatives

To fully realize the potential of HTPNCs, future research can focus on:

1. Improving the compatibility and stability of nanofillers.
2. Enhancing cost-effective and scalable production techniques.
3. Developing standardized methods for evaluating thermal, mechanical, and environmental performance.
4. Designing smart nanocomposites with features like self-healing, shape memory, and adaptive responses.
5. Addressing recyclability and end-of-life sustainability to align with circular economy goals.

4.10 Conclusion

HTPNCs constitute a cutting-edge class of materials that satisfy the expanding demands of advanced industries working under harsh thermal, mechanical, and chemical environments. By incorporating nanomaterials such as CNTs, graphene, nanoclays, metal, and ceramic nanoparticles into high-performance polymer matrices, significant improvements have been achieved in terms of thermal stability, mechanical strength, wear resistance, flame retardancy, and electrical conductivity. The performance of these nanocomposites is vitally dependent on both the structure and loading concentration of the nanoparticles. Optimized nanoparticle loading promotes interfacial bonding, stress transmission, and heat dissipation, whereas excessive loading can lead to agglomeration and processing problems. The shape and dimensionality of nanoparticles (e.g., 0D NDs, 1D CNTs, 2D graphene) influence their dispersion and reinforcing efficiency inside the polymer matrix.

Equally significant are the synthesis and processing methods, including solution blending, melt blending, in situ polymerization, electrospinning, and LbL assembly, which play a crucial role in shaping the microstructure and final properties of the nanocomposite. Each process offers particular advantages depending on the type of polymer, targeted property upgrades, and target applications. The integration of high-temperature nanocomposites has already altered important sectors such as aerospace, automotive, electronics, energy, and biomedicine, offering lightweight, durable, and versatile alternatives to standard materials. Looking ahead, the development of multifunctional hybrids, ecofriendly production methods, smart and recyclable nanocomposites, and AI-assisted material design will further broaden the frontiers of these materials.

However, for widespread commercial application, difficulties such as homogeneous nanoparticle dispersion, interface compatibility, scalable manufacturing, cost reduction, and environmental safety must be systematically addressed. Collaborative, interdisciplinary research will be required to overcome these hurdles and fully exploit the transformative promise of HTPNCs in next-generation technology.

References

Adem, A.A., Panjiar, H., Daniel, B.S.S., 2024. The effect of nanocarbon inclusion on mechanical, tribological, and thermal properties of phenolic resin-based composites: an overview. Engineering Reports 6 (4), e12861. Available at https://doi.org/10.1002/eng2.12861.

Albdiry, M., 2024. Effect of melt blending processing on mechanical properties of polymer nanocomposites: a review. Polymer Bulletin 81 (7), 5793–5821. Available at https://doi.org/10.1007/s00289-023-05012-z.

Ali, A., Rahimian Koloor, S.S., Alshehri, A.H., Arockiarajan, A., 2023. Carbon nanotube characteristics and enhancement effects on the mechanical features of polymer-based materials and structures – a review. Journal of Materials Research and Technology 24, 6495–6521. Available at https://doi.org/10.1016/j.jmrt.2023.04.072.

Al-Mutairi, N.H., Mehdi, A.H., Kadhim, B.J., 2022. Nanocomposite materials: definitions, types, and some of their applications: a review. European Journal of Research Development and Sustainability 3, 102–108.

Alqutaibi, A.Y., Alghauli, M.A., Algabri, R.S., Hamadallah, H.H., Aboalrejal, A.N., Zafar, M.S., et al., 2025. Applications, modifications, and manufacturing of polyetheretherketone (PEEK) in dental implantology: a comprehensive critical review. International Materials Reviews 70 (2), 103–136. Available at https://doi.org/10.1177/09506608251314251.

Alshammari, B.A., Alsuhybani, M.S., Almushaikeh, A.M., Alotaibi, B.M., Alenad, A.M., Alqahtani, N.B., et al., 2021. Comprehensive review of the properties and modifications of carbon fiber-reinforced thermoplastic composites. Polymers 13 (15). Available at https://doi.org/10.3390/polym13152474.

Apmann, K., Fulmer, R., Soto, A., Vafaei, S., 2021. Thermal conductivity and viscosity: review and optimization of effects of nanoparticles. Materials 14 (5), 1291. Available at https://doi.org/10.3390/ma14051291.

Arshad, M.U., Wei, C., Li, Y., Li, J., Khakzad, M., Guo, C., et al., 2023. Mechanics – microstructure relations in 1D, 2D and mixed dimensional carbon nanomaterials. Carbon 204, 162–190. Available at https://doi.org/10.1016/j.carbon.2022.12.042.

Balasubramanian, K., Tirumalai, M., 2013. High Temperature Polymer Nanocomposites. In: Njuguna, J. (Ed.), Structural Nanocomposites. Engineering Materials. Springer, Berlin, Heidelberg, pp. 165–186. Available at https://doi.org/10.1007/978-3-642-40322-4_7.

Basha, S.I., Shah, S.S., Helal, A., Aziz, M.A., Yoo, D.Y., 2024. Unveiling the limitless potential: exploring metal-organic frameworks (MOFs)/MXene based construction materials. Case Studies in Construction Materials 21. Available at https://doi.org/10.1016/j.cscm.2024.e03586.

Bershtein, V.A., Yakushev, P.N., 2023. High-Temperature Polymer Nanocomposites Based on Heterocyclic Networks from Nitrile Monomers. Springer, Cham, pp. 1–168. Available at https://doi.org/10.1007/978-3-031-32943-2.

Dantas de Oliveira, A., Augusto Gonçalves Beatrice, C., 2019. Polymer nanocomposites with different types of nanofiller. In: Sivasankaran, S. (Ed.), Nanocomposites - Recent Evolutions. IntechOpen. Available at https://doi.org/10.5772/intechopen.81329.

Bhat, A., Budholiya, S., Raj, S.A., Sultan, M.T.H., Hui, D., Shah, A.U.M., et al., 2021. Review on nanocomposites based on aerospace applications. Nanotechnology Reviews 10 (1), 237–253. Available at https://doi.org/10.1515/ntrev-2021-0018.

Bokobza, L., 2023. Elastomer nanocomposites: effect of filler–matrix and filler–filler interactions. Polymers 15 (13), 2900. Available at https://doi.org/10.3390/polym15132900.

Cai, Z., Wang, X., Luo, B., Hong, W., Wu, L., Li, L., 2017. Dielectric response and breakdown behavior of polymer-ceramic nanocomposites: the effect of nanoparticle distribution. Composites Science and Technology 145, 105–113. Available at https://doi.org/10.1016/j.compscitech.2017.03.039.

Camargo, P.H.C., Satyanarayana, K.G., Wypych, F., 2009. Nanocomposites: synthesis, structure, properties and new application opportunities. Materials Research 12 (1), 1–39. Available at https://doi.org/10.1590/S1516-14392009000100002.

Castro, K.C.D., Costa, J.M., Campos, M.G.N., 2022. Drug-loaded polymeric nanoparticles: a review. International Journal of Polymeric Materials and Polymeric Biomaterials 71 (1), 1–13. Available at https://doi.org/10.1080/00914037.2020.1798436.

Chen, J., Huang, X., Sun, B., Jiang, P., 2019. Highly thermally conductive yet electrically insulating polymer/boron nitride nanosheets nanocomposite films for improved thermal management capability. ACS Nano 13 (1), 337–345. Available at https://doi.org/10.1021/acsnano.8b06290.

Chitoria, A.K., Mir, A., Shah, M.A., 2023. A review of ZrO_2 nanoparticles applications and recent advancements. Ceramics International 49 (20), 32343–32358. Available at https://doi.org/10.1016/j.ceramint.2023.06.296.

Chow, W.S., Mohd Ishak, Z.A., 2020. Smart polymer nanocomposites: a review. Express Polymer Letters 14 (5), 416–435. Available at https://doi.org/10.3144/expresspolymlett.2020.35.

Chung, T.-S., Calundann, G.W., East, A.J., 2023. Liquid crystal polymers and their applications. In: Cheremisinoff (Ed.), Handbook of Polymer Science and Technology. CRC Press, Boca Raton.

Darwish, M.S.A., Mostafa, M.H., Al-Harbi, L.M., 2022. Polymeric nanocomposites for environmental and industrial applications. International Journal of Molecular Sciences 23 (3). Available at https://doi.org/10.3390/ijms23031023.

Das, A., Chatham, C.A., Fallon, J.J., Zawaski, C.E., Gilmer, E.L., Williams, C.B., et al., 2020. Current understanding and challenges in high temperature additive manufacturing of engineering thermoplastic polymers. Additive Manufacturing 34. Available at https://doi.org/10.1016/j.addma.2020.101218.

Deepa, C., Rajeshkumar, L., Ramesh, M., 2022. Preparation, synthesis, properties and characterization of graphene-based 2D nano-materials for biosensors and bioelectronics. Journal of Materials Research and Technology 19, 2657–2694. Available at https://doi.org/10.1016/j.jmrt.2022.06.023.

Dlamini, D.S., Mamba, B.B., Li, J., 2019. The role of nanoparticles in the performance of nano-enabled composite membranes - a critical scientific perspective. Science of the Total Environment 656, 723–731. Available at https://doi.org/10.1016/j.scitotenv.2018.11.421.

Dossin Zanrosso, C., Piazza, D., Lansarin, M.A., 2020. PVDF/ZnO composite films for photocatalysis: a comparative study of solution mixing and melt blending methods. Polymer Engineering and Science 60 (6), 1146–1157. Available at https://doi.org/10.1002/pen.25368.

Ekeocha, J., Ellingford, C., Pan, M., Wemyss, A.M., Bowen, C., Wan, C., 2021. Challenges and opportunities of self-healing polymers and devices for extreme and hostile environments. Advanced Materials 33 (33). Available at https://doi.org/10.1002/adma.202008052.

El Hawary, A., Hasan, S., El Sttar, R.A., Mohamed, S., Bassyouni, M., 2019. A review on processing, and applications of nanocomposites. Journal of Composites and Biodegradable Polymers 7, 40–50. Available at https://doi.org/10.12974/2311-8717.2019.07.6.

Fedullo, N., Sorlier, E., Sclavons, M., Bailly, C., Lefebvre, J.M., Devaux, J., 2007. Polymer-based nanocomposites: overview, applications and perspectives. Progress in Organic Coatings 58 (2-3), 87–95. Available at https://doi.org/10.1016/j.porgcoat.2006.09.028.

Gao, Y., Zhou, X., Zhang, M., Lyu, L., Li, Z., 2022. Polyphenylene sulfide-based membranes: recent progress and future perspectives. Membranes 12 (10), 924. Available at https://doi.org/10.3390/membranes12100924.

Gay, D., 2022. Composite Materials: Design and Applications. CRC Press. Available at 10.1201/9781003195788.

Gerasin, V.A., Antipov, E.M., Karbushev, V.V., Kulichikhin, V.G., Karpacheva, G.P., Talroze, R.V., et al., 2013. New approaches to the development of hybrid nanocomposites: from structural materials to high-tech applications. Russian Chemical Reviews 82 (4), 303–332. Available at https://doi.org/10.1070/RC2013v082n04ABEH004322.

Glover, R.D., Miller, J.M., Hutchison, J.E., 2011. Generation of metal nanoparticles from silver and copper objects: nanoparticle dynamics on surfaces and potential sources of nanoparticles in the environment. ACS Nano 5 (11), 8950–8957. Available at https://doi.org/10.1021/nn2031319.

Gonzalez-Ortiz, D., Salameh, C., Bechelany, M., Miele, P., 2020. Nanostructured boron nitride–based materials: synthesis and applications. Materials Today Advances 8, 100107. Available at https://doi.org/10.1016/j.mtadv.2020.100107.

Guerrero, J.M., Loh, P.C., Lee, T.L., Chandorkar, M., 2013. Advanced control architectures for intelligent microgridsPart II: power quality, energy storage, and AC/DC microgrids. IEEE Transactions on Industrial Electronics 60 (4), 1263–1270. Available at https://doi.org/10.1109/TIE.2012.2196889.

Gupta, A., Patel, P., Shah, S., Patel, K., 2025. Nanocomposites in focus: tailoring drug delivery for enhanced therapeutic outcomes. Future Journal of Pharmaceutical Sciences 11 (1). Available at https://doi.org/10.1186/s43094-025-00789-4.

Hanemann, T., Szabó, D.V., 2010. Polymer-nanoparticle composites: from synthesis to modern applications. Materials 3 (6), 3468–3517. Available at https://doi.org/10.3390/ma3063468.

Hassan, T., Salam, A., Khan, A., Khan, S.U., Khanzada, H., Wasim, M., et al., 2021. Functional nanocomposites and their potential applications: a review. Journal of Polymer Research 28 (2). Available at https://doi.org/10.1007/s10965-021-02408-1.

Hong, R.Y., Chen, Q., 2014. Dispersion of inorganic nanoparticles in polymer matrices: challenges and solutions. In: Kalia, S., Haldorai, Y.S., Haldorai, Y. (Eds.), Organic-Inorganic Hybrid Nanomaterials. Advances in Polymer Science, vol 267. Springer, Cham, pp. 1–38. Available at https://doi.org/10.1007/12_2014_286.

Hu, D., Wang, S., Zhang, C., Yi, P., Jiang, P., Huang, X., 2021. Ultrathin MXene-aramid nanofiber electromagnetic interference shielding films with tactile sensing ability withstanding harsh temperatures. Nano Research 14 (8), 2837–2845. Available at https://doi.org/10.1007/s12274-021-3297-z.

Huang, X., Zhi, C., Lin, Y., Bao, H., Wu, G., Jiang, P., et al., 2020. Thermal conductivity of graphene-based polymer nanocomposites. Materials Science and Engineering: R: Reports 142, 100577. Available at https://doi.org/10.1016/j.mser.2020.100577.

Huang, X., Su, S., Xu, Z., Miao, Q., Li, W., Wang, L., 2023. Advanced composite materials for structure strengthening and resilience improvement. Buildings 13 (10), 2406. Available at https://doi.org/10.3390/buildings13102406.

Ibrahim, O., Oluwadunsin, J., Antwi, M., Mekunye, F., Oluwatobi, M., Olorunfemi, A., et al., 2025. The evolution of self-healing electrodes: a critical review of nanomaterial contributions. American Journal of Nanosciences 9 (1), 8–31. Available at https://doi.org/10.11648/j.ajn.20250901.12.

Jain, P., Verma, C., Raman, A.P.S., Kumari, K., Singh, P., 2024. Biosensors Based on Graphene, Graphene Oxide and Graphynes for Early Detection of Cancer. CRC Press, India, pp. 1–281. Available at https://doi.org/10.1201/9781003491361.

Ji, D., Lin, Y., Guo, X., Ramasubramanian, B., Wang, R., Radacsi, N., et al., 2024. Electrospinning of nanofibers. Nature Reviews Methods Primers 4, 1. Available at https://doi.org/10.1038/s43586-023-00278-z.

Kalita, M., Baruah, S., Deka, R., Dutta, G.K., Barua, S., 2025.

Kambhampati, P., 2021. Nanoparticles, nanocrystals, and quantum dots: what are the implications of size in colloidal nanoscale materials? The Journal of Physical Chemistry Letters 12 (20), 4769–4779. Available at https://doi.org/10.1021/acs.jpclett.1c00754.

Kausar, A., 2024. Polymer/Nanodiamond Nanocomposites: Fundamentals, Properties and Applications. Elsevier, Pakistan, pp. 1–293. Available at https://doi.org/10.1016/C2022-0-02205-4.

Khan, S., Hossain, M.K., 2022. 2 - Classification and properties of nanoparticles. In: Rangappa, S.M., Parameswaranpillai, J., Yashas Gowda, T.G., Siengchin, S., Ozgur Seydibeyoglu, M. (Eds.), Nanoparticle-based Polymer Composites. Woodhead Publishing, pp. 15–54. Available at https://doi.org/10.1016/B978-0-12-824272-8.00009-9.

Khdary, N.H., Almuarqab, B.T., El Enany, G., 2023. Nanoparticle-embedded polymers and their applications: a review. Membranes 13 (5). Available at https://doi.org/10.3390/membranes13050537.

Kong, D., Li, J., Guo, A., Xiao, X., 2021. High temperature electromagnetic shielding shape memory polymer composite. Chemical Engineering Journal 408, 127365. Available at https://doi.org/10.1016/j.cej.2020.127365.

Król-Morkisz, K., Pielichowska, K., 2019. 13 - Thermal decomposition of polymer nanocomposites with functionalized nanoparticles. In: Pielichowski, K., Majka, T.M. (Eds.), Polymer Composites with Functionalized Nanoparticles: Synthesis, Properties, and Applications. Elsevier BV, pp. 405–435. Available at https://doi.org/10.1016/b978-0-12-814064-2.00013-5.

Kumar, A., Singh, R.C., Chaudhary, R., 2020. Recent progress in production of metal matrix composites by stir casting process: an overview. Materials Today: Proceedings 21, 1453–1457. Available at https://doi.org/10.1016/j.matpr.2019.10.079.

Kumar, D., Kanchan, D.K., 2024. Functionalization and surface modification strategies of nanofillers: dispersion state and specific interaction between functionalized nanofillers and polymeric systems. Advances in Functionalized Polymer Nanocomposites: From Synthesis to Applications. Elsevier, India, pp. 35–64. Available at https://doi.org/10.1016/B978-0-443-18860-2.00019-0.

Kumar, A., Dixit, S., Singh, S., Sreenivasa, S., Bains, P.S., Sharma, R., 2025. Recent developments in the mechanical properties and recycling of fiber-reinforced polymer composites. Polymer Composites 46 (5), 3883–3908. Available at https://doi.org/10.1002/pc.29261.

Kutvonen, A., Rossi, G., Puisto, S.R., Rostedt, N.K.J., Ala-Nissila, T., 2012. Influence of nanoparticle size, loading, and shape on the mechanical properties of polymer nanocomposites. Journal of Chemical Physics 137 (21). Available at https://doi.org/10.1063/1.4767517.

Lalire, T., Longuet, C., Taguet, A., 2024. Electrical properties of graphene/multiphase polymer nanocomposites: a review. Carbon 225, 119055. Available at https://doi.org/10.1016/j.carbon.2024.119055.

Li, H., Zhou, Y., Liu, Y., Li, L., Liu, Y., Wang, Q., 2021. Dielectric polymers for high-temperature capacitive energy storage. Chemical Society Reviews 50 (11), 6369–6400. Available at https://doi.org/10.1039/d0cs00765j.

Li, H., Hou, L., Liu, Y., Yao, Z., Liang, L., Tian, D., et al., 2024. Balanced thermal insulation, flame-retardant and mechanical properties of PU foam constructed via cost-effective EG/APP/SA ternary synergistic modification. Polymers 16 (3), 330. Available at https://doi.org/10.3390/polym16030330.

Liang, Y., Sun, W., Li, T., Chen, Y., Cheng, Y., Zhang, L., 2023. All organic polymer dielectrics for high-temperature energy storage from the classification of heat-resistant insulation grades. Journal of Polymer Science 61 (22), 2777–2795. Available at https://doi.org/10.1002/pol.20230334.

Lipton, J., Weng, G.M., Röhr, J.A., Wang, H., Taylor, A.D., 2020. Layer-by-layer assembly of two-dimensional materials: meticulous control on the nanoscale. Matter 2 (5), 1148–1165. Available at https://doi.org/10.1016/j.matt.2020.03.012.

Liu, Y., Yu, Z., Wang, B., Li, P., Zhu, J., Ma, S., 2022. Closed-loop chemical recycling of thermosetting polymers and their applications: a review. Green Chemistry 24 (15), 5691–5708. Available at https://doi.org/10.1039/d2gc00368f.

Luo, J.J., Daniel, I.M., 2003. Characterization and modeling of mechanical behavior of polymer/clay nanocomposites. Composites Science and Technology 63 (11), 1607–1616. Available at https://doi.org/10.1016/S0266-3538(03)00060-5.

Ma, J., Liu, X., Wang, R., Lu, C., Wen, X., Tu, G., 2023. Research progress and application of polyimide-based nanocomposites. Nanomaterials 13 (4), 656. Available at https://doi.org/10.3390/nano13040656.

Markandan, K., Lai, C.Q., 2023. Fabrication, properties, and applications of polymer composites additively manufactured with filler alignment control: a review. Composites Part B: Engineering 256, 110661. Available at https://doi.org/10.1016/j.compositesb.2023.110661.

Marquis, D.M., Guillaume, E., Chivas-Joly, C., 2011. Properties of nanofillers in polymer. In: Cuppoletti, J. (Ed.), Nanocomposites and Polymers with Analytical Methods. InTech. Available at https://doi.org/10.5772/21694.

Mat Yazik, M.H., Ismail, I., 2025. Chapter Fifteen - Aerospace structures and engines from polymer composites. In: Hameed Sultan, M.T., Uthayakumar, M., Korniejenko, K., Mashinini, P.M., Najeeb, M.I., Krishnamoorthy, R.R. (Eds.), Aerospace Materials: Novel Technologies and Practical Applications. Elsevier, Malaysia, pp. 349–383. Available at https://doi.org/10.1016/B978-0-443-22118-7.00015-4.

Mehta, J., Das, P.P., Mehta, S., Chaudhary, V., Gupta, S., Gupta, N., et al., 2022. Effect of nanofillers and nanotoxicity on the performance of composites: influencing factors, future scope, challenges and applications. Polymer Composites 43 (6), 3335–3349. Available at https://doi.org/10.1002/pc.26635.

Mercante, L.A., Andre, R.S., Facure, M.H.M., Correa, D.S., Mattoso, L.H.C., 2023. Recent progress in conductive electrospun materials for flexible electronics: energy, sensing, and electromagnetic shielding applications. Chemical Engineering Journal 465. Available at https://doi.org/10.1016/j.cej.2023.142847.

Merhari, L., 2009. Hybrid Nanocomposites for Nanotechnology: Electronic, Optical, Magnetic and Biomedical Applications. Springer, New York, NY, pp. 1–847. Available at https://doi.org/10.1007/978-0-387-30428-1.

Mohajerani, A., Burnett, L., Smith, J.V., Kurmus, H., Milas, J., Arulrajah, A., et al., 2019. Nanoparticles in construction materials and other applications, and implications of nanoparticle use. Materials 12 (19), 3052. Available at https://doi.org/10.3390/ma12193052.

Mohd, A., Ortjohann, E., Schmelter, A., Hamsic, N., Morton, D., 2008. Challenges in integrating distributed energy storage systems into future smart grid. 2008 IEEE International Symposium on Industrial Electronics, Cambridge, UK. IEEE, pp. 1627–1632. Available at https://doi.org/10.1109/ISIE.2008.4676896.

Monteiro, B., Simões, S., 2024. Recent advances in hybrid nanocomposites for aerospace applications. Metals 14 (11), 1283. Available at https://doi.org/10.3390/met14111283.

Mousavi, S.R., Estaji, S., Paydayesh, A., Arjmand, M., Jafari, S.H., Nouranian, S., et al., 2022. A review of recent progress in improving the fracture toughness of epoxy-based composites using carbonaceous nanofillers. Polymer Composites 43 (4), 1871–1886. Available at https://doi.org/10.1002/pc.26518.

Nguyen, T.P., 2011. Polymer-based nanocomposites for organic optoelectronic devices. A review. Surface and Coatings Technology 206 (4), 742–752. Available at https://doi.org/10.1016/j.surfcoat.2011.07.010.

Njuguna, J., Ansari, F., Sachse, S., Rodriguez, V.M., Siqqique, S., Zhu, H., 2021. 1 - Nanomaterials, nanofillers, and nanocomposites: types and properties. In: Njuguna, J., Pielichowski, K., Zhu, H. (Eds.), Health and Environmental Safety of Nanomaterials: Polymer Nanocomposites and Other Materials Containing

Nanoparticles. Woodhead Publishing, pp. 3–37. Available at https://doi.org/10.1016/B978-0-12-820505-1.00011-0.

Rahimi-Ahar, Z., Rahimi Ahar, L., 2024. Thermal, optical, mechanical, dielectric, and electrical properties of nanocomposites. European Polymer Journal 218, 113337. Available at https://doi.org/10.1016/j.eurpolymj.2024.113337.

Sabet, M., 2024. Unveiling advanced self-healing mechanisms in graphene polymer composites for next-generation applications in aerospace, automotive, and electronics. Polymer-Plastics Technology and Materials 63 (15), 2032–2059. Available at https://doi.org/10.1080/25740881.2024.2365282.

Saha, R., Munshi, M.H., Akter, M., Shikder, A.A.R., Islam, T., 2025. Synthesis techniques, fundamental properties, and emerging applications of nanocomposites: a comprehensive review. SPE Polymers 6 (2). Available at https://doi.org/10.1002/pls2.70004.

Saxena, A., Mehta, A., Vasudev, H., Prashar, G., Yasir Jasim, N., 2023. A short review on machine learning for the purpose of optimizing and predicting the properties of polymeric nanocomposites. Materials Today: Proceedings. Available at https://doi.org/10.1016/j.matpr.2023.07.347.

Sedlmeier, A., Gorris, H.H., 2015. Surface modification and characterization of photon-upconverting nanoparticles for bioanalytical applications. Chemical Society Reviews 44 (6), 1526–1560. Available at https://doi.org/10.1039/c4cs00186a.

Selvaraj, S.K., Ramesh, R., Narendhra, T.M.V., Agarwal, I.N., Chadha, U., Paramasivam, V., et al., 2021. New developments in carbon-based nanomaterials for automotive brake pad applications and future challenges. Journal of Nanomaterials 2021. Available at https://doi.org/10.1155/2021/6787435.

Shabaev, A.S., Shakhmurzova, K.T., Zhansitov, A.A., Slonov, A.L., Fomicheva, I.N., Dolbin, I.V., et al., 2024. Thermal transformation and mechanical properties of high-temperature-resistant matrix based polyetherketones. Institute for Problems in Mechanical Engineering, Russian Academy of Sciences, Russian Federation Materials Physics and Mechanics 52 (2), 123–132. Available at https://doi.org/10.18149/MPM.5222024_13.

Sharifzadeh, E., Cheraghi, K., 2021. Temperature-affected mechanical properties of polymer nanocomposites from glassy-state to glass transition temperature. Mechanics of Materials 160, 103990. Available at https://doi.org/10.1016/j.mechmat.2021.103990.

Sharma, G., Kaur, M., Punj, S., Singh, K., 2020. Biomass as a sustainable resource for value-added modern materials: a review. Biofuels, Bioproducts and Biorefining 14 (3), 673–695. Available at https://doi.org/10.1002/bbb.2079.

Sheikh, T., Behdinan, K., 2023. Insight of discrete scale and multiscale methods for characterization of composite and nanocomposite materials. Archives of Computational Methods in Engineering 30 (2), 1231–1265. Available at https://doi.org/10.1007/s11831-022-09840-6.

Shunmugasamy, V.C., Xiang, C., Gupta, N., 2015. Clay/polymer nanocomposites: processing, properties, and applications. In: Kim, C.S., Randow, C., Sano, T. (Eds.), Hybrid and Hierarchical Composite Materials. Springer, Cham, pp. 161–200. Available at https://doi.org/10.1007/978-3-319-12868-9_5.

Simões, S., 2024. High-performance advanced composites in multifunctional material design: state of the art, challenges, and future directions. Materials 17 (23), 5997. Available at https://doi.org/10.3390/ma17235997.

Singh, M., Tiwary, S.K., Karim, A., 2024. Sub-nano fillers for high-temperature storage. Nature Energy 9 (2), 113–114. Available at https://doi.org/10.1038/s41560-023-01446-x.

Sui, Y., Cui, Y., Meng, X., Zhou, Q., 2022. Research progress on the correlation between properties of nanoparticles and their dispersion states in polymer matrix. Journal of Applied Polymer Science 139 (19), 52096. Available at https://doi.org/10.1002/app.52096.

Sun, X., Huang, C., Wang, L., Liang, L., Cheng, Y., Fei, W., et al., 2021. Recent progress in graphene/polymer nanocomposites. Advanced Materials 33 (6), 2001105. Available at https://doi.org/10.1002/adma.202001105.

Thakur, A., Thakur, P., Baccar, S., 2023. Structural properties of nanoparticles. In: Suhag, D., Thakur, A., Thakur, P. (Eds.), Integrated Nanomaterials and their Applications. Springer, Singapore, pp. 57–80. Available at https://doi.org/10.1007/978-981-99-6105-4_4.

Thippeswamy, B.H., Maligi, A.S., Hegde, G., 2021. Roadmap of effects of biowaste-synthesized carbon nanomaterials on carbon nano-reinforced composites. Catalysts 11 (12), 1485. Available at https://doi.org/10.3390/catal11121485.

Treccani, L., 2022. Introduction to ceramic materials. In: Treccani, L., Meder, F. (Eds.), Surface-Functionalized Ceramics: For Biotechnological and Environmental Applications. Wiley, pp. 1–46. Available at https://doi.org/10.1002/9783527698042.ch1.

Uddin, F., 2008. Clays, nanoclays, and montmorillonite minerals. Metallurgical and Materials Transactions A 39 (12), 2804–2814. Available at https://doi.org/10.1007/s11661-008-9603-5.

Wang, L., Wang, T., Peng, L., Wang, Y., Zhang, M., Zhou, J., et al., 2022. The promises, challenges, and pathways to room-temperature sodium-sulfur batteries. National Science Review 9 (3), nwab050. Available at https://doi.org/10.1093/nsr/nwab050.

Wani, I.N., Shukla, P.K., Gulzar, O., Kumar, M., 2024. Techniques and Considerations of Materials for Thermal Protection and Temperature Regulation in Spacecraft. Tuijin Jishu/Journal of Propulsion Technology 45 (3), 2969–2990. Available at https://doi.org/10.52783/tjjpt.v45.i03.7693.

Xu, X., Chen, J., Zhou, J., Li, B., 2018. Thermal conductivity of polymers and their nanocomposites. Advanced Materials 30 (17), 1705544. Available at https://doi.org/10.1002/adma.201705544.

Xu, H., Hao, Z., Wang, C., Deng, J., Wang, T., Zhang, J., 2023. In situ emulsion polymerization to multifunctional polymer nanocomposites: a review. Macromolecular Chemistry and Physics 224 (20), 2300185. Available at https://doi.org/10.1002/macp.202300185.

Xu, L.Y., Wang, X.Y., Lin, Y.Z., Huang, Y., Tao, C.C., Zhang, D.W., 2025. Experimental and numerical investigations of carbon-based nanoparticle reinforcement on microstructure and mechanical properties of epoxy coatings. Chinese Journal of Polymer Science (English Edition) 43 (1), 211–224. Available at https://doi.org/10.1007/s10118-025-3252-7.

Yasmin, H., Giwa, S.O., Noor, S., Sharifpur, M., 2023. Thermal conductivity enhancement of metal oxide nanofluids: a critical review. Nanomaterials 13 (3), 597. Available at https://doi.org/10.3390/nano13030597.

Yazman, Ş., Uyaner, M., Karabörk, F., Akdemir, A., 2021. Effects of nano reinforcing/matrix interaction on chemical, thermal and mechanical properties of epoxy nanocomposites. Journal of Composite Materials 55 (28), 4257–4272. Available at https://doi.org/10.1177/00219983211037059.

Yilmaz, O., Kucuk, M., Yilmaz, C.N., 2024. 4 - Polyimide nanocomposites and blends for textiles. In: Damaceanu, M-D., Darie-Nita, R.N. (Eds.), Polyimides: Advances in Blends and Nanocomposites. Elsevier BV, pp. 145–188. Available at https://doi.org/10.1016/b978-0-323-90294-6.00011-4.

Zafar, M., Muhammad Imran, S., Iqbal, I., Azeem, M., Chaudhary, S., Ahmad, S., et al., 2024. Graphene-based polymer nanocomposites for energy applications: recent advancements and future prospects. Results in Physics 60, 107655. Available at https://doi.org/10.1016/j.rinp.2024.107655.

Zhang, X., Xu, Y., Zhang, X., Wu, H., Shen, J., Chen, R., et al., 2019. Progress on the layer-by-layer assembly of multilayered polymer composites: strategy, structural control and applications. Progress in Polymer Science 89, 76–107. Available at https://doi.org/10.1016/j.progpolymsci.2018.10.002.

Zhou, T., Zhao, C., Liu, Y., Huang, J., Zhou, H., Nie, Z., et al., 2022. Large-area ultrastrong and stiff layered mxene nanocomposites by shear-flow-induced alignment of nanosheets. ACS Nano 16 (8), 12013–12023. Available at https://doi.org/10.1021/acsnano.2c02062.

5

Polymer blend–based high temperature composite materials

Chibueze Godwin Achi, Williams Kehinde Kupolati, Julius Musyoka Ndambuki, Jacques Snyman

DEPARTMENT OF CIVIL ENGINEERING, TSHWANE UNIVERSITY OF TECHNOLOGY, PRETORIA, GAUTENG, SOUTH AFRICA

5.1 Introduction

5.1.1 Overview of high-temperature polymer-based composites

High-temperature polymer composites are well-developed materials that possess outstanding mechanical, thermal, and chemical characteristics, which make them useful in conditions that cannot be sustained by regular materials. These composites usually include a polymer base, in combination with high-performance fibers or particles (Oladele et al., 2020; Parveez et al., 2022).

The polymer matrix used in high-temperature composites is chosen to provide thermostability and resist degradation at high temperatures. Some examples of polymer matrices that are widely employed in the preparation of composite materials are polyimides (PIs), polyetheretherketone (PEEK), polybenzimidazole, and polyphenylene sulfide (PPS). These polymers have high thermal stability, high T_g, and good mechanical properties at high temperature (Byerlin and Pakhomova, 1990; Kemmish, 2011; Kyriacos, 2017; Seferis, 1986; Wilson et al., 2013).

The reinforcing fibers or particles, incorporated into high-temperature composites, provide enhanced mechanical properties and heat resistance. Reinforcement materials can include carbon, ceramic, aramid, or metallic fibers, depending on the specific application requirements. These fibers offer high strength, stiffness, and thermal stability, allowing the composites to maintain their mechanical integrity at elevated temperatures (Rajak et al., 2019; Yang et al., 2019).

High-temperature polymer-based composites find applications in various industries, including aerospace, automotive, energy, and chemical processing. In the aerospace domain, they are used for components, such as engine parts, aircraft interiors, and thermal protection systems (Oladele et al., 2020). The automotive industry utilizes these composites for components that require high-temperature resistance, such as engine components and exhaust systems. They are also employed in the energy sector for applications such as turbine blades, heat exchangers, and insulation materials (Das et al., 2021).

High-Temperature Polymeric Materials. DOI: https://doi.org/10.1016/B978-0-443-32792-6.00001-4

The advantages of high-temperature polymer-based composites include their lightweight nature, excellent thermal stability, corrosion resistance, and design flexibility. These materials offer improved performance and durability in extreme environments, enabling the development of more efficient and reliable products (Parin and Demirci, 2024).

However, challenges such as the need for specialized manufacturing techniques, high costs associated with raw materials, and limited availability of large-scale fabrication technologies exist in the production and processing of high-temperature composites (Alkunte et al., 2024; Das et al., 2020; Koumoulos et al., 2019). Ongoing research and development efforts aim to overcome these challenges and further enhance the properties and affordability of high-temperature polymer-based composites.

Researchers have been exploring various strategies to improve the properties of high-temperature polymer composites. For example, efforts have been made to fabricate organic–inorganic composites using alternative polymers to achieve higher operational temperatures (Cinquino et al., 2024; Sanchez et al., 2005). Furthermore, the interfacial interaction between carbon fibers and polymer matrices has been a focal point of research aimed at manufacturing innovative carbon fiber-reinforced composites using high-temperature polymers, for example, polysulfone(Chukov et al., 2018). Understanding the mechanical behavior of high-temperature polymer composites is essential for their utilization in demanding environments. Research has focused on analyzing the temperature and damage-dependent tensile strength of fiber/polymer composites, indicating a growing interest in investigating the high-temperature mechanical properties of these materials (Li et al., 2022). Additionally, the impact of the cure cycle on high-temperature polymer composite structures has been identified as an area that necessitates further exploration to fully exploit the potential of these composites (Collinson et al., 2022).

In summary, high-temperature polymer-based composites offer a promising solution for applications requiring materials that withstand elevated temperatures. By harnessing the unique properties of polymers and reinforcing materials, for example, carbon fibers, researchers have continued to strive to develop composites with enhanced mechanical strength and thermal stability for diverse industrial applications.

5.1.2 Importance of polymer blends in composite development

Polymer blends are indeed crucial in the development of composites, offering a cost-effective and advantageous method for fabricating novel structural materials (Dutta et al., 2019). The compatibility and performance of combined materials, such as poly(lactic) acid (PLA) and starch, are analyzed through blending or multilayer strategies, highlighting the importance of polymer compatibility in achieving good performance (Muller et al., 2017). However, it is noted that most biodegradable polymer blends exhibit insufficient performance due to thermodynamic immiscibility and poor compatibility between the blended components (Muthuraj et al., 2018). Nevertheless, polymer blends and composites have been successfully utilized in the development of consumer products, such as shopping bags and packaging, with short lifetimes (Easton et al., 2023). The rheological behavior of polymer composites is influenced by factors such as particle-particle and particle-matrix interactions, which are essential considerations in polymer blending (Zhou et al., 2008). Additionally, the application of quaternary ammonium compounds as compatibilizers for polymer blends and composites has been identified as a promising area for further research in this field (Shamsuri and Jamil, 2021). The study of the nanometer-scale structures and morphologies of block copolymers and their blends, is essential for the understanding of phase separation, miscibility, adhesion, and interface phenomena in polymers (Bar et al., 1998). The blending of important biodegradable polymers, such as PLA and polybutylene adipate terephthalate, with silk powder has been explored, emphasizing the significance of compatibilization and performance evaluation in such blends (Nakayama et al., 2018). Moreover, the thermal stability of polymer blends is greatly influenced by blending, with the interaction between the individual polymers playing a crucial role in determining the thermal stability of the blends (Rao et al., 1999). In addition, the use of polymer blends as matrix resin can lead to multiple percolation, resulting in changes in conductivity and the electrical properties of the composites (Zhang et al., 1998). In summary, polymer blends are essential in the development of composites, offering a wide processing window for the manufacturing of electrically conductive composites, multiple shape-memory behaviors, and improved mechanical and thermal properties. The compatibility and performance of combined materials, the influence of nanometer-scale structures, and the use of compatibilizers are crucial factors in the successful development of polymer blends for composite applications.

5.1.3 Objectives of using polymer blends for high-temperature applications

The utilization of polymer blends for high-temperature applications serves to enhance thermal stability, mechanical properties, and functionality, making them essential items for industries that require materials that are capable of withstanding extreme temperatures and harsh conditions. Polymer blends offer a versatile approach to tailor properties such as thermal stability, mechanical strength, and chemical resistance, rendering them ideal for high temperature applications (Gumyusenge et al., 2018). By combining semicrystalline conjugated polymers with high glass-transition temperature insulating matrices, thermally stable high-temperature semiconducting polymer blends can be developed, ensuring stable charge transport even at elevated temperatures (Gumyusenge et al., 2018). Furthermore, the use of polymer blends enables the creation of materials with improved mechanical stability and ionic conductivity, crucial for high-performance applications like battery membranes (Gul and Mahmood, 2023). The compatibility and miscibility of polymer blends significantly influence the thermal and mechanical properties of the final composite, making them suitable for use in high-temperature environments (Lin et al., 2013). In addition, the segmental dynamics and relaxation behavior of polymer blends at high temperatures are critical factors that impact their performance in various applications (Leroy et al., 2003). Moreover, the development of scalable thermochromic composites based on polymer blends allows for temperature-adaptive solar heat management, highlighting the versatility and potential of polymer blends in high-temperature settings (Hirai et al., 2023). The modification of the behavior of polymers, through blending, provides a cost-effective and efficient method to expand the application range of polymeric materials for high-temperature conditions (Vedamurthy et al., 2023).

5.2 Fundamentals of polymer blends

5.2.1 Definition and types of polymer blends

Polymer blends are defined by the physical mixtures of structurally different homopolymers and/or copolymers, thereby offering a versatile approach to combine the properties of individual polymers (Barzegari et al., 2004). These blends typically consist of at least two polymeric species, allowing for the creation of materials with tailored properties that can be customized for specific applications. The types of polymer blends can vary based on the nature of the polymers being combined and the desired characteristics of the final material. The common classification of polymer blends includes thermoplastic–thermoplastic blends, thermoplastic–rubber blends, thermoplastic–thermosetting blends, rubber–thermosetting blends, and polymer–filler blends, each of which has been extensively studied.

Thermoplastic/thermoplastic blends involve the combination of two thermoplastic polymers, hence, offering a wide range of properties depending on the specific polymers used (Adeniyi et al., 2016). Thermoplastic–rubber blends combine the flexibility of rubber with the processability of thermoplastics, creating materials with enhanced elasticity and toughness. Thermoplastic/thermosetting blends combine the properties of thermoplastics with the heat resistance of thermosetting polymers, thereby providing materials that are suitable for high-temperature applications. Rubber/thermosetting blends combine the resilience of rubber with the durability of thermosetting polymers, thereby offering materials with unique mechanical properties (Biron, 2015). Polymer/filler blends incorporate fillers into polymers to enhance specific characteristics such as strength, stiffness, or conductivity.

Another classification of polymer blends includes miscible blends, where the constituent polymers exhibit maximum interaction, leading to superior performance (Hema and Sureshan, 2019). Miscible blends are characterized by a high degree of compatibility between the polymers, resulting in materials with homogeneous properties throughout the blend (Fig. 5–1). This type of blend is particularly advantageous when a seamless integration of properties from different polymers is desired. Polymer blends play a crucial role in various industries, since they offer a cost-effective and efficient route to fabricate materials with tailored properties. By combining different polymers, researchers and engineers can develop materials that exhibit unique combination of characteristics, such as mechanical strength, thermal stability, chemical resistance, and electrical conductivity. The versatility of polymer blends makes them valuable for a wide range of applications, from packaging and consumer goods to high-performance materials used in the aerospace and automotive industries.

Generally, polymer blends are essential components in the realm of materials science, since they provide a platform for the fabrication of innovative materials with diverse

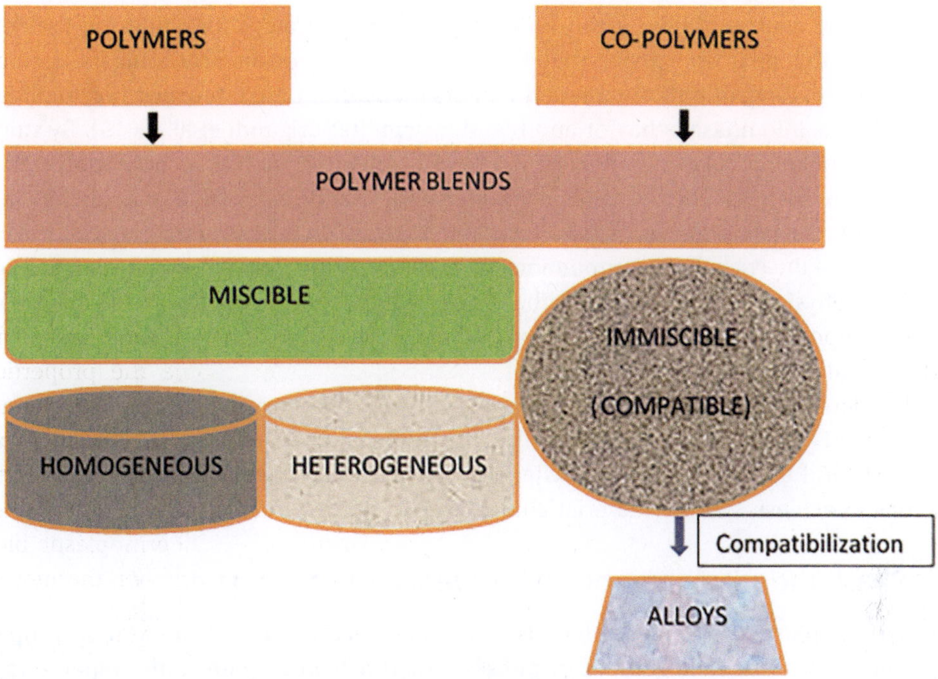

FIGURE 5–1 Classifications of polymer blends based on miscibility.

properties. The classification of polymer blends into different types allows for a systematic understanding of their characteristics and applications, hence enabling researchers to design materials that meet specific requirements for various industries.

5.2.2 Thermodynamics of polymer blending

Thermodynamics significantly influences the behavior of polymer blends, affecting their miscibility, phase separation, and properties. Parameters, such as the Gibbs free energy and the Flory–Huggins parameter (X), are commonly used to describe the thermodynamic interactions between polymer components in a blend, aiding in determining blend miscibility (Muthuraj et al., 2018). These parameters are crucial for evaluating the compatibility of different polymers and predicting their behavior when combined. The morphology of polymer blends is determined by a combination of thermodynamic and kinetic effects, as well as the localization of nanoparticles within the blend (Lin et al., 1996). The distribution of fillers in polymer blends is influenced by factors such as interfacial tensions between the fillers and the polymer phases, reflecting the thermodynamic preferences that are essential for the control of filler dispersion and localization of inclusions, such as carbon nanotubes or graphene, within the blend. In immiscible polymer blends, thermodynamic factors can lead to the selective localization of fillers at the interface between the two polymer phases, driven by the minimization of interfacial energy (Penning and St. John Manley, 1996). This thermodynamic control of filler localization is an effective strategy for achieving the desired properties in the final composite material. Small-angle neutron scattering is a valuable technique for investigating thermodynamic interactions in multicomponent polymer blends, offering insights into phase behavior and blend miscibility (Brigandi et al., 2015). By studying the thermodynamic interactions between different polymers, including homopolymers and block copolymers, researchers can gain a better understanding of the blend's behavior and properties. In miscible polymer blends, favorable thermodynamic interactions can lower the melting point of the crystalline components, demonstrating the impact of thermodynamics on the blend's physical properties. The thermodynamic control of conductive networks in polymer composites can be achieved by inducing the self-assembled structures through thermodynamically driven processes.

Thermodynamics is fundamental in shaping the behavior, properties, and morphology of polymer blends. The understanding of the thermodynamic interactions between polymer components and fillers enables researchers to customize blend composition and processing conditions to achieve specific material characteristics and performance.

5.2.3 Phase morphology and its influence on properties

The phase morphology of polymer blends is a critical factor in determining their properties and performance. The arrangement of phases within a blend significantly impacts various characteristics, such as mechanical strength, thermal behavior, and barrier properties. The morphology of polymer blends is influenced by factors such as blend composition, processing conditions, and the presence of compatibilizers or additives/modifiers. Research has shown that phase morphology greatly influences the properties of polymer blends (Wu et al., 2017). The phase structure, whether cocontinuous, dispersed, or core–shell morphology, directly affects the mechanical, thermal, and barrier properties of the blend (Dou et al., 2013). For example, in a study on polyamide 6 (PA6), maleic anhydride grafted ethylene-propylene--diene monomer (EPDM-*g*-MA), high-density polyethylene (HDPE), PA6/EPDM-*g*-MA/HDPE ternary blends, the phase morphology observation revealed a core–shell structure, where HDPE core was encapsulated by the EPDM-*g*-MA phase within the PA6 matrix (Dou et al., 2013). The morphology of polymer blends can exhibit various structures, such as droplet, fiber, lamellar, and cocontinuous microstructures (Lopez-Barron and Macosko, 2009). The evolution of these morphologies depends on factors such as the rheological properties of each component, volume fractions, and interfacial tension between phases (Daux et al., 2013). In addition, the presence of nanoparticles or fillers can influence the phase morphology, leading to unique structures that can affect the overall properties of the blend (Huang et al., 2016; Mochane et al., 2020). The understanding and control of the phase structure, through processing techniques, blend composition, and compatibilization strategies, are essential for tailoring the properties of polymer blends to meet specific application requirements. By optimizing the phase morphology, researchers can enhance the mechanical strength, thermal stability, and other key properties of polymer blends.

5.2.4 Compatibilization strategies for polymer blends

Polymer blending often involves the combination of immiscible polymers, which can lead to phase separation and poor mechanical properties. Compatibilization strategies play a crucial role in improving the compatibility between the different polymer phases, enhancing the properties of the blend, and achieving a more uniform morphology. Several research studies have explored various methods of compatibilizing polymer blends to optimize their performance. One effective strategy for compatibilizing immiscible polymer blends involves using graphene oxide sheets to optimize interfacial tension and stabilize the morphology against high stresses during forming (Cao et al., 2011). Another approach utilizes the in situ production route and forms the Janus nanomicelles by reactive blending to enhance the compatibility of immiscible polymer blends, thereby achieving a highly refined and stabilized morphology (Wang et al., 2015). In addition, the formation of interfacial Janus nanomicelles, via the reactive blending, has been shown to lower interfacial tension and enhance the phase adhesion, leading to homogeneous dispersion of the crosslinking and compatibilizing agents in polymer blends (Wang et al., 2016). Reactive processing is another common method of compatibilizing polymer blends. It involves the use of suitably functionalized blend constituents, polymeric compatibilizers, or low molecular weight compounds to enhance the compatibility between the phases (Xanthos and Dagli, 1991). The addition of reactive nanofillers has been shown to thermodynamically compatibilize polymer blends, thereby resulting in advanced blended nanocomposites with improved properties (Zhao et al., 2018). Furthermore, the use of ionic liquid-grafted PA6 as a compatibilizer has been demonstrated to enhance the interfacial adhesion in polymer blends, hence, improving their overall properties (Zheng et al., 2022). The incorporation of a third component into a polymer blend, with the aim of inducing a chemical interaction at the interface, is a state-of-the-art method for achieving compatibilization (El Sayed, 2007). This approach allows for the enhancement of interfacial adhesion and the stabilization of the blend morphology. Additionally, the use of reactive copolymers has been shown to improve the stability of polymer blends at the interface, thereby leading to enhanced compatibility and improved properties (Dong et al., 2015). Compatibilization strategies are essential for optimizing the properties of polymer blends by improving interfacial adhesion, stabilizing morphology, and enhancing overall product performance. Various methods, such as the use of nanomaterials, reactive blending, and the incorporation of a third component, offer effective routes to achieve compatibilization and create high-performance polymer blends.

5.3 High-temperature polymer blends

5.3.1 Selection of high-temperature polymers for blending

The selection of high-temperature polymers for blending purposes involves careful consideration of several factors to achieve the desired material properties and performance characteristics. High-temperature polymers are essential in industries such as aerospace, automotive, electronics, and manufacturing, where components are subjected to elevated temperatures and harsh operating conditions. Some of the factors to consider during the selection process are as follows:

1. **Temperature resistance**: The primary consideration is the polymer's ability to withstand high temperatures without undergoing significant thermal degradation or loss of mechanical properties (Ray and Cooney, 2018). Polymers, such as PEEK, PIs, PPS, and polyetherimide, are known for their excellent thermal stability, and they can withstand continuous operating temperatures, ranging from 150°C to over 300°C.
2. **Chemical compatibility**: The polymer blend should be chemically compatible with the environment in which it will be used (Paul and Barlow, 1980). This includes exposure to fuels, lubricants, solvents, acids, and other chemicals. Compatibility ensures that the material retains its integrity and performance over time.
3. **Mechanical properties**: Polymers operating in high-temperature environments should possess adequate mechanical properties, including strength, stiffness, toughness, and dimensional stability, even at elevated temperatures. Blending polymers with complementary mechanical properties can enhance the overall performance, such as an improvement in impact resistance or a reduction in creep deformation (Das et al., 2020).
4. **Processing conditions**: Consideration should be given to the conditions required for polymer blending, such as melt temperature, melt viscosity, and compatibility with processing methods (e.g., injection molding, extrusion, or compression molding). Compatibility between polymers is crucial to ensure uniform blending and homogeneity of the final material.
5. **Cost and availability**: The cost of raw materials and the availability of high-temperature polymers can influence the selection process. While certain polymers may offer superior performance, they may also come at a high cost. The evaluation of the trade-offs between performance and cost is essential for selecting the most suitable polymer blend for a specific application (Hasegawa et al., 2016).
6. **Additive compatibility**: If additives, such as fillers, reinforcements, or flame retardants, are required to enhance certain properties of the polymer blend, their compatibility with the base polymers should be assessed. Additives should not adversely affect the polymer's thermal stability or mechanical properties (Pfaendner, 2006).
7. **End-use requirements**: Understanding the specific requirements of the end-use application is critical in selecting high-temperature polymers for blending. Factors, such as regulatory compliance, dimensional tolerances, electrical conductivity, flame resistance, and wear resistance, should be considered to ensure the final material meets the performance standards and customer expectations.

When selecting high-temperature polymers for blending, it is crucial to consider polymer compatibility, processing conditions, and the envisaged properties of the final blend. Overall, selecting high-temperature polymers for blending involves a comprehensive evaluation of their thermal stability, chemical compatibility, mechanical properties, processing conditions, cost, additive compatibility, and end-use requirements to achieve optimal performance in often demanding applications.

5.3.2 Miscibility and immiscibility considerations

Miscibility in polymer blends refers to the ability of two or more polymers to mix at the molecular level, forming a homogeneous mixture with uniform properties throughout the material. Immiscibility occurs when polymers do not mix homogeneously, thereby resulting in a phase-separated blend, with distinct regions of each constituent polymer, often leading to varied properties in different parts of the material.

Polymer blends play a crucial role in material science, where the miscibility or immiscibility of the components significantly impacts the final properties of the resulting blend. Miscible polymer blends, characterized by a single phase, exhibit properties that are a combination of the individual components, while immiscible blends show phase separation, leading to distinct domains within the materials in the blend (Ali and Eisa, 2014). The degree of miscibility directly influences the homogeneity and thermodynamic behavior of the blend, with reactive blending being a method employed to enhance miscibility by heating and mixing the polymers at temperatures that are well above their individual melting points (Poulopoulou et al., 2019). The phase behavior in copolymer blends has been extensively studied, with the Mean Field theories being developed to understand the miscibility in polymer systems, especially in copolymer blends (Brinke et al., 1983). The miscibility of polymer blends is not solely determined by the glass-transition temperatures of the components. Even thermodynamically miscible blends can exhibit dynamic heterogeneity, thereby showcasing multiple glass-transition temperatures (Poulopoulou et al., 2018). The miscibility of polymers is a critical factor that governs the applicability and performance of the resulting blends. In cases where immiscible polymer blends are formed, the addition of suitable compatibilizers can render them compatible, thereby expanding their potential applications. The understanding of the thermodynamics of associative polymer blends is essential for the quantification of the conditions that lead to miscibility, partial miscibility, or immiscibility in blends (Prusty et al., 2018). The effect of miscibility on the elongational viscosity of polymer blends has been investigated, highlighting the differences in the behavior between homogeneous, miscible, and immiscible states. For instance, the addition of ultrahigh-molecular-weight polymer in blends can shed light on the impact of miscibility on elongational viscosity (Takahashi et al., 1999). In ternary polymer blends, the phase behavior and hydrogen bonding interactions can lead to unique phenomena, e.g., closed immiscibility loops in phase diagrams, demonstrating the complex nature of miscibility in polymer systems (Kuo et al., 2002). By tailoring the components in polyamide blends, the properties under various conditions can be improved, showcasing the versatility of blending miscible and immiscible components to achieve the desired material characteristics (Hirai et al., 2020). The miscibility in polymer blends that involve high-temperature polymers, for example, polystyrene and polyether, has been extensively researched, hence emphasizing the importance of comprehending the phase behavior and miscibility in such systems (Chang and Woo, 2003). The determination of miscibility in polymer blends can vary, depending on the method used and the temperature range of the study, necessitating the need for comprehensive investigations to assess, accurately, the blend compatibility (Kumaraswamy and Ranganathaiah, 2006). The interaction energy in polymer blends, as described by the Flory–Huggins theory, can be inferred from the extent of the melting point depression, thereby providing insights into the miscibility of the components (Papageorgiou and Bikiaris, 2006). Dipole-dipole interactions have been identified as key factors that influence the miscibility phenomenon in polyester/chlorinated polymer blends, hence underscoring the role of specific interactions in blend compatibility (Prud'homme, 1982). The thermal and rheological properties of polymer blends, such as mLLDPE/LDPE blends, can be indicative of their miscibility, with techniques like Cole–Cole plots being utilized to determine immiscibility in blends (Fang et al., 2005). The shape and size asymmetry of monomers can lead to new patterns of miscibility in polymer blends, showcasing the impact of molecular structure on blend compatibility (Dudowicz et al., 2002). Factors, for example, the conductivity of polymer blends can be influenced by their miscibility, hence highlighting the importance of the blend compatibility in functional applications (Zakaria et al., 2012). The effects of the blend ratio and miscibility on elongational viscosity have been studied in polymer blends, emphasizing the role of composition in determining the rheological behavior of the material (Takahashi et al., 1999). In homologous semicrystalline aryl polyesters, slight variations in chemical structures can lead to phase separation or immiscibility in the blend systems, illustrating the delicate balance required for achieving blend compatibility (Kuo and Woo, 2003). The rheological properties of blends can deviate from additivity predictions, based on their miscibility; with immiscible blends showing distinct phase-separated regions with unique viscosities (Kim et al., 1993). The miscibility of polymers in bicomponent blends is a critical parameter for successful spinning processes, underscoring the importance of blend compatibility in fiber formation (Kim et al., 2009). The rheological properties and miscibility behavior of biodegradable aliphatic polyester and poly(vinyl acetate) blends have been investigated, revealing obvious immiscibility and distinct glass transition temperatures in the blend system (Shin et al., 2000). Reactive compatibilization can be employed to enhance the compatibility of immiscible blends, leading to the formation of crosslinks and improved properties in the resulting materials (Ramesh and De, 1993). Ternary miscibility in blends involving three polymers, with balanced interactions, demonstrates the complexity of blend systems and the interplay of multiple components in determining miscibility (Woo et al., 2003).

5.3.3 Processing techniques for high-temperature polymer blends

The processing techniques for high-temperature polymer blends are crucial for achieving desired material properties and performance. Various studies have explored innovative methods to process polymer blends at elevated temperatures to enhance their characteristics. For instance, Gumyusenge et al. (2018) introduced a strategy to develop thermally stable high-temperature semiconducting polymer blends by combining semicrystalline conjugated polymers with high glass-transition temperature insulating matrices. This approach was aimed at maintaining a stable charge transport in polymer blends at elevated temperatures, thereby expanding their potential for applications that require reasonable thermal stability.

In the study of polymer blends with multiple hydrogen-bonded end groups, researchers have examined the effectiveness of these interactions in the melt. They cast chain-end-functionalized linear polymers into films, annealed them at temperatures above the glass-transition temperature, and analyzed their structures by using techniques such as transmission electron microscopy and differential scanning calorimetry (DSC) (Feldman et al., 2008).

This study shed light on the processing techniques that can influence the structure and properties of polymer blends with specific end-group interactions. Moreover, the processing of ternary photovoltaic blends, as explored by Mulherin et al. (2011), incorporated all-conjugated donor–acceptor diblock copolymers, under conditions that would typically lead to phase separation in regular blends, such as high-temperature annealing to enhance polymer crystallinity and charge transport (Mulherin et al., 2011). This research highlighted the importance of specialized processing techniques in achieving the desired blend morphologies and functionalities, especially in complex ternary systems. Zhao et al. (2017) delved into the melt-processing of complementary semiconducting polymer blends for high-performance organic transistors. The study demonstrated the potential to achieve exceptional charge-carrier mobility and on/off current ratios through melt-processing, setting new records for organic field-effect transistors. This study emphasized the significance of processing conditions in optimizing the performance of polymer blends for advanced electronic applications. Furthermore, the effects of moisture content and heat treatment on the physical properties of starch and poly(lactic acid) blends were investigated by Ke and Sun (2001), who prepared the blends by using a laboratory-scale twin-screw extruder and subsequently fabricated some tensile bars for mechanical testing through compression and injection molding. This research emphasized the influence of processing parameters on the properties of polymer blends, underscoring the importance of precise processing techniques for achieving the desired material characteristics. In the context of exploring next-generation engineering bioplastics, reactive blending was highlighted as a method that enhances homogeneity and thermodynamic miscibility in polymer blends by heating and mixing the polymers at temperatures well above their melting points (Poulopoulou et al., 2019). This approach underscores the importance of tailored processing techniques for optimizing blend compatibility and performance in advanced polymer systems.

5.3.4 Characterization methods for high-temperature polymer blends

Characterization methods for high-temperature polymer blends are crucial for the understanding of their properties and behavior. Various techniques have been utilized to analyze several aspects, such as composition, miscibility, crystallinity, and other key features of these blends. For example, X-ray diffraction (XRD), polarized optical microscopy (POM), Fourier transform infrared spectroscopy (FTIR), and electrical impedance spectroscopy have been employed to characterize polymer blend films, thereby providing insights into their structure and electrical properties (Ahmed and Abdullah, 2019). Dynamic thermomechanical analysis has been recognized as a valuable method for studying the thermal and flow characteristics of blends of polymers with similar glass transition temperatures, thereby offering a deep understanding of their miscibility and dynamic behavior (Bikiaris et al., 2004). In addition, techniques, such as DSC, thermogravimetric analysis (TGA), and ultraviolet-visible (UV–vis) spectroscopy, have been used to optimize the glass-transition temperatures of polymer blend films and to investigate the interactions between parental polymers, thermal stability, and optical properties (Yazie et al., 2023). Impedance spectroscopy has emerged as a novel approach to probe the phase transitions and the microstructures in polymer blends, with UV–vis spectrophotometry being employed to confirm specific characteristics in blended samples (Aziz et al., 2018). Moreover, techniques, such as XRD, FTIR, DSC, and impedance analyses, have been instrumental in characterizing the lithium-ion conducting solid polymer blend electrolytes, thereby shedding light on their structure and conductivity (Rajeswari et al., 2013). The miscibility and crystallization behavior of polymer blends have been investigated by using DSC, wide-angle XRD, and polarizing light microscopy (PLM), hence providing valuable insights into the blend compatibility and its phase behavior (Papageorgiou and Bikiaris, 2006). Furthermore, the dilute solution viscometry technique, based on the Flory–Huggins lattice theory, has been utilized to elucidate the miscibility characteristics of polymer blends, thereby offering a quantitative approach to understanding blend interactions (Siddamallappa et al., 2018).

In summary, a combination of advanced analytical techniques, such as XRD, FTIR, DSC, TGA, UV–vis spectroscopy, impedance spectroscopy, and microscopy methods, such as POM and PLM, has been instrumental in characterizing high-temperature polymer blends. These methods/techniques provide valuable information on the structure, properties, and behavior of polymer blends, enabling researchers to optimize their composition and performance for various applications.

5.4 Influence of crosslinking on composite properties

5.4.1 Types of crosslinking reactions and mechanisms

Crosslinking reactions are essential for the modification of the properties of polymer blends. Such reactions often impact positively the mechanical strength, thermal stability, and the overall performance of the blends. Various mechanisms and types of crosslinking reactions have been investigated to comprehend their influence on the structure and properties of polymer blends. One common type of crosslinking reaction involves amine-catalyzed aldol condensation/dehydration, as demonstrated in the study by Boyars et al. (2007). This mechanism results in an increase in the crosslink density of polymer blends, thereby improving their mechanical properties. Another study by Hu et al. (2017), showcased the self-catalysis of trans-esterification to produce partially crosslinked thermoplastic polymer blends, thereby presenting an alternative mechanism for achieving desirable crosslinking feat through the interaction of linear polymers with specific functional groups. In a specific polymer system context, Manoj et al. (1993), explored the crosslinking reaction between poly(vinyl chloride) and acrylonitrile-butadiene rubber, emphasizing the role of allylic chlorine sites and the nitrile groups in the crosslinking process. This study elucidated the chemical interactions that led to crosslink formation in polymer blends. Reactive compatibilization has been studied as a method to enhance the interfacial adhesion and the morphology of polymer-polymer interfaces through crosslinking reactions, as discussed by Jiang et al. (2010). By generating block or graft copolymers at the interface, crosslinking reactions reduce interfacial tension and stabilize the blend's morphology, thereby enhancing its overall properties. Radiation-based crosslinking techniques were employed by Lee et al. (2021) to boost the thermal and mechanical properties of polymer blends. By inducing the crosslinking reactions between the polymer chains by using the gamma-ray radiation, the interfacial adhesion and the properties of the blends were significantly enhanced. Moreover, the formation of the chemical crosslinks, such as the β-sheet crosslinks and epoxy polymer network crosslinks, has been investigated in silk fibroin-epoxy resin blend films (Yang et al., 2019). These crosslinks played a crucial role in stabilizing the silk polymer structure and enhancing the mechanical performance of the blend.

The scheme of crosslinking classifications is shown in Fig. 5–2, and the schematics of the various mechanisms of crosslinking, as presented in Fig. 5–3 are adapted from Grabska-Zielińska and Sionkowska (2021). The diverse mechanisms and types of crosslinking reactions applicable to polymer blends underscore the importance of understanding these processes to

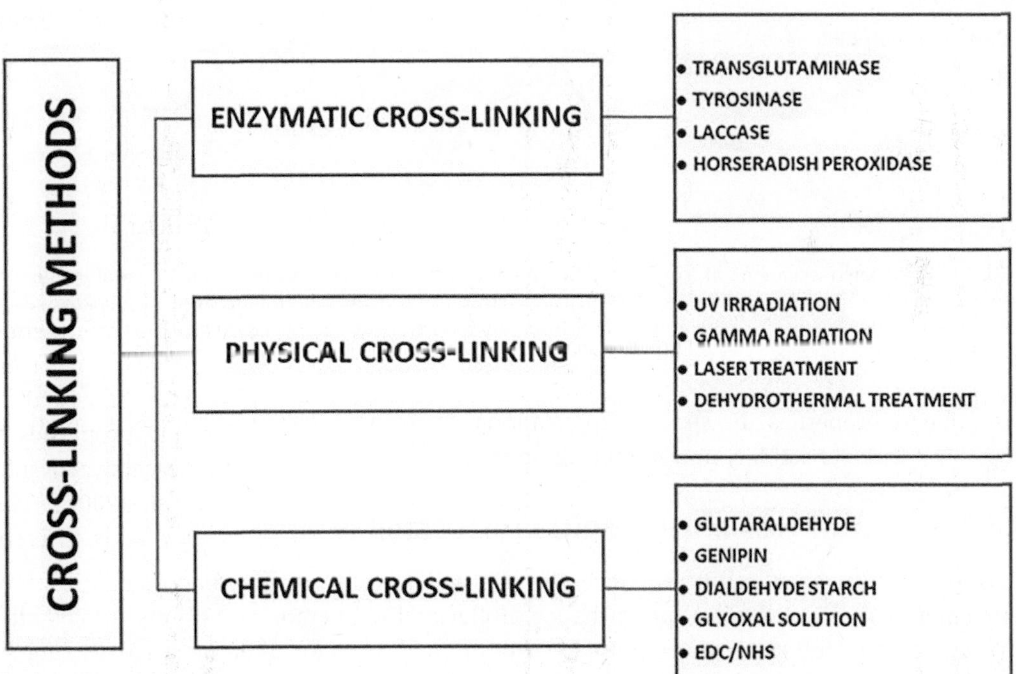

FIGURE 5–2 Crosslinking classifications. *From Grabska-Zielińska, S., Sionkowska, A., 2021. How to improve physico-chemical properties of silk fibroin materials for biomedical applications?—blending and cross-linking of silk fibroin—a review. Materials, 14(6). https://doi.org/10.3390/ma14061510.*

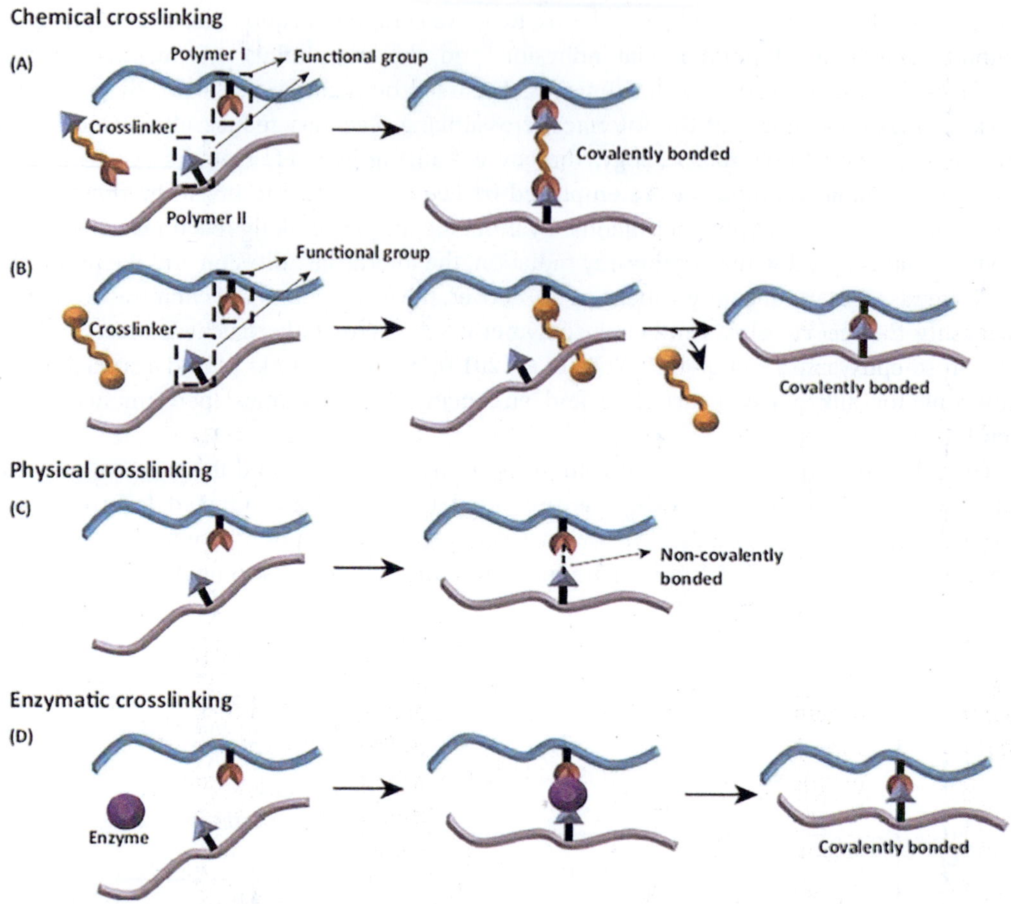

FIGURE 5–3 Various mechanisms of crosslinking. (A) Chemical crosslinking with polymer I; (B) Chemical crosslinking with polymer II; (C) Physical crosslinking; (D) Enzymatic crosslinking. *From Grabska-Zielińska, S., Sionkowska, A., 2021. How to improve physico-chemical properties of silk fibroin materials for biomedical applications?—mlending and cross-linking of silk fibroin—a review. Materials, 14(6). https://doi.org/10.3390/ma14061510.*

tailor blend properties for specific applications, from enhancing mechanical strength to improving thermal stability and interfacial adhesion.

5.4.2 Impact of crosslinking on thermal stability

Crosslinking can significantly enhance the thermal stability of polymer blends, thereby impacting their properties and performance at elevated temperatures. Various studies have investigated the influence of crosslinking on thermal stability, thereby underscoring its importance in enhancing the properties of polymeric materials. In their study, they (Kim et al., 2009) reported on photo-crosslinkable polythiophenes. They found that introducing crosslinking via photo-crosslinkable materials resulted in excellent thermal stability in organic photovoltaic devices. The materials exhibited high power conversion efficiencies and maintained reasonable stability, even at elevated temperatures. Similarly, reported that the incorporation of the basic oxide of boron, that is, diboron trioxide B_2O_3, into polylactic acid/thermoplastic polyethylene glycol shape-memory composites improved the thermal properties by crosslinking the matrix, thereby enhancing the thermal stability of the polymer blend. Furthermore, their research (Rashid et al., 2024) highlighted the fact that an increase in the crosslinked network influenced the thermal stability of polymers by reducing the rate of volatile product evaporation generated during heating, hence emphasizing the role of crosslinking in enhancing thermal properties. Laskoski et al. (2017) reported that by increasing the crosslinking density of thermosetting polymers, they directly enhanced the thermal stability and mechanical properties of the polymers, thereby illustrating the positive impact of crosslinking on material performance. In addition, Kwon et al. (2021) reported that crosslinking reactions improved the stability of all-polymer solar cells, emphasizing the effectiveness of crosslinking in enhancing the stability and functions of polymer films at high temperatures. Lee et al. (2021) also found that the radiation-based crosslinking techniques significantly enhanced the thermal stability of HDPE–PU–EVA blends, therefore, exhibiting a substantial improvement in the material's thermal stability when compared to the non-crosslinked blends. In summary, the studies so far discussed underscore the crucial role of crosslinking in enhancing the thermal stability of polymer blends. Crosslinking reactions contribute to improved properties and performance, particularly at elevated temperatures.

5.4.3 Effect of crosslinking on mechanical properties

Crosslinking significantly influences the mechanical properties of polymer blends, and it affects parameters, such as tensile strength, elongation-at-break, toughness, and stiffness. Various studies have investigated the impact of crosslinking on these properties, hence shedding light on the mechanisms and outcomes of this process. Jin et al. (2019) discussed the development of covalent adaptable network (CAN) polymers, which merge the mechanical properties of thermosets with the reprocessability of thermoplastics. This innovative approach enhances the mechanical properties of polymers while allowing for reprocessing and demonstrating the potential of crosslinking in enhancing material performance. Fulati et al. (2022) studied the effects of crystallinity and crosslinking density on the shape recovery force in shape-memory polymer blends. By adjusting the crystallinity gradient and the crosslinking density, the researchers achieved polymer blends with customizable mechanical properties, including stretchability and toughness, while maintaining excellent shape-memory characteristics of the materials. Martinez et al. (2014) explored how the different crosslinking strategies affect the mechanical properties, drug release, and cytocompatibility of protein polymers. They found that distinct crosslinking methods can adjust the stiffness, strain-at-failure, and ultimate tensile strength of protein-based polymer blends, underscoring the significant influence of crosslinking on mechanical performance. In addition, Lee et al. (2021) demonstrated that radiation-based crosslinking techniques can enhance the thermal and mechanical properties of polymer blends. By implementing a crosslinking procedure in polyolefin-based polymers, the researchers observed a notable enhancement in both thermal stability and mechanical properties, therefore, highlighting the effectiveness of crosslinking in improving material performance. These studies underscore the crucial role of crosslinking in enhancing the mechanical properties of polymer blends. By manipulating crosslink density, crystallinity, and specific crosslinking strategies, researchers can tailor the mechanical properties of polymer blends to meet specific application requirements.

5.4.4 Influence of crosslinking on chemical resistance

Crosslinking is a fundamental process that significantly enhances the chemical resistance of polymer materials, thereby impacting their durability and performance across various applications. Numerous studies have delved into the effects of crosslinking on chemical resistance, underscoring its pivotal role in bolstering the resistance of polymer blends against degradation and environmental factors. For example, the report of Sung et al. (2003) illustrated the fact that the chemical crosslinking of biological tissues by using genipin and/or carbodiimide substantially heightened their resistance to enzymatic degradation and diminished antigenicity, hence underscoring the efficacy of crosslinking in augmenting chemical resistance. Similarly, McGann et al. (2015) observed that the use of genipin in the crosslinking of cartilage amplified its resistance to both chemical degradation and mechanical wear, thereby emphasizing the significance of crosslinking in enhancing the chemical stability of the resulting product. Furthermore, Baker et al. (1999) scrutinized the fatigue resistance of medical-grade ultrahigh-molecular-weight polyethylene, crosslinked through chemical and radiation methods, hence elucidating the role of crosslinking in fortifying the material's resistance to wear and degradation. Ren et al. (2021) explored polyphenol-assisted chemical crosslinking as a means to achieve a highly crosslinked ultrahigh-molecular-weight polyethylene for total joint replacement, thereby accentuating the improved wear resistance and the longevity of the product, as conferred by crosslinking. In addition, Zadhoush et al. (2009) expounded on the efficacy of crosslinking in ameliorating the properties of plasticized poly(vinyl chloride), encompassing the enhancements in mechanical, chemical, and thermal resistance. Yoo et al. (2009) observed that by augmenting the content of interparticle crosslinking agents in acrylic copolymers, they bolstered the tensile strength, water resistance, and chemical resistance of the film, thereby emphasizing the favorable impact of crosslinking on the chemical stability of the product. Summarily, the amalgamated findings from these studies unequivocally demonstrate the indispensable role of crosslinking in enhancing the chemical resistance of polymeric materials, thereby contributing to their longevity, durability, and performance across diverse applications.

5.5 Influence of immiscibility on composite properties

5.5.1 Role of interfacial interactions in immiscible blends

Interfacial interactions are fundamental in immiscible polymer blends, since they significantly impact the properties and performance of the blends. Various studies have investigated the influence of interfacial interactions on immiscible blends, hence underscoring the importance of compatibilization and interfacial reinforcement. Wang et al. (2016) aptly illustrated the fact that the creation of the interfacial Janus nanomicelles effectively enhanced the miscibility of polymer blends, thereby boosting the performance of immiscible blends. This research highlights the pivotal role of interfacial interactions in the advancement of blend compatibility and the overall properties of polymer blends. López-Barrón and Tsou (2017) explored the strain hardening of polyethylene/polypropylene blends through the interfacial reinforcement with poly(ethylene-*cb*-propylene) comb block copolymers. The study emphasized that interfacial compatibilization not only diminished the interfacial tension and stabilized the morphology but also altered the viscoelastic properties of immiscible blends. Sinha Ray and Bousmina (2005) demonstrated the impact of organic modification on the efficiency of compatibilization of clay in immiscible polymer blends, thereby showcasing the significance of organoclay interfacial activity in enhancing the properties of polystyrene/polypropylene blends. This study underscored the importance of interfacial interactions in ameliorating the blend properties. Mao et al. (2017) discussed the selective localization of fillers in immiscible polymer blends, thereby emphasizing that interfacial interactions between fillers and the blend components dictated the fillers' dispersion. This study highlights the critical role of interfacial interactions that govern the filler localization and blend properties.

The reviewed studies collectively stressed the crucial role of interfacial interactions in immiscible polymer blends. Compatibilization, interfacial reinforcement, and selective filler localization are key strategies that harness interfacial interactions that are desired to enhance the properties and performance of immiscible polymer blends.

5.5.2 Effect of phase morphology on mechanical properties

The phase morphology in polymer blends is a critical factor in determining their mechanical properties, such as tensile strength, ductility, toughness, and stiffness. Various studies have researched this relationship, shedding light on how phase structure impacts material performance. Xu et al. (2006) examined the influence of Mg-Zn-Y phase characteristics on the mechanical properties of Mg-Zn-Y-Zr alloys, emphasizing the significant role of phase morphology in determining the material behavior. Li et al. (2009) studied the impact of using recycled polyethylene terephthalate glycol (PETG) plastic on the tensile strength and ductility of acrylonitrile butadiene styrene, highlighting the effect of the processing temperature on the mechanical properties and phase morphology. The research concluded that changes in phase morphology, due to polyethylene terephthalate (PET) degradation, could lead to a substantial loss in mechanical properties. Long et al. (2015) discussed the impact of phase morphology and interface properties on the mechanical properties of polyarylene ether nitrile blends, underscoring the importance of phase structure in the determination of the material behavior. Rohde et al. (2016) investigated thermoset blends of an epoxy resin and polydicyclopentadiene, finding that the mechanical properties of these phase-separated blends aligned reasonably well with the predictions by the rule of mixtures for blend tensile strength, modulus, and fracture toughness. This study highlighted the significant influence of phase morphology on the mechanical properties of polymer blends.

The collective findings of these studies underscore the crucial role of phase morphology in shaping the mechanical properties of polymer blends. The understanding and management of phase structure are vital for the customization of the mechanical behavior of polymer materials for specific applications.

5.5.3 Influence of immiscibility on thermal and electrical properties

The immiscibility of polymer blends significantly influences their thermal and electrical properties, hence impacting their behavior and performance. Several studies have explored the relationship between immiscibility and these properties, shedding light on the effects of phase separation and interfacial interactions. Dell'Erba et al. (2001) investigated the thermal properties and phase morphology of poly(L-lactic acid (PLLA)/polycaprolactone immiscible blends, highlighting the immiscibility of the components. The phase morphology resulting from immiscibility can affect the thermal conductivity and heat transfer properties in such blends. Gholami et al. (2022) discussed how microphase separation arising from thermodynamic immiscibility affects the microstructure and characteristics of polyurethanes, thereby influencing their thermal and electrical properties. The immiscibility between the hard and soft segments in polyurethanes can lead to variations in their thermal and electrical behaviors. Fina et al. (2012) studied the morphology and the conduction properties of graphite-filled immiscible polyvinylidene fluoride / poly(propylene-graft-maleic anhydride) (PPgMA) blends, with the aim of improving the electrical and thermal conductive properties through a double-percolation structure. The phase morphology in immiscible blends plays a crucial role in the determination of the electrical and thermal conductivities of the material. Salehiyan and Ray (2019) discussed the impact of immiscibility on the conductivity of nanocomposites, emphasizing the importance of understanding the fundamental interactions in immiscible polymer blends. Immiscible blends can exhibit poor interfacial adhesion, which can affect electrical properties.

In summary, the immiscibility of polymer blends can have a significant impact on their thermal and electrical properties. An understanding of the phase morphology and the interfacial interactions in immiscible blends is crucial for the tailoring of these properties to meet specific application requirements.

5.5.4 Tailoring immiscibility for specific applications

To optimize immiscibility for specific applications, researchers can employ various strategies and techniques to manipulate the phase morphology and interfacial interactions in polymer blends. By selecting suitable compatibilizers, adjusting the processing conditions, and optimizing the blend compositions, immiscible polymer blends can be tailored to exhibit the desired properties for specific applications. One practical approach involves the use of reactive blending, as discussed by Dong et al. (2015), where in situ formed Janus nanomicelles acted as compatibilizers to enhance the interfacial interactions in immiscible polymer blends.

This method facilitates the creation of well-dispersed phases and improved blend properties, making it suitable for applications that require enhanced compatibility and performance. Another strategy is the incorporation of block or graft copolymers at the polymer–polymer interfaces, as explored by Wang et al. (2015), which can serve as an effective compatibilization method for immiscible polymer blends. These copolymers promote adhesion between phases and control the domain size, thereby optimizing the mechanical, thermal, and electrical properties for applications, where blend compatibility is crucial. Moreover, the utilization of immiscible polymer blends with selective filler localization, as investigated by Marischal et al. (2019), can be advantageous for applications that require specific material properties. By controlling the filler content and localization in cocontinuous polymer phases, the overall properties of the blend can be tailored to meet the requirements of the heating applications. In the realm of smart textiles, Eutionnat-Diffo et al. (2020) demonstrated the development of flexible and conductive immiscible thermoplastic/elastomer monofilaments for smart textiles applications by using the 3D printing technique. This approach enabled the customization of the mechanical and electrical properties, thereby making it suitable for applications where flexibility, stress, strain, and electrical conductivity are essential. In conclusion, through the strategic selection of blending techniques, compatibilizers, and processing conditions, researchers can customize immiscible polymer blends to fulfill specific application needs, spanning from packaging materials to smart textiles, membranes, and conductive materials. Some examples of successful high-temperature polymer blend composites are presented in Table 5–1.

5.6 Conclusion, challenges, and future perspectives

5.6.1 Technical challenges in developing high-temperature polymer blend composites

The development of high-temperature polymer blend composites poses several technical challenges that researchers are currently addressing to optimize their properties for specific applications. Some of the key challenges in this area include the following:

1. Preservation of intermolecular interactions: The maintenance of close intermolecular interactions and packing motifs at elevated temperatures is crucial, especially for semiconducting polymers, to ensure the desired properties are retained (Gumyusenge et al., 2018).
2. Thermodynamic immiscibility: Immiscibility between polymer components can pose challenges in achieving certain desired properties, such as thermal and electrical conductivity, in polymer blends. Strategies to address phase separation and immiscibility are essential for optimizing blend properties (Liu et al., 2024).
3. Interfacial interactions: The achievement of effective interfacial interactions between immiscible polymer phases is critical for the enhancement of the mechanical and electrical properties of polymer blend composites. Strategies, such as compatibilization

Table 5–1 Case studies of high-temperature polymer blend composites.

Blends	Properties	Application	References
PEEK/PPS	PEEK and PPS blends offer excellent thermal stability, mechanical strength, and chemical resistance	Used in demanding applications, such as aerospace components, automotive parts, and electrical insulation	Cao et al. (2016), Zuo et al. (2019)
PEI/PAEK	PEI and PAEK blends combine the high-temperature performance of PAEK with the processability of PEI	Suitable for applications that require high strength, stiffness, and dimensional stability at elevated temperatures	Korycki et al. (2023)
PEEK/PTFE	PEEK/PTFE blends combine the excellent tribological properties of PTFE with the high-temperature performance of PEEK	Used in bearings, seals, and other wear-resistant components operating at high temperatures	Nunez et al. (2011), Zhu et al. (2015)
PI/PPS	PI and PPS blends offer high thermal stability, good mechanical properties, and excellent electrical insulation	Used in structural components, automotive parts, and electronic devices requiring high dimensional stability	Tcherdyntsev et al. (2018)
PEEK/LCP	PEEK/LCP blends combine the high strength and stiffness of LCP with the processability of PEEK	Used in structural components, automotive parts, and electronic devices requiring high dimensional stability	Al Christopher et al. (2021)

LCP, Liquid crystal polymer; *PAEK*, polyaryletherketone; *PEEK*, polyetheretherketone; *PEI*, polyetherimide; *PI*, polyimide; *PPS*, polyphenylene sulfide; *PTFE*, polytetrafluoroethylene.

and interface engineering, are essential for improving blend performance (Liu et al., 2024).

4. Nanoparticle dispersion: Ensuring uniform dispersion, distribution, and localization of nanoparticles in immiscible binary blends can be challenging, but it is crucial for the determination of the mechanical and electrical performances of the composites.
5. Phase morphology control: The manipulation of the phase morphology in immiscible polymer blends to achieve the desired properties, such as self-healing behavior or thermal conductivity, presents a technical challenge that requires precise control over the blend composition and processing conditions (Gebrekrstos and Ray, 2023).
6. Temperature sensing materials: The development of temperature-sensitive materials by using polymer blends requires a deep understanding of the composition–temperature relationship and the electrical conductivity behavior of the blends. Combinatorial studies and inkjet printing techniques can aid in the evaluation and optimization of these properties (Sabzi et al., 2019).
7. Multicomponent blends: The design of multicomponent polymer blends with tailored properties for specific applications involves certain challenges that are related to material compatibility, phase behavior, and processing conditions. High-throughput experimentation methods can accelerate the discovery of high-performance materials (Bolar et al., 2024).
8. Compatibilization: The incorporation of compatibilizers in polymer blends to improve mechanical, thermal, and thermomechanical behavior can be challenging due to the need

for precise control over the blend composition and the addition of compatibilizing agents (Mahmud et al., 2024).

9. Miscibility control: The achievement of an adequate and desirable miscibility control in polymer blends for enhanced properties, such as transparency, conductivity, and mechanical strength, requires careful selection of blend components and processing conditions (Hirai et al., 2023; Ismail et al., 2022).
10. Reactive blending: The utilization of reactive blending techniques to enhance the properties of polymer blends can be challenging due to the need for effective compatibilization and control over the reaction kineticsto achieve the desired properties (Li et al., 2022; Roy et al., 2023).

Addressing these technical challenges through innovative approaches, advanced characterization techniques, and precise control over the blend composition and processing parameters can lead to the development of high-temperature polymer blend composites with tailored properties for a wide range of applications.

5.6.2 Economic and environmental considerations

Economic and environmental considerations associated with high-temperature polymer blend composites are crucial factors to consider in the development and application of these materials. Some of the key economic and environmental considerations include the following:

1. **Cost-effectiveness**: The economic viability of producing high-temperature polymer blend composites is essential for commercial applications. Other utilization of cost-effective raw materials, efficient processing techniques, and scalable manufacturing processes can help to reduce production costs and enhance the economic feasibility of these materials (Adamu et al., 2023).
2. **Sustainability**: The consideration of the environmental impact of polymer blend composites is crucial for sustainable material development. Using biobased or recycled polymers, reducing energy consumption during processing, and minimizing waste generation can contribute to the environmental sustainability of high-temperature polymer blends (Lodge and McLeish, 2000).
3. **Energy efficiency**: The optimization of the processing conditions and energy consumption during the production of polymer blend composites can lead to energy-efficient manufacturing processes. Utilizing energy-efficient equipment and technologies can help reduce the environmental footprint of these materials (Chen et al., 2018).
4. **Recyclability**: Designing polymer blend composites with recyclability in mind can enhance their environmental sustainability. Developing materials that are easily recyclable or biodegradable at the end of their lifecycle can reduce waste generation and promote a circular economy approach (Kurniawan et al., 2024; Velvizhi et al., 2020).
5. **Life cycle assessment (LCA)**: Conducting an LCA of high-temperature polymer blend composites can provide valuable insights into their environmental impact throughout their

entire life cycle. Assessing factors such as raw material sourcing, production, use, and end-of-life disposal can help identify opportunities for improving sustainability (Sikorska et al., 2021).

6. **Material efficiency**: The maximization of material efficiency, by reducing material waste, optimizing material usage, and minimizing excess production, can contribute to cost savings and environmental benefits. The design of polymer blend composites with high material efficiency can enhance their economic and environmental performance (Lloyd and Lave, 2003; Hirai et al., 2022).
7. **Regulatory compliance**: Ensuring compliance with environmental regulations and standards is essential for the development and use of high-temperature polymer blend composites. Adhering to regulations on material sourcing, production processes, emissions, and waste management is critical for sustainable and responsible material development (Yu et al., 2006).

By addressing these economic and environmental considerations in the development and application of high-temperature polymer blend composites, researchers and industry professionals can formulate and manufacture materials that are cost-effective, sustainable, energy-efficient, and compliant with regulatory requirements.

5.6.3 Emerging trends and future research directions

Emerging trends and future research directions for high-temperature polymer blend composites encompass a wide range of topics and advancements in materials science. Some potential areas of focus for future research in this field include the following:

Advanced compatibilization strategies: Development of innovative compatibilization strategies to enhance the interfacial interactions and phase morphology in high-temperature polymer blend composites. This could involve the use of novel compatibilizers, reactive blending techniques, and surface modification approaches to improve blend properties (Al-Amiery et al., 2024; Podara et al., 2024).
Tailored phase morphologies: Investigation into the impact of phase morphology control on the mechanical and thermal properties of high-temperature polymer blend composites (He et al., 2024).

AI disclosure

During the preparation of this work, the author(s) used Scite Assistant to gather relevant literature on the subject on my research. After using this tool/service, the author(s) reviewed and edited the content as needed and take(s) full responsibility for the content of the publication.

References

Adamu, Y., Bello, T.K., Shehu, U., Bello, A., Adamu, S., Isa, M.T., 2023. Mechanical, Physical and Thermal Properties of Waste Polypropylene/Polystyrene/Natural Rubber (wPP/PS/NR) Blends. Research Square, Nigeria Research Square. Available at https://doi.org/10.21203/rs.3.rs-2966999/v1.

Adeniyi, A., Agboola, O., Sadiku, E.R., Durowoju, M.O., Olubambi, P.A., Babul Reddy, A., et al., 2016. Thermoplastic-thermoset nanostructured polymer blends. Design and Applications of Nanostructured Polymer Blends and Nanocomposite Systems. Elsevier Inc., South Africa, pp. 15–38. Available at https://doi.org/10.1016/B978-0-323-39408-6.00002-9.

Ahmed, H.T., Abdullah, O.G., 2019. Preparation and composition optimization of PEO: MC polymer blend films to enhance electrical conductivity. Polymers 11 (5). Available at https://doi.org/10.3390/polym11050853.

Al-Amiery, A.A., Fayad, M.A., Abdul Wahhab, H.A., Al-Azzawi, W.K., Mohammed, J.K., Majdi, H.S., 2024. Interfacial engineering for advanced functional materials: surfaces, interfaces, and applications. Results in Engineering 22. Available at https://doi.org/10.1016/j.rineng.2024.102125.

Ali, Z.I., Eisa, W.H., 2014. Characterization of electron beam irradiated poly vinyl alcohol/poly ethylene glycol blends. Journal of Scientific Research 6 (1), 29–42. Available at https://doi.org/10.3329/jsr.v6i1.13071.

Alkunte, S., Fidan, I., Naikwadi, V., Gudavasov, S., Ali, M.A., Mahmudov, M., et al., 2024. Advancements and challenges in additively manufactured functionally graded materials: a comprehensive review. Journal of Manufacturing and Materials Processing 8 (1). Available at https://doi.org/10.3390/jmmp8010023.

Aziz, S.B., Faraj, M.G., Abdullah, O.G., 2018. Impedance spectroscopy as a novel approach to probe the phase transition and microstructures existing in CS:PEO based blend electrolytes. Scientific Reports 8 (1). Available at https://doi.org/10.1038/s41598-018-32662-1.

Baker, D.A., Hastings, R.S., Pruitt, L., 1999. Study of fatigue resistance of chemical and radiation crosslinked medical grade ultrahigh molecular weight polyethylene. Journal of Biomedical Materials Research 46 (4), 573–581. Available at https://doi.org/10.1002/(SICI)1097-4636 (19990915)46:4<573::AID-JBM16>3.3.CO;2-1.

Bar, G., Thomann, Y., Whangbo, M.-H., 1998. Characterization of the morphologies and nanostructures of blends of poly(styrene)- block -poly(ethene- co -but-1-ene)- block -poly(styrene) with isotactic and atactic polypropylenes by tapping-mode atomic force microscopy. Langmuir 14 (5), 1219–1226. Available at https://doi.org/10.1021/la9711544.

Barzegari, M.R., Yousefi, A.A., Zeynali, M.E., Rabii, A., Rodrigue, D., 2004. Rheology of co-continuous phase in physical thermoplastic elastomers. Polymer International 53 (10), 1448–1455. Available at https://doi.org/10.1002/pi.1559.

Bikiaris, D., Prinos, J., Botev, M., Betchev, C., Panayiotou, C., 2004. Blends of polymers with similar glass transition temperatures: a DMTA and DSC study. Journal of Applied Polymer Science 93 (2), 726–735. Available at https://doi.org/10.1002/app.20531.

Biron, M., 2015. Material Selection for Thermoplastic Parts: Practical and Advanced Information. William Andrew.

Bolar, S., Ito, Y., Fujita, T., 2024. Future prospects of high-entropy alloys as next-generation industrial electrode materials. Chemical Science 15, 8664–8722. Available at https://doi.org/10.1039/d3sc06784j.

Boyars, B., Daniels, E.S., Storer, R., Klein, A., 2007. The influence of latex blend composition on crosslinking and mechanical properties. Journal of Applied Polymer Science 104 (6), 3774–3779. Available at https://doi.org/10.1002/app.25947.

Brigandi, P.J., Cogen, J.M., Wolf, C.A., Reffner, J.R., Pearson, R.A., 2015. Kinetic and thermodynamic control in conductive PP/PMMA/EAA carbon black composites. Journal of Applied Polymer Science 132 (25). Available at https://doi.org/10.1002/app.42134.

Brinke, G.T., Karasz, F.E., MacKnight, W.J., 1983. Phase behavior in copolymer blends: poly(2,6-dimethyl-l,4-phenylene oxide) and halogen-substituted styrene copolymers. Macromolecules 16 (12), 1827–1832. Available at https://doi.org/10.1021/ma00246a006.

Byerlin, A.A., Pakhomova, L.K., 1990. Polymeric matrices for high-strength reinforced composites. Review. Polymer Science U.S.S.R. 32 (7), 1275–1311. Available at https://doi.org/10.1016/0032-3950(90)90094-M.

Cao, L., Deng, S., Lin, Z., 2016. Enhancement of carbon nanotube particle distribution in PPS/PEEK/carbon nanotube ternary composites with sausage-like structure. Polymers 8 (2), 50. Available at https://doi.org/10.3390/polym8020050.

Cao, Y., Zhang, J., Feng, J., Wu, P., 2011. Compatibilization of immiscible polymer blends using graphene oxide sheets. ACS Nano 5 (7), 5920–5927. Available at https://doi.org/10.1021/nn201717a.

Chang, L.L., Woo, E.M., 2003. Thermal, morphology, and NMR characterizations on phase behavior and miscibility in blends of isotactic polystyrene with poly(cyclohexyl methacrylate). Journal of Polymer Science, Part B: Polymer Physics 41 (8), 772–784. Available at https://doi.org/10.1002/polb.10431.

Chen, C.H., Chiang, C.L., Chen, W.J., Shen, M.Y., 2018. The effect of mbs toughening for mechanical properties of wood-plastic composites under environmental ageing. Polymers and Polymer Composites 26 (1), 45–57. Available at https://doi.org/10.1177/096739111802600106.

Christopher, Al, Silva, Í.G., Pangilinan, K.D., Chen, Q., Caldona, E.B., Advincula, R.C., 2021). High performance polymers for oil and gas applications. Reactive and Functional Polymers 162, 104878. Available at 10.1016/j.reactfunctpolym.2021.104878.

Chukov, D.I., Stepashkin, A.A., Tcherdyntsev, V.V., Olifirov, L.K., Kaloshkin, S.D., 2018. Structure and properties of composites based on polyphenylene sulfide reinforced with Al-Cu-Fe quasicrystalline particles. Federation Journal of Thermoplastic Composite Materials 31 (7), 882–895. Available at https://doi.org/10.1177/0892705717729015.

Cinquino, M., Prontera, C.T., Giuri, A., Pugliese, M., Giannuzzi, R., Maggiore, A., et al., 2024. Thermochromic printable and multicolor polymeric composite based on hybrid organic–inorganic perovskite. Advanced Materials 36 (2). Available at https://doi.org/10.1002/adma.202307564.

Collinson, M.G., Bower, M.P., Swait, T.J., Atkins, C.P., Hayes, S.A., Nuhiji, B., 2022. Novel composite curing methods for sustainable manufacture: a review. Composites Part C: Open Access 9, 100293. Available at https://doi.org/10.1016/j.jcomc.2022.100293.

Wilson, D., Stenzenberger, H.D., Hergenrother, P.M., 2013. Polymides. Springer Science & Business Media.

Das, A., Chatham, C.A., Fallon, J.J., Zawaski, C.E., Gilmer, E.L., Williams, C.B., et al., 2020. Current understanding and challenges in high temperature additive manufacturing of engineering thermoplastic polymers. Additive Manufacturing 34. Available at https://doi.org/10.1016/j.addma.2020.101218.

Das, P.P., Chaudhary, V., Mishra, S., 2021. Emerging trends in green polymer based composite materials: properties, fabrication and applications. Graphene Based Biopolymer Nanocomposites 1–24.

Daux, V., Prochazka, F., Carrot, C., 2013. Cocontinuous morphology in vinylidene fluoride based polymers/poly(ethylene oxide) blends. Journal of Applied Polymer Science 128 (1), 265–274. Available at https://doi.org/10.1002/app.38167.

Dell'Erba, R., Groeninckx, G., Maglio, G., Malinconico, M., Migliozzi, A., 2001. Immiscible polymer blends of semicrystalline biocompatible components: thermal properties and phase morphology analysis of PLLA/PCL blends. Polymer 42 (18), 7831–7840. Available at https://doi.org/10.1016/s0032-3861(01)00269-5.

Dong, W., Wang, H., He, M., Ren, F., Wu, T., Zheng, Q., et al., 2015. Synthesis of reactive comb polymers and their applications as a highly efficient compatibilizer in immiscible polymer blends. Industrial and Engineering Chemistry Research 54 (7), 2081–2089. Available at https://doi.org/10.1021/ie503645a.

Dou, R., Wang, W., Zhou, Y., Li, L.-peng, Gong, L., Yin, B., et al., 2013. Effect of core-shell morphology evolution on the rheology, crystallization, and mechanical properties of PA6/EPDM- g -MA/HDPE ternary blend. Journal of Applied Polymer Science 129 (1), 253–262. Available at https://doi.org/10.1002/app.38733.

Dudowicz, J., Freed, K.F., Douglas, J.F., 2002. New patterns of polymer blend miscibility associated with monomer shape and size asymmetry. Journal of Chemical Physics 116 (22), 9983–9996. Available at https://doi.org/10.1063/1.1476696.

Dutta, S.D., Patel, D.K., Lim, K.T., 2019. Functional cellulose-based hydrogels as extracellular matrices for tissue engineering. Journal of Biological Engineering 13 (1). Available at https://doi.org/10.1186/s13036-019-0177-0.

Easton, Z.H.W., Essink, M.A.J., Rodriguez Comas, L., Wurm, F.R., Gojzewski, H., 2023. Acceleration of biodegradation using polymer blends and composites. Macromolecular Chemistry and Physics 224 (6). Available at https://doi.org/10.1002/macp.202200421.

El Sayed, A.M., 2007. Evaluation of new conducting polymeric membrane based on sulfonation of compatibilized NR/EPDM blend. Journal of Applied Polymer Science 104 (6), 3804–3812. Available at https://doi.org/10.1002/app.25957.

Eutionnat-Diffo, P.A., Cayla, A., Chen, Y., Guan, J., Nierstrasz, V., Campagne, C., 2020. Development of flexible and conductive immiscible thermoplastic/elastomer monofilament for smart textiles applications using 3D printing. Polymers 12 (10), 2300. Available at https://doi.org/10.3390/polym12102300.

Fang, Y., Carreau, P.J., Lafleur, P.G., 2005. Thermal and rheological properties of mLLDPE/LDPE blends. Polymer Engineering and Science 45 (9), 1254–1264. Available at https://doi.org/10.1002/pen.20401.

Feldman, K.E., Kade, M.J., De Greef, T.F.A., Meijer, E.W., Kramer, E.J., Hawker, C.J., 2008. Polymers with multiple hydrogen-bonded end groups and their blends. Macromolecules 41 (13), 4694–4700. Available at https://doi.org/10.1021/ma800375r.

Fina, A., Han, Z., Saracco, G., Gross, U., Mainil, M., 2012. Morphology and conduction properties of graphite-filled immiscible PVDF/PPgMA blends. Polymers for Advanced Technologies 23 (12), 1572–1579. Available at https://doi.org/10.1002/pat.3031.

Fulati, A., Uto, K., Ebara, M., 2022. Influences of crystallinity and crosslinking density on the shape recovery force in poly(ε-caprolactone)-based shape-memory polymer blends. Polymers 14 (21), 4740. Available at https://doi.org/10.3390/polym14214740.

Gebrekrstos, A., Ray, S.S., 2023. Superior electrical conductivity and mechanical properties of phase-separated polymer blend composites by tuning the localization of nanoparticles for electromagnetic interference shielding applications. Journal of Polymer Science 61 (21), 2567–2584. Available at https://doi.org/10.1002/pol.20230059.

Gholami, M., Haddadi-Asl, V., Jouibari, I.S., 2022. A review on microphase separation measurement techniques for polyurethanes. Journal of Plastic Film and Sheeting 38 (4), 502–541. Available at https://doi.org/10.1177/87560879221088939.

Grabska-Zielińska, S., Sionkowska, A., 2021. How to improve physico-chemical properties of silk fibroin materials for biomedical applications?—Blending and cross-linking of silk fibroin—a review. Materials 14 (6), 1510. Available at https://doi.org/10.3390/ma14061510.

Gul, R., Mahmood, W.A.K., 2023. Studies on a novel microporous electrospun poly(vinyl alcohol)-based polymer electrolyte membranes for high-performance battery applications. Journal of the Chinese Chemical Society 70 (11), 1972–1985. Available at https://doi.org/10.1002/jccs.202300122.

Gumyusenge, A., Tran, D.T., Luo, X., Pitch, G.M., Zhao, Y., Jenkins, K.A., et al., 2018. Semiconducting polymer blends that exhibit stable charge transport at high temperatures. Science 362 (6419), 1131–1134. Available at https://doi.org/10.1126/science.aau0759.

Hasegawa, M., Kaneki, T., Tsukui, M., Okubo, N., Ishii, J., 2016. High-temperature polymers overcoming the trade-off between excellent thermoplasticity and low thermal expansion properties. Polymer 99, 292–306. Available at https://doi.org/10.1016/j.polymer.2016.07.010.

He, L., Zhang, W., Liu, X., Tong, L., 2024. Substantial improvement of thermal conductivity and mechanical properties of polymer composites by incorporation of boron nitride nanosheets and modulation of thermal curing reaction. Polymer Composites 45 (3), 2215–2231. Available at https://doi.org/10.1002/pc.27915.

Hema, K., Sureshan, K.M., 2019. Solid-state synthesis of two different polymers in a single crystal: a miscible polymer blend from a topochemical reaction. Angewandte Chemie 131 (9), 2780–2785. Available at https://doi.org/10.1002/ange.201813198.

Hirai, T., Onochi, Y., Kawada, J., 2020. Multifaceted property tailoring of polyamide 6 by blending miscible and immiscible components: ternary blends of polyamide 6/polyethylene terephthalate/phenol novolac. RSC Advances 10 (26), 15132–15138. Available at https://doi.org/10.1039/d0ra02344b.

Hirai, T., Muraoka, Y., Okamoto, H., 2022. Strong, transparent composites based on glass-fiber textile and a polycarbonate-polycaprolactone blend with matching refractive indices. Journal of Applied Polymer Science 139 (42). Available at https://doi.org/10.1002/app.52925.

Hirai, T., Kugimoto, K., Oyama, S., Takeda, Y., 2023. Scalable thermochromic composite based on a ternary polymer blend for temperature-adaptive solar heat mManagement. ACS Applied Materials & Interfaces 15 (15), 19427–19434. Available at https://doi.org/10.1021/acsami.3c00397.

Hu, X., Wang, Y., Yu, J., Zhu, J., Hu, Z., 2017. The mechanical and fatigue properties of flowable crosslink thermoplastic polymer blends based on self-catalysis of transesterification. Journal of Applied Polymer Science 134 (24). Available at https://doi.org/10.1002/app.44964.

Huang, S., Bai, L., Trifkovic, M., Cheng, X., Macosko, C.W., 2016. Controlling the morphology of immiscible cocontinuous polymer blends via silica nanoparticles jammed at the interface. Macromolecules 49 (10), 3911–3918. Available at https://doi.org/10.1021/acs.macromol.6b00212.

Ismail, A.S., Jawaid, M., Hamid, N.H., Yahaya, R., Hassan, A., Asim, M., et al., 2022. Effect of curing temperature on mechanical properties of bio-phenolic/epoxy polymer blends. Journal of Polymers and the Environment 30 (3), 878–885. Available at https://doi.org/10.1007/s10924-021-02244-w.

Jiang, G., Wu, H., Guo, S., 2010. Reinforcement of adhesion and development of morphology at polymer–polymer interface via reactive compatibilization: a review. Polymer Engineering & Science 50 (12), 2273–2286. Available at https://doi.org/10.1002/pen.21686.

Jin, Y., Lei, Z., Taynton, P., Huang, S., Zhang, W., 2019. Malleable and recyclable thermosets: the next generation of plastics. Matter 1 (6), 1456–1493. Available at https://doi.org/10.1016/j.matt.2019.09.004.

Ke, T., Sun, X., 2001. Effects of moisture content and heat treatment on the physical properties of starch and poly(lactic acid) blends. Journal of Applied Polymer Science 81 (12), 3069–3082. Available at https://doi.org/10.1002/app.1758.

Kemmish, D.J., 2011. Practical Guide to High Performance Engineering Plastics. Smithers Rapra.

Kim, B.J., Miyamoto, Y., Ma, B., Fréchet, J.M.J., 2009. Photocrosslinkable polythiophenes for efficient, thermally stable, organic photovoltaics. Advanced Functional Materials 19 (14), 2273–2281. Available at https://doi.org/10.1002/adfm.200900043.

Kim, Y.J., Shin, G.S., Lee, I.T., Kim, B.K., 1993. Miscible and immiscible blends of ABS with PMMA. I. Morphology and rheology. Journal of Applied Polymer Science 47 (2), 295–304. Available at https://doi.org/10.1002/app.1993.070470209.

Korycki, A., Carassus, F., Tramis, O., Garnier, C., Djilali, T., Chabert, F., 2023. Polyaryletherketone based blends: a review. Polymers 15 (19), 3943. Available at https://doi.org/10.3390/polym15193943.

Koumoulos, E.P., Trompeta, A.F., Santos, R.M., Martins, M., Santos, C.M.D., Iglesias, V., et al., 2019. Research and development in carbon fibers and advanced high-performance composites supply chain in Europe: a roadmap for challenges and the industrial uptake. Journal of Composites Science 3 (3). Available at https://doi.org/10.3390/jcs3030086.

Kumaraswamy, G.N., Ranganathaiah, C., 2006. Free volume microprobe studies on poly(methyl methacrylate)/poly(vinyl chloride) and poly(vinyl chloride)/polystyrene blends. Polymer Engineering & Science 46 (9), 1231–1241. Available at https://doi.org/10.1002/pen.20566.

Kuo, S.W., Lin, C.L., Chang, F.C., 2002. Phase behavior and hydrogen bonding in ternary polymer blends of phenolic resin/poly(ethylene oxide)/poly(ε-caprolactone). Macromolecules 35 (1), 278–285. Available at https://doi.org/10.1021/ma011255f.

Kuo, Y.H., Woo, E.M., 2003. Miscibility in two blend systems of homologous semicrystalline aryl polyesters involving poly(trimethylene terephthalate). Polymer Journal 35 (3), 236–244. Available at https://doi.org/10.1295/polymj.35.236.

Kurniawan, T.A., Meidiana, C., Goh, H.H., Zhang, D., Othman, M.H.D., Aziz, F., et al., 2024. Unlocking synergies between waste management and climate change mitigation to accelerate decarbonization through circular-economy digitalization in Indonesia. Sustainable Production and Consumption 46, 522–542. Available at https://doi.org/10.1016/j.spc.2024.03.011.

Kwon, N.Y., Park, S.H., Kang, H., Kim, Y.U., Chau, H.D., Harit, A.K., et al., 2021. Improved stability of all-polymer solar cells using crosslinkable donor and acceptor polymers bearing vinyl moieties in the side-chains. ACS Applied Materials and Interfaces 13 (14), 16754–16765. Available at https://doi.org/10.1021/acsami.1c00960.

Kyriacos, D., 2017. Chapter 21 - High-temperature engineering thermoplastics. In: Gilbert, M. (Ed.), Brydson's Plastics Materials. Butterworth-Heinemann, pp. 545–615. Available at https://doi.org/10.1016/b978-0-323-35824-8.00021-9.

Laskoski, M., Clarke, J.S., Neal, A., Keller, T.M., 2017. Phthalonitrile blends: simple way to improve physical properties by increasing crosslinking density. Macromolecular Chemistry and Physics 218 (22). Available at https://doi.org/10.1002/macp.201700207.

Lee, J.G., Jeong, J.O., Jeong, S.I., Park, J.S., 2021. Radiation-based crosslinking technique for enhanced thermal and mechanical properties of hdpe/eva/pu blends. Polymers 13 (16). Available at https://doi.org/10.3390/polym13162832.

Leroy, E., Alegría, A., Colmenero, J., 2003. Segmental dynamics in miscible polymer blends: modeling the combined effects of chain connectivity and concentration fluctuations. Macromolecules 36 (19), 7280–7288. Available at https://doi.org/10.1021/ma034144k.

Li, B., Zhang, X., Zhang, Q., Chen, F., Fu, Q., 2009. Synergistic enhancement in tensile strength and ductility of ABS by using recycled PETG plastic. Journal of Applied Polymer Science 113 (2), 1207–1215. Available at https://doi.org/10.1002/app.30002.

Li, Y., Li, W., Zhang, X., Shao, J., Deng, Y., 2022. Modeling and influencing factor analysis for the temperature and damage dependent tensile strength of fiber/polymer composites. Polymer Composites 43 (8), 5573–5581. Available at https://doi.org/10.1002/pc.26869.

Lin, C.C., Jonnalagadda, S.V., Balsara, N.P., Han, C.C., Krishnamoorti, R., 1996. Thermodynamic interactions in multicomponent polymer blends. Macromolecules 29 (2), 661–669. Available at https://doi.org/10.1021/ma951189f.

Lin, Y., Shangguan, Y., Chen, F., Zuo, M., Zheng, Q., 2013. Nonlinear phase-separation behavior of poly(methyl methacrylate)/poly(styrene- co -maleic anhydride) blends. Polymer International 62 (4), 676–683. Available at https://doi.org/10.1002/pi.4349.

Liu, C., Zhu, C.Y., Zhang, C., Yang, H.C., Xu, Z.K., 2024. Thermodynamic and kinetic understanding for managing the controllability of interfacial polymerization. Progress in Polymer Science 152. Available at https://doi.org/10.1016/j.progpolymsci.2024.101815.

Lloyd, S.M., Lave, L.B., 2003. Life cycle economic and environmental implications of using nanocomposites in automobiles. Environmental Science and Technology 37 (15), 3458–3466. Available at https://doi.org/10.1021/es026023q.

Lodge, T.P., McLeish, T.C.B., 2000. Self-concentrations and effective glass transition temperatures in polymer blends. Macromolecules 33 (14), 5278–5284. Available at https://doi.org/10.1021/ma9921706.

Long, C., Huang, X., Wei, J., Feng, M., Jia, K., Liu, X., 2015. Rheology, morphology, and properties of polyarylene ether nitrile blends. High Performance Polymers 27 (8), 1016–1023. Available at https://doi.org/10.1177/0954008315606200.

Lopez-Barron, C.R., Macosko, C.W., 2009. Characterizing interface shape evolution in immiscible polymer blends via 3D image analysis. Langmuir 25 (16), 9392–9404. Available at https://doi.org/10.1021/la803450y.

López-Barrón, C.R., Tsou, A.H., 2017. Strain hardening of polyethylene/polypropylene blends via interfacial reinforcement with poly(ethylene-cb-propylene) comb block copolymers. Macromolecules 50 (7), 2986–2995. Available at https://doi.org/10.1021/acs.macromol.7b00264.

Mahmud, S., Hasan, K.M.F., Dai, J., Xiong, Z., Wang, J., Zhu, J., 2024. Thermo-blending of biobased poly (propylene 2,5-furan dicarboxylate) and poly(lactic acid) with improved mechanical performance. ACS Applied Polymer Materials 6 (6), 3073–3081. Available at https://doi.org/10.1021/acsapm.3c02386.

Manoj, N.R., De, P.P., De, S.K., 1993. Self-crosslinkable plastic-rubber blend system based on poly (vinyl chloride) and acrylonitrile- co -butadiene rubber. Journal of Applied Polymer Science 49 (1), 133–142. Available at https://doi.org/10.1002/app.1993.070490116.

Mao, H., Zhang, T., Huang, T., Zhang, N., Wang, Y., Yang, J., 2017. Fabrication of high- k poly(vinylidene fluoride)/Nylon 6/carbon nanotube nanocomposites through selective localization of carbon nanotubes in blends. Polymer International 66 (4), 604–611. Available at https://doi.org/10.1002/pi.5302.

Marischal, L., Cayla, A., Lemort, G., Campagne, C., Devaux, É., 2019. Selection of immiscible polymer blends filled with carbon nanotubes for heating applications. Polymers 11 (11), 1827. Available at https://doi.org/10.3390/polym11111827.

Martinez, A.W., Caves, J.M., Ravi, S., Li, W., Chaikof, E.L., 2014. Effects of crosslinking on the mechanical properties, drug release and cytocompatibility of protein polymers. Acta Biomaterialia 10 (1), 26–33. Available at https://doi.org/10.1016/j.actbio.2013.08.029.

McGann, M.E., Bonitsky, C.M., Jackson, M.L., Ovaert, T.C., Trippel, S.B., Wagner, D.R., 2015. Genipin crosslinking of cartilage enhances resistance to biochemical degradation and mechanical wear. Journal of Orthopaedic Research 33 (11), 1571–1579. Available at https://doi.org/10.1002/jor.22939.

Mochane, M.J., Sefadi, J.S., Motsoeneng, T.S., Mokoena, T.E., Mofokeng, T.G., Mokhena, T.C., 2020. The effect of filler localization on the properties of biopolymer blends, recent advances: a review. Polymer Composites 41 (7), 2958–2979. Available at https://doi.org/10.1002/pc.25590.

Mulherin, R.C., Jung, S., Huettner, S., Johnson, K., Kohn, P., Sommer, M., et al., 2011. Ternary photovoltaic blends incorporating an all-conjugated donor-acceptor diblock copolymer. Nano Letters 11 (11), 4846–4851. Available at https://doi.org/10.1021/nl202691n.

Muller, J., González-Martínez, C., Chiralt, A., 2017. Combination of poly (lactic) acid and starch for biodegradable food packaging. Materials 10, 2017.

Muthuraj, R., Misra, M., Mohanty, A.K., 2018. Biodegradable compatibilized polymer blends for packaging applications: a literature review. Journal of Applied Polymer Science 135 (24). Available at https://doi.org/10.1002/app.45726.

Nakayama, D., Wu, F., Mohanty, A.K., Hirai, S., Misra, M., 2018. Biodegradable composites developed from PBAT/PLA binary blends and silk powder: compatibilization and performance evaluation. ACS Omega 3 (10), 12412–12421. Available at https://doi.org/10.1021/acsomega.8b00823.

Nunez, E.E., Yeo, S.M., Polychronopoulou, K., Polycarpou, A.A., 2011. Tribological study of high bearing blended polymer-based coatings for air-conditioning and refrigeration compressors. Surface and Coatings Technology 205 (8-9), 2994–3005. Available at https://doi.org/10.1016/j.surfcoat.2010.11.008.

Oladele, I.O., Omotosho, T.F., Adediran, A.A., 2020. Polymer-based composites: an indispensable material for present and future applications. International Journal of Polymer Science 2020. Available at https://doi.org/10.1155/2020/8834518.

Papageorgiou, G.Z., Bikiaris, D.N., 2006. Biodegradable poly(alkylene succinate) blends: thermal behavior and miscibility study. Journal of Polymer Science, Part B: Polymer Physics 44 (3), 584–597. Available at https://doi.org/10.1002/polb.20727.

Parin, F.N., Demirci, F., 2024. Durability of polymer composite materials for high-temperature applications. Aging and Durability of FRP Composites and Nanocomposites. Elsevier, Turkey, pp. 135–170. Available at https://doi.org/10.1016/B978-0-443-15545-1.00017-2.

Parveez, B., Kittur, M.I., Badruddin, I.A., Kamangar, S., Hussien, M., Umarfarooq, M.A., 2022. Scientific advancements in composite materials for aircraft applications: a review. Polymers 14 (22), 5007. Available at https://doi.org/10.3390/polym14225007.

Paul, D.R., Barlow, J.W., 1980. POLYMER BLENDS (OR ALLOYS). Journal of Macromolecular Science. Reviews in Macromolecular Chemistry C18 (1), 109–168.

Penning, J.P., St. John Manley, R., 1996. Miscible blends of two crystalline polymers. 1. Phase behavior and miscibility in blends of poly(vinylidene fluoride) and poly(1,4-butylene adipate). Macromolecules 29 (1), 77–83. Available at https://doi.org/10.1021/ma950651t.

Pfaendner, R., 2006. How will additives shape the future of plastics? Polymer Degradation and Stability 91 (9), 2249–2256. Available at https://doi.org/10.1016/j.polymdegradstab.2005.10.017.

Podara, C., Termine, S., Modestou, M., Semitekolos, D., Tsirogiannis, C., Karamitrou, M., et al., 2024. Recent trends of recycling and upcycling of polymers and composites: a comprehensive review. Recycling 9 (3). Available at https://doi.org/10.3390/recycling9030037.

Poulopoulou, N., Kasmi, N., Bikiaris, D.N., Papageorgiou, D.G., Floudas, G., Papageorgiou, G.Z., 2018. Sustainable polymers from renewable resources: polymer blends of furan-based polyesters. Macromolecular Materials and Engineering 303 (8). Available at https://doi.org/10.1002/mame.201800153.

Poulopoulou, N., Kasmi, N., Siampani, M., Terzopoulou, Z., Bikiaris, D., Achilias, D., et al., 2019. Exploring next-generation engineering bioplastics: poly(alkylene furanoate)/Poly(alkylene terephthalate) (PAF/PAT) Blends. Polymers 11 (3), 556. Available at https://doi.org/10.3390/polym11030556.

Prud'homme, R.E., 1982. Miscibility phenomena in polyester/chlorinated polymer blends. Polymer Engineering & Science 22 (2), 90–95. Available at https://doi.org/10.1002/pen.760220210.

Prusty, D., Pryamitsyn, V., Olvera de la Cruz, M., 2018. Thermodynamics of associative polymer blends. Macromolecules 51 (15), 5918–5932. Available at https://doi.org/10.1021/acs.macromol.8b00661.

Rajak, D.K., Pagar, D.D., Kumar, R., Pruncu, C.I., 2019. Recent progress of reinforcement materials: a comprehensive overview of composite materials. Journal of Materials Research and Technology 8 (6), 6354–6374. Available at https://doi.org/10.1016/j.jmrt.2019.09.068.

Rajeswari, N., Selvasekarapandian, S., Prabu, M., Karthikeyan, S., Sanjeeviraja, C., 2013. Lithium ion conducting solid polymer blend electrolyte based on bio-degradable polymers. Bulletin of Materials Science 36 (2), 333–339. Available at https://doi.org/10.1007/s12034-013-0463-2.

Ramesh, P., De, S.K., 1993. Carboxylated nitrile rubber as a reactive compatibilizer for immiscible blends of poly(vinyl chloride) and epoxidized natural rubber. Journal of Applied Polymer Science 50 (8), 1369–1377. Available at https://doi.org/10.1002/app.1993.070500808.

Rao, V., Latha, P., Ashokan, P.V., Shridhar, M.H., 1999. Thermal degradation of poly(N-vinylpyrrolidone)–Poly (vinyl alcohol) Blends. Polymer Journal 31 (10), 887–889. Available at https://doi.org/10.1295/polymj.31.887.

Rashid, F.J., Mohamed, N., Daik, R., Ehsan, A.A., 2024. Customizing ultraviolet curable polymer-ink properties via reactive diluent modification and viscosity control. Journal of Applied Polymer Science 141 (3). Available at https://doi.org/10.1002/app.54809.

Ray, S., Cooney, R.P., 2018. Chapter 9 - Thermal degradation of polymer and polymer composites. In: Kutz, M. (Ed.), Handbook of Environmental Degradation of Materials. William Andrew, pp. 185–206. Available at https://doi.org/10.1016/b978-0-323-52472-8.00009-5.

Ren, Y., Wang, F.Y., Lan, R.T., Fu, W.Q., Chen, Z.J., Lin, H., et al., 2021. Polyphenol-assisted chemical crosslinking: a new strategy to achieve highly crosslinked, antioxidative, and antibacterial ultrahigh-molecular-weight polyethylene for total joint replacement. ACS Biomaterials Science and Engineering 7 (1), 373–381. Available at https://doi.org/10.1021/acsbiomaterials.0c01437.

Rohde, B.J., Le, K.M., Krishnamoorti, R., Robertson, M.L., 2016. Thermoset blends of an epoxy resin and polydicyclopentadiene. Macromolecules 49 (23), 8960–8970. Available at https://doi.org/10.1021/acs.macromol.6b01649.

Roy, A.L., Beaumont, C., Leclerc, M., Walus, K., 2023. Evaluating polythiophenes as temperature sensing materials using combinatorial inkjet printing. Flexible and Printed Electronics 8 (1), 014002. Available at https://doi.org/10.1088/2058-8585/acc07f.

Sabzi, M., Ranjbar-Mohammadi, M., Zhang, Q., Kargozar, S., Leng, J., Akhtari, T., et al., 2019. Designing triple-shape memory polymers from a miscible polymer pair through dual-electrospinning technique. Journal of Applied Polymer Science 136 (19). Available at https://doi.org/10.1002/app.47471.

Salehiyan, R., Ray, S.S., 2019. Tuning the conductivity of nanocomposites through nanoparticle migration and interface crossing in immiscible polymer blends: a review on fundamental understanding. Macromolecular Materials and Engineering 304 (2). Available at https://doi.org/10.1002/mame.201800431.

Sanchez, C., Julián, B., Belleville, P., Popall, M., 2005. Applications of hybrid organic-inorganic nanocomposites. Journal of Materials Chemistry 15 (35-36), 3559–3592. Available at https://doi.org/10.1039/b509097k.

Seferis, J.C., 1986. Polyetheretherketone (PEEK): processing-structure and properties studies for a matrix in high performance composites. Polymer Composites 7 (3), 158–169. Available at https://doi.org/10.1002/pc.750070305.

Shamsuri, A.A., Jamil, S.N.A.M., 2021. Application of quaternary ammonium compounds as compatibilizers for polymer blends and polymer composites—a concise review. Applied Sciences 11 (7), 3167. Available at https://doi.org/10.3390/app11073167.

Shin, T.K., Kim, J., Choi, H.J., Jhon, M.S., 2000. Miscibility of biodegradable aliphatic polyester and poly(vinyl acetate) blends. Journal of Applied Polymer Science 77 (6), 1348–1352. Available at https://doi.org/10.1002/1097-4628 (20000808)77:6<1348::AID-APP20>3.0.CO;2-Z.

Siddamallappa, N.G., Basavaraju, M., Dase Gowda, C.G., 2018. Elastin-based polymer: synthesis, characterization and examination of its miscibility characteristics with poly(vinyl alcohol) and electrospinning of the miscible blends. Polymer International 67 (11), 1511–1522. Available at https://doi.org/10.1002/pi.5669.

Sikorska, W., Musioł, M., Zawidlak-Węgrzyńska, B., Rydz, J., Leonard, D., 2021. Tijing, end-of-life options for (Bio)degradable polymers in the circular economy. Advances in Polymer Technology 2021, 1–18. Available at https://doi.org/10.1155/2021/6695140.

Sinha Ray, S., Bousmina, M., 2005. Effect of organic modification on the compatibilization efficiency of clay in an immiscible polymer blend. Macromolecular Rapid Communications 26 (20), 1639–1646. Available at https://doi.org/10.1002/marc.200500447.

Sung, H.-W., Chang, W.-H., Ma, C.-Y., Lee, M.-H., 2003. Crosslinking of biological tissues using genipin and/or carbodiimide. Journal of Biomedical Materials Research Part A 64A (3), 427–438. Available at https://doi.org/10.1002/jbm.a.10346.

Takahashi, T., Takimoto, J.I., Koyama, K., 1999. Elongational viscosity for miscible and immiscible polymer blends. II. Blends with a small amount of UHMW polymer. Journal of Applied Polymer Science 72 (7), 961–969. Available at https://doi.org/10.1002/(SICI)1097-4628 (19990516)72:7<961::AID-APP12>3.0.CO;2-T.

Tcherdyntsev, V.V., Olifirov, L.K., Kaloshkin, S.D., Zadorozhnyy, M.Y., Danilov, V.D., 2018. Thermal and mechanical properties of fluorinated ethylene propylene and polyphenylene sulfide-based composites obtained by high-energy ball milling. Journal of Materials Science 53 (19), 13701–13712. Available at https://doi.org/10.1007/s10853-018-2508-9.

Vedamurthy, T., Lai, C.H., Vamhindi, B.S.D.R., 2023. Preparation of hydrophobic nylon 6-phenol resin derivative polymer blends for the dielectric application and theoretical evaluation of their hydrophobic property. Journal of Polymer Research 30 (9). Available at https://doi.org/10.1007/s10965-023-03738-y.

Velvizhi, G., Shanthakumar, S., Das, B., Pugazhendhi, A., Priya, T.S., Ashok, B., et al., 2020. Biodegradable and non-biodegradable fraction of municipal solid waste for multifaceted applications through a closed loop integrated refinery platform: paving a path towards circular economy. Science of the Total Environment 731, 138049. Available at https://doi.org/10.1016/j.scitotenv.2020.138049.

Wang, H., Dong, W., Li, Y., 2015. Compatibilization of immiscible polymer blends using in situ formed janus nanomicelles by reactive blending. ACS Macro Letters 4 (12), 1398–1403. Available at https://doi.org/10.1021/acsmacrolett.5b00763.

Wang, H., Fu, Z., Dong, W., Li, Y., Li, J., 2016. Formation of interfacial janus nanomicelles by reactive blending and their compatibilization effects on immiscible polymer blends. The Journal of Physical Chemistry B 120 (34), 9240–9252. Available at https://doi.org/10.1021/acs.jpcb.6b06761.

Woo, E.M., Mao, Y.-J., Wu, P.-L., Su, C.-C., 2003. Ternary miscibility in blends of three polymers with balanced binary interactions. Polymer Engineering & Science 43 (3), 543–557. Available at https://doi.org/10.1002/pen.10044.

Wu, Y., Zhang, H., Shentu, B., Weng, Z., 2017. Relationship between structure and properties in high-performance PA 6/ SEBS - g - MA /(PPO / PS) blends: the role of PPO and PS. Journal of Applied Polymer Science 134 (36). Available at https://doi.org/10.1002/app.45281.

Xanthos, M., Dagli, S.S., 1991. Compatibilization of polymer blends by reactive processing. Polymer Engineering & Science 31 (13), 929–935. Available at https://doi.org/10.1002/pen.760311302.

Xu, D.K., Liu, L., Xu, Y.B., Han, E.H., 2006. The influence of element Y on the mechanical properties of the as-extruded Mg–Zn–Y–Zr alloys. Journal of Alloys and Compounds 426 (1-2), 155–161. Available at https://doi.org/10.1016/j.jallcom.2006.02.035.

Yang, K., Yazawa, K., Tsuchiya, K., Numata, K., Guan, J., 2019. Molecular interactions and toughening mechanisms in silk fibroin–epoxy resin blend films. Biomacromolecules 20 (6), 2295–2304. Available at https://doi.org/10.1021/acs.biomac.9b00260.

Yazie, N., Worku, D., Gabbiye, N., Alemayehu, A., Getahun, Z., 2023. Optimization of the glass transition temperature of polyvinyl pyrrolidone/polyacrylamide blend films using response surface methodology. Journal of Applied Polymer Science 140 (38). Available at https://doi.org/10.1002/app.54445.

Yoo, Y., Hong, G.H., Hur, S.R., Kim, Y.S., Lee, S.G., Kim, H.J., et al., 2009. Preparation of acrylic copolymers and crosslinking agents and properties as a film. Journal of Applied Polymer Science 112 (3), 1587–1594. Available at https://doi.org/10.1002/app.29564.

Yu, L., Dean, K., Li, L., 2006. Polymer blends and composites from renewable resources. Progress in Polymer Science 31 (6), 576–602. Available at https://doi.org/10.1016/j.progpolymsci.2006.03.002.

Zadhoush, A., Esmaeili, M., Ghaeli, I., 2009. Crosslinking of plasticized PVC used in coated fabrics. Journal of Vinyl and Additive Technology 15 (2), 108–112. Available at https://doi.org/10.1002/vnl.20178.

Zakaria, N.A., Isa, M.I.N., Mohamed, N.S., Subban, R.H.Y., 2012. Characterization of polyvinyl chloride/polyethyl methacrylate polymer blend for use as polymer host in polymer electrolytes. Journal of Applied Polymer Science 126 (S2). Available at https://doi.org/10.1002/app.36940.

Zhang, M.Q., Yu, G., Zeng, H.M., Zhang, H.B., Hou, Y.H., 1998. Two-step percolation in polymer blends filled with carbon black. Macromolecules 31 (19), 6724–6726. Available at https://doi.org/10.1021/ma9806918.

Zhao, X., Wang, H., Fu, Z., Li, Y., 2018. Enhanced interfacial adhesion by reactive carbon nanotubes: new route to high-performance immiscible polymer blend nanocomposites with simultaneously enhanced toughness, tensile strength, and electrical conductivity. ACS Applied Materials & Interfaces 10 (10), 8411–8416. Available at https://doi.org/10.1021/acsami.8b01704.

Zhao, Y., Zhao, X., Roders, M., Gumyusenge, A., Ayzner, A.L., Mei, J., 2017. Melt-processing of complementary semiconducting polymer blends for high performance organic transistors. Advanced Materials 29 (6). Available at https://doi.org/10.1002/adma.201605056.

Zheng, X., Li, Y., Tang, J., Yu, G., 2022. Structure and properties of PVDF/PA6 blends compatibilized by ionic liquid-grafted PA6. ACS Omega 7 (15), 12772–12778. Available at https://doi.org/10.1021/acsomega.1c07341.

Zhou, Z., Zhang, Y., Zhang, Y., Yin, N., 2008. Rheological behavior of polypropylene/octavinyl polyhedral oligomeric silsesquioxane composites. Journal of Polymer Science Part B: Polymer Physics 46 (5), 526–533. Available at https://doi.org/10.1002/polb.21386.

Zhu, S., Zhang, Y., Li, Q., Wei, L., Guan, S., 2015. Influence of polytetrafluoroethylene (PTFE) content on mechanical and tribological properties of poly(ether ether ketone)/PTFE coatings prepared by electrostatic powder spraying technique. High Performance Polymers 27 (1), 3–9. Available at https://doi.org/10.1177/0954008314537098.

Zuo, P., Tcharkhtchi, A., Shirinbayan, M., Fitoussi, J., Bakir, F., 2019. Overall investigation of poly (Phenylene Sulfide) from synthesis and process to applications—a review. Macromolecular Materials and Engineering 304 (5). Available at https://doi.org/10.1002/mame.201800686.

6

Fiber-reinforced high-temperature composite materials

Stephen C. Agwuncha[1], Tochukwu Okonkwo[2], Chioma Grace Anusionwu[3], Emmanuel Rotimi Sadiku[4]

[1]DEPARTMENT OF INDUSTRIAL CHEMISTRY, FEDERAL UNIVERSITY LOKOJA, LOKOJA, KOGI STATE, NIGERIA
[2]DEPARTMENT OF CHEMISTRY, IBRAHIM BADAMASI BABANGIDA UNIVERSITY, LAPAI, NIGER STATE, NIGERIA
[3]DEPARTMENT OF CHEMISTRY, UNIVERSITY OF JOHANNESBURG, JOHANNESBURG, SOUTH AFRICA
[4]DEPARTMENT OF CHEMICAL, METALLURGICAL AND MATERIALS ENGINEERING, TSHWANE UNIVERSITY OF TECHNOLOGY, PRETORIA, GAUTENG, SOUTH AFRICA

6.1 Introduction

The demand for polymers that can withstand operating conditions in the aerospace and electronic industries sparked the need for high temperatures in the 1950s (Hergenrother, 2003). Since then, many high-temperature polymers have been commercially used and researched to improve existing ones, and this is an active research area. These high-temperature polymers are vital to many applications, especially those with high dielectric breakdown strength (Zhang et al., 2021). The question that naturally comes to mind is, what is a high-temperature polymer? According to Hergenrother (2003), polymers that can withstand a temperature range between 150°C and 350°C for a defined period of time before degrading, that can withstand long-term durability at 177°C, or that have a thermal decomposition temperature above 450°C are considered high-temperature polymers. Also, according to Meador (1998), high-temperature polymers have high T_g, which is 25°C higher than their operating temperature. Due to their chemical nature or design, not all polymers are suited for high-temperature environments. This presented a significant challenge that required a deeper understanding of the polymer material's properties and behavior.

The definition given above can be considered relative and depends more on the fields of application. High-temperature polymers for application as fuel cell membranes can be considered anything above 180°C since their operating temperature ranges from 120°C to 180°C (Søndergaard et al., 2017). Therefore according to Hergenrother (1990), high-temperature polymer matrices are polymers that retain useful properties after thousands of hours of exposure to 200°C in air under static conditions.

High-temperature polymer composites can be defined as polymer composites prepared using a polymer matrix with high T_g, as described above. Generally, composite materials are created by combining two or more materials into a final product with properties superior to those of the individual materials (Dagdag et al., 2020, 2019; Hsissou et al., 2019, 2021; Hsissou et al., 2020; Hsissou, Dagdag, et al., 2019; Khalil et al., 2013). These composite materials have

High-Temperature Polymeric Materials. DOI: https://doi.org/10.1016/B978-0-443-32792-6.00007-5

fillers embedded in the selected polymer matrix to ensure cohesion, load orientation, and stress transmission (Asim et al., 2017; Hsissou et al., 2019; Werken et al., 2020). Composite materials are typically heterogeneous and anisotropic, with their properties influenced by many factors, such as the chemical nature of the matrix and reinforcement, the shape and properties of the reinforcement, the quality of the interface, and the production process (Amrollahi et al., 2019; Asim et al., 2018; Khalil et al., 2013; Le Guen et al., 2016; Nagarajan et al., 2019; Nobile, 2011; Rod et al., 2019; Venkateshwaran and Elayaperumal, 2010; Zhou et al., 2019). The matrix and reinforcement components of composite materials can be made from metallic, ceramic, or plastic materials, and they can also be organic or inorganic. This versatility allows for a wide range of combinations, as illustrated in Fig. 6–1 (Chen and Gao,

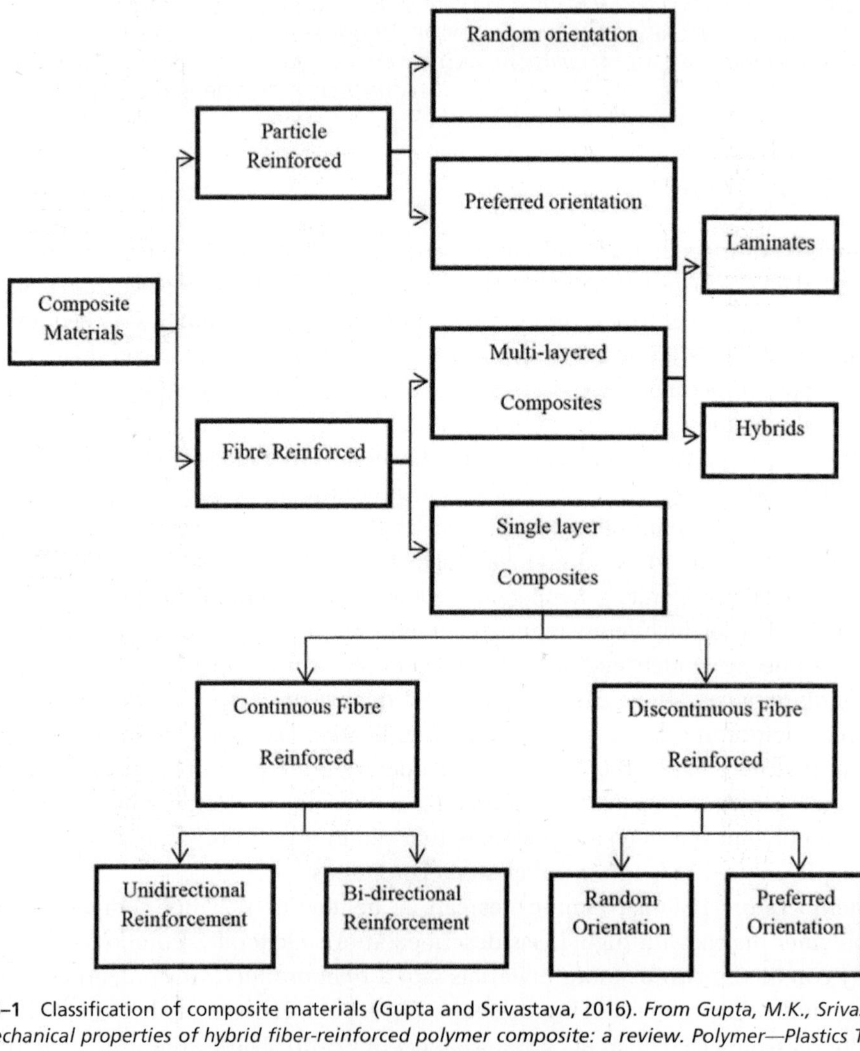

FIGURE 6–1 Classification of composite materials (Gupta and Srivastava, 2016). *From Gupta, M.K., Srivastava, R.K., 2016. Mechanical properties of hybrid fiber-reinforced polymer composite: a review. Polymer—Plastics Technology and Engineering 55(6), 626–642. https://doi.org/10.1080/03602559.2015.1098694.*

2019; Cheng et al., 2020; Devnani and Sinha, 2019; Li et al., 2018). When there are multiple discontinuous phases of different natures, the composite is classified as a hybrid (Ahmadijokani et al., 2020; Arabpour et al., 2020; Dong et al., 2019). The continuous phase is known as the matrix, while the discontinuous phase is the reinforcement. By combining components with complementary physical and mechanical properties, composite materials can exhibit enhanced characteristics (Baek et al., 2020; De et al., 2020). For example, adding reinforcements with high tensile strength and high modulus to a polymer matrix can improve the mechanical and thermal qualities (Hsissou et al., 2021).

Fiber-reinforced high-temperature polymer composites are a type of composite material that incorporates fibers as reinforcing agents. These fibers can be either organic or inorganic, as shown in Fig. 6-2 (Kharbanda et al., 2021). Fiber-reinforced composites (FRCs) can be

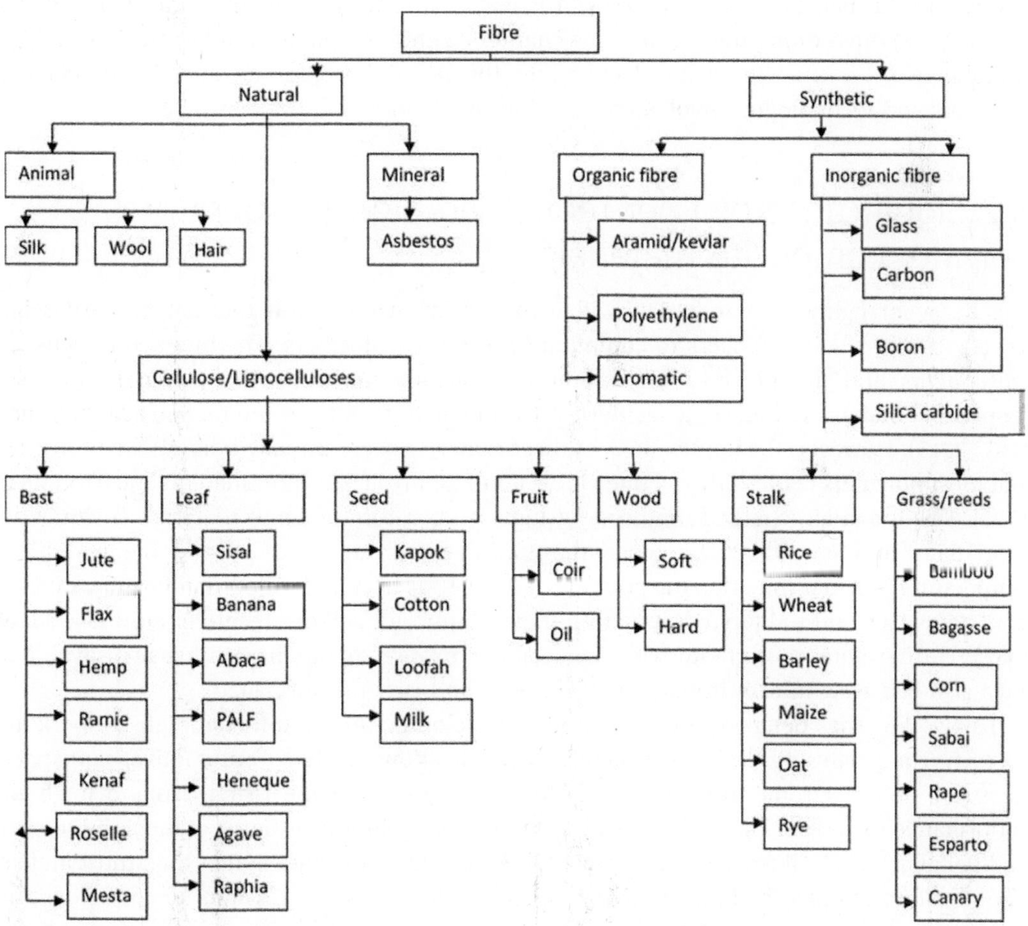

FIGURE 6–2 A comprehensive classification of natural and synthetic fibers (Gupta and Srivastava, 2016). *From Gupta, M.K., Srivastava, R.K., 2016. Mechanical properties of hybrid fiber-reinforced polymer composite: a review. Polymer—Plastics Technology and Engineering 55(6), 626–642. https://doi.org/10.1080/03602559.2015.1098694.*

categorized based on their fiber length or structure, as depicted in Fig. 6–2. In terms of fiber length, FRC can be continuous, discontinuous, or discrete. In relation to fiber structure or arrangement, FRC can be unidirectional, bidirectional, aligned, or randomly oriented. Each type has its own set of advantages and disadvantages, and the selection depends on the specific polymer matrix involved and the intended application. This reinforcement serves as the framework that imparts mechanical strength, including rigidity and tensile strength (Hsissou et al., 2021).

FRCs represented a significant advancement in material science and engineering due to their lightweight nature and numerous advantages (Mangalgiri, 2005). High-temperature polymer composites (HTPC), particularly in the aerospace and electronic industries, have revolutionized the field. According to Mangalgiri (2005), the simultaneous development of high-temperature polymers has been a crucial advancement in technological progress. To sustain and further these developments, it is essential to foster new and innovative ways of thinking and envisioning the future. This chapter examines past accomplishments in HTPC, stimulating researchers to delve deeper into the possibilities for advancing this class of materials and technologies involved for the future.

6.2 Significance of fiber-reinforced high-temperature polymer composite materials

The quest for lightweight materials with superior mechanical and thermal properties has driven scientists and engineers to compound polymers with fibers. The interaction between these fibers and the polymer matrix can significantly influence the material's physical properties, making it crucial to achieve an optimal fiber–matrix interaction (Mohammed et al., 2015). However, the applications of these materials were initially limited due to the fact that most polymers could only operate at low temperatures (Balasubramanian and Tirumalai, 2013). This limitation sparked the need for high-temperature polymer materials (Asim et al., 2017; Saba and Jawaid, 2018). Since the 1950s when Dupole developed the first high-temperature polymer material, the potential applications have expanded significantly, ranging from aerospace and military equipment to navy ships and facilities (Pinto et al., 1999; Shaw, 1987). Furthermore, these materials have facilitated satellite launches and space travel, and have also had a significant impact the electronics industry (Strauß, 2020).

Lately, there has been a growing interest in developing fiber-reinforced high-temperature polymer composites for various engineering applications such as automotive, aerospace, electrical, and electronic industries. This interest stems from the rising demand for high-performance materials. These composite materials offer solutions to some of the limitations of traditional materials like metals and ceramics. For instance, they are lighter than most metals, can be tailored to provide high strength in specific directions, are resistant to corrosion, and can be manufactured with minimal secondary operations. In comparison to ceramics, high-temperature polymer composites exhibit higher fracture toughness, vibration-damping capacity, and ductility. The key advantage lies in their ability to be designed for strength, stiffness, toughness, and applicability to diverse conditions by carefully selecting the matrix and reinforcements. While both thermoplastics and thermosetting polymers can serve as matrix materials in these composites, thermosetting polymers such as epoxies, phenolics, polyimides (PIs), bismaleimides (BMIs), and cyanate esters are the most commonly used due to their superior property retention at elevated temperatures.

A variety of reinforcing fibers, including glass, aramid, carbon, boron, and silicon carbide, are readily available for high-temperature polymer composites. The continuous advancement of matrix materials, reinforcements, and fabrication technology is expected to further drive the development of high-temperature polymer composites. These composites are increasingly replacing metals in advanced structural applications, particularly in the aerospace and aircraft industries, due to their lightweight nature, customizable properties, high strength, and stiffness. Additionally, polymers are being chosen over conventional materials for their ease of fabrication, good fatigue resistance, and relatively low cost. Compared to ceramics, polymer composites offer much higher impact resistance and are less likely to fracture when overloaded. Exploring the high temperature capabilities of polymer composites is essential to expand their potential applications in fields such as automotive and marine, which also require advanced materials for high temperature use.

6.3 Types of high-temperature polymer matrices

As defined earlier, high-temperature polymer matrices are utilized in applications where materials are needed to retain their mechanical properties and structural integrity at elevated temperatures. Polymers with excellent thermal stability, such as cyanate esters, PIs, and phthalonitriles, have been widely accessible for an extended period and have been effectively employed in the electronics, energy, machinery, and aerospace industries (DeMeuse, 2014; Dominguez and Keller, 2006; Hergenrother, 1990; Mittal, 2018; Tsampas et al., 2018). Efforts to explore and characterize polymers as potential matrices for HT FRCs, allowing for the use of fiber-reinforced polymers at elevated temperatures, have been documented in various studies (Mangalgiri, 2005). Most of these studies have focused on composites based on phenylethynyl terminated imide (PETI), following the development of Polymerized Monomeric Reactant formulations by NASA and the U.S. Air Force in the 1970s (Bain et al., 2007; Chuang et al., 2003; Connell et al., 2003; Gupta et al., 1998). It is also well reported that the properties of matrices that affect the mechanical behavior of composites are (1) modulus of elasticity, (2) stress–strain behavior, (3) stain capacity, (4) strengths in tension, and (5) fiber–matrix bond strength (Papakonstantinou et al., 2001). This section will highlight the unique properties of each of the standard high-temperature polymer matrices.

Thermal stability and glass transition temperature are some of the most important material requirements to consider when choosing a material for an application that involves high temperatures. However, most polymeric materials have limited applications due to their T_g and material stability (Meador, 1998). High-temperature polymeric materials such as PIs are in high demand due to their low density and high specific strength, which make them applicable in aerospace systems, marine, oil and gas industries, and electronics (Meador, 1998). However, there are other types of polymeric materials that fall into this category. This section focuses on types of high-temperature polymeric materials, their characteristic properties, and their unique applications. High-temperature polymers can either be thermoset or thermoplastic. Several commonly used high temperature polymeric materials include the following:

6.3.1 Polyimide

PIs which comprise an acyl group linked to nitrogen as shown in Fig. 6-3, were first discovered by researchers at DuPont. They modified Nylon chemistry in two reaction steps that comprise the condensation reaction between a diamine and a dianhydride to produce amic acid (Form poly) and cyclodehydration reaction to produce PI as shown in Fig. 6-3. PIs in general are naturally resistant to flame and have passed the UL95 V-0 rating for flame retardance. However, PIs can be classified either as thermosetting or thermoplastic. Thermosetting PI (Fig. 6-3) is thermally stable material up to 500°C, displaying excellent dielectric and mechanical properties. They also have good chemical resistance. These unique properties make them applicable to aerospace systems, microelectronics, and medicine.

PIs' limitations are their hygroscopicity (water adsorption) and poor processability. Over the years, scientists have modified PI by adding materials such as carbon or glass fiber (GF), reinforcing the polymer composites. PI is known for its poor thermal conductivity, as are many other polymers (Haruki, 2021). However, PIs have been used to prepare high-performance polymer composites with remarkable properties. Dong and coresearchers prepared carbon fiber (CF)-reinforced PI composites using the hot-pressing technique. The mechanical properties of the PI composites were optimal with 20 wt.% of CF loading, displaying excellent improvements (Dong et al., 2018). PI composites with prepreg fibers can be prepared using most conventional methods available (Fernberg et al., 2018).

Ke et al. (2020) reported the preparation of PI with improved mechanical and thermal properties. This group used thermoplastic PI (TPI) to toughen thermosetting PIs by blending the two polymers to produce toughened PI (TPI/PI) blends. The modified PI resin system exhibited good thermal stability, heat resistance, and toughness. The findings showed that the

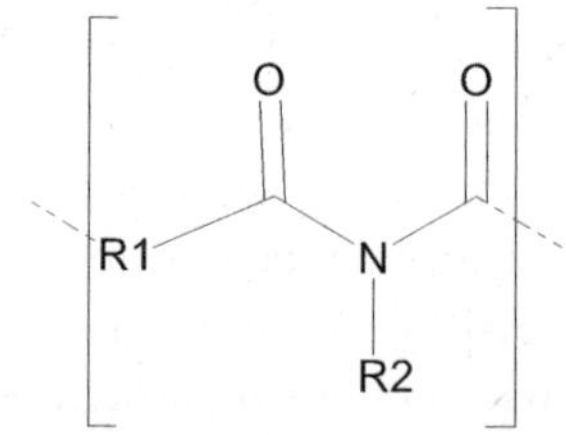

FIGURE 6–3 Molecular framework of polyimides showing the linkage between the two acyl groups and the nitrogen atom (Wu and Koo, 2022). *From Wu, H., Koo, J.H., 2022. Characterization of high-temperature polymers for extreme environments. In: Analysis of Flame Retardancy in Polymer Science. Elsevier, pp. 299–331. https://doi.org/10.1016/B978-0-12-824045-8.00008-3.*

TPI/PI blends preserved the curing behavior and characteristics of the PI oligomer. The T_g of the cured TPI/PI blend exceeded 395°C, and the $T_{5\%}$ values ranged from 533°C to 563°C, indicating outstanding thermal stability and heat resistance. The maximum impact strength increased by 46% compared to pure PI, demonstrating the exceptional toughening effect of the TPI. CF-reinforced PI composites were developed using this toughening system as a matrix. The compression-after-impact values of the CF-reinforced PI composites reached up to 190 MPa, signifying the exceptional toughness of the materials (Ke et al., 2020).

These reinforced PIs exhibit improved thermal and mechanical properties (Wu and Koo, 2022). However, these improved mechanical properties often have the limitation of high cost, and poor thermal and solvent processability (Xu et al., 2021).

6.3.2 Polyetherimide

Polyetherimide (PEI) is a thermoplastic material with great potential for engineering applications based on its lightweight, high strength, flame retardancy, low toxicity, and chemical resistance (Vakharia et al., 2023). The processability of PI was significantly improved when a particular class of PIs known as PEI was developed (Fig. 6–4) (Eastmond and Paprotny, 1996). PEI is an amorphous high-temperature aromatic thermoplastic polymer with unique chemical properties and good thermal and mechanical properties. PEI exhibits a T_g of 186°C (Gebisa and Lemu, 2018). They are used for special applications such as wire coating and automobiles where high temperatures and harsh environmental conditions are encountered, mainly due to their unique properties (Santhosh Kumar and Reghunadhan Nair, 2014). PEI composites have been applied to electronics, medicine, automobiles, aerospace, and nuclear plants. PEI is a polymer of choice for high-temperature applications due to its multifunctional molecular structure and superior mechanical properties (Khanbareh et al., 2017). PEI has been compounded with CF, GF, AF, and nanoparticle. PEI is widely recognized as an exceptional high-performance thermoplastic with the potential to replace metal components in cutting-edge applications like aerospace engineering. However, several studies have underscored the necessity of enhancing its mechanical properties to fully realize its potential (Lin et al., 2022; Sun et al., 2021).

6.3.3 Polyetherketoneketone

Polyetherketoneketone (PEKK) is a semicrystalline thermoplastic polymer belonging to the polyaryletherketone (PAEK) family. It is known for its high-temperature stability (Rivière et al.,

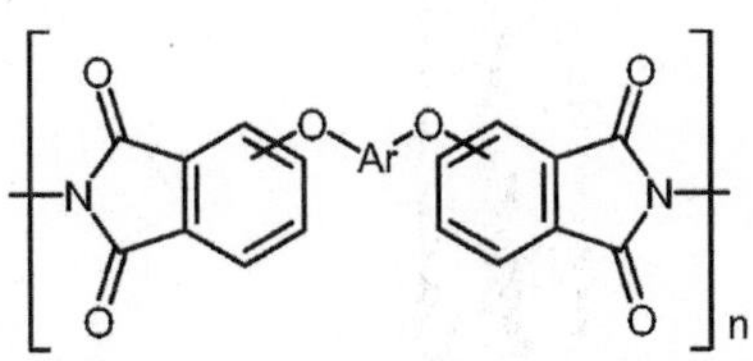

FIGURE 6–4 Chemical structure of polyetherimide (Ramgobin et al., 2019). *From Ramgobin, A., Fontaine, G., Bourbigot, S., 2019. Thermal degradation and fire behavior of high performance polymers. Polymer Reviews 59(1), 55–123. https://doi.org/10.1080/15583724.2018.1546736.*

2016; Tadini et al., 2017). PEKK has exceptional mechanical properties with low flammability, high thermal stability, glass transition temperature, and good corrosion resistance over a wide range of temperatures (Díez-Pascual et al., 2012; Garcia-Leiner et al., 2021; Harsha and Tewari, 2004). PEKK is one of the most utilized polymers among the PAEK polymer family with high T_g of 160°C and melting temperature (T_m) of 345°C (Karnik et al., 2008; Chang et al., 2021; Sukur et al., 2022; Mazur et al., 2013). Composites of PEKK have been reported in journal publications using conventional processing methods (Sukur et al., 2022; Veazey et al., 2017; Wang et al., 2022).

6.3.4 Polyetheretherketone

PEEK is also a member of the PAEK family. A polymer with excellent thermoplastic properties and good mechanical properties over a wide range of temperatures (Sukur et al., 2022). The main difference between PEKK and polyetheretherketone (PEEK) is in their ether and ketone groups, as shown in Fig. 6–5, even though they are related. Because of this structural difference, PEKK is stiffer than PEEK and has a higher glass transition temperature (T_g). PEKK is frequently used in additive manufacturing (AM) because of its amorphous nature, making it easy to process (Wu and Koo, 2022). Fig. 6–6 presents the TGAs of PEI, PEKK, and PEEK showing their thermal stability at elevated temperatures. PEEK resin can be compression molded, extruded, or injection molded using conventional equipment and has a high processing temperature, up to 350°C, with no degradation of material (Alsayed et al., 2012; Zalaznik et al., 2016).

6.3.5 Phenolic and polybenzoxazine

Phenolic resins find extensive commercial use in applications such as adhesives and composites. This is due to their cost-effectiveness in manufacturing, ability to maintain dimensional stability, and resistance to chemicals (Da-Peng and Hong, 2008). The primary process of phenolic resin formation is the reaction of phenols or their derivatives with formaldehyde. The main phenolic resin types, novolacs and resoles, are made by additional polymerization processes that are base- and acid-catalyzed, as shown in Fig. 6–7 (Wu and Koo, 2022). The resole method is primarily used to manufacture phenolic resins. It was

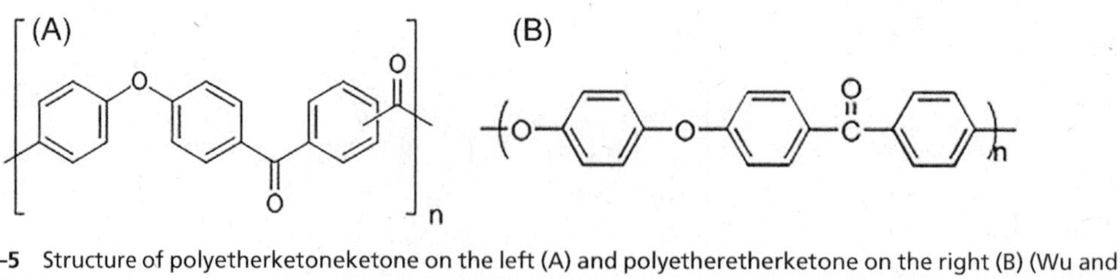

FIGURE 6–5 Structure of polyetherketoneketone on the left (A) and polyetheretherketone on the right (B) (Wu and Koo, 2022). *From Wu, H., Koo, J.H., 2022. Characterization of high-temperature polymers for extreme environments. In: Analysis of Flame Retardancy in Polymer Science. Elsevier, pp. 299–331. https://doi.org/10.1016/B978-0-12-824045-8.00008-3.*

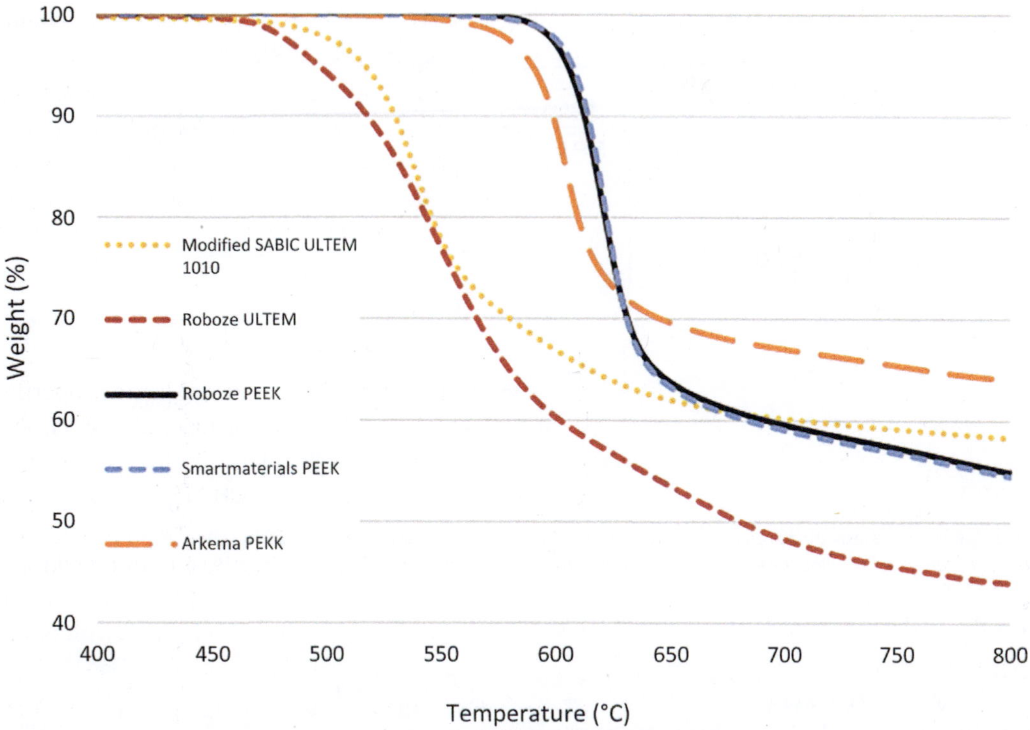

FIGURE 6–6 Thermogravimetric analysis of polyetherimide, polyetherketoneketone, and polyetheretherketone in nitrogen gas and at a rate of 20°C min^{-1} (Wu et al., 2020). *From Wu, H., Kim, S., Fahy, W., Koo, J.H., Kafi, A., Bateman, S. et al., 2020. Ablation performances of additively manufactured high-temperature thermoplastic polymers. In: AIAA Scitech 2020 Forum. American Institute of Aeronautics and Astronautics (AIAA), United States, 1. American Institute of Aeronautics and Astronautics, https://doi.org/10.2514/6.2020-1125.*

chosen for this process due to its noteworthy qualities, which include excellent dimensional stability, cost-effectiveness, and high char yield (Jwo et al., 1992), as shown in Fig. 6-8.

Phenol and its derivatives are widely used in many different industries, such as composite materials, adhesives, coatings, and molding. These reinforced composites, which include materials like silica or carbon, are mostly used as TPS materials in space applications (Venkateshwaran and Elayaperumal, 2010).

Among the challenges phenolic resins face are the requirement for elevated temperatures and pressures, a brief shelf life, and the possibility of void formation (Santhosh Kumar and Reghunadhan Nair, 2014). One of the newly discovered phenolic derivatives is polybenzoxazine. This polymer has improved properties, such as it has no volatile by-products and therefore low void formation, it can cure without the use of a catalyst, it has low hygroscopic properties, high char yield, good stability, and good mechanical properties, among others (Santhosh Kumar and Reghunadhan Nair, 2014). Table 6–1 compares the properties of phenolics with those of polybenzoxazine.

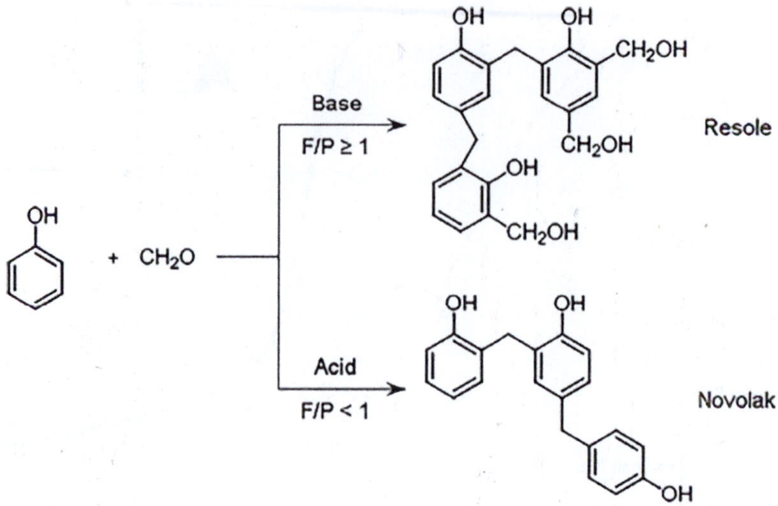

FIGURE 6–7 Synthesis of novolac and resole resins (Pilato, 2013). *From Pilato, L., 2013. Phenolic resins: 100 Years and still going strong. Reactive and Functional Polymers. 73 (2), 270–277. https://doi.org/10.1016/j.reactfunctpolym.2012.07.008.*

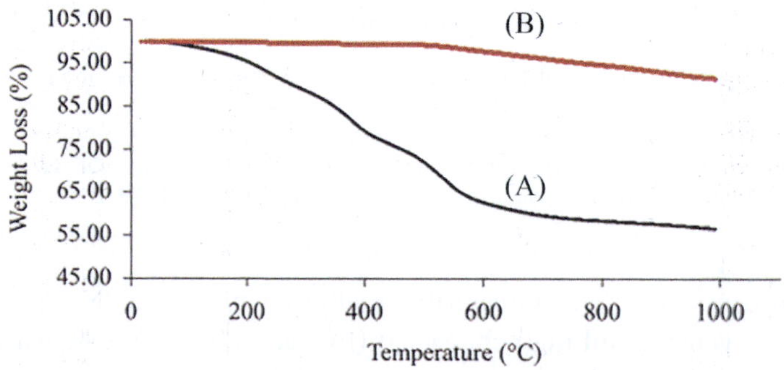

FIGURE 6–8 The results of thermal gravimetric analysis for phenolic resin (A) and carbon fiber resin composite (B) at a heating rate of 10°C min^{-1} in an atmosphere of nitrogen gas (Souza et al., 2015). *From De Souza, W.O., Garcia, K., De Avila Von Dollinger, C.F., Pardini, L.C., 2015. Electrical behavior of carbon fiber/phenolic composite during pyrolysis. Materials Research 18(6), 1209–1216. https://doi.org/10.1590/1516-1439.000515.*

CF and PEEK were recently used to strengthen polybenzoxazine, and the resulting novel polymer composite has better mechanical and thermal characteristics. The tensile strength, flexural modulus, and Young modulus of this polymer composite are 50.1-78.6 MPa, 13.1-27.4 GPa, 62.1-88.3 MPa, and 10.2-24.3 GPa, respectively. Its coefficient of friction is stable. The wear resistance (1.79×10^{-7} cm^3 Nm^{-1}) and thermal conductivity of the composites are improved by the inclusion of CF. However, the incorporation of CF reduces the linear coefficient of thermal expansion (Guo et al., 2023).

Table 6–1 The comparative properties of phenolics and polybenzoxazine.

Properties	Phenolics	Polybenzoxazine
Cure temperature (°C)	150–190	160–220
Maximum use temperature (°C)	200	130–280
Cure shrinkage	0.002	~0
T_g (°C)	170	170–340
Tensile strength (MPa)	24–45	100–125
Tensile modulus (GPa)	5-Mar	3.8–4.5
Elongation (%)	0.3	2.3–2.9
Dielectric constant (1 MHz)	10-Apr	3–3.5
Degradation onset temperature (°C)	300–360	380–400

Source: Adapted from Santhosh Kumar, K.S., Reghunadhan Nair, C.P., 2014. Polybenzoxazine-new generation phenolics. In: Handbook of Thermoset Plastics. Elsevier Inc., pp. 45–73. https://doi.org/10.1016/B978-1-4557-3107-7.00003-8.

6.3.6 Polyphenylene sulfide

Polyphenylene sulfide (PPS), a kind of thermoplastic polymer, consists of repeat units of an aromatic ring and a sulfur atom. One of the easiest ways to synthesize PPS is the polymerization of diphenyl disulfide with oxidants, that is, Lewis acids such as $SbCl_5$, $MoCl_5$, and $TiCl_4$. These Lewis acids enhanced the structure, morphology, and transport properties of the polymer composites according to Lenz et al. (1962). PPS offers several unique qualities, such as a high tensile strength of around 65 MPa, accurate moldability, thermal stability up to 500°C, low solubility in organic solvent below 200°C, 1.1% elongation, and chemical resistance. The glass transition (T_g) and melting temperatures (T_m) of the semicrystalline PPS are 88°C–93°C and 280°C–285°C, respectively (Oyama et al., 2011). Due to its special qualities, PPS may be used in a variety of settings, such as electronics, composite materials, and ball valves (Lopez and Wilkes, 1989).

PPS is a common material option for engineering applications because of its improved mechanical, chemical, thermal, and electrical characteristics. It is frequently reinforced with fillers, fibers, or other polymers (Chen et al., 2021). Melted glass, mineral fillers, CFs, polymer fibers, and metal particles are often used as fillers for PPS composite reinforcement. Because they provide better mechanical properties to the polymer composites, GF fillers and GFs are especially common in reinforcing PPS composites (Barbosa et al., 2019; Lohr et al., 2019; Zhao et al., 2019).

Jiang et al. investigated the mechanical characteristics of PPS that had been improved by carbon nanotubes. Fig. 6–9 shows how the homogeneous dispersion of carbon nanotubes throughout the polymer matrix led to better mechanical characteristics (Jiang et al., 2012). However, the prepared PPS composites exhibited a lot of issues. For example, the carbon fillers' short length prevents them from continuously reinforcing the polymer composites. Thus more investigation is required to attain high mechanical characteristics in fiber textiles and their structural optimization (Chen et al., 2021).

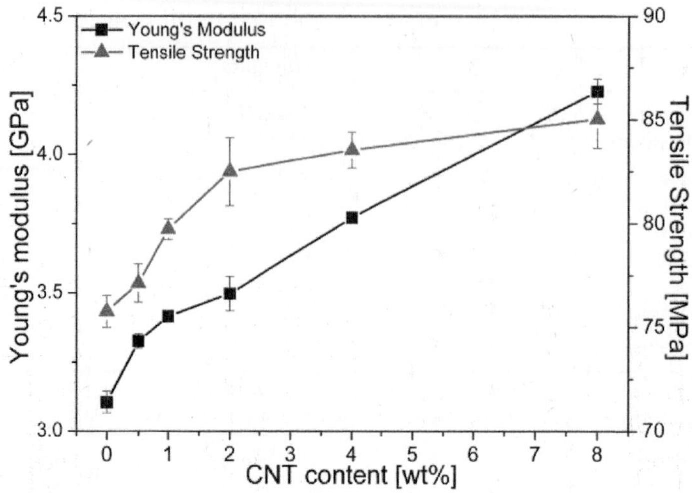

FIGURE 6–9 Mechanical properties of polyphenylene sulfide with different amounts of carbon nanotubes (Jiang et al., 2012). *From Jiang, Z., Hornsby, P., McCool, R., Murphy, A., 2012. Mechanical and thermal properties of polyphenylene sulfide/multiwalled carbon nanotube composites. Journal of Applied Polymer Science 123(5), 2676–2683. https://doi.org/10.1002/app.34669.*

6.3.7 Thermoset

Thermoset, or highly crosslinked covalent network polymers, provide superior mechanical, thermal, and chemical resistance as well as dimensional stability. They are frequently utilized in many different applications, including surface coatings and kettle handles. However, thermosets cannot be recycled or reprocessed after failure. Also, "creep," a shape-changing phenomenon brought on by reversible bond-exchange events, such as disulfide crosslinks, is perceived as a disadvantage of these polymeric materials (Shaw, 1987). As a result, thermoset networks are made to be basically nonrecyclable and irreversible. Other examples of high-temperature polymers include polyamide-imide, polyethersulfone, polyetherketone, Liquid crystal polymers, polytetrafluoroethylene, ethylene tetrafluoroethylene, perfluoroalkoxy, and fluorinated ethylene propylene.

6.4 Types of fiber reinforcements

6.4.1 Carbon fibers

CFs are usually obtained from polyacrylonitrile carbonization and then graphitized, as shown in Fig. 6–10. (Hsissou et al., 2021). CFs have extensive applications in high-temperature scenarios due to their remarkable strength and rigidity at elevated temperatures (Yao et al., 2018). CFs have notable mechanical properties and resilience to high temperatures. They are primarily composed of carbon atoms arranged in a crystalline structure; these fibers exhibit exceptional strength, stiffness, and thermal stability, rendering them well-suited for applications in elevated-temperature environments (Zheng et al., 2022). The thermal resistance of

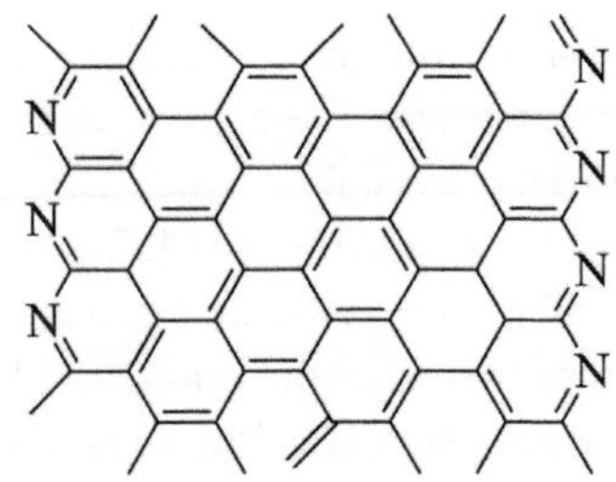

FIGURE 6–10 Structure of carbon fiber (Hsissou et al., 2021). *From Hsissou, R., Seghiri, R., Benzekri, Z., Hilali, M., Rafik, M., Elharfi, A., 2021. Polymer composite materials: a comprehensive review. Composite Structures 262. https://doi.org/10.1016/j.compstruct.2021.113640.*

CFs can maintain integrity even at temperatures surpassing 1000°C. CFs are lightweight, possess strong fatigue resistance, and are noncorrosive (Li et al., 2022). Their ability to withstand high temperatures makes them extensively used in the production of thermosetting composite materials. They are used across different industrial sectors such as space construction, aeronautics, aviation, and automobile industries (Bello, 2021).

A recent review by Zheng et al. (2022) critically analyzed the possible theories of the interphases in CF polymer composites. They listed some theories they believe are the likely reasons why the fiber–resin interphase in CF polymer composites is so strong. These include (1) Mechanical interlocking, (2) chemical interaction, (3) infiltration, (4) diffusion, (5) adsorption, and (6) transition layer. Also, since the surface of CF can be modified, modification of the CF improves the interphase interaction significantly.

Carpier et al. (2020) evaluated the thermomechanical behavior of some thermoplastic-based polymer composites compounded with CFs when subjected to a combination of mechanical loading under tension and homogeneous temperature conditions. Their results showed that the decomposition onset temperature was about 500°C, regardless of the testing atmosphere. Based on the results of isothermal tensile tests conducted at various temperatures from room temperature to 520°C, the quasiisotropic CFs reinforced PPS laminates retain a significant portion of their axial stiffness. Specifically, they retain about 70% of their initial stiffness beyond the decomposition temperature. In another work, Benoit and coresearchers evaluated CF composites' postfire behavior for aeronautical applications. The work involved comparing the properties of epoxy and PPS CF laminates (Benoit et al., 2014). Examining the tensile properties after a fire indicates that carbon/epoxy laminates are more adversely affected by severe fire exposure than carbon/PPS laminates. Despite the highly ductile behavior of PPS matrix at temperatures exceeding the glass transition temperature, it is evident that the decline in tensile properties of PPS-based composites is significantly slower than that observed in carbon/epoxy laminates under similar severe fire conditions.

6.4.2 Glass fibers

GFs are synthetic fibers made from finely ground quartz, sand, limestone, kaolin, and colemanite (Veit, 2023). They offer excellent strength, are lightweight and corrosion-resistant, and have durability, thermal stability, impact resistance, chemical resistance, friction, and wear properties; however, recycling these fibers is usually difficult. They are among the cheapest lightweight fiber materials available with comparatively higher strength and outstanding ability to adhere to polymer matrices (Chalmers, 1991; Prakash, 2019).

In a study conducted by Torabizadeh (2013), the behavior and material properties of unidirectional glass FRCs were investigated under static and low-temperature conditions. The study revealed that as the temperature decreases, the tensile strength and Young's modulus of the laminates increase until breakage. Moreover, brittleness increases significantly at cryogenic temperatures. The shear stress–strain relation was found to be somewhat nonlinear, and in all cases, ultimate failure occurred catastrophically. The imperfect bonding between the fibers and matrix is responsible for this behavior. At low temperatures, compressive loading causes bonding imperfections due to the curing of unidirectional laminates, which is associated with the plastic deformation of the polymeric matrix. The results indicate highly nonlinear behavior of in-plane shear stress at all temperatures.

Sathishkumar et al. (2014) published a review study on GF-reinforced polymer composites, which evaluated a variety of processing technologies. The study reported on the mechanical, thermal, water absorption, and vibrational characteristics of different GF-reinforced polymers. Various GFRP composites were created using different manufacturing techniques, including hand-lay process, hydraulic press, hot press technique, compression molding, and matrix mixing and molding methods. The study found that the maximum flexural strength and tensile strength were observed at 25% V_f of fiber, while the highest impact strength was found at 0% V_f of fiber. Furthermore, the ductility of the composite material was observed to increase with increasing curing pressure for nonsymmetrical arrangement and decrease for symmetrical arrangement.

Ye et al. (2020) conducted a study on the failure of FICs under thermomechanical processing. The findings revealed that significant ambient temperature fluctuations generally lead to stress degradation and heightened failure intensity in both uniaxially and biaxially loaded laminates. The highest stresses in composites are typically located near the interface under both uniaxial and biaxial tensile loads, regardless of considering temperature differences. GF polymer composites can be manufactured using various methods depending on (1) intended matrix, (2) the temperature requirements for processing, including melting, forming, and curing of the material (Morampudi et al., 2020; Rajak et al., 2019; Rajak et al., 2021). The mechanical and thermal properties of some class of GF are given in Table 6–2.

6.4.3 Aramid fibers

Aramid fiber was the first manmade organic fiber employed as reinforcement in advanced composites because it possesses high tensile modulus, high thermal stability, excellent wear resistance, low density, and strength high enough for various applications due to the increased crystallization level as well as strong interchain bonding. Aramid fibers are fibers from aromatic polyamide (aramids) with at least 85% of the amide group (CO–NH) directly bonded to two aromatic rings (Jassal et al., 2020; Seydibeyoğlu et al., 2017). These fibers offer

Table 6–2 The properties of different classes of glass fibers as reported by Rajak et al. (2021). A—alkali-lime glass with little or no boron oxide, C—alkali-lime glass with high boron oxide content, D—borosilicate glass, E—alumino-borosilicate glass with less than 1 wt.% alkali oxides, AR-Glass resistant to alkali environment, R—alumino-silicate glass without MgO and CaO content, S—alumino-silicate glass without CaO but with high MgO content.

Properties		Type of glass fiber						
		A	C	D	E	AR	R	S
Physical	Density (g cm^{-3})	2.44	2.52–2.56	2.11–2.14	2.54–2.6	2.7	2.54	2.48–2.49
Mechanical	Tensile strength (MPa)	3310	3310	2415	3450	3241	4135	4585
	Elongation break (%)	4.8	4.8	4.6	4.8	4.4	4.8	5.4
	Young modulus (GPa)	68.9	68.9	51.7	72.4	73.1	85.5	85.5-46.9
	Poisson's ratio (–)	0.183	0.276		0.2			0.22
	Shear modulus (GPa)	29.1	27		30			35
Electrical	Electrical resistivity (Ω cm)	10^8	–		4.02×10^{12}		2.03×10^{12}	9.05×10^{10}
	Dielectric constant (–)	6.2	6.9	3.8	5.9–6.4	8.1	6.4	5.1–5.34
	Dissipation constant (–)	–	0.0085		0.0025			0.0034
	Dielectric strength (kV mm^{-1})				10.3			13
Thermal	CTE linear (μm m^{-1} °C)	9	6.3	2.5	5	6.5	3.3	5.2–5.6
	Specific heat capacity (J g^{-1} °C)	0.796	0.787	0.733	0.81		0.732	0.737
	Softening point (°C)	727	750	771	840.6	773	952	
	Thermal conductivity (W m^{-1} K)	–	1.1		1.3			
	Thermal expansion coefficient (x10^{-7})	73	63	25	54	65	33	
	Melting point (°C)				≥1725			≥1725
	Annealing point (°C)		588	521	657			
	Strain point (°C)		522	477	615		736	
Optical	Refractive index (–)	1.534	1.533	1.465	1.558	1.562	1.546	1.525

Source: From Rajak, D.K., Wagh, P.H., Linul, E., 2021. Manufacturing technologies of carbon/glass fiber-reinforced polymer composites and their properties: a review. Polymers 13(21). https://doi.org/10.3390/polym13213721.

significantly superior mechanical properties that are 5%–10% higher compared to other synthetic fibers, as well as glass and steel fibers with an equal weight. Additionally, aramid fibers are naturally heat and flame-resistant, retaining these properties even at elevated temperatures. These excellent properties of the aramids make them applicable in the manufacturing of aircraft, marine, and cars. They are used to make bulletproof vests and protective clothing for chemical, high-temperature environments, and radiation purposes. They are used as finishing materials for flame-resistant purposes, industrial filters, and desalination fibers in reverse osmosis (Cai et al., 2015; Jassal and Ghosh, 2002). Examples of aramids include, but are not limited to, Kevlar, Nomex, Technora, etc. Kevlar fiber is a type of aramid fiber used to reinforce the beam-column joint. This type of aramid fiber is obtained from the reaction between terephthalic diacid chloride and *para*-phenylene diamine in hexamethyl phosphoramide and *N*-methyl pyrrolidone (Granata and Parvin, 2001).

In recent years, aramid nanofibers have become increasingly prevalent in the fields of polymer reinforcement and toughening. This is attributed to their substantial specific surface area, high aspect ratio, and ability to maintain the exceptional strength, modulus, and high-temperature resistance of aramid micron fibers (Zhang et al., 2020). In AF-reinforced polymers, the adhesion between the fiber–polymer directly impacts the safety and durability of the composites during use. However, AF's chemical inertness, high crystallinity, and smooth surface often require surface modification to enhance the chemical bonding and mechanical interlocking with the polymer matrix (Lertwassana et al., 2019; Qi et al., 2017; Raphael et al., 2018; Zhang et al., 2021). Al-Quraishi et al. (2020) used plasma treatment to modify the aramid surface and increase the mechanical properties by improving the interface between the fibers and the polymer matrix. A modified AF can be processed with most conventional methods used for high-temperature polymers.

6.4.4 Ceramic fibers

Tressler (1999) reported that Ceramic fibers are engineered for high-temperature stability. They are frequently employed in applications necessitating resilience against thermal shock. They can withstand temperatures up to 1000°C or higher, depending on their specific composition. Ceramic fibers have higher tensile strength than CFs. They are capable of withstanding high temperatures due to their high strengths, high Young modulus, better chemical stabilities, and resistance to oxidation. Ceramic fibers for high-temperature applications are grouped into two categories, monoxide fibers or oxide fibers. The monoxide fibers are mainly based on silicon carbide, which could be either amorphous or nanocrystalline fibers. The oxide fibers are mainly alumina-based. Silicon carbide fiber (SiC): SiC fibers such as Nicalon boast exceptional thermal stability and mechanical properties in high-temperature environments and can resist oxidation. They can endure temperatures reaching up to 1600°C in inert atmospheres, exhibit commendable resistance to oxidation, and retain excellent chemical, physical, and mechanical properties in this severe environment. They can resist oxidation at high temperatures due to the SiO_2 passivation layer formed in an oxidation environment (Ohnabe et al., 1999; Yilmaz et al., 2024). They are often produced by chemical vapor deposition on CF (Hsu and Speyer, 1992). Alumina fibers, composed predominantly of aluminum oxide, prove suitable for high-temperature applications up to 1600°C. They offer outstanding thermal stability along with chemical resistance. Koch et al. (2008) reported that ceramic fiber reinforced matrix composite offers higher strength, stiffness, and they are developed for different applications at high temperatures.

6.4.5 Other specialty fibers

Other specialty fibers include, but are not limited to, natural fibers, cellulosic fibers, etc. Natural fibers are either plant-based, animal-based, or mineral-based (Rajak et al., 2019). They are readily accessible and widely available in nature. They exhibit exceptional properties such as biodegradability, low cost per unit volume, excellent strength, and specific stiffness. Composites reinforced with natural fibers offer several advantages over synthetic fibers, including weight, price, toxicity, and enhanced recyclability. These economic and environmental benefits make natural fiber composites preferable to synthetic FICs for modern applications (Rajak et al., 2019; Suresh et al., 2020). Natural fibers are readily available in large quantities and with a large specific surface area and can be chemically treated to rival synthetic fibers (Suresh et al., 2020). Despite the numerous advantages of natural fibers, their limitations are dimensional changes due to moisture absorption of these fibers, which could occur, resulting in weak interfacial adhesion of the compound (Bisanda and Ansell, 1991). Natural fibers show limited thermal stability, which may result in degradation during the processing at high temperatures; hence the need to treat or add other composites to the natural fibers to improve their water resistance, improve surface wetting, and improve the interfacial adhesion (Buitrago et al., 2015). The following are some examples of natural fibers commonly utilized in the industry, this include sisal fiber, jute fiber, hemp fiber, kenaf fiber, flax, fiber, etc.

6.5 Manufacturing processes

The final performance of a composite depends not only on the choice of the matrix and the fiber but also on the manufacturing process by which they are made. From the inception of the composite's life, it is important to consider the presence of imperfections resulting from manufacturing. These imperfections can readily damage the composite piece or lead to rapid deterioration. The rapid expansion of manufacturing industries has created a demand for materials with improved properties such as strength, stiffness, density, and cost-effectiveness, while also being more sustainable. Composite materials have emerged as a promising solution, offering enhanced properties and finding applications across various industries (Chukov, Nematulloev, Zadorozhnyy, et al., 2019; Linul et al., 2019; Sherif et al., 2019; Yashas Gowda et al., 2018). The incorporation of both natural and synthetic fibers in the production of composite materials has demonstrated extensive potential in a wide range of industries, including construction, mechanical engineering, automotive, aerospace, biomedical, and marine applications, as referenced in sources (Monteiro et al., 2018; Movahedi and Linul, 2017; Zagho et al., 2018).

Research studies from the past two decades have presented composites as an alternative to many conventional materials. FRC material significantly enhances the structural, mechanical, and tribological properties (Chukov et al., 2019; Linul et al., 2018). The strength depends on the fiber–matrix bonding, each fiber's properties, the microvoids, and the optimum fabricating method (Muralidhara et al., 2019).

6.5.1 Resin transfer molding

The resin transfer molding (RTM) process involves using a rigid closed mold. In a basic scenario, a preformed fiber preform is initially placed in an open mold. Typically, the RTM process utilizes a two-sided metal mold with a channel carved into it (Muralidhara et al., 2019). Once the mold is closed and clamped, resin with low viscosity is injected under positive gradient pressure through the gate points, displacing any air trapped within the preform. Typically, a vacuum is used at specific vents to assist in the removal of air from the mold (Adhikari et al., 2021). Once the resin reaches the vents, the gates are sealed, and the preform is saturated. This marks the beginning of the curing phase (Laurenzi and Marchetti, 2012; Miranda Campos et al., 2022; Ornaghi et al., 2010). After the manufacturing process, the mold is opened to remove the part. For structural applications in aerospace, an additional postcuring phase may be required to ensure the complete polymerization of the matrix and the release of internal thermal stress (Laurenzi and Marchetti, 2012).

The RTM process is governed by variables and parameters that are dependent on each other. Their combination affects the process and the quality of the finished product. Consequently, they need to be carefully determined. The most important parameters, which can not be neglected in the design, are pressure, temperature, viscosity, permeability, volume fraction, and filling time of the process (Ageyeva et al., 2019; Lee et al., 2020; Sun et al., 2018).

RTM has the following advantages over other processing techniques: (1) less manual operations and low labor cost, since the mold can be designed to be a closed one and the resin injection can be mechanized, (2) good dimensional tolerance and reproducibility, because the volume of the part and necessary materials have been precisely defined by the mold design and the preform, (3) smooth surface finish without need of secondary finishing operations (Gong and Yang, 2010). However, RTM also has its inherent limitations. For example, the cycle time is relatively long; it will easily take several minutes for each cycle, so it is not suitable for mass production. And the mechanical property of the product is highly dependent on the fiber content and fiber orientation (Ornaghi et al., 2010). Also, RTM is only suitable for fiber reinforced polymers with thermosetting matrix, since the polymer has to crosslink during the curing process to form a solid part. RTM is widely used to manufacture large and relatively thin structural parts for aerospace, automotive, marine, and wind energy applications. These parts include radomes, aerospace fairings, automotive body panels, wind turbine blades, etc.

Many research works have been focusing on improving the manufacturing efficiency and mechanical performance of RTM products. For example, intelligent sensors can be integrated into the mold to monitor the resin flow front in real time and optimize mold designs using improved resin flow simulation methods. Fernberg and his colleagues produced carbon FICs using RTM in a stainless-steel tool with a flow and pressure-controlled injection machine from ISOJET. The composites were typically 350 mmx350 mmx3 mm in size and featured a multidirectional [(-45/45)/(90/0)]2S quasiisotropic layup with a fiber volume fraction (V_f) of 58%±1%. Two different curing schedules were utilized at various stages of the development process, and the mechanical and thermal properties were assessed (Fernberg et al., 2018).

The effect of fiber length was investigated by Ornaghi and Coworkers, who prepared hybrid composites using RTM (Ornaghi et al., 2010). The analysis of flexural and impact results showed that, overall, the properties tended to be higher with increased reinforcement content. The study observed that the storage and loss modulus increased with higher glass loading and overall fiber volume. Additionally, there was a shift to higher values in these moduli with increasing glass loading and overall fiber volume. The filler showed improved efficiency, and the calculated activation energy for the relaxation process in the glass transition region also increased. Additionally, the fiber length did not have a significant impact on the results observed in all analyses conducted in this study (Hsiao and Heider, 2012).

Also, a combination of RTM with other processing techniques like filament winding and injection molding has been studied to fulfill the specific requirements for different applications. Fu and his coresearchers used a vacuum-assisted RTM (VARTM) processing method to prepare high-temperature PETIs (Fu et al., 2011). This process has proven to be cost-effective for manufacturing composite structures compared with prepreg/autoclave and traditional RTM processes. Two processing methods, in-plane and through-thickness resin flow, were used. These methods were confirmed to be capable of fabricating PI matrix composites, and the CF laminates yield good fiber–resin interfacial bonding and comparable mechanical properties to those laminates fabricated using RTM (see Fig. 6–11).

According to Hong et al. (2021), phenylethynyl-terminated aromatic PIs, which exhibit the requirements for RTM and process high glass transition temperature (T_g). In addition, the relationship between the PI backbones structure and their melting stability was investigated. These oligoimides exhibit excellent melting flowability with wide processing temperature window and low minimum melt viscosities (<1 Pa s). Two of the oligoimides display good melting stability at 280°C–290°C, which is a requirement of the RTM process. The oligoimide powders were compression molded in stainless steel (80x80 mm^2) with a hot-press machine. They were melted at 305°C, gradually raised to 380°C in 50 min, and cured for 2 h under 4 MPa pressure. When the cured resins were cooled to about 250°C, the pressure was released, and the sample was removed from the mold and stored under ambient temperature. After thermally curing, all resins show high glass transition temperatures (T_g, 363°C–391°C)

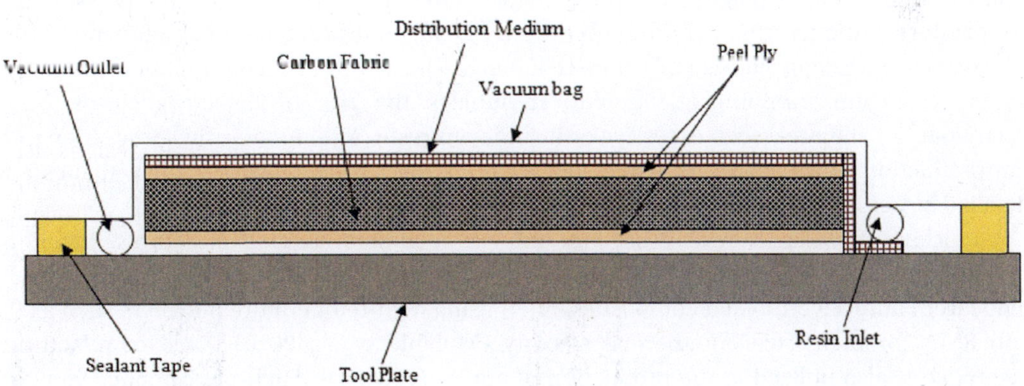

FIGURE 6–11 Schematic representation of a single-bag vacuum-assisted resin transfer molding process by Menta et al. (2013). *From Menta, V., Vuppalapati, R., Chandrashekhara, K., Schuman, T., Sha, J., 2013. Elevated-temperature vacuum-assisted resin transfer molding process for high performance aerospace composites. Polymer International 62(10), 1465–1476. https://doi.org/10.1002/pi.4444.*

and good tensile strength (51–66 MPa). The cure kinetics studied by the differential scanning calorimetry, 13C nuclear magnetic resonance characterization, and density functional theory definitely confirmed that the electron-withdrawing ability of oligoimide backbone can tremendously affect the curing reactivity of terminated phenylethynyl groups.

6.5.2 Autoclave molding

Autoclave molding is a highly specialized and widely used process for manufacturing high-temperature polymer composites, particularly in the aerospace and automotive industries (Laurenzi and Marchetti, 2012). Preimpregnated fibers with resin, known as composite material prepregs, are meticulously laid up in the mold following a specified orientation and sequence. This process can be carried out manually for precision or by automated machines for enhanced efficiency and consistency. Subsequently, a release film is placed over the lay-up, followed by a breather fabric to ensure uniform pressure distribution. The entire assembly is then enclosed in a vacuum bag, which is sealed to the tool/mold to create an airtight environment (Adhikari et al., 2021).

In the manufacturing process, a vacuum is utilized to eliminate air and volatiles, preventing any voids in the final composite material and ensuring thorough consolidation of the layers (Muralidhara et al., 2019). The vacuum-bagged lay-up is then placed inside an autoclave, a pressure vessel that applies both heat and pressure. The autoclave is heated to the curing temperature specified for the resin system, typically between 120°C and 180°C for standard epoxies, but possibly exceeding 400°C for advanced polymers like PIs. The curing cycle duration varies from a few hours to over 24 h, depending on the complexity of the part and the resin system used. Subsequently, the part is gradually cooled to prevent thermal shocks. Following this, the vacuum bagging (VB) materials are removed, and the cured composite part is demolded.

The autoclave molding process yields composites with outstanding mechanical properties, characterized by high strength and stiffness, attributed to the application of high pressure and temperature during curing (Abdurohman et al., 2018). Autoclave methods offer numerous advantages, including consistency and reliability. The controlled environment within the autoclave ensures uniform quality and minimizes the risk of defects such as voids, delamination, and areas with excessive or insufficient resin. Additionally, this process enables the production of intricate geometries, allowing for the fabrication of complex shapes and detailed structures that are challenging to achieve using other molding methods.

Autoclave molding is widely used in the production of aircraft structures, including fuselage panels, wing components, and control surfaces. These components are typically made from autoclave-molded composites, often using high-temperature polymers such as PIs and BMIs for their excellent thermal stability (Mujahid et al., 2020). This manufacturing technique is also utilized in the production of automobile parts. High-performance vehicles, such as sports cars and racing vehicles, use autoclave-molded composites for body panels, chassis components, and interior parts to create lightweight yet strong structures. Additionally, autoclave-molded composites find extensive use in industrial and military applications, including radomes, satellite structures, and other high-performance components that require exceptional material performance in demanding conditions.

Autoclave molding is a widely employed technique in the manufacturing of aircraft structures, encompassing fuselage panels, wing components, and control surfaces. These parts are commonly crafted from autoclave-molded composites, often utilizing high-temperature polymers like PIs and BMIs for their superb thermal stability. This method is also utilized in the production of automotive components. High-performance vehicles, such as sports cars and racing vehicles, employ autoclave-molded composites for body panels, chassis components, and interior parts to achieve lightweight yet durable structures. Moreover, autoclave-molded composites are extensively utilized in industrial and military applications, including radomes, satellite structures, and other high-performance components that necessitate exceptional material performance in challenging conditions. Composites for industrial/mass production applications are typically produced by autoclave process under elevated temperature and pressure (Grunenfelder et al., 2017).

6.5.3 Vacuum bagging

Vacuum infusion or vacuum-assisted resin transfer mold (VARTM) is a recent development in which perform fibers are placed on a mold, and a perforated tube is positioned between vacuum bag (VB) and resin container (Laurenzi and Marchetti, 2012). Vacuum force causes the resin to be sucked through the perforated tubes over the fibers to consolidate the laminate structure, as shown in Fig. 6–12 (Abdurohman et al., 2018). VB is a standard method of manufacturing composite material (Menta et al., 2013). The purpose of VB is to squeeze out air from between the layers of dry fiber and between the layers of fiber and resin in a wet lay-

FIGURE 6–12 Vacuum bagging process (Abdurohman et al., 2018). *From Abdurohman, K., Satrio, T., Muzayadah, N.L., Teten, 2018. A comparison process between hand lay-up, vacuum infusion and vacuum bagging method toward e-glass EW 185/lycal composites. Journal of Physics: Conference Series. 1130, 012018. https://doi.org/10.1088/1742-6596/1130/1/012018.*

up, which helps to reduce void content. This process leaves no room for excess air in the composite structure, making it popular for manufacturing large objects, such as boat hulls and wind turbine blades (Mujahid et al., 2020).

VB is ideal rather than the traditional hand lay-up process for large, concave, and deep composite parts. The VB technique can be used as a repair method for composites. When utilized, it is common to refer to it as a "wet vacuum bag" repair. This means that the composite part is repaired with a wet resin lay-up, and a vacuum bag is applied to it. When properly managed, VB can be safer and offer a higher quality solution to many manufacturing needs. It helps create a better quality laminate and decreases waste because an excess of resin is less likely to accumulate with vacuum pressure. Quality assurance and consistency are also improved by the use of VB, making it a highly regarded composite manufacturing method. VB is a promising technique for manufacturing advanced laminated composites and sandwich structures (Mujahid et al., 2020). Vacuum bag autoclave process (VBA) is commonly known to produce high-quality composites with very minimal void content. However, manufacturing process for composites using VBA process is extremely costly because of high procurement, operation, and maintenance costs involved in a specialized autoclave vessel.

The nonautoclave VB processes offer the potential for reduced cost and cycle time and have been receiving increased attention as an alternative for producing aerospace-quality parts (Menta et al., 2013). Compared with RTM and out-of-autoclave vacuum-bag-only (VBO) prepreg processes, VARTM is rapidly emerging as a competitive low-cost alternative for autoclave processes (Menta et al., 2013). The VARTM process uses a one-sided vacuum-sealed mold and vacuum pressure to draw the resin into the mold.

Abdurohman et al. (2018) compared hand lay-up, vacuum infusion, and VB methods using plain weave type woven glass fabric as reinforcement and epoxy resin. The results obtained showed that the vacuum process was unable to eliminate the gas and voids that were already stuck in the laminate because the curing started before the vacuum. This is a potential reason for failure in the material. However, those on the surface were effectively removed, improving the overall material's mechanical properties. Complex composite structures have been reported to be very susceptible to defects when manufactured using a low-pressure VBO method when compared to flat laminates. Mujahid et al. (2020) examined how various processing parameters, including different VB techniques (single VBO, modified single-VBO, and double VBO), curing profiles (manufacturer's recommended curing cycle, extended manufacturer's recommended curing cycle, and direct cure cycle), and laminate structures (concave corner and convex corner), impact the shape nonconformity of complex-shaped laminated composites. The findings demonstrated strong consistency in the uniformity of laminate thickness with the MSVB technique. This method not only maximizes shape conformation by avoiding the bridging effect in concave corners of the mold to prevent significant thickening but also prevents significant thinning in convex corners. The research laid the groundwork for minimizing variations in thickness for laminated composites with intricate shapes. This was achieved through the precise selection of processing parameters for prepreg processing.

Zhang et al. (2017) investigated the void reduction mechanism in oven vacuum bag processing of high CF composites. The researchers found that through-thickness air diffusion and in-plane flow to the laminate edges occur through the permeable interlayer regions. Therefore laminates with edges open to a vacuum, which allows the diffusion of air through a single layer and flow through the permeable interlayer, leading to essentially void-free laminates.

6.5.4 Filament winding

Filament winding is another highly specialized technique for manufacturing high-temperature polymer composites. It is particularly useful for creating cylindrical or spherical components with excellent strength-to-weight ratios (Boon et al., 2021). The mandrel, which defines the internal shape of the wound composite, can be made from various materials, including metals, plastic, or soluble materials for easy removal. The mandrel surface is treated or coated to facilitate easy removal of the finished part. Release agents are often applied (Liang et al., 2023). Continuous fibers (e.g., carbon, glass, aramid) impregnated with high-temperature polymer resins (e.g., epoxy, PI) are used. Fibers can be preimpregnated (prepreg) or wet-impregnated during the winding process using a resin bath (Fig. 6–13). Fibers are wound around the mandrel in specific patterns (hoop, helical, polar) to achieve desired mechanical properties. The angle and sequence of winding are critical for optimizing strength and performance (Sciti et al., 2021). Maintaining consistent tension in the fibers during winding is crucial to ensure uniformity and avoid defects. The wound mandrel is placed in an oven or autoclave to cure the resin. The temperature and pressure conditions depend on the resin system used (Sobrinho et al., 2011). High-temperature polymers require curing temperatures ranging from 120°C to 400°C. For some high-performance composites, an autoclave is used to apply additional pressure during curing, enhancing fiber consolidation and resin distribution. After curing, the part is gradually cooled to room temperature. The mandrel is removed, which can involve dissolving a soluble mandrel, collapsing an inflatable mandrel, or extracting a reusable mandrel.

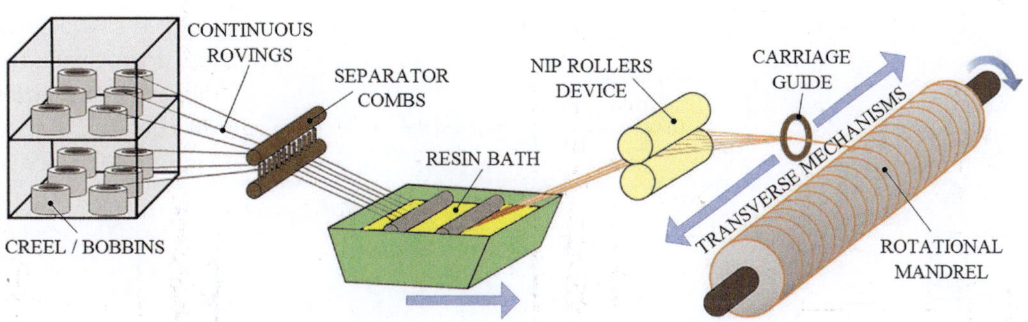

FIGURE 6–13 Schematic diagram of a wet winding process for filament production (Rajak et al., 2021). *From Rajak, D.K., Wagh, P.H., Linul, E., 2021. Manufacturing technologies of carbon/glass fiber-reinforced polymer composites and their properties: A review. Polymers 13(21). https://doi.org/10.3390/polym13213721.*

Filament winding produces composites with superior mechanical properties, particularly in the direction of the fibers (high strength-to-weight ratio), making them ideal for high-performance applications. Filament winding produces composites with good uniformity and precision: The controlled winding process ensures consistent fiber placement and tension, resulting in uniform mechanical properties and minimal defects. The process is well-suited for creating complex cylindrical or spherical shapes, such as pipes, tanks, and pressure vessels (Asyraf et al., 2023).

Filament winding molding is used to manufacture rocket motor casings, pressure vessels, and fuel tanks that require high strength and thermal stability. In the automobile industry, it is used in the production of drive shafts, pressure vessels, and other components that benefit from high strength and lightweight properties. Filament molding is used to manufacture pipes, tanks, and pressure vessels for chemical processing, oil and gas, and other industries. It is also used to produce high-performance sporting goods, such as bicycle frames and golf club shafts.

Some of the challenges in filament molding methods include mandrel removal, initial setup costs, and material waste. However, advancements have been made to improve the process, such as in automated filament winding, advanced resin systems, and multiaxis winding. Filament winding is a versatile and highly effective method for producing high-temperature polymer composites with excellent mechanical properties. Its ability to create complex shapes with precise fiber placement makes it indispensable in industries where performance and reliability are critical.

6.5.5 Pultrusion

Pultrusion is a continuous manufacturing process used to produce high-strength, high-temperature polymer composites with constant cross-sectional profiles (Minchenkov et al., 2021). This process is particularly well-suited for creating structural components that require consistent properties along their length. Continuous fibers such as carbon, glass, or aramid are used, often in the form of rovings (bundles of fibers), mats, or fabrics (Fig. 6-14). High-

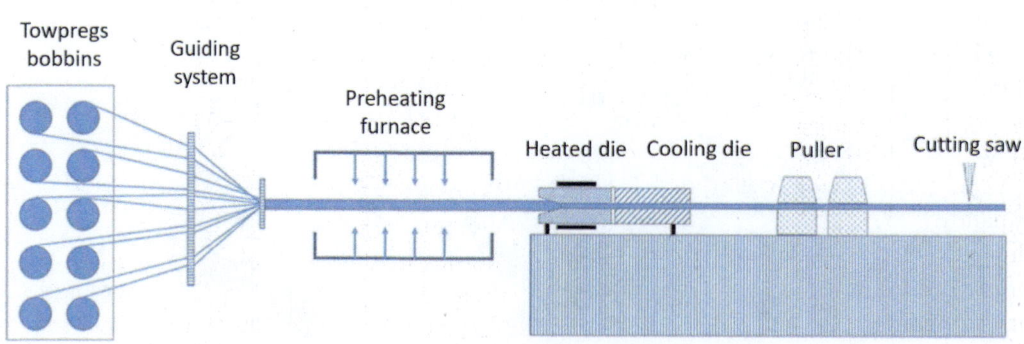

FIGURE 6–14 A schematic diagram of a nonreactive thermoplastic pultrusion line (Minchenkov et al., 2021). *From Minchenkov, K., Vedernikov, A., Safonov, A., Akhatov, I., 2021. Thermoplastic pultrusion: a review. Polymers 13(2), 1–36. https://doi.org/10.3390/polym13020180.*

temperature resin systems, such as epoxy, PI, BMI, or phenolic resins, are chosen based on the required thermal and mechanical properties. Fibers are pulled through a resin bath where they are thoroughly impregnated with the resin (Asensio et al., 2020). Alternatively, preimpregnated fibers (prepregs) can be used. Ensuring proper resin impregnation and control is critical for achieving the desired composite properties (Ranganayakulu et al., 2019). The impregnated fibers pass through a series of guides and preforming dies, shaping them into the desired cross-sectional profile. The material is then pulled through a heated die, which sets the final shape and initiates the curing process. The temperature of the die is carefully controlled to ensure complete curing of the resin.

High-temperature resins require precise temperature control during the curing process, which can range from 120°C to over 400°C, depending on the resin system. Because pultrusion is a continuous process, the curing happens progressively as the material moves through the heated die (Miller et al., 1998; Costa Dias et al., 2018). Continuous pulling of the composite material is achieved using a pulling mechanism that maintains constant tension and speed. After exiting the die and completing the curing process, the continuous profile is cut to the desired length using saws or other cutting tools.

Pultrusion is highly efficient for producing long lengths of composite materials with consistent cross-sectional shapes. The alignment of fibers in the pultrusion process results in composites with high unidirectional strength and stiffness (Correia et al., 2013). The automated and controlled nature of the process ensures uniform quality and properties throughout the length of the composite. For high-volume production, pultrusion offers significant cost savings due to its continuous and automated nature.

Pultruded composites are used in bridges, walkways, handrails, and utility poles due to their high strength, corrosion resistance, and durability. Structural profiles, beams, and panels made by pultrusion are used in building construction for their lightweight and high-strength properties (Irfan et al., 2017; Xu et al., 2020). In aerospace and automotive, pultruded components are used in applications requiring lightweight and high-strength materials, such as aircraft and vehicle structural elements. Wind turbine blades and other renewable energy components benefit from the high strength-to-weight ratio of pultruded composites.

The pultrusion process is best suited for constant cross-sectional profiles, which can limit design flexibility. Setting up pultrusion equipment and dies can be capital-intensive, although it is cost-effective for high-volume production. Processing high-temperature resin systems requires precise thermal control and can be more challenging than standard resins (Irfan et al., 2017; Joshi, 2012).

The development of new high-temperature resin systems and fiber reinforcements has expanded the range of applications for pultruded composites. Furthermore, combining different types of fibers (e.g., glass and carbon) within the same profile can optimize properties for specific applications. Also, advances in automation and monitoring technologies enhance the precision, efficiency, and quality control of the pultrusion process (Tena et al., 2016). Pultrusion is a powerful and efficient method for producing high-temperature polymer composites with uniform properties and high structural performance. Its ability to continuously produce long, consistent profiles makes it ideal for various applications in infrastructure, construction, aerospace, automotive, and energy sectors.

6.5.6 Automated fiber placement

Automated fiber placement (AFP) is a cutting-edge manufacturing technology used to fabricate high-temperature polymer composites, particularly in the aerospace, defense, and high-performance automotive industries. This process involves precisely placing composite material tapes onto a mold or mandrel, layer by layer, using robotic or computer-controlled machinery (Rajak et al., 2021). AFP uses narrow, preimpregnated (prepreg) composite tapes, typically made from high-performance fibers such as carbon, glass, or aramid embedded in a high-temperature resin matrix like epoxy, BMI, or PI. These materials must be stored under controlled conditions to maintain their properties until use. A robotic arm equipped with a fiber placement head lays down the composite tape onto a mold or mandrel according to a programmed path. The placement head can rotate and pivot to follow complex contours and shapes accurately.

The process is often conducted in a temperature- and humidity-controlled environment to ensure material consistency and quality (Rajak et al., 2019). Some AFP systems include mechanisms for applying heat (using infrared heaters, hot gas torches, or lasers) and compaction pressure during tape placement to ensure good bonding and minimize voids. AFP can be used in conjunction with out of autoclave curing methods, although traditional autoclave curing may still be employed for certain high-performance applications. The robot places the tape layer by layer, following precise paths and orientations to build up the desired thickness and achieve optimal mechanical properties. Sensors and feedback systems monitor the process in real-time to ensure accuracy and detect any defects or irregularities. After lay-up, the composite structure undergoes a curing cycle, typically in an autoclave or oven, to fully polymerize the resin and consolidate the composite material.

AFP offers exceptional precision in fiber placement, allowing for the creation of complex and highly optimized composite structures with consistent quality. The process minimizes material waste by precisely placing material only where it is needed, which is especially important for expensive high-performance fibers and resins. AFP allows for the fabrication of complex shapes and structures that would be difficult or impossible to achieve with traditional hand lay-up techniques. Automation reduces the reliance on skilled manual labor and can significantly decrease production time for large and complex parts.

AFP is extensively used in the aerospace industry to manufacture large and complex components such as fuselage sections, wing skins, spars, and fairings. High-temperature polymer composites are essential for parts that must withstand extreme thermal and mechanical loads. Military applications include the production of advanced missile casings, radar domes, and structural components for high-performance aircraft and drones. The high-end automotive and motorsport sectors utilize AFP to produce lightweight and strong structural components, contributing to overall vehicle performance and efficiency. Wind turbine blades and other renewable energy structures benefit from the strength and durability of AFP-manufactured composites.

The cost of AFP equipment and the complexity of setup can be substantial, making it more suitable for high-volume or high-value production runs. Operating AFP systems and programming the robotic lay-up paths require specialized skills and knowledge. Managing and storing high-temperature prepregs properly to maintain their properties can be challenging. Ensuring consistent quality and dealing with potential issues such as fiber wrinkling, gaps, or overlaps requires sophisticated monitoring and control systems.

Continuous improvements in robotic systems and software are enhancing AFP's precision, speed, and flexibility. The development of new high-temperature resin systems and fiber materials is expanding AFP's capabilities and applications. Integration with advanced simulation, design, and quality control software is improving the overall efficiency and reliability of the AFP process. AFP is a transformative technology for producing high-temperature polymer composites with superior mechanical properties and complex geometries (Morampudi et al., 2020). Its precision, efficiency, and ability to handle advanced materials make it an essential tool in modern composite manufacturing, particularly in demanding sectors like aerospace and defense.

6.5.7 Additive manufacturing

AM, also known as 3D printing, is revolutionizing the production of high-temperature polymer composites by offering new design freedoms, rapid prototyping, and customized manufacturing capabilities (Vakharia et al., 2023). While traditionally associated with metals and plastics, AM is increasingly being applied to advanced polymer composites, including those suitable for high-temperature applications. High-temperature polymer composites suitable for AM are typically reinforced with fibers such as carbon, glass, or aramid (Zhang et al., 2021). The polymer matrix may consist of thermoplastics or thermosets capable of withstanding elevated temperatures, such as PEEK, PPS, or PIs (Zalaznik et al., 2016).

Composite filaments or powders are deposited layer by layer according to a digital design file. Some AM processes allow for controlled fiber orientation within each layer, enabling tailored mechanical properties and enhanced performance in specific directions (Fig. 6–15). Various energy sources, including lasers, electron beams, or heat, are used to fuse or solidify the deposited material. After printing, additional processing steps may be required, such as heat treatment to improve material properties, machining to achieve final dimensions, or surface finishing for desired esthetics or functionality.

AM enables the creation of complex geometries and internal structures that are difficult or impossible to achieve with traditional manufacturing methods. AM processes are typically more material-efficient compared to subtractive manufacturing techniques, as they only use the material necessary to build the part. AM allows for the production of customized parts tailored to specific applications or individual requirements (Kafi et al., 2020). AM facilitates rapid prototyping and iteration cycles, speeding up the development process and reducing time-to-market for new products. The ability to manufacture parts on demand and in small batch sizes offers flexibility and reduces inventory costs.

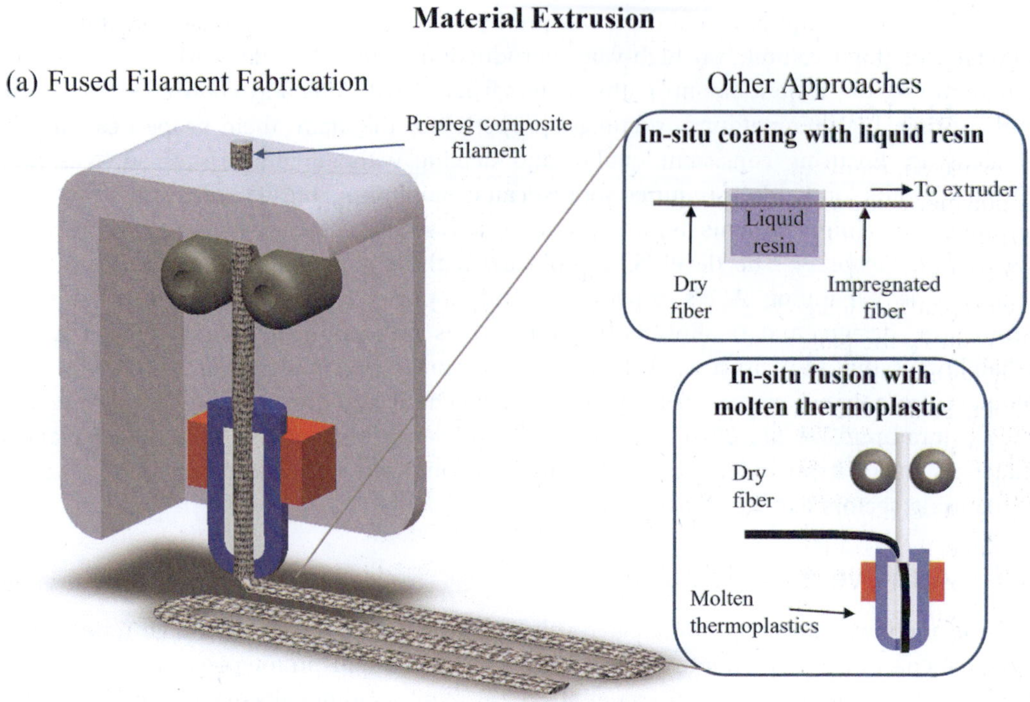

FIGURE 6–15 Schematic representations of a fused filament extrusion process, an example of additive manufacturing. *From Goh, G.D., Yap, Y.L., Agarwala, S., Yeong, W.Y., 2019. Recent progress in additive manufacturing of fiber reinforced polymer composite. Advanced Materials Technologies 4(1). https://doi.org/10.1002/admt.201800271.*

AM is used to produce lightweight, high-strength components for aircraft and spacecraft, such as engine components, brackets, ducting, and interior panels. High-temperature polymer composites manufactured via AM are also used in automotive applications, including engine components, exhaust systems, and lightweight structural parts. AM enables the production of patient-specific implants and medical devices made from biocompatible high-temperature polymer composites. Composite tooling and fixtures with high-temperature resistance can be 3D printed to support various manufacturing processes, such as composite lay-up or metal forming.

Some challenges often encountered with AM include material selection and compatibility, print quality and consistency, postprocessing requirements, and cost and scalability. However, advancements in new materials, process optimization, and hybrid manufacturing have made AM attractive (Kafi et al., 2020; Rajak et al., 2021; Vakharia et al., 2023, 2022). AM for high-temperature polymer composites holds great promise for unlocking new possibilities in design, performance, and customization across various industries. As materials, processes, and technologies continue to evolve, the adoption of AM for producing high-temperature polymer composites is expected to grow, enabling innovative solutions for demanding applications.

6.6 Properties of fiber-reinforced high-temperature polymer composites

6.6.1 Mechanical properties of HTPC

As expected, many articles have reported deterioration in mechanical and thermal properties of polymer composites at elevated temperatures (Bazli and Abolfazli, 2020; Khotbehsara et al., 2019; Li et al., 2017). The tensile, compression, flexural, impact, and stiffness strengths are all affected at elevated temperature. As the temperature is raised, the polymer resin mollify and creeps, causing a significant decrease in the mechanical properties (Jiangtao et al., 2017). At the resin T_g, the material changes from glassy state to rubbery state. This change in state of the polymer material at temperatures above its T_g defines its physical and application properties. Zhang et al. (2023) reported that increasing the temperature of a CF reinforced polymer composite from 20°C to 200°C and then to 600°C saw the material loss 32% and 84% of its tensile strength, respectively, when compared to room temperature value. The presence of the fibers led to improved mechanical and thermal properties at lower temperatures. However, at elevated temperatures when the material has transitioned to a rubbery state, the polymer resin loses the ability to effectively transmit externally applied stress to the fiber particles, leading to the overall failure of the composite material. The stress-displacement curve of CFRP plates prepared and tested by Wang et al. (2011) from 20°C to 700°C showed linear relationship at temperatures below 530°C. However, the strength decreased to 50% of room temperature value at 300°C due to decomposition of resin and oxidation of CF. A 60% volume loading of ceramic fiber onto PI matrix gave a composite that maintains good flexibility and increases the thermal conductivity significantly (Sato et al., 2010).

At low temperatures, the mechanical properties of composites are primarily influenced by the characteristics of the fibers. However, at high temperatures, the properties of the matrix take precedence. Therefore when selecting materials for applications across a wide temperature range, it is crucial to consider the glass transition temperature and the time- and temperature-dependent mechanical behavior of the polymer matrix. Additionally, high-temperature polymer composites exhibit excellent chemical resistance to a variety of aggressive chemicals, including acids, bases, and organic solvents.

6.6.2 Thermal properties

According to Chen et al. (2016), thermal management is critical to the performance, lifetime, and reliability of conductive polymers, especially electronic devices. The thermal resistance of a material, which is defined as the long-term thermal stability of the material, is the most important thermal property. With the miniaturization, thermal dissipation becomes a challenging problem. Especially with the integration and functionalization of electronics and the emergence of new applications such as light-emitting diodes. The glass transition temperature (T_g) and the melting temperature (T_m) are the two most important parameters to describe the behavior of a polymer material under load, and they are related to the heat resistance. The inclusion of fillers in composites enhances their heat conductivity, which plays a crucial role in boosting their ability to withstand high temperatures. Consequently, the choice of fiber employed in the production of composites significantly influences their thermal characteristics. The thermal properties of high-temperature polymer composites have been reported to be improved by compounding with fibers, and/or blending two polymers together (Adak et al., 2023; Tadini et al., 2017). Ke et al. enhanced the toughness of thermosetting PIs composites by incorporating TPI. The resulting blends displayed a T_g exceeding 395°C and $T_{5\%}$ within the range of 533°C–563°C, signifying exceptional thermal stability and heat resistance. The mechanical and thermal properties were significantly improved when the enhanced blends were compounded with CFs (Ke et al., 2020). Zegaoui et al. successfully produced cyanate ester/benzoxazine blend composites by compounding them with silane-surface modified Kevlar and GFs. The compounded samples exhibited noteworthy enhancements in thermal stability and heat resistance compared to the unfilled blend (Zegaoui et al., 2019). In general, the degree of improvement targeted can be achieved by varying the vol% fraction of the various components used in preparing the material (Goh et al., 2019).

6.6.3 Electrical properties

High-temperature polymer composites have demonstrated enhanced electrical properties in comparison to neat polymers or their blends, largely due to the incorporation of conducting fibers as fillers (Li et al., 2004). The utilization of graphene and carbon black as fillers in the preparation of polymer composites has been shown to improve the electrical conductivity of the resulting materials (Li et al., 2015; Wang et al., 2023). Additionally, the functionalization of fibers with inorganic particles has been reported to further enhance electrical conductivities (Li et al., 2021; Santhosh Kumar and Reghunadhan Nair, 2014; Zhou et al., 2019).

6.7 Challenges and future directions for high-temperature polymer composites

Developing novel high-temperature polymer matrices is a complex endeavor. High-temperature polymer composites offer significant advantages in various industries but also come with challenges that need to be addressed for broader adoption and improved performance. Listed below are some key challenges and future directions for high-temperature polymer composites:

1. Achieving reliable and consistent mechanical properties, stable thermal characteristics, and resistance to environmental factors such as moisture and chemicals continues to be difficult, particularly for intricate shapes and long-lasting durability.
2. Efficiently and cost-effectively developing advanced manufacturing processes capable of handling high-temperature polymers, while maintaining quality, repeatability, and scalability, is of paramount importance.

3. As mentioned earlier in Section 6.5, high-temperature polymer composites are often quite costly due to the expense of raw materials, the intricacy of manufacturing processes, and the need for specialized equipment. To encourage wider use of these materials, finding ways to lower production costs without sacrificing performance is crucial.
4. Sophisticated design and simulation tools are imperative for optimizing the performance of high-temperature polymer composites. These tools play a crucial role in accurately predicting the material's behavior under a wide range of loading conditions, thermal gradients, and aging effects.
5. Developing reliable and efficient methods for joining high-temperature polymer composite components, such as adhesives, welding, or mechanical fastening, is important for assembly and structural integrity.
6. Establishing industry standards and certification processes for high-temperature polymer composites is crucial for ensuring quality, safety, and regulatory compliance, particularly in sectors like aerospace and automotive.
7. Improving recycling methods and developing sustainable alternatives for high-temperature polymer composites to reduce environmental impact and promote circular economy principles (Goh et al., 2019).

To mitigate the above-listed challenges, the following should be the future directions in high-temperature polymer composite design and development: (1) Continued research into new polymer matrices, reinforcement fibers, and nanomaterial additives to enhance the mechanical, thermal, and chemical properties of high-temperature polymer composites. (2) Integrating additional functionalities into high-temperature polymer composites, such as self-healing capabilities, sensing capabilities, electromagnetic shielding, or thermal management properties, to expand their potential applications. (3) Integration of digital manufacturing technologies, such as AM, AFP, and advanced robotics, to enable efficient and customizable production of high-temperature polymer composite parts with complex geometries. (4) Leveraging data analytics, IoT sensors, and machine learning algorithms to optimize manufacturing processes, monitor material performance in real time, and enable predictive maintenance for high-temperature polymer composite components (smart manufacturing). (5) Collaboration between industry stakeholders, regulatory bodies, and research institutions to develop standardized testing methods, performance metrics, and certification processes for high-temperature polymer composites. (6) Encouraging collaboration and knowledge exchange between industries, academia, and government agencies to accelerate innovation, address common challenges, and explore new applications for high-temperature polymer composites. (7) Investing in education and training programs to develop a skilled workforce capable of designing, manufacturing, and implementing high-temperature polymer composite solutions across various industries.

It is crucial to tackle these obstacles and follow these upcoming paths to unleash the full potential of high-temperature polymer composites. This will facilitate their widespread use in various applications, ranging from aerospace and automotive to energy, infrastructure, and more.

6.8 Conclusion

High-temperature polymer composites are described as polymer composites that can function at elevated temperatures for a given period of time. They are in high demand around the world for advanced technological applications. To keep pace with today's technological advancements, including space travel and satellites, among others, continuous research into high-temperature polymer resin and the improved fiber surface is required. Their uncommon properties, such as lightweight, chemical, and corrosion resistance, thermal stability at elevated temperatures, and availability of manufacturing techniques, have made high-temperature composites more attractive. These classes of polymer composites can be fabricated with the existing manufacturing methods. However, the problems associated with the processing conditions, use, and waste management generated by these materials have limited their wider application. Even in the face of these shortcomings, high-temperature polymer composites have remained a very important material for the advancement of technology in many human endeavors.

References

Abdurohman, K., Satrio, T., Muzayadah, N.L., Teten, 2018. A comparison process between hand lay-up, vacuum infusion and vacuum bagging method toward e-glass EW 185/lycal composites. Journal of Physics: Conference Series 1130, 012018. Available at 10.1088/1742-6596/1130/1/012018.

Adak, N.C., Lee, G.H., Tung, H.T., Lim, S., Lingappan, N., Kang, H.W., et al., 2023. Superior high-temperature mechanical and thermal performance of carbon fiber/epoxy composites by incorporating highly dispersed aramid nanofibers. Applied Materials Today 35. Available at https://doi.org/10.1016/j.apmt.2023.101956.

Adhikari, D., Gururaja, S., Hemchandra, S., 2021. Resin infusion in porous preform in the presence of HPM during VARTM: flow simulation using level set and experimental validation. Composites Part A: Applied Science and Manufacturing 151. Available at https://doi.org/10.1016/j.compositesa.2021.106641.

Ageyeva, T., Sibikin, I., Kovács, J.G., 2019. Review of thermoplastic resin transfer molding: process modeling and simulation. Polymers 11 (10). Available at https://doi.org/10.3390/polym11101555, https://res.mdpi.com/d_attachment/polymers/polymers-11-01555/article_deploy/polymers-11-01555-v2.pdf.

Ahmadijokani, F., Shojaei, A., Dordanihaghighi, S., Jafarpour, E., Mohammadi, S., Arjmand, M., 2020. Effects of hybrid carbon-aramid fiber on performance of non-asbestos organic brake friction composites. Wear 452-453, 203280. Available at https://doi.org/10.1016/j.wear.2020.203280.

Al-Quraishi, K.K., Wang, M., Liang, J., Jia, S., Kauppila, W., Wang, J., et al., 2020. Strengthening the interface between individual aramid fibers and polymer at room and elevated temperatures. Materials Today Communications 24. Available at https://doi.org/10.1016/j.mtcomm.2020.101254, http://www.journals.elsevier.com/materials-today-communications/.

Alsayed, S., Al-Salloum, Y., Almusallam, T., El-Gamal, S., Aqel, M., 2012. Performance of glass fiber reinforced polymer bars under elevated temperatures. Composites Part B: Engineering 43 (5), 2265–2271. Available at https://doi.org/10.1016/j.compositesb.2012.01.034.

Amrollahi, S., Ramezanzadeh, B., Yari, H., Ramezanzadeh, M., Mahdavian, M., 2019. Synthesis of polyaniline-modified graphene oxide for obtaining a high performance epoxy nanocomposite film with excellent UV blocking/anti-oxidant/ anti-corrosion capabilities. Composites Part B: Engineering 173. Available at https://doi.org/10.1016/j.compositesb.2019.05.015, https://www.journals.elsevier.com/composites-part-b-engineering.

Arabpour, A., Shockravi, A., Rezania, H., Farahati, R., 2020. Investigation of anticorrosive properties of novel silane-functionalized polyamide/GO nanocomposite as steel coatings. Surfaces and Interfaces 18. Available at https://doi.org/10.1016/j.surfin.2020.100453, http://www.journals.elsevier.com/surfaces-and-interfaces.

Asensio, M., Esfandiari, P., Núñez, K., Silva, J.F., Marques, A., Merino, J.C., et al., 2020. Processing of pre-impregnated thermoplastic towpreg reinforced by continuous glass fibre and recycled PET by pultrusion. Composites Part B: Engineering 200. Available at https://doi.org/10.1016/j.compositesb.2020.108365, https://www.journals.elsevier.com/composites-part-b-engineering.

Asim, M., Jawaid, M., Nasir, M., Saba, N., 2018. Effect of fiber loadings and treatment on dynamic mechanical, thermal and flammability properties of pineapple leaf fiber and kenaf phenolic composites. Journal of Renewable Materials 6 (4), 383–393. Available at https://doi.org/10.7569/JRM.2017.634162, http://www.techscience.com/JRM/index.html.

Asim, M., Saba, N., Jawaid, M., Nasir, M., Pervaiz, M., Alothman, O.Y., 2017. A review on phenolic resin and its composites. Current Analytical Chemistry 13 (3). Available at https://doi.org/10.2174/1573411013666171003154410.

Asyraf, M.R.M., Ishak, M.R., Syamsir, A., Amir, A.L., Nurazzi, N.M., Norrrahim, M.N.F., et al., 2023. Filament-wound glass-fibre reinforced polymer composites: potential applications for cross arm structure in transmission towers. Polymer Bulletin 80 (2), 1059–1084. Available at https://doi.org/10.1007/s00289-022-04114-4, https://www.springer.com/journal/289.

Baek, Y.M., Shin, P.S., Kim, J.H., Park, H.S., DeVries, K.L., Park, J.M., 2020. Thermal transfer, interfacial, and mechanical properties of carbon fiber/polycarbonate-CNT composites using infrared thermography. Polymer Testing 81. Available at https://doi.org/10.1016/j.polymertesting.2019.106247, https://www.journals.elsevier.com/polymer-testing.

Bain, S., Ozawa, H., Criss, J.M., 2007. Mechanical properties of thermally aged PETI-330 laminates fabricated by resin transfer molding. *International SAMPE Symposium and Exhibition Proceedings*. 52, 08910138.

Balasubramanian, K., Tirumalai, M., 2013. High temperature polymer nanocomposites. Engineering Materials. Springer Science and Business Media B.V., India, pp. 165–186. Available at 10.1007/978-3-642-40322-4_7.

Barbosa, L.C.M., de Souza, S.D.B., Botelho, E.C., Cândido, G.M., Rezende, M.C., 2019. Fractographic evaluation of welded joints of PPS/glass fiber thermoplastic composites. Engineering Failure Analysis 102, 60–68. Available at https://doi.org/10.1016/j.engfailanal.2019.04.032.

Bazli, M., Abolfazli, M., 2020. Mechanical properties of fibre reinforced polymers under elevated temperatures: an overview. Polymers 12 (11), 1–31. Available at https://doi.org/10.3390/polym12112600.

Bello, S.A., 2021. Carbon-fiber composites: development, structure, properties, and applications. Handbook of Nanomaterials and Nanocomposites for Energy and Environmental Applications. Springer, Cham, pp. 63–84. Available at https://doi.org/10.1007/978-3-030-36268-3_86.

Benoit, V., Cédric, L., Alexis, C., 2014. Post fire behavior of carbon fibers polyphenylene sulfide- and epoxy-based laminates for aeronautical applications: a comparative study. Materials and Design 63, 56–68. Available at https://doi.org/10.1016/j.matdes.2014.05.046.

Bisanda, E.T.N., Ansell, M.P., 1991. The effect of silane treatment on the mechanical and physical properties of sisal-epoxy composites. Composites Science and Technology 41 (2), 165–178. Available at https://doi.org/10.1016/0266-3538(91)90026-L.

Boon, Y.D., Joshi, S.C., Bhudolia, S.K., 2021. Review: filament winding and automated fiber placement with in situ consolidation for fiber reinforced thermoplastic polymer composites. Polymers 13 (12). Available at https://doi.org/10.3390/polym13121951.

Buitrago, B., Jaramillo, F., Gómez, M., 2015. Some properties of natural fibers (Sisal, Pineapple, and Banana) in comparison to man-made technical fibers (Aramide, Glass, Carbon). Journal of Natural Fibers 12 (4), 357–367. Available at https://doi.org/10.1080/15440478.2014.929555.

Cai, P., Li, Z., Wang, T., Wang, Q., 2015. Effect of aspect ratios of aramid fiber on mechanical and tribological behaviors of friction materials. Tribology International 92, 109–116. Available at https://doi.org/10.1016/j.triboint.2015.05.024.

Carpier, Y., Vieille, B., Coppalle, A., Barbe, F., 2020. About the tensile mechanical behaviour of carbon fibers fabrics reinforced thermoplastic composites under very high temperature conditions. Composites Part B: Engineering 181. Available at https://doi.org/10.1016/j.compositesb.2019.107586.

Chalmers, D.W., 1991. Experience in design and production of FRP marine structures. Marine Structures 4 (2), 93–115. Available at https://doi.org/10.1016/0951-8339(91)90015-4.

Chang, W., Rose, L.R.F., Islam, M.S., Wu, S., Peng, S., Huang, F., Kinloch, A.J., Wang, C.H., 2021. Strengthening and toughening epoxy polymer at cryogenic temperature using cupric oxide nanorods. Composites Science and Technology 208, 108762. Available at https://doi.org/10.1016/j.compscitech.2021.108762.

Chen, G., Mohanty, A.K., Misra, M., 2021. Progress in research and applications of polyphenylene sulfide blends and composites with carbons. Composites Part B: Engineering 209. Available at https://doi.org/10.1016/j.compositesb.2020.108553.

Chen, H., Ginzburg, V.V., Yang, J., Yang, Y., Liu, W., Huang, Y., et al., 2016. Thermal conductivity of polymer-based composites: fundamentals and applications. Progress in Polymer Science 59, 41–85. Available at https://doi.org/10.1016/j.progpolymsci.2016.03.001.

Chen, J., Gao, X., 2019. Thermal and electrical anisotropy of polymer matrix composite materials reinforced with graphene nanoplatelets and aluminum-based particles. Diamond and Related Materials 100. Available at https://doi.org/10.1016/j.diamond.2019.107571.

Cheng, Q., Jiang, H., Li, Y., 2020. Effect of fiber content and orientation on the scratch behavior of short glass fiber reinforced PBT composites. Tribology International 146. Available at https://doi.org/10.1016/j.triboint.2020.106221.

Chuang, K.C., Bowman, C.L., Tsotsis, T.K., Arendt, C.P., 2003. 6F-Polyimides with Phenylethynyl Endcap for 315-370 °C Applications. High Performance Polymers 15 (4), 459–472. Available at https://doi.org/10.1177/0954008303015004005.

Chukov, D., Nematulloev, S., Torokhov, V., Stepashkin, A., Sherif, G., Tcherdyntsev, V., 2019. Effect of carbon fiber surface modification on their interfacial interaction with polysulfone. Results in Physics 15, 102634. Available at https://doi.org/10.1016/j.rinp.2019.102634.

Chukov, D., Nematulloev, S., Zadorozhnyy, M., Tcherdyntsev, V., Stepashkin, A., Zherebtsov, D., 2019. Structure, mechanical and thermal properties of polyphenylene sulfide and polysulfone impregnated carbon fiber composites. Polymers 11 (4). Available at https://doi.org/10.3390/polym11040684.

Connell, J.W., Smith, J.G., Hergenrother, P.M., Criss, J.M., 2003. High temperature transfer molding resins: laminate properties of PETI-298 and PETI-330. High Performance Polymers 15 (4), 375–394. Available at https://doi.org/10.1177/0954008303015004001.

Correia, J.R., Gomes, M.M., Pires, J.M., Branco, F.A., 2013. Mechanical behaviour of pultruded glass fibre reinforced polymer composites at elevated temperature: experiments and model assessment. Composite Structures 98, 303–313. Available at https://doi.org/10.1016/j.compstruct.2012.10.051.

Costa Dias, Rd.C., Santos, Ld.S., Ouzia, H., Schledjewski, R., 2018. Improving degree of cure in pultrusion process by optimizing die-temperature. Materials Today Communications 17, 362–370. Available at https://doi.org/10.1016/j.mtcomm.2018.08.017.

Dagdag, O., Safi, Z., Hsissou, R., Erramli, H., El Bouchti, M., Wazzan, N., et al., 2019. Epoxy pre-polymers as new and effective materials for corrosion inhibition of carbon steel in acidic medium: computational and experimental studies. Scientific Reports 9 (1). Available at https://doi.org/10.1038/s41598-019-48284-0.

Dagdag, O., Hsissou, R., El Harfi, A., Safi, Z., Berisha, A., Verma, C., et al., 2020. Epoxy resins and their zinc composites as novel anti-corrosive materials for copper in 3% sodium chloride solution: experimental and computational studies. Journal of Molecular Liquids 315. Available at https://doi.org/10.1016/j.molliq.2020.113757.

Da-Peng, Z., Hong, F., 2008. Mechanical and high-temperature properties of glass fibers reinforced phenolic composites. Journal of Reinforced Plastics and Composites 27 (13), 1449–1460. Available at https://doi.org/10.1177/0731684407086613.

De, S., Fulmali, A.O., Shivangi, P.N., Choudhury, S., Prusty, R.K., Ray, B.C., 2020. Interface modification of carbon fiber reinforced epoxy composite by hydroxyl/carboxyl functionalized carbon nanotube. Materials Today: Proceedings 27, 1473–1478. Available at https://doi.org/10.1016/j.matpr.2020.02.970.

DeMeuse, M.T., 2014. Polysulfones as a reinforcement in high temperature polymer blends. High Temperature Polymer Blends. Elsevier Inc., United States, pp. 165–173. Available at 10.1533/9780857099013.165.

Devnani, G.L., Sinha, S., 2019. Effect of nanofillers on the properties of natural fiber reinforced polymer composites. Materials Today: Proceedings 18, 647–654. Available at https://doi.org/10.1016/j.matpr.2019.06.460.

Díez-Pascual, A.M., Guan, J., Simard, B., Gómez-Fatou, M.A., 2012. Poly(phenylene sulphide) and poly(ether ether ketone) composites reinforced with single-walled carbon nanotube buckypaper: I - Structure, thermal stability and crystallization behaviour. Composites Part A: Applied Science and Manufacturing 43 (6), 997–1006. Available at https://doi.org/10.1016/j.compositesa.2011.11.002.

Dominguez, D.D., Keller, T.M., 2006. Low-melting phthalonitrile oligomers: preparation, polymerization and polymer properties. High Performance Polymers 18 (3), 283–304. Available at https://doi.org/10.1177/0954008306060143.

Dong, F., Hou, G., Liu, H., Liu, L., Cao, F., Wang, J., et al., 2018. An investigation on the mechanical and tribological properties of carbon fiber/polyimide composites at elevated temperatures. Polymer Composites 39. Available at https://doi.org/10.1002/pc.24290 E869.

Dong, Z.J., Zhou, T., Luan, H., Williams, R.C., Wang, P., Leng, Z., 2019. Composite modification mechanism of blended bio-asphalt combining styrene-butadiene-styrene with crumb rubber: a sustainable and environmental-friendly solution for wastes. Journal of Cleaner Production 214, 593–605. Available at https://doi.org/10.1016/j.jclepro.2019.01.004.

Eastmond, G.C., Paprotny, J., 1996. Scope in the synthesis and properties of poly(ether imide)s. Reactive and Functional Polymers 30 (1-3), 27–41. Available at https://doi.org/10.1016/1381-5148(95)00114-x.

Fernberg, P., Gong, G., Mannberg, P., Tsampas, S., 2018. Development of novel high Tg polyimide-based composites. Part I: RTM processing properties. Journal of Composite Materials 52 (2), 253–260. Available at https://doi.org/10.1177/0021998317705705.

Fu, X., Zhang, C., Liang, R., Wang, B., Fielding, J.C., 2011. High temperature vacuum assisted resin transfer molding of phenylethynyl terminated imide composites. Polymer Composites 32 (1), 52–58. Available at https://doi.org/10.1002/pc.21015.

Garcia-Leiner, M., Streifel, B., Başgül, C., MacDonald, D.W., Kurtz, S.M., 2021. Characterization of polyaryletherketone (PAEK) filaments and printed parts produced by extrusion-based additive manufacturing. Polymer International 70 (8), 1128–1136. Available at https://doi.org/10.1002/pi.6231.

Gebisa, A., Lemu, H., 2018. Investigating Effects of Fused-Deposition Modeling (FDM) Processing Parameters on Flexural Properties of ULTEM 9085 using Designed Experiment. Materials 11 (4), 500. Available at https://doi.org/10.3390/ma11040500.

Goh, G.D., Yap, Y.L., Agarwala, S., Yeong, W.Y., 2019. Recent progress in additive manufacturing of fiber reinforced polymer composite. Advanced Materials Technologies 4 (1). Available at https://doi.org/10.1002/admt.201800271.

Gong, Y., Yang, G., 2010. All-polyamide composites prepared by resin transfer molding. Journal of Materials Science 45 (19), 5237–5243. Available at https://doi.org/10.1007/s10853-010-4565-6.

Granata, P.J., Parvin, A., 2001. An experimental study on Kevlar strengthening of beam-column connections. Composite Structures 53 (2), 163–171. Available at https://doi.org/10.1016/S0263-8223(00)00187-2.

Grunenfelder, L.K., Dills, A., Centea, T., Nutt, S., 2017. Effect of prepreg format on defect control in out-of-autoclave processing. Composites Part A: Applied Science and Manufacturing 93, 88–99. Available at https://doi.org/10.1016/j.compositesa.2016.10.027.

Guo, Y., Chen, X., Wu, G., Huang, S., 2023. Tunable thermal, mechanical, and tribological properties of polybenzoxazine-based composite for vehicle applications. Coatings 13 (7). Available at https://doi.org/10.3390/coatings13071147.

Gupta, M.K., Srivastava, R.K., 2016. Mechanical properties of hybrid fibers-reinforced polymer composite: a review. Polymer - Plastics Technology and Engineering 55 (6), 626–642. Available at https://doi.org/10.1080/03602559.2015.1098694.

Gupta, V., Anand, K., Grape, J., 1998. Failure of woven carbon-polyimide laminates under off-axis compression loading. Acta Materialia 46 (2), 711–718. Available at https://doi.org/10.1016/S1359-6454(97)00175-4.

Harsha, A.P., Tewari, U.S., 2004. Tribological studies on glass fiber reinforced polyetherketone composites. Journal of Reinforced Plastics and Composites 23 (1), 65–82. Available at https://doi.org/10.1177/0731684404029349.

Haruki, M., 2021. Thermal conductivity for polymer composite materials: recent advances in polyimide materials. Journal of Chemical Engineering of Japan 54 (5), 186–194. Available at https://doi.org/10.1252/jcej.20we136.

Hergenrother, P./M., 1990. Perspectives in the Development of High-Temperature Polymers. Angewandte Chemie International Edition in English 29, 1262–1268. Available at https://doi.org/10.1002/anie.199012621.

Hergenrother, P.M., 2003. The use, design, synthesis, and properties of high performance/high temperature polymers: an overview. High Performance Polymers 15 (1), 3–45. Available at https://doi.org/10.1177/095400830301500101.

Hong, W., Yuan, L., Ma, Y., Cui, C., Zhang, H., Yang, S., et al., 2021. Resin transfer moldable fluorinated phenylethynyl-terminated imide oligomers with high tg: structure–melt stability relationship. Polymers 13 (6). Available at https://doi.org/10.3390/polym13060903.

Hsiao, K.T., Heider, D., 2012. Vacuum assisted resin transfer molding (VARTM) in polymer matrix composites. Manufacturing Techniques for Polymer Matrix Composites (PMCs). Elsevier Inc, United States, pp. 310–347. Available at 10.1016/B978-0-85709-067-6.50010-9.

Hsissou, R., Abbout, S., Berisha, A., Berradi, M., Assouag, M., Hajjaji, N., et al., 2019. Experimental, DFT and molecular dynamics simulation on the inhibition performance of the DGDCBA epoxy polymer against the corrosion of the E24 carbon steel in 1.0 M HCl solution. Journal of Molecular Structure 1182, 340–351. Available at https://doi.org/10.1016/j.molstruc.2018.12.030.

Hsissou, R., Dagdag, O., Abbout, S., Benhiba, F., Berradi, M., El Bouchti, M., et al., 2019. Novel derivative epoxy resin TGETET as a corrosion inhibition of E24 carbon steel in 1.0 M HCl solution. Experimental and computational (DFT and MD simulations) methods. Journal of Molecular Liquids 284, 182–192. Available at https://doi.org/10.1016/j.molliq.2019.03.180.

Hsissou, R., Abbout, S., Seghiri, R., Rehioui, M., Berisha, A., Erramli, H., et al., 2020. Evaluation of corrosion inhibition performance of phosphorus polymer for carbon steel in [1 M] HCl: computational studies (DFT, MC and MD simulations). Journal of Materials Research and Technology 9 (3), 2691–2703. Available at https://doi.org/10.1016/j.jmrt.2020.01.002.

Hsissou, R., Abbout, S., Safi, Z., Benhiba, F., Wazzan, N., Guo, L., et al., 2021. Synthesis and anticorrosive properties of epoxy polymer for CS in [1 M] HCl solution: electrochemical, AFM, DFT and MD simulations. Construction and Building Materials 270. Available at https://doi.org/10.1016/j.conbuildmat.2020.121454.

Hsissou, R., Seghiri, R., Benzekri, Z., Hilali, M., Rafik, M., Elharfi, A., 2021. Polymer composite materials: a comprehensive review. Composite Structures 262. Available at https://doi.org/10.1016/j.compstruct.2021.113640.

Hsu, J.Y., Speyer, R.F., 1992. Fabrication and properties of SiC fibre-reinforced Li2O·Al2O3·6SiO2 glass-ceramic composites. Journal of Materials Science 27 (2), 381–390. Available at https://doi.org/10.1007/BF00543927.

Irfan, M.S., Shotton-Gale, N., Paget, M.A., Machavaram, V.R., Leek, C., Wootton, S., et al., 2017. A modified pultrusion process. Journal of Composite Materials 51 (13), 1925–1941. Available at https://doi.org/10.1177/0021998316666653.

Jassal, M., Ghosh, S., 2002. Aramid fibres - an overview. Indian Journal of Fibre and Textile Research 27 (3), 290–306.

Jassal, M., Agrawal, A.K., Gupta, D., Panwar, K., 2020. Aramid fibers. In: Hu, J., Kumar, B., Lu, J. (Eds.), Handbook of Fibrous Materials. Wiley, pp. 207–231. Available at https://doi.org/10.1002/9783527342587.ch8.

Jiang, Z., Hornsby, P., McCool, R., Murphy, A., 2012. Mechanical and thermal properties of polyphenylene sulfide/multiwalled carbon nanotube composites. Journal of Applied Polymer Science 123 (5), 2676–2683. Available at https://doi.org/10.1002/app.34669.

Jiangtao, Y., Yichao, W., Kexu, H., Kequan, Y., Jianzhuang, X., 2017. The performance of near-surface mounted CFRP strengthened RC beam in fire. Fire Safety Journal 90, 86–94. Available at https://doi.org/10.1016/j.firesaf.2017.04.031.

Joshi, S.C., 2012. The pultrusion process for polymer matrix composites. Manufacturing Techniques for Polymer Matrix Composites (PMCs). Elsevier Inc, Singapore, pp. 381–413. Available at https://doi.org/10.1016/B978-0-85709-067-6.50012-2.

Jwo, S.L., Chen, J.C., Chen, W.Y., Hsu, S.E., Shan, C., 1992. Preparation of high char-yield phenolic resins. Composites and Advanced Materials. Available at https://doi.org/10.1177/096369359200100206.

Karnik, S.R., Gaitonde, V.N., Rubio, J.C., Correia, A.E., Abrão, A.M., Davim, J.P., 2008. Delamination analysis in high speed drilling of carbon fiber reinforced plastics (CFRP) using artificial neural network model. Materials & Design 29, 1768–1776. Available at https://doi.org/10.1016/j.matdes.2008.03.014.

Kafi, A., Wu, H., Langston, J., Atak, O., Kim, H., Kim, S., et al., 2020. Evaluation of additively manufactured ultraperformance polymers to use as thermal protection systems for spacecraft. Journal of Applied Polymer Science 137 (37). Available at https://doi.org/10.1002/app.49117.

Ke, H., Zhao, L., Zhang, X., Qiao, Y., Wang, G., Wang, X., 2020. Performance of high-temperature thermosetting polyimide composites modified with thermoplastic polyimide. Polymer Testing 90. Available at https://doi.org/10.1016/j.polymertesting.2020.106746.

Khalil, H.P.S.A., Tehrani, M.A., Davoudpour, Y., Bhat, A.H., Jawaid, M., Hassan, A., 2013. Natural fiber reinforced poly(vinyl chloride) composites: a review. Journal of Reinforced Plastics and Composites 32 (5), 330–356. Available at https://doi.org/10.1177/0731684412458553.

Khanbareh, H., Hegde, M., Bijlevold, J.C., Van Der Zwaag, S., Groen, P., 2017. Functionally graded ferroelectric polyetherimide composites for high temperature sensing. Journal of Materials Chemistry C 5 (36), 9389–9397. Available at https://doi.org/10.1039/c7tc02649h.

Kharbanda, S., Bhadury, T., Gupta, G., Fuloria, D., Pati, P.R., Mishra, V.K., et al., 2021. Polymer composites for thermal applications - a review. Materials Today: Proceedings 47, 2839–2845. Available at https://doi.org/10.1016/j.matpr.2021.03.609.

Khotbehsara, M.M., Manalo, A., Aravinthan, T., Reddy, K.R., Ferdous, W., Wong, H., et al., 2019. Effect of elevated in-service temperature on the mechanical properties and microstructure of particulate-filled epoxy polymers. Polymer Degradation and Stability 170. Available at https://doi.org/10.1016/j.polymdegradstab.2019.108994.

Koch, D., Tushtev, K., Grathwohl, G., 2008. Ceramic fiber composites: experimental analysis and modeling of mechanical properties. Composites Science and Technology 68 (5), 1165–1172. Available at https://doi.org/10.1016/j.compscitech.2007.06.029.

Laurenzi, S., Marchetti, M., 2012. Advanced composite materials by resin transfer molding for aerospace applications. In: Hu, N. (Ed.), Composites and Their Properties. IntechOpen. Available at 10.5772/48172.

Le Guen, M.J., Newman, R.H., Fernyhough, A., Emms, G.W., Staiger, M.P., 2016. The damping-modulus relationship in flax-carbon fibre hybrid composites. Composites Part B: Engineering 89, 27–33. Available at https://doi.org/10.1016/j.compositesb.2015.10.046.

Lee, J., Lim, J.W., Kim, M., 2020. Effect of thermoplastic resin transfer molding process and flame surface treatment on mechanical properties of carbon fiber reinforced polyamide 6 composite. Polymer Composites 41 (4), 1190–1202. Available at https://doi.org/10.1002/pc.25445.

Lertwassana, W., Parnklang, T., Mora, P., Jubsilp, C., Rimdusit, S., 2019. High performance aramid pulp/carbon fiber-reinforced polybenzoxazine composites as friction materials. Composites Part B: Engineering 177. Available at https://doi.org/10.1016/j.compositesb.2019.107280.

Li, H., Zhou, Y., Liu, Y., Li, L., Liu, Y., Wang, Q., 2021. Dielectric polymers for high-temperature capacitive energy storage. Chemical Society Reviews 50 (11), 6369–6400. Available at https://doi.org/10.1039/d0cs00765j.

Li, Q., He, R., Jensen, J.O., Bjerrum, N.J., 2004. PBI-based polymer membranes for high temperature fuel cells - preparation, characterization and fuel cell demonstration. Fuel Cells 4 (3), 147–159. Available at https://doi.org/10.1002/fuce.200400020.

Li, Q., Chen, L., Gadinski, M.R., Zhang, S., Zhang, G., Li, H., et al., 2015. Flexible higherature dielectric materials from polymer nanocomposites. Nature 523 (7562), 576–579. Available at https://doi.org/10.1038/nature14647.

Li, X., Bao, Y., Wu, L., Yan, Q., Ma, H., Chen, G., et al., 2017. Thermal and mechanical properties of high-performance fiber-reinforced cementitious composites after exposure to high temperatures. Construction and Building Materials 157, 829–838. Available at https://doi.org/10.1016/j.conbuildmat.2017.09.125.

Li, Y., Yi, X., Yu, T., Xian, G., 2018. An overview of structural-functional-integrated composites based on the hierarchical microstructures of plant fibers. Advanced Composites and Hybrid Materials 1 (2), 231–246. Available at https://doi.org/10.1007/s42114-017-0020-3.

Li, Y.F., Wang, H.F., Syu, J.Y., Ramanathan, G.K., Tsai, Y.K., 2022. Investigating the mechanical performance on static and shock wave loading of aramid fiber-reinforced concrete. Fibers 10 (10). Available at https://doi.org/10.3390/fib10100082.

Liang, J., Liu, L., Qin, Z., Zhao, X., Li, Z., Emmanuel, U., et al., 2023. Experimental study of curing temperature effect on mechanical performance of carbon fiber composites with application to filament winding pressure vessel design. Polymers 15 (4). Available at https://doi.org/10.3390/polym15040982.

Lin, Qh, Liu, Qq, Dai, Zx, Qi, Xd, Yang, Jh, Wang, Y., 2022. Enhanced high-temperature dielectric performances of polyetherimide-based composites via surface functionalized silicon carbide whiskers. Applied Surface Science 599. Available at https://doi.org/10.1016/j.apsusc.2022.153991.

Linul, E., Vălean, C., Linul, P.A., 2018. Compressive behavior of aluminum microfibers reinforced semi-rigid polyurethane foams. Polymers 10 (12). Available at https://doi.org/10.3390/polym10121298.

Linul, E., Lell, D., Movahedi, N., Codrean, C., Fiedler, T., 2019. Compressive properties of zinc syntactic foams at elevated temperatures. Composites Part B: Engineering 167, 122–134. Available at https://doi.org/10.1016/j.compositesb.2018.12.019.

Lohr, C., Beck, B., Henning, F., Weidenmann, K.A., Elsner, P., 2019. Mechanical properties of foamed long glass fiber reinforced polyphenylene sulfide integral sandwich structures manufactured by direct thermoplastic foam injection molding. Composite Structures 220, 371–385. Available at https://doi.org/10.1016/j.compstruct.2019.03.056.

Lopez, L.C., Wilkes, G.L., 1989. Poly(p-Phenylene Sulfide]-an overview of an important engineering thermoplastic. Journal of Macromolecular Science, Part C 29 (1), 83–151. Available at https://doi.org/10.1080/07366578908055165.

Mangalgiri, P.D., 2005. Polymer-matrix composites for high-temperature applications. Defence Science Journal 55 (2), 175–193. Available at https://doi.org/10.14429/dsj.55.1980.

Mazur, R.L., Oliveira, P.C., Rezende, M.C., Botelho, E.C., 2013. Environmental effects on viscoelastic behavior of carbon fiber/PEKK thermoplastic composites. Journal of Reinforced Plastics and Composites 33, 749–757. Available at https://doi.org/10.1177/0731684413515955.

Meador, M.A., 1998. Recent advances in the development of processable high-temperature polymers. Annual Review of Materials Science 28 (1), 599–630. Available at https://doi.org/10.1146/annurev.matsci.28.1.599.

Menta, V., Vuppalapati, R., Chandrashekhara, K., Schuman, T., Sha, J., 2013. Elevated-temperature vacuum-assisted resin transfer molding process for high performance aerospace composites. Polymer International 62 (10), 1465–1476. Available at https://doi.org/10.1002/pi.4444.

Miller, A.H., Dodds, N., Hale, J.M., Gibson, A.G., 1998. High speed pultrusion of thermoplastic matrix composites. Composites Part A: Applied Science and Manufacturing 29 (7), 773–782. Available at https://doi.org/10.1016/S1359-835X(98)00006-2.

Minchenkov, K., Vedernikov, A., Safonov, A., Akhatov, I., 2021. Thermoplastic pultrusion: a review. Polymers 13 (2), 1–36. Available at https://doi.org/10.3390/polym13020180.

Miranda Campos, B., Bourbigot, S., Fontaine, G., Bonnet, F., 2022. Thermoplastic matrix-based composites produced by resin transfer molding: a review. Polymer Composites 43 (5), 2485–2506. Available at https://doi.org/10.1002/pc.26575.

Mittal, K., 2018. Polyimides and Other High Temperature Polymers: Synthesis, Characterization and Applications. Taylor & Francis Group.

Mohammed, L., Ansari, M.N.M., Pua, G., Jawaid, M., Islam, M.S., 2015. A review on natural fiber reinforced polymer composite and its applications. International Journal of Polymer Science 2015. Available at https://doi.org/10.1155/2015/243947.

Monteiro, S.N., de Assis, F.S., Ferreira, C.L., Simonassi, N.T., Weber, R.P., Oliveira, M.S., et al., 2018. Fique fabric: a promising reinforcement for polymer composites. Polymers 10 (3). Available at https://doi.org/10.3390/polym10030246.

Morampudi, P., Namala, K.K., Gajjela, Y.K., Barath, M., Prudhvi, G., 2020. Review on glass fiber reinforced polymer composites. Materials Today: Proceedings 43, 314–319. Available at https://doi.org/10.1016/j.matpr.2020.11.669.

Movahedi, N., Linul, E., 2017. Quasi-static compressive behavior of the ex-situ aluminum-alloy foam-filled tubes under elevated temperature conditions. Materials Letters 206, 182–184. Available at https://doi.org/10.1016/j.matlet.2017.07.018.

Mujahid, Y., Sallih, N., Abdullah, M.Z., 2020. A comparison of single-vacuum-bag and double-vacuum-bag methods for manufacturing high-quality laminated composites. Advances in Manufacturing Engineering. Lecture Notes in Mechanical Engineering. Springer, Singapore., pp. 457–467. Available at https://doi.org/10.1007/978-981-15-5753-8_42.

Mujahid, Y., Sallih, N., Mustapha, M., Abdullah, M.Z., Mustapha, F., 2020. Effects of processing parameters for vacuum-bagging-only method on shape conformation of laminated composites. Processes 8 (9). Available at https://doi.org/10.3390/PR8091147.

Muralidhara, B., Kumaresh Babu, S.P., Suresha, B., 2019. Utilizing vacuum bagging process to prepare carbon fiber/epoxy composites with improved mechanical properties. Materials Today: Proceedings 27, 2022–2028. Available at https://doi.org/10.1016/j.matpr.2019.09.051.

Nagarajan, B., Arshad, M., Ullah, A., Mertiny, P., Qureshi, A.J., 2019. Additive manufacturing ferromagnetic polymers using stereolithography – materials and process development. Manufacturing Letters 21, 12–16. Available at https://doi.org/10.1016/j.mfglet.2019.06.003.

Nobile, M.R., 2011. Rheology of polymer-carbon nanotube composites melts. Polymer-Carbon Nanotube Composites: Preparation, Properties and Applications. Elsevier Ltd, Italy, pp. 428–481. Available at 10.1533/9780857091390.2.428.

Ohnabe, H., Masaki, S., Onozuka, M., Miyahara, K., Sasa, T., 1999. Potential application of ceramic matrix composites to aero-engine components. Composites Part A: Applied Science and Manufacturing 30 (4), 489–496. Available at https://doi.org/10.1016/S1359-835X(98)00139-0.

Ornaghi, H.L., Bolner, A.S., Fiorio, R., Zattera, A.J., Amico, S.C., 2010. Mechanical and dynamic mechanical analysis of hybrid composites molded by resin transfer molding. Journal of Applied Polymer Science 118 (2), 887–896. Available at https://doi.org/10.1002/app.32388.

Oyama, H.T., Matsushita, M., Furuta, M., 2011. High performance reactive blends composed of poly(p-phenylene sulfide) and ethylene copolymers. Polymer Journal 43 (12), 991–999. Available at https://doi.org/10.1038/pj.2011.106.

Papakonstantinou, C.G., Balaguru, P., Lyon, R.E., 2001. Comparative study of high temperature composites. Composites Part B:Engineering 32 (8), 637–649. Available at https://doi.org/10.1016/S1359-8368(01)00042-7.

Pilato, L., 2013. Phenolic resins: 100 Years and still going strong. Reactive and Functional Polymers 73 (2), 270–277. Available at https://doi.org/10.1016/j.reactfunctpolym.2012.07.008.

Pinto, G., López-González, C., Jiménez-Martín, A., 1999. Polymer composites prepared by compression molding of a mixture of carbon black and nylon 6 powder. Polymer Composites 20 (6), 804–808. Available at https://doi.org/10.1002/pc.10404.

Prakash, S., 2019. Experimental investigation of surface defects in low-power CO_2 laser engraving of glass fiber-reinforced polymer composite. Polymer Composites 40 (12), 4704–4715. Available at https://doi.org/10.1002/pc.25339.

Qi, G., Zhang, B., Du, S., Yu, Y., 2017. Estimation of aramid fiber/epoxy interfacial properties by fiber bundle tests and multiscale modeling considering the fiber skin/core structure. Composite Structures 167, 1–10. Available at https://doi.org/10.1016/j.compstruct.2017.01.047.

Rajak, D.K., Wagh, P.H., Linul, E., 2021. Manufacturing technologies of carbon/glass fiber-reinforced polymer composites and their properties: a review. Polymers 13 (21). Available at https://doi.org/10.3390/polym13213721.

Rajak, D.K., Pagar, D.D., Menezes, P.L., Linul, E., 2019. Fiber-reinforced polymer composites: manufacturing, properties, and applications. Polymers 11 (10). Available at https://doi.org/10.3390/polym11101667.

Ramgobin, A., Fontaine, G., Bourbigot, S., 2019. Thermal degradation and fire behavior of high performance polymers. Polymer Reviews 59 (1), 55–123. Available at https://doi.org/10.1080/15583724.2018.1546736.

Ranganayakulu, N.S., Singh, A.D., Babu, D.S., Vardhan T, B.V., 2019. Experimental development of FRP pultrusion moulds. Materials Today: Proceedings 19, 354–360. Available at https://doi.org/10.1016/j.matpr.2019.07.612.

Raphael, N., Namratha, K., Chandrashekar, B.N., Sadasivuni, K.K., Ponnamma, D., Smitha, A.S., et al., 2018. Surface modification and grafting of carbon fibers: a route to better interface. Progress in Crystal Growth and Characterization of Materials 64 (3), 75–101. Available at https://doi.org/10.1016/j.pcrysgrow.2018.07.001.

Rivière, L., Caussé, N., Lonjon, A., Dantras, É., Lacabanne, C., 2016. Specific heat capacity and thermal conductivity of PEEK/Ag nanoparticles composites determined by modulated-temperature differential scanning calorimetry. Polymer Degradation and Stability 127, 98–104. Available at https://doi.org/10.1016/j.polymdegradstab.2015.11.015.

Rod, K.A., Nguyen, M.T., Elbakhshwan, M., Gills, S., Kutchko, B., Varga, T., et al., 2019. Insights into the physical and chemical properties of a cement-polymer composite developed for geothermal wellbore applications. Cement and Concrete Composites 97, 279–287. Available at https://doi.org/10.1016/j.cemconcomp.2018.12.022.

Saba, N., Jawaid, M., 2018. A review on thermomechanical properties of polymers and fibers reinforced polymer composites. Journal of Industrial and Engineering Chemistry 67, 1–11. Available at https://doi.org/10.1016/j.jiec.2018.06.018.

Santhosh Kumar, K.S., Reghunadhan Nair, C.P., 2014. Polybenzoxazine-new generation phenolics. Handbook of Thermoset Plastics. Elsevier Inc, India, pp. 45–73. Available at 10.1016/B978-1-4557-3107-7.00003-8.

Sathishkumar, T.P., Satheeshkumar, S., Naveen, J., 2014. Glass fiber-reinforced polymer composites - a review. Journal of Reinforced Plastics and Composites 33 (13), 1258–1275. Available at https://doi.org/10.1177/0731684414530790.

Sato, K., Horibe, H., Shirai, T., Hotta, Y., Nakano, H., Nagai, H., et al., 2010. Thermally conductive composite films of hexagonal boron nitride and polyimide with affinity-enhanced interfaces. Journal of Materials Chemistry 20 (14), 2749–2752. Available at https://doi.org/10.1039/b924997d.

Sciti, D., Galizia, P., Reimer, T., Schoberth, A., Gutiérrez-Gonzalez, C.F., Silvestroni, L., et al., 2021. Properties of large scale ultra-high temperature ceramic matrix composites made by filament winding and spark plasma sintering. Composites Part B: Engineering 216. Available at https://doi.org/10.1016/j.compositesb.2021.108839.

Seydibeyoğlu, M.Ö., Mohanty, A.K., Misra, M., 2017. Fiber Technology for Fiber-Reinforced Composites. Elsevier, Turkey, pp. 1–325.

Shaw, S.J., 1987. High-temperature polymers for adhesive and composite applications. Materials Science and Technology 3 (8), 589–599. Available at 10.1179/mst.1987.3.8.589.

Sherif, G., Chukov, D., Tcherdyntsev, V., Torokhov, V., 2019. Effect of formation route on the mechanical properties of the polyethersulfone composites reinforced with glass fibers. Polymers 11 (8), 1364. Available at https://doi.org/10.3390/polym11081364.

Sobrinho, L.L., De Araújo Calado, V.M., Bastian, F.L., 2011. Development and characterization of composite materials for production of composite risers by filament winding. Materials Research 14 (3), 287–298. Available at https://doi.org/10.1590/S1516-14392011005000062.

Søndergaard, T., Cleemann, L.N., Becker, H., Aili, D., Steenberg, T., Hjuler, H.A., et al., 2017. Long-term durability of HT-PEM fuel cells based on thermally cross-linked polybenzimidazole. Journal of Power Sources 342, 570–578. Available at https://doi.org/10.1016/j.jpowsour.2016.12.075.

Souza, W.O.D., Garcia, K., Von Dollinger, C.F.D.A., Pardini, L.C., 2015. Electrical behavior of carbon fiber/phenolic composite during pyrolysis. Materials Research 18 (6), 1209–1216. Available at https://doi.org/10.1590/1516-1439.000515.

Strauß, S., 2020. Development of a flexible injection and impregnation chamber for pultrusion of high reactive resins. Procedia Manufacturing 47, 956–961. Available at https://doi.org/10.1016/j.promfg.2020.04.294.

Sukur, E.F., Elmas, S., Seyyednourani, M., Eskizeybek, V., Yildiz, M., Sas, H.S., 2022. A rational study on the hydrothermal aging of AFP manufactured CF/polyetherketoneketone composites with in situ consolidation supported by acoustic emission inspection. Journal of Applied Polymer Science 139 (29). Available at https://doi.org/10.1002/app.52480

Sun, Z., Xiao, J., Tao, L., Wei, Y., Wang, S., Zhang, H., et al., 2018. Preparation of high-performance carbon fiber-reinforced epoxy composites by compression resin transfer molding. Materials 12 (1). Available at https://doi.org/10.3390/ma12010013.

Sun, Z., Li, Y.Q., Huang, P., Cao, H.J., Zeng, W., Li, J., et al., Fu, S.Y., 2021. Temperature-dependent mechanical properties of polyetherimide composites reinforced by graphene oxide-coated short carbon fibers. Composite Structures 270. Available at https://doi.org/10.1016/j.compstruct.2021.114075.

Suresh, S., Sudhakara, D., Vinod, B., 2020. Investigation on industrial waste eco-friendly natural fiber-reinforced polymer composites. Journal of Bio- and Tribo-Corrosion 6 (2). Available at https://doi.org/10.1007/s40735-020-00339-w.

Tadini, P., Grange, N., Chetehouna, K., Gascoin, N., Senave, S., Reynaud, I., 2017. Thermal degradation analysis of innovative PEKK-based carbon composites for high-temperature aeronautical components. Aerospace Science and Technology 65, 106–116. Available at https://doi.org/10.1016/j.ast.2017.02.011.

Tena, I., Sarrionandia, M., Torre, J., Aurrekoetxea, J., 2016. The effect of process parameters on ultraviolet cured out of die bent pultrusion process. Composites Part B: Engineering 89, 9–17. Available at https://doi.org/10.1016/j.compositesb.2015.11.027.

Torabizadeh, M.A., 2013. Tensile, compressive and shear properties of unidirectional glass/epoxy composites subjected to mechanical loading and low temperature services. Indian Journal of Engineering and Materials Sciences 20 (4), 299–309.

Tressler, R.E., 1999. Recent developments in fibers and interphases for high temperature ceramic matrix composites. Composites Part A: Applied Science and Manufacturing 30 (4), 429–437. Available at https://doi.org/10.1016/S1359-835X(98)00131-6.

Tsampas, S., Fernberg, P., Joffe, R., 2018. Development of novel high T g polyimide-based composites. Part II: Mechanical characterisation. Journal of Composite Materials 52 (2), 261–274. Available at https://doi.org/10.1177/0021998317705706.

Vakharia, V.S., Leonard, H., Singh, M., Halbig, M.C., 2023. Multi-material additive manufacturing of high temperature polyetherimide (pei)-based polymer systems for lightweight aerospace applications. Polymers 15 (3). Available at https://doi.org/10.3390/polym15030561.

Vakharia, V.S., Singh, M., Salem, A., Halbig, M.C., Salem, J.A., 2022. Effect of reinforcements and 3-D printing parameters on the microstructure and mechanical properties of acrylonitrile butadiene styrene (abs) polymer composites. Polymers 14 (10). Available at https://doi.org/10.3390/polym14102105.

Veazey, D., Hsu, T., Gomez, E.D., 2017. Next generation high-performance carbon fiber thermoplastic composites based on polyaryletherketones. Journal of Applied Polymer Science 134 (6). Available at https://doi.org/10.1002/app.44441.

Veit, D., 2023. Fibers: History, Production, Properties, Market. Springer International Publishing, Germany, pp. 1–1040. Available at https://doi.org/10.1007/978-3-031-15309-9.

Venkateshwaran, N., Elayaperumal, A., 2010. Banana fiber reinforced polymer composites - a review. Journal of Reinforced Plastics and Composites 29 (15), 2387–2396. Available at https://doi.org/10.1177/0731684409360578.

Wang, K., Young, B., Smith, S.T., 2011. Mechanical properties of pultruded carbon fibre-reinforced polymer (CFRP) plates at elevated temperatures. Engineering Structures 33 (7), 2154–2161. Available at https://doi.org/10.1016/j.engstruct.2011.03.006.

Wang, L., Yang, C., Wang, X., Shen, J., Sun, W., Wang, J., et al., 2023. Advances in polymers and composite dielectrics for thermal transport and high-temperature applications. Composites Part A: Applied Science and Manufacturing 164. Available at https://doi.org/10.1016/j.compositesa.2022.107320.

Wang, Q.B., Jia, D.L., Pei, X.H., Wu, X.L., Xu, F., Ye, Z.H., et al., 2022. Mechanical performance of graphenex/poly(ether ketone ketone) composite sheets by hot pressing. Scientific Reports 12 (1). Available at https://doi.org/10.1038/s41598-022-08221-0.

Werken, H., Tekinalp, H., Khanbolouki, P., Ozcan, S., Williams, A., Tehrani, M., 2020. Additively manufactured carbon fiber-reinforced composites: State of the art and perspective. Additive Manufacturing 31, 100962. Available at 10.1016/j.addma.2019.100962.

Wu, H., Koo, J.H., 2022. Characterization of high-temperature polymers for extreme environments. Analysis of Flame Retardancy In Polymer Science. Elsevier, United States, pp. 299–331. Available at 10.1016/B978-0-12-824045-8.00008-3.

Wu, H., Kim, S., Fahy, W., Koo, J.H., Kafi, A., Bateman, S., et al., 2020. Ablation performances of additively manufactured high-temperature thermoplastic polymers. AIAA Scitech 2020 Forum. American Institute of Aeronautics and Astronautics (AIAA), United States, 1. American Institute of Aeronautics and Astronautics. Available at https://doi.org/10.2514/6.2020-1125.

Xu, J., Wang, W., Han, Q., 2020. Mechanical properties of pultruded high-temperature-resistant carbon-fiber-reinforced polymer tendons at elevated temperatures. Construction and Building Materials 258, 119526. Available at https://doi.org/10.1016/j.conbuildmat.2020.119526.

Xu, Z., Croft, Z.L., Guo, D., Cao, K., Liu, G., 2021. Recent development of polyimides: synthesis, processing, and application in gas separation. Journal of Polymer Science 59 (11), 943–962. Available at https://doi.org/10.1002/pol.20210001.

Yao, S.S., Jin, F.L., Rhee, K.Y., Hui, D., Park, S.J., 2018. Recent advances in carbon-fiber-reinforced thermoplastic composites: a review. Composites Part B: Engineering 142, 241–250. Available at https://doi.org/10.1016/j.compositesb.2017.12.007.

Yashas Gowda, T.G., Sanjay, M.R., Subrahmanya Bhat, K., Madhu, P., Senthamaraikannan, P., Yogesha, B., 2018. Polymer matrix-natural fiber composites: an overview. Cogent Engineering 5 (1). Available at https://doi.org/10.1080/23311916.2018.1446667.

Ye, J., Wang, Y., Li, Z., Saafi, M., Jia, F., Huang, B., et al., 2020. Failure analysis of fiber-reinforced composites subjected to coupled thermo-mechanical loading. Composite Structures 235. Available at https://doi.org/10.1016/j.compstruct.2019.111756.

Yilmaz, S., Theodore, M., Ozcan, S., 2024. Silicon carbide fiber manufacturing: cost and technology. Composites Part B: Engineering 269. Available at https://doi.org/10.1016/j.compositesb.2023.111101.

Zagho, M.M., Hussein, E.A., Elzatahry, A.A., 2018. Recent overviews in functional polymer composites for biomedical applications. Polymers 10 (7). Available at https://doi.org/10.3390/polym10070739.

Zalaznik, M., Kalin, M., Novak, S., 2016. Influence of the processing temperature on the tribological and mechanical properties of poly-ether-ether-ketone (PEEK) polymer. Tribology International 94, 92–97. Available at https://doi.org/10.1016/j.triboint.2015.08.016.

Zegaoui, A., Derradji, M., Dayo, A.Q., Medjahed, A., Zhang, H.Y., Cai, W.A., et al., 2019. High-performance polymer composites with enhanced mechanical and thermal properties from cyanate ester/benzoxazine resin and short Kevlar/glass hybrid fibers. High Performance Polymers 31 (6), 719–732. Available at https://doi.org/10.1177/0954008318793181.

Zhang, B., Wang, W., Tian, M., Ning, N., Zhang, L., 2020. Preparation of aramid nanofiber and its application in polymer reinforcement: a review. European Polymer Journal 139. Available at https://doi.org/10.1016/j.eurpolymj.2020.109996.

Zhang, B., Jia, L., Tian, M., Ning, N., Zhang, L., Wang, W., 2021. Surface and interface modification of aramid fiber and its reinforcement for polymer composites: a review. European Polymer Journal 147. Available at https://doi.org/10.1016/j.eurpolymj.2021.110352.

Zhang, C., Li, Y., Wu, J., 2023. Mechanical properties of fiber-reinforced polymer (FRP) composites at elevated temperatures. Buildings 13 (1). Available at https://doi.org/10.3390/buildings13010067.

Zhang, D., Heider, D., Gillespie, J.W., 2017. Void reduction of high-performance thermoplastic composites via oven vacuum bag processing. Journal of Composite Materials 51 (30), 4219–4230. Available at https://doi.org/10.1177/0021998317700700.

Zhang, Q., Chen, X., Zhang, B., Zhang, T., Lu, W., Chen, Z., et al., 2021. High-temperature polymers with record-high breakdown strength enabled by rationally designed chain-packing behavior in blends. Matter 4 (7), 2448–2459. Available at https://doi.org/10.1016/j.matt.2021.04.026.

Zhao, L., Yu, Y., Huang, H., Yin, X., Peng, J., Sun, J., et al., 2019. High-performance polyphenylene sulfide composites with ultra-high content of glass fiber fabrics. Composites Part B: Engineering 174. Available at https://doi.org/10.1016/j.compositesb.2019.05.001.

Zheng, H., Zhang, W., Li, B., Zhu, J., Wang, C., Song, G., et al., 2022. Recent advances of interphases in carbon fiber-reinforced polymer composites: a review. Composites Part B: Engineering 233. Available at https://doi.org/10.1016/j.compositesb.2022.109639.

Zhou, W., Kou, Y., Yuan, M., Li, B., Cai, H., Li, Z., et al., 2019. Polymer composites filled with core@double-shell structured fillers: effects of multiple shells on dielectric and thermal properties. Composites Science and Technology 181. Available at https://doi.org/10.1016/j.compscitech.2019.107686.

7

Hybrid polymer–based high-temperature composite materials

Teboho Clement Mokhena[1,2], Katekani Shingange[3], Sivuyisiwe Mapukata[4], Jonas Mochane[5], Bulelwa Ntsendwana[1,4], Mathato Motaung[1,4], William Moloto[1,4], Nonkululeko Miya[1,4], Amos Nhlapo[6], Teboho Patrick Mokoena[7], Mokae Bambo[8], Themba Ntuli[1,4]

[1]DSTI/MINTEK-NANOTECHNOLOGY INNOVATION CENTRE, RANDBURG, SOUTH AFRICA [2]SCHOOL OF CHEMICAL AND PHYSICAL SCIENCES, UNIVERSITY OF MPUMALANGA, MBOMBELA, SOUTH AFRICA [3]SCHOOL OF PHYSICS, UNIVERSITY OF THE WITWATERSRAND, JOHANNESBURG, SOUTH AFRICA [4]MINTEK, ADVANCED MATERIALS DIVISION, RANDBURG, SOUTH AFRICA [5]DEPARTMENT OF CHEMISTRY, UNIVERSITY OF SOUTH AFRICA, JOHANNESBURG, SOUTH AFRICA [6]DEPARTMENT OF PHYSICS, MEDUNSA, SEFAKO MAKGATHO HEALTH SCIENCES UNIVERSITY, SOUTH AFRICA [7]DEPARTMENT OF PHYSICS, UNIVERSITY OF THE FREE STATE, PHUTHADITJHABA, SOUTH AFRICA [8]DEPARTMENT OF NATURAL SCIENCES, VAAL UNIVERSITY OF TECHNOLOGY, VANDERBIJLPARK, SOUTH AFRICA

7.1 Introduction

There is an ever-increasing demand for lightweight, environmentally friendly, strong, and durable structures (Matumba et al., 2024; Asim et al., 2015; Sethupathi et al., 2024; Raheem and Subbaya, 2023; Kaleni et al., 2022). This demand has sparked interest in using fiber-reinforced polymer composites as suitable alternatives to conventional materials like metals and glass. Currently, glass fiber (GF) dominates the market, accounting for 95% of total fiber-reinforced polymer composites (Kanitkar et al., 2017). However, natural fibers offer attractive attributes such as lightweight, biodegradability, excellent damping properties, no damage to processing equipment, abundant availability, and excellent health safety. Numerous studies have confirmed that natural fibers are more environmentally friendly compared to GFs (Sethupathi et al., 2024; Rashid et al., 2024; Mokhothu et al., 2011).

Despite these advantages, natural fibers have inherent drawbacks, including hydrophilicity, poor wettability, low thermal degradation temperatures, and poor chemical resistance (Asim et al., 2015; Todkar and Patil, 2019; Baysal et al., 2023). These issues often need to be addressed for natural fibers to be used in high-end applications. One common solution is the combination of both natural and synthetic fibers, a process known as hybridization.

Hybridization involves incorporating additional materials to resolve the issues associated with the fillers added to the matrix, resulting in multifunctional, superior composite products (Jagadeesh Chandra et al., 2019; Todkar and Patil, 2019). To optimize the resultant properties,

High-Temperature Polymeric Materials. DOI: https://doi.org/10.1016/B978-0-443-32792-6.00010-5

the materials used in preparing hybrids must be carefully considered. These materials are typically categorized into the following systems:

- Systems with two reinforcing fibers (Baysal et al., 2023)
- Systems composed of fibers and micron-nanosized particles (Rashid et al., 2024; Adhikari et al., 2018)
- Systems composed of two nanoparticles (Rashid et al., 2024)

In this context, hybridization of at least two reinforcing fillers is briefly defined. Hybrid composites are crucial in various fields, from household items to automotive applications, due to their unique features such as low cost, lightweight, and high specific and mechanical strength. In this chapter, preparation and application of hybrid polymer composites in different fields are discussed. Emphasis is based on the preparation of these materials to meet high-end applications. The challenges and advantages of hybrid composites are also highlighted.

7.2 Hybrid composites: an overview

Hybrid composites consist of polymeric materials or their blends mixed with fibers that act as a reinforcing phase (Jagadeesh Chandra et al., 2019; Mochane et al., 2019). The choice of preparation method is often influenced by economic considerations (Mochane et al., 2019). These composites are characterized by unique features such as lightweight, superior mechanical performance, and excellent thermal properties. Polymeric materials serve as the continuous phase that surrounds the fibers, transferring stress to the fillers and protecting the fibers from environmental conditions. Typically, polymeric materials make up more than 40% of the composite's overall weight. A wide range of polymeric materials is used as matrices for fillers, including thermosets (e.g., epoxy resins and phenolic resins) and thermoplastics (e.g., polypropylene and polyethylene). Common fillers include synthetic fibers like GF, carbon fiber (CF), and aramid, as well as natural fibers, depending on the intended application (Todkar and Patil, 2019). Epoxy resins are the most frequently used polymeric resins due to their excellent properties, such as low shrinkage, minimal moisture absorption, quick curing, good mechanical properties, and ease of processing (Ding et al., 2019).

7.3 Fibers

Polymer composites are typically reinforced with either natural or synthetic fibers (Sethupathi et al., 2024; Todkar and Patil, 2019; Rajak et al., 2019b; Singh et al., 2022). Natural fiber-reinforced composites are particularly attractive due to their inherent properties, such as low density, exceptional specific strength, reduced wear during processing, cost-effectiveness, low energy consumption, renewability, and biodegradability (Gomez-Caturla et al., 2023). These advantages have driven researchers to focus on balancing the benefits and drawbacks of these materials. However, natural fibers have limitations, including high moisture absorption and low thermal stability. To address these issues, researchers and industrialists have explored integrating natural fibers with other fillers that can complement their properties (Sethupathi et al., 2024).

On the other hand, synthetic fiber-reinforced composites are categorized based on the type of fiber used as the reinforcing phase (Singh et al., 2022). These man-made fibers are often associated with higher costs but offer distinctive properties such as excellent thermal and mechanical performance. Common synthetic fibers include aramid, carbon, and GFs, which are widely used in various industries, including aerospace, automotive, construction, and leisure.

Natural fiber composites are gaining popularity due to their environmental benefits and cost savings (Sethupathi et al., 2024). They are particularly useful in applications where biodegradability and sustainability are important. However, their susceptibility to moisture and thermal degradation can limit their use in high-performance applications. By combining natural fibers with synthetic fillers or other materials, it is possible to enhance their properties and expand their range of applications.

Synthetic composites, despite their higher cost, are valued for their superior performance characteristics (Kanitkar et al., 2017; Rajak et al., 2019b). Aramid fibers, for example, are known for unique features, including high strength and heat resistance, which afford them applicability in aerospace and military applications. CFs offer excellent stiffness and strength-to-weight ratios, which are crucial for high-performance automotive and sporting goods. GFs are widely used due to their good mechanical properties and cost-effectiveness, which make afford a broad range of applications from construction to consumer products.

In summary, both natural and synthetic fiber-reinforced composites have their unique advantages and challenges (Sethupathi et al., 2024). The choice between them relies on the anticipated application, including performance, cost, and environmental considerations. Current research efforts and development aim to optimize these materials to meet the growing demands of various industries.

7.3.1 Natural fibers

Natural fibers are often sourced from various materials, including waste residues and natural resources (Adhikari et al., 2018; Rajak et al., 2019b; Verma and Goh, 2019). They can be extracted from different parts of plants, such as bast stems, leaves, and seeds, leading to their classification, as shown in Fig. 7–1 (Todkar and Patil, 2019; Asim et al., 2017). Beyond plant-based fibers, animals also produce natural fibers like hair, wool, and silk. The use of natural fibers as reinforcement in polymeric materials has grown due to concerns about the sustainability of petroleum reserves. This has sparked interest in green, sustainable alternatives that can replace petroleum-based fibers.

Natural fiber-reinforced composites offer numerous advantages, including being light-weight, abundantly available, and cost-effective (Todkar and Patil, 2019; Singh et al., 2022; Verma and Goh, 2019; Bocz et al., 2020). However, the main limitation of using natural fibers as reinforcement for polymers is the compatibility between the fiber and the host polymeric materials. This often results in poor interfacial interaction and high moisture absorption. Over the years, researchers have developed strategies to improve fiber adhesion with various

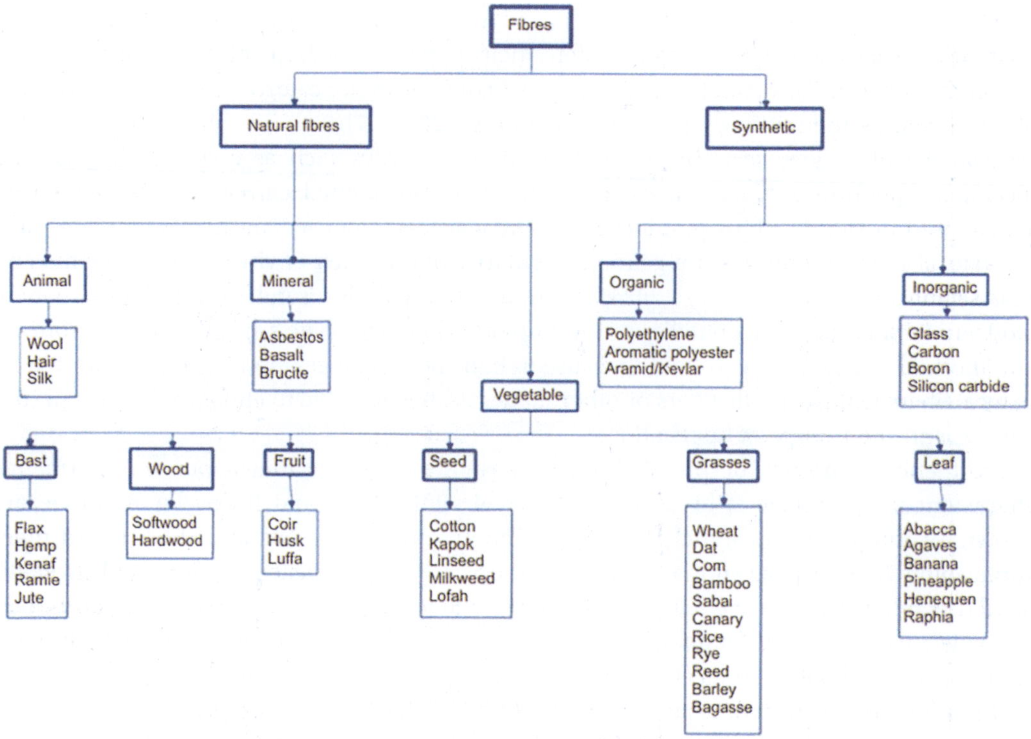

FIGURE 7–1 Classification of fiber. *Reproduced from Mochane, M.J., Mokhena, T.C., Mokhothu, T.H., Mtibe, A., Sadiku, E.R., Ray, S.S., et al., 2019. Recent progress on natural fiber hybrid composites for advanced applications: a review. Express Polymer Letters 13 (2), 159–198. https://doi.org/10.3144/expresspolymlett.2019.15. Categorization of fibres according to their origin.*

matrices. These strategies include treating natural fibers with different chemical modifications to reduce moisture absorption and enhance interaction with the host matrix.

Several chemical and physical modifications have been employed to improve the interaction and overall properties of the resulting composite products (Kumar et al., 2021). These modifications include treatments with sodium hydroxide, sodium chlorite, methacrylate, isocyanate, silane, acetylation, etherification, enzymatic treatments, peroxide treatment, benzoylation, dicumyl peroxide, and plasma treatments (Mokhothu et al., 2011; Todkar and Patil, 2019).

7.3.2 Synthetic fibers

Synthetic fiber-reinforced composites typically offer high thermal stability, mechanical strength, and stiffness, which make afford application in the aerospace and automotive industries (Kanitkar et al., 2017; Todkar and Patil, 2019; Rajak et al., 2019b). These fibers include GFs, CFs, aramid fibers, and many others. The combination of synthetic and natural fibers is often developed with the primary goal of enhancing both the thermal and mechanical properties of the resulting composites (Todkar and Patil, 2019). This complementary approach addresses the limitations of each type of fiber rather than replacing one with the other. Synthetic fibers, such as glass, carbon, and aramid, are known for their excellent performance characteristics. GFs provide good mechanical properties and are cost-effective, making them widely used in various applications. CFs offer exceptional stiffness and strength-to-weight ratios, which are crucial for high-performance applications. Aramid fibers are valued for their high strength and heat resistance, making them ideal for aerospace and military uses. On the other hand, natural fibers, derived from plants and animals, are lightweight, renewable, and biodegradable. They are increasingly used as reinforcements in polymeric materials due to their environmental benefits and cost savings. However, natural fibers often face challenges such as high moisture absorption and lower thermal stability compared to synthetic fibers. By combining synthetic and natural fibers, researchers aim to create hybrid composites that leverage the strengths of both types of fibers. This approach can improve the overall performance of the composites, making them suitable for a broader range of applications. For instance, the high mechanical strength and thermal stability of synthetic fibers can complement the lightweight and ecofriendly nature of natural fibers, resulting in composites that are both strong and sustainable. Additionally, the integration of synthetic and natural fibers can lead to innovative solutions in various industries. In the automotive sector, hybrid composites can be used to manufacture lightweight and durable components, contributing to fuel efficiency and reduced emissions. In the aerospace industry, these composites can help create strong yet lightweight structures, enhancing performance and reducing costs. The construction industry can also benefit from hybrid composites, using them to develop sustainable building materials with improved mechanical properties. Overall, the development of hybrid composites combining synthetic and natural fibers represents a promising direction for creating high-performance, sustainable materials that meet the demands of modern applications (Todkar and Patil, 2019).

7.4 Polymers

The choice of host polymeric materials plays a critical role in determining the properties of the resultant composites (Sethupathi et al., 2024; Mokhena and Mochane, 2018). Researchers and material scientists select these materials based on the intended application. Polymeric materials can be categorized based on their source into natural or synthetic polymers. Natural polymers include cellulose, lignin, chitin, starch, and their derivatives (Sethupathi et al., 2024; Mokhena et al., 2015). These polymers are derived from natural sources and are often used for their biodegradability and renewable nature. Synthetic polymers, on the other hand, are produced through chemical reactions using monomers as precursors. These reactions, known as polymerization, result in high molecular weight materials with diverse behaviors and properties, making them suitable for various applications. Synthetic polymers are further classified into thermoplastics and thermosets, as discussed below (Ding et al., 2019).

7.4.1 Thermosets

Thermosets are durable resins known for their desirable properties, such as high modulus, strength, and resistance to thermal and chemical degradation, due to their crosslinking density (Ding et al., 2019; Rajak et al., 2019b; Asim et al., 2017). Once cured or polymerized, thermosets become brittle and cannot be reshaped. Examples of thermosets include epoxy resins, phenol-formaldehyde, polyesters, and polyurethanes (Ding et al., 2019). Due to their brittleness, thermosets are often used in high-performance and impact-resistant industrial applications, including construction and transportation. Hybrid composites made from thermosets are typically manufactured using hand layup techniques and compression molding methods.

7.4.2 Thermoplastics

Thermoplastics are produced in large quantities through chemical processes (Sethupathi et al., 2024; Mtibe et al., 2023; Moothoo et al., 2021). Some common examples of thermoplastics include polybutylene succinate, poly lactic acid, polyethylene, polycaprolactone, poly(butylene succinate-*co*-adipate), polypropylene, poly(vinyl chloride), poly(vinyl alcohol), polyethylene terephthalate, and polystyrene (Mokhena et al., 2024; Pérez-Fonseca et al., 2025; Gomez-Caturla et al., 2023; Mtibe et al., 2023; Van de Perre et al., 2024). These polymers are highly versatile and can be melted and remelted multiple times at suitable temperatures, allowing them to be shaped and reshaped as needed. This property makes thermoplastics ideal for a wide range of applications, from packaging and consumer goods to automotive and aerospace components.

7.5 Preparation methods

7.5.1 Hand layup

Hand layup is one of the simplest methods for manufacturing hybrid composites (Devendra and Rangaswamy, 2012; Ravi Raj and Vijaya Ramnath, 2021; Seretis et al., 2017). In this process, fibers are laid down layer by layer into a mold of the desired shape, with each layer being coated with the appropriate amount of polymer resin. To prevent the composite from sticking to the mold, various nonstick materials are applied before placing the fibers and polymer matrix. Curing takes time, and the accuracy, shape, and quality of the resulting composites depend on the skill of the processor. Ravi Raj and Vijaya Ramnath (2021) investigated the influence of silicon carbide on polymethyl methacrylate (PMMA) toughened GF epoxy hybrid composites. They used the hand layup method to disperse PMMA and silicon carbide (SiC) filler in an epoxy matrix at room temperature. The prepared hybrid composites were then postcured at 80°C to enhance performance. The incorporation of the filler improved the overall mechanical and thermal properties of the composites, making them suitable for applications in various fields such as automotive, aerospace, drones, electronic gadgets, and domestic appliances. Similarly, Uma Devi et al. (2012) prepared hybrid composites from GFs and pineapple leaf fibers (PALF) by impregnating them with polyester resin. The impregnated fibers were kept in the mold under pressure for 24 hours, followed by postcuring at 50°C for 24 hours. The resulting hybrid composites exhibited outstanding mechanical performance.

7.5.2 Compression molding

It is one of the simplest manufacturing processes, operating on a similar principle to bulk composite production through compression molding (Uma Devi et al., 2012; Nurul and Mariatti, 2013; Corbridge et al., 2017). The machine is designed with one part as male and the other as female (Nurul and Mariatti, 2013). The fibers and polymeric resin are placed in the female part, and the male part is positioned above. Desired pressure is then applied to allow curing at room temperature. After curing, the resulting composite is removed from the mold. This method is often suitable for manufacturing automotive components. The design of compression molding can be optimized based on the host polymeric materials to ensure that the resulting hybrid composites exhibit excellent overall performance (Corbridge et al., 2017). For thermoplastic reinforced hybrid composites, material flow during processing can cause distortions in the continuous fiber reinforcement, such as fiber waviness, ply splitting, or ply migration. For instance, Mallick (1986) reported that fabricating hybrid composites using GF-based sheet molding and continuous GF plies resulted in reduced overall properties as the initial charge size decreased. This reduction was attributed to increased fiber waviness in the continuous filler with greater flow. This behavior is also observed in thermoplastic-based composites, where discontinuous fibers tend to drag and penetrate the continuous fiber reinforcement.

7.5.3 Resin transfer molding technique

In this process, the mold is equipped with an inlet to introduce the resin/catalyst mixture into the fibers already laid in the mold (Rafiee et al., 2019; Bhanuprakash et al., 2019). An outlet vent allows air to escape. The catalyst and resin are thoroughly mixed in appropriate ratios and then pumped into the mold containing dry fillers. The resin cures under atmospheric pressure. This technique has been modified to use vacuum to draw the resin/catalyst mixture into the mold (Rafiee et al., 2019). Vacuum bags are used along the mold to improve resin flow and reduce void fractions (Rafiee et al., 2019). This process is cleaner and releases fewer volatiles, making it suitable for producing autobody panels, swimming pool panels, and sandwich panels. Rafiee et al. (2019) investigated the thermal properties of epoxy/GF composites in the presence of carbon-based nanoparticles, namely, reduced graphene oxide, graphene nanoplatelets, and graphene oxide. They used vacuum-assisted resin transfer molding (RTM) to create these multifunctional hybrid composites. The study concluded that the overall properties of the resulting hybrid composites can be enhanced depending on the type and amount of filler, as well as its interaction with the host polymeric material. The thermal properties of the composites were found to improve because of better filler dispersion and strong interfacial interaction with the matrix.

7.5.4 Pultrusion technique

This manufacturing method is a continuous and highly automated process, commonly used to produce large quantities of hybrid composites (Asim et al., 2017; Rajak et al., 2019a). The process involves aligning fibers to match the die, pulling them through a resin bath, and then onto a heated metal pultrusion die. The die is maintained at a desired temperature to ensure proper heat transfer to the matrix and fibers. The solid part is then pulled out of the die and cut into appropriate lengths. This method is often employed in the production of materials for construction, transportation, and other industries. As a highly automated manufacturing technique, pultrusion is considered efficient, consistent, cost-effective, scalable, and versatile. The precise control over fiber alignment and resin distribution allows for the optimization of material properties, thereby enhancing the overall properties of the resulting composites.

7.5.5 Melt processing

Thermoplastics are crucial materials due to their unique properties, including low density, excellent water and chemical resistance, easy processability, and cost-effectiveness (Dvir et al., 2003; Bledzki et al., 2015). Their affordability has driven their widespread industrial applications, from automotive to construction. These materials are typically shaped by melting the polymer and mixing it with fillers using various melt processing techniques (Table 7-1).

7.5.5.1 Melt extrusion

Melt extrusion is a widely used melt processing technique for manufacturing thermoplastic-based products (Dvir et al., 2003; Bledzki et al., 2015). This process involves melting the material and mixing it under shear stress from rotating screws, then forcing it through a die to produce items such as pipes, plastic sheets, and pellets. These pellets can be further processed using other techniques, such as blow molding, compression molding, or injection molding, depending on the intended application (Bledzki et al., 2015). Fu et al. (2000) produced polypropylene (PP) composites reinforced with short GFs and short CFs using melt extrusion followed by injection molding. Their study demonstrated that the overall performance of the resulting hybrids depends on the fiber ratio. This indicates that melt extrusion is often combined with other processing techniques to achieve desired shapes and structures for specific applications.

7.5.5.2 Injection molding

This process involves melting polymeric material and injecting it into a mold to create desired items (Bledzki et al., 2015; Holmström et al., 2020; Mortazavian and Fatemi, 2015; Lee et al., 2013). It can be used to produce various products, such as automotive components and packaging materials. Despite the technique's advantages, such as repeatability and the ability to easily produce complex, intricate shapes with fine details, the design and production of molds can be expensive and time-consuming. Consequently, small production runs are often costly due to the significant investment required for setup, design, and any subsequent design

Table 7–1 Summary of preparation methods used to fabricate hybrid composites.

Matrix	Filler-type	Preparation method	Verdict	References
PMMA/Epoxy	Silicon carbide (SiC)/GF	Hand layup	The resultant composites had balanced mechanical properties with improved thermal stability	Ravi Raj and Vijaya Ramnath (2021)
Epoxy	GF/aluminum oxide, magnesium hydroxide, silicon carbide, hematite	Hand layup	The improvements of the properties depended on the type of filler used	Devendra and Rangaswamy (2012)
Epoxy	GF/graphene nanoplatelets	Hand layup	The addition of graphene nanoplatelets into epoxy/GF composites enhanced the overall mechanical performance	Seretis et al. (2017)
Polypropylene PP	GF/pentabromobenzyl acrylate (PBBMA)/magnesium hydroxide	Melt extrusion/ injection molding	GFs enhanced modulus with limited influence on impact strength; meanwhile, PBBMA and magnesium hydroxide negatively affected impact strength, meanwhile enhanced flame resistance property	Dvir et al. (2003)
PP/polylactic acid	Cellulose fibers/softwood flour	Melt extrusion/ injection molding	Hybrid exhibited excellent performance, indicating that it can serve as an alternative to GF-based hybrid composites	Bledzki et al. (2015)
Epoxy	GF/graphene nanoplatelets, graphene oxide, reduced-graphene oxide	Vacuum-assisted RTM	Better dispersion of the nanoparticles and good interfacial interaction are crucial for improving their overall properties	Rafiee et al. (2019)
Epoxy	CF/graphene oxides	Vacuum-assisted RTM	The coating of GOs onto CF enhanced interfacial interaction between fibers and the host matrix	Bhanuprakash et al. (2019)

CF, Carbon fiber; *GF*, glass fiber; *PMMA*, polymethyl methacrylate; *RTM*, resin transfer molding.

changes. Recent advancements in injection molding have enabled the creation of multifunctional composites (Wang et al., 2025; Valente et al., 2023; Azam and Lee, 2022). In their study, Wang et al. (2025) prepared hybrid continuous and discontinuous fiber reinforced products using insert-injection molding. Here, premolded composites act as continuous inserts placed in mold cavities, allowing them to bond directly with injected discontinuous fiber-reinforced composites, as illustrated in Fig. 7–2. These premolded composites can be made from randomly oriented fibers, woven fibers, unidirectional prepreg tapes, or fiber-roving-based composites. This method facilitates the production of complex geometries while reducing costs and enhancing esthetics. Additionally, the continuous reinforcement can be tailored to the expected application, with high-stress areas in the composites being reinforced with more fibers to prevent failure. This technique has been used to manufacture components like bumpers and seat beams. However, variations in processing conditions can affect the melting and crystallization behavior of the thermoplastic, influencing interfacial interactions and the overall properties of the hybrid product. By modifying inserts, polymeric materials, filler types, and filler orientations, multifunctional materials can be produced for high-performance applications, such as in aeronautics.

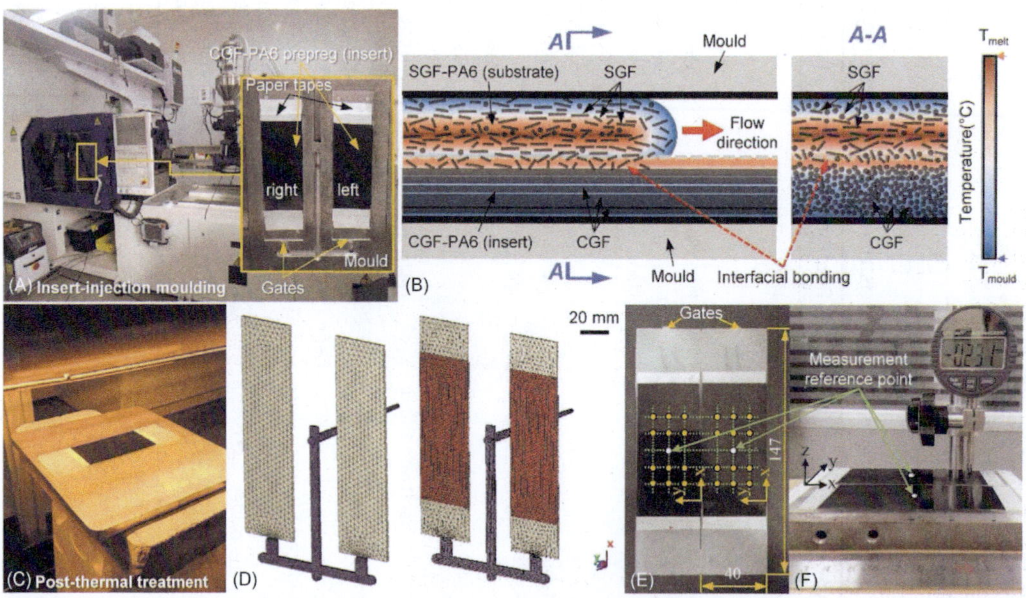

FIGURE 7–2 Illustration of insert injection molding: (A) typical injection machine with the mold cavities with paper-taped composites; (B) schematic presentation of insert injection-molding; (C) postthermal treatment; (D) 3D mesh models of composites; (E) as-prepared samples and warping measurement positions; and (F) Warpage measuring device. Injection moulding as a thermal technique for processing polymer composites. *Reproduced from Wang, J., Li, H., Li, F., Fan, C., Liu, T., Wang, D., 2025. Insert-injection moulding and post-thermal treatment of hybrid continuous and discontinuous glass-fibre-reinforced polyamide composite products. Composites Part A: Applied Science and Manufacturing 188, 108534. https://doi.org/10.1016/j.compositesa.2024.108534; Mochane, M.J., Mokhena, T.C., Mokhothu, T.H., Mtibe, A., Sadiku, E.R., Ray, S.S., et al., 2019. Recent progress on natural fiber hybrid composites for advanced applications: a review. Express Polymer Letters 13 (2), 159–198. https://doi.org/10.3144/expresspolymlett.2019.15.*

In their research, the authors used unidirectional continuous GF reinforced polyamide-6 (CFG/PA) prepreg tapes as inserts fixed in the mold cavity (Fig. 7-2A) (Wang et al., 2025). Molten short GF reinforced PA6 was then introduced to fill the cavity and integrate with the prepreg tapes. The process of impregnating the insert is shown in Fig. 7-2B. During this process, the prepreg polymeric materials remelt and solidify with the injected short GF-PA6, creating strong interfacial interactions. Optimized preparation conditions resulted in improved mechanical performance, which was further enhanced through postthermal treatment of the hybrid composite (Fig. 7-2C).

Although GF is commonly used to reinforce various polymeric materials for engineering applications, some studies have explored replacing GF with natural materials to create more cost-effective and ecofriendly products. For instance, Bledzki et al. (2015) developed hybrid biocomposites reinforced with cellulose fibers and softwood flour using injection molding. These hybrid biocomposites demonstrated mechanical performance comparable to those reinforced with GF.

7.5.5.3 Blow molding

In this process, hollow objects are created by inflating a molten preform within a mold (Balthes, 2013). It is commonly used to produce bottles for packaging. Blow molding is one of the more cost-effective processing methods with high production rates. It allows for the creation of a variety of products with different shapes and uniform wall thickness. However, its disadvantages include material limitations and the fact that it can only produce hollow objects (Balthes, 2013).

7.6 Properties and performance

7.6.1 Thermal properties

Reinforced polymer composites with conventional fiber have been proven to be somewhat not enough for many applications, and this has prompted the use of hybrid polymer composites. In one of the studies, basalt-reinforced sheep wool, paddy straw, (BPS) and bagasse-based (BBS) hybrid polymer composites were prepared by compression molding (Hajare et al., 2022). The hybrid composites undergone thermal test (Fig. 7-3) (Hajare et al., 2022).

Interestingly, the composites improved their thermal stability by ~10 degrees as degradation temperature was increased (Hajare et al., 2022). Based on the thermogravimetric analysis (TGA) analysis, the BPS hybrid composite showed more thermal stability than BBS. The increased char yield in the BPS hybrid suggested that degradation had slowed. The presence of keratin in the wool fiber, which contains amino acid cysteine that results in bridges and crosslinking, and thus facilitates thermal insulation and hence stability.

It must be noted that polymers are usually susceptible to low thermal and heat resistance. Similar to the above example, other materials can be used in conjunction with polymers to form hybrids and thereby increase their thermal resistance. Commonly, to increase thermal properties of polymers, silicon carbides, GF, zinc oxide fillers, etc., are added (Cheewawuttipong et al., 2013). The important ingredient in the polymer reinforcement is

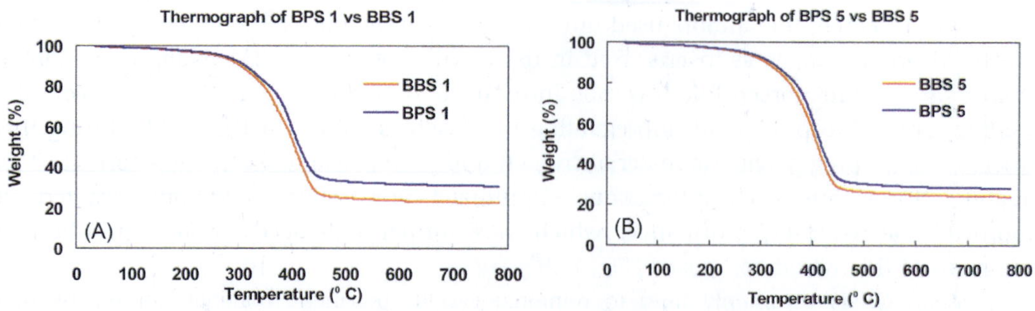

FIGURE 7–3 Thermogravimetric analysis of unsaturated polyester basalt-reinforced with (A) BBS1 (10/90 sheep wool/paddy straw) versus BPS1 (10/90 sheep wool/baggasse) and (B) BBP5 (50/50 sheep wool/paddy straw) versus BPS5 (50/50 sheep wool/baggasse). Thermogravimetric analysis comparing BBS1 and BBP5 with their respective counterparts, BPS and BPS-5. *Hajare, B., Patil, O., Barpande, G.S., Radhakrishnan, S., Kulkarni, M.B., 2022. Tensile, impact and thermal properties of farm-waste based hybrid basalt polymer composites. Cleaner Materials 6, 100157. https://doi.org/10.1016/j.clema.2022.100157.*

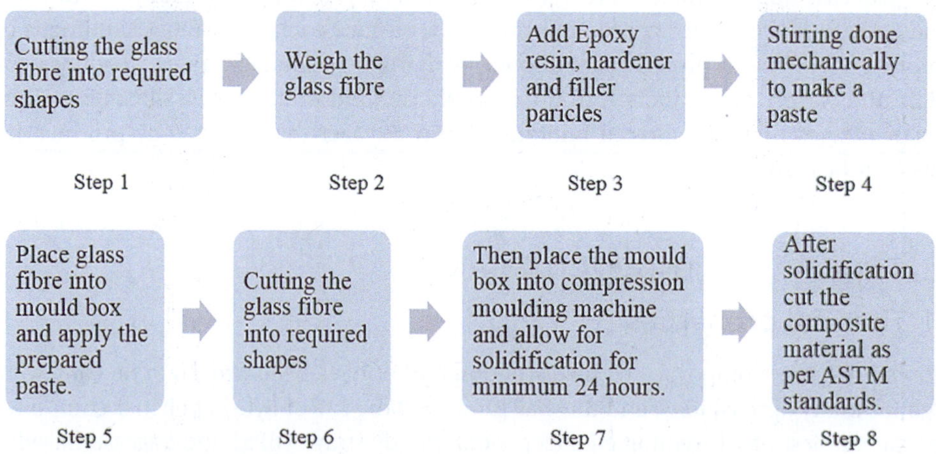

FIGURE 7–4 Overall steps involved in the manufacturing of hybrid polymer composites. Key stages in the production process of hybrid polymer composites. *Vaggar, G.B., Kamate, S.C., Shetty, R., 2021. A review on thermal properties of hybrid polymer matrix composites. IOP Conference Series: Materials Science and Engineering 1065 (1), 012030. https://doi.org/10.1088/1757-899x/1065/1/012030.*

either the epoxy resin, polyester, phenolic, or polyimide resins. The most commonly used reinforcement ingredient due to its low moisture absorption, low shrinkage, outstanding stiffness, good mechanical properties, and excellent interface. It is also important to note that identifying and predicting the temperature behaviors of hybrid polymers is very difficult due to the variation in temperature. Fig. 7–4 shows a summary of important steps involved in the fabrication of hybrid polymer composites (Vaggar et al., 2021).

Studies have found that to improve the thermal stability of pristine epoxy polymer composites, silicon carbide fillers must be added. In fact, the increase in 20% and 30% filler

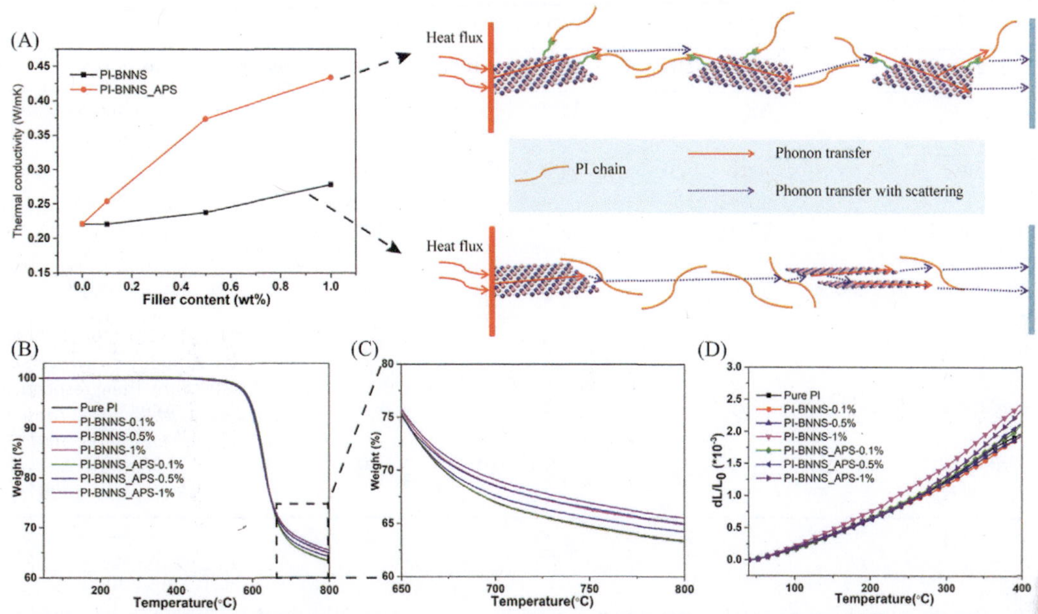

FIGURE 7–5 (A) Thermal conductivities of the PI-based hybrid system containing different filler contents and (B) schematic presentation of heat transfer processes of two series of hybrid polymer PIs. thermogravimetric analysis (TGA) curves of the PI-based hybrid system in nitrogen and (C) magnification images between 650°C and 800°C. (D) TMA curves of the hybrid PIs. (A) The thermal conductivities of the PI-based hybrid system with varying filler contents; (B) a schematic representation of heat transfer mechanisms in two series of hybrid polymer PIs; (C) TGA curves of the PI-based hybrid system in a nitrogen atmosphere, including magnified views between 650°C and 800°C; and (D) TMA curves of the hybrid PIs. *Ou, X., Lu, X., Chen, S., Lu, Q., 2020. Thermal conductive hybrid polyimide with ultrahigh heat resistance, excellent mechanical properties and low coefficient of thermal expansion. European Polymer Journal 122, 109368. https://doi.org/10.1016/j.eurpolymj.2019.109368.*

particles has the potential to increase the thermal conductivity by up to 0.41 and 0.51 W m^{-1} K^{-1}, respectively (Srinivas and Bhagyashekar, 2015). Another study revealed that amino-functionalized boron nitride nanosheets hybrid polyamide (PI-B-NNS), which was fabricated through a simple polymerization method (Ou et al., 2020). The hybrid (PI-BNNS-APS) showed higher heat resistance and thermal conductivity. In addition, the glass transition temperature increases to ~473°C and thermal conductivity by 100% when compared to the pristine PI (Fig. 7-5). The influence of filler content is evident in an increase in thermal conductivity, as a low content of fillers, no filler–filler networks could form. In fact, the dispersion of BNN-APS matrix in PI led to a better filler–filler transfer network, resulting in high thermal conductivity. In addition, the high crystallinity of PI matrix contributed to the higher thermal conductivity (Ou et al., 2020).

7.6.2 Mechanical properties

The mechanical properties of hybrid polymers are distinct and enhanced compared to those of individual metal alloys and composite materials (Sinha et al., 2020). By combining different metal layers (such as aluminum, magnesium, and titanium alloys) with glass, aramid, or

Table 7–2 Mechanical properties of hybrid polymers.

Composites	Tensile strength	Flexural strength	Impact resistance	References
Abaca/GFRP	44.5	12.5	16	Vijaya Ramnath et al. (2013)
Jute/GFRP	46.5	11.9	15	Vijaya Ramnath et al. (2013)
Abaca/Jute/GFRP	57.0	12.1	12	Vijaya Ramnath et al. (2013)
Pineapple leaf fiber/epoxy resin	71.66	239.37	–	Indra Reddy et al. (2018)
Coir/Bagasse (30:20 wt.%)	23.2	44.2	4.2	Sivaraj and d Rajeshkumar (2014)
Coir/Bagasse (30:20 wt.%)	18.72	37.8	4.0	Sivaraj and Rajeshkumar (2014)
Coir/Bagasse (30:20 wt.%)	12.52	30.84	3.0	Sivaraj and Rajeshkumar (2014)

CF-reinforced polymer matrix composites, and by orienting individual layers, it is possible to create laminate packs with specific thicknesses and a defined number of layers (both metal and polymer composite). Various surface treatment methods also enable the precise modification of the strength properties of these hybrid polymers. To achieve the desired composite properties, it is crucial to ensure a proper interface between the reinforcement and the host matrix (Sinha et al., 2020; Seoane-Rivero et al., 2023; Afolabi and Ndou, 2024). Table 7–2 represents the mechanical properties of various hybrid polymers (Bhanuprakash et al., 2019).

7.6.2.1 Tensile strength

The incorporation of strong reinforcement fibers (e.g., CFs, GFs, or aramid fibers) within the polymer matrix can significantly enhance the tensile strength of the composite (Shrivastava et al., 2017). This makes it suitable for load-bearing applications at high temperatures. Engineering materials with hybrid composite properties have significant amounts of potential for use in various applications. Choosing the right fibers and matrix for a hybrid polymer composite material enables designers to precisely achieve the desired properties. Using various fiber types to include in a single resin matrix allows the material to have certain qualities. A study on epoxy hybrid composite reinforced with carbon and GFs investigated the tensile properties of the composite material. The process of vacuum bagging was used to produce hybrid composite materials. ASTM standards were followed in determining the mechanical characteristics of the hybrid composites, including hardness, tensile strength, tensile modulus, ductility, and peak load. When the amount of fiber reinforcement in the matrix increased, the mechanical characteristics also improved (Zikmundová and Eliášová, 2022). Shrivastava et al. (2017) have made efforts to make use of the characteristics of both natural and synthetic fibers to enhance the mechanical qualities. Mats made of synthetic GF and naturally occurring coir fiber were employed as reinforcement for this purpose, while thermosetting polymeric resin epoxy was used as the matrix material. Sodium hydroxide was chemically applied to coir fiber to reduce its moisture absorption behavior. Improved matrix reinforcement interface bonding was the outcome of this chemical treatment, which also enhances surface roughness and lowers the fiber's affinity for the material. The hand-layup method was used to produce the hybrid composite of short coir fiber and GF mat. The study demonstrates that the volume fraction and fiber length of the composites affect their mechanical properties (Shrivastava et al., 2017). Researchers have investigated how high temperatures affect the mechanical properties of materials with sodium hydroxide and silane adhesion treatments between metal and prepreg layers, as well as vacuum molding techniques (Shrivastava et al., 2017). Tensile strength of the fabricated fiber metal laminates (FMLs) is significantly higher than that of the aramid-fiber-reinforced aluminum alloy laminates and glass-reinforced aluminum laminates, as stated by the European Space Agency (ESA). In addition, FMLs that were produced at 175°C maintained 4% of their tensile strain at ambient temperature and more than 70% of their tensile strength and modulus. Their work has led to the remarkable finding that good candidates have been identified for a variety of applications, including but not limited to aeronautical and space applications (Seoane-Rivero et al., 2023).

7.6.2.2 Flexural strength

Hybrid polymer composites can exhibit excellent flexural strength due to the combined stiffness of the polymer matrix and the reinforcing fibers (Indra Reddy et al., 2018). This is critical in applications where bending and flexural loads are common (Indra Reddy et al., 2018; Krishnakumari et al., 2018). The flexural strength of polymer composites indicates the maximum bending stress they can endure under bending load conditions. This property is crucial for analyzing the bending performance of hybrid polymer composites, which is vital for structural applications (Sinha et al., 2020; Sivaraj and Rajeshkumar, 2014). Researchers have explored the tensile and flexural properties of polymer hybrid composites reinforced with jute, PALF, and GF in a 1:1:1 ratio within an epoxy resin. The fiber content varied from 0.18 to 0.42 by volume fraction, and the mechanical properties, including tensile and flexural qualities, were examined. Results showed that tensile strength improved with increased fiber content, and similarly, flexural strength also increased with higher fiber content. Sinha et al. (2019) reported the mechanical modeling and characterization of hybrid abaca epoxy composites that are filled with red mud. These hybrid composites were prepared using the hand layup technique, with experiments designed using a full factorial approach. The three control parameters were the weight percentage of abaca (2.6, 5.26, and 7.9 wt.%), the weight percentage of red mud (4, 8, and 12 wt.%), and the particle size of red mud (68, 82, and 98 μm). The composites were evaluated for flexural and impact strength. The response surface method was employed to develop mathematical models for these properties, and analysis of variance was used to determine the significance of the control parameters on the mechanical properties. The study also examined the effects of these input variables on the flexural and impact strength of the hybrid composites. The developed model enables the prediction of the mechanical properties of hybrid composites (Sinha et al., 2019).

7.6.2.3 Impact resistance

The hybrid nature allows for a tailored toughness, with some fibers contributing to high strength and others improving the energy absorption capacity. This leads to better impact resistance even at high temperatures. Impact strength measures a hybrid composite's ability to resist impact loads without failing. Therefore it is crucial for design engineers and scientists to understand the impact properties of a component before its use in structural applications (Wang et al., 2020). This is particularly important in the automotive industry for parts such as fenders, underfloor panels, and doors. Research has shown that the impact strength of hybrid composites is influenced by fiber length and fiber loading. The impact strength of a hybrid composite affects its ability to withstand impacts without failure. Thus, when a component is used in any structural application, it is evident that a design engineer or scientist should be aware of its impact qualities. These qualities are very important, particularly in the automotive industry for parts (Bieniaś et al., 2017; Nguyen et al., 2017). Impact resistance can result from both low- and high-velocity sources, including aircraft ground handling, maintenance work, take-off and landing (such as solid objects thrown up by the wheels, wind-lifted debris, bird strikes, and hail), and collisions with ballistic items (Bieniaś et al., 2017; Nguyen et al., 2017). Shih et al. (2022) assess the impact resistance of aramid fabric reinforced with shear thickening fluids (STFs), epoxy, or polyurea elastomers using ballistic experiments. According to ballistic test findings, aramid composite structures coated with polyurea elastomers exhibit the highest impact resistance and absorb the most impact energy per unit area density. However, the impact resistance of the epoxy-treated aramid composite structure was reduced by the concentration of stress that occurs during ballistic impact. Aramid fabric coated with STFs enhances structural protection but increases the composite's weight and decreases specific energy absorption. These findings evaluate the energy absorption properties, deformation characteristics, and damage mechanisms of various aramid composites, providing valuable insights for future research (Shih et al., 2022).

7.6.3 Flame resistance performance

Polymeric materials are flammable in nature due to their chemical composition, which limits their applications in areas where high flame resistance is required (Bocz et al., 2020; Devendra and Rangaswamy, 2012; Mokhena et al., 2021). The addition of flame-retardant fillers to polymer composites serves as an alternative to enhance the overall fire resistance of the resulting hybrid composites. Devendra and Randaswamy investigated the impact of various fillers, aluminum oxide (Al_2O_3), magnesium hydroxide ($Mg(OH)_2$), silicon carbide (SiC), and hematite powder (HP), on the fire resistance of epoxy/GF composites. They reported that increasing the amounts of Al_2O_3, $Mg(OH)_2$, and HP accelerated the time to ignition and reduced flame propagation. The authors concluded that Al_2O_3 and $Mg(OH)_2$ are superior flame-retardant additives compared to the other fillers. Notably, all samples exhibited self-extinguishing times of more than 30 seconds. Dvir et al. (2003) studied the influence of pentabromobenzyl acrylate (PBBMA) as a primary flame retardant and $Mg(OH)_2$ as a secondary flame retardant on the flame resistance of PP/GF composites. The inclusion of PBBMA and magnesium hydroxide enhanced the flame-retardant properties of the resulting hybrid composites. It was demonstrated that an optimal formulation can be achieved to improve flame resistance without compromising the mechanical properties of the composites.

7.7 Applications

7.7.1 Aerospace and aviation

The aerospace and aviation industry contributes immensely to the development and advancement of technology and is a major economic contributor for several countries, making it a super industry (Gonçalves et al., 2023). Central to the aerospace and aviation industry are the materials used to manufacture tools, aircraft, spacecraft, parts, and other related equipment (Steinberg, 1986). The desired materials have characteristics such as lightweight, high strength, heat resistance, and oxidation resistance (Steinberg, 1986). Using Boeing aircraft as an example, aluminum alloys take up most components, followed by steel and titanium alloys, though other materials are used composite materials are gaining attention (Zhang et al., 2018). The most used metal alloys comprise aluminum (62%), steel (13%), titanium (11%), magnesium (8%), and a combination of aluminum and lithium (6%), as shown in Fig. 7–6 (Trzepieciński et al., 2021). From this, it can be seen that the combination of different alloys, such as aluminum-lithium alloys, presents an opportunity to use the synergistic effect of different materials.

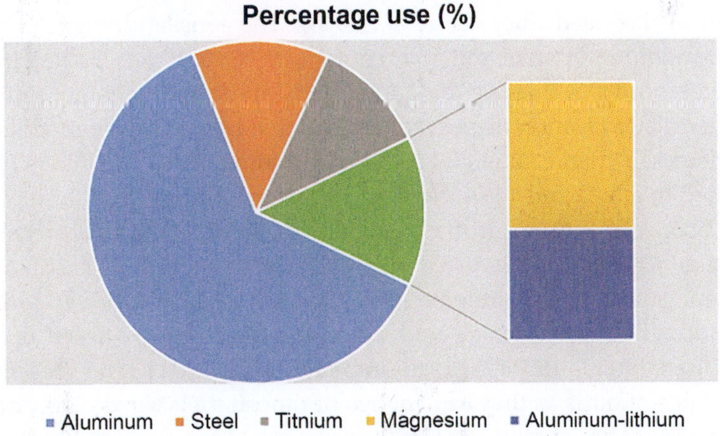

FIGURE 7–6 Mostly used metal alloys in the aerospace and aviation industry. Commonly used metal alloys in the aerospace and aviation sectors. *Trzepieciński, T., Najm, S.M., Sbayti, M., Belhadjsalah, H., Szpunar, M., Lemu, H.G., 2021. New advances and future possibilities in forming technology of hybrid metal–polymer composites used in aerospace applications. Journal of Composites Science 5 (8), 217. https://doi.org/10.3390/jcs5080217.*

Polymers and their hybrid equivalents have found increasing application in the aerospace and aviation industry, particularly in their thermal properties, which allow them to withstand high temperatures. They include polyether–ether–ketone (Jiang et al., 2020), polyimide (Sezer Hicyilmaz and Celik Bedeloglu, 2021), fluoropoly(ether amide) (Vora, 2010), polychlorotrifluoroethylene (Gardiner, 2015), polytetrafluoroethylene (Riba et al., 2022). This is because hybrid polymers not only come with the prospect of improved performance through synergy but also have attractive properties. These include high specific strength, modulus, thermal conductivity, fracture toughness, hardness, Vickers, and high-temperature tolerance (Zhang et al., 2018). Chemical or physical modification has proven to enhance the properties of different materials. However, depending on the type of modification, the thermal stability of the hybrid polymers could decrease. For example, Li et al. (2019) demonstrated that the addition of Ni-CF decreases the degradation temperature from 547°C (polyether–ether–ketone) to 535°C (polyether–ether–ketone modified with Ni-CF). This was attributed to the decomposition of the CF. The final residual solid of the modified hybrid polymer was 23% of the weight of the polymer, making the modified hybrid polymer suitable for high-temperature applications and as a flame-retardant material. However, the modification of the polyether–ether–ketone with Ni only was not presented, making it difficult to predict the thermal degradation of the polymer. The results do give an indication that modification provides a platform to change the properties of hybrid polymers. The modification with Ni-CF did not only alter the thermal properties of the hybrid polymer but also increased the thermal conductivity to almost 2.5 times more than the unmodified polymer. The Ni was added via electron spinning, creating a Ni coating, and the combination of the CF and the polyether–ether–ketone improved the electrical conductivity, making the hybrid polymer suitable for varies application and has an appeal in the aerospace and aviation industry (Li et al., 2019).

Another example, the properties of high-density polyethene (HDPE) have been improved by the addition of flax seed fiber (Li et al., 2006). The tensile strength of the HDPE was increased by the addition of increased fiber content, were 5% fiber resulting in a 1% increase in tensile strength, and 175 increase by adding 30% fiber. As the content of the fibers was increased, the tensile elongation decreased, indicating a higher tensile modulus. The flexural strength was increased from 23.29 to 33.53 MPa by adding 5% to 30% fibers. The addition of 30% fiber resulted in a 51% and 128% increase in flexural strength and flexural modulus. The addition of the fibers also influenced the melting point of the HDPE by decreasing it by almost 2 degrees (Li et al., 2006).

Hybrid polymers have found prominent applications in aircraft design to improve safety, performance, and efficiency. Dilshad et al. presented an overview based on polymorphing skins that are used for morphing aircraft (Ahmad et al., 2024). The idea emanated from observing the flights of birds as they can change or morph their wings as the birds take flight, as shown in Fig. 7-7 (Ahmad et al., 2024).

The hybrid polymers are suitable for morphing aircraft due to their properties that include enhanced flexibility, durability, low density, fatigue resistance, high-temperature tolerance, high elasticity, and self-healing. Key to the polymer materials is the electroactive and shape

FIGURE 7–7 (A) The flight of birds compared to the (B) flight of an aircraft (Ahmad et al., 2024). Exploring the similarities and differences between bird flight and aircraft flight. *Ahmad, D., Parancheerivilakkathil, M.S., Kumar, A., Goswami, M., Ajaj, R.M., Patra, K., et al., 2024. Recent developments of polymer-based skins for morphing wing applications. Polymer Testing 135, 108463. https://doi.org/10.1016/j.polymertesting.2024.108463.*

memory, which allows the material to undergo different shape changes without damaging the structural integrity. Other hybrid polymers are centered on polymer reinforcement, including epoxies used to enhance the thermomechanical properties for aerospace and aviation applications. Phenolic resins in combination with fiber materials have also been used to enhance the flame-retardancy of aerospace equipment, tools, and aircrafts. To enhance the thermal properties of components used for the manufacturing and support of engines and airframes, hybrid polymers incorporated with polyimides have found application. For purposes of increasing the thermal properties of hybrid polymers, cyanate esters have been incorporated (Dyer and Kumru, 2023). The use of hybrid polymers has proven to be an attractive complementary field in the aerospace and aviation industry in providing smart, flexible, durable, strong, and thermally stable materials.

7.7.2 Automotive industry

In recent years, there has been a paradigm shift of the global automotive industry toward zero emission transportation (Asim et al., 2015; Mortazavian and Fatemi, 2015; Wang et al., 2025). This has been translated to the rapid growth of electric vehicles (EVs) with a target of reaching high performance of these EVs. Polymer composites are of essential importance for the production of lightweight components not only for EVs but also for other conventional vehicles (Todkar and Patil, 2019; Wang et al., 2025). Their ecofriendliness, reusability, and recyclability provide comfort when it comes to sustainable manufacturing from environmental and economic viewpoints. The easy processing of hybrid composites makes them ideal materials to be used in the automotive industry as engine mounts, clutch pedals, front-end carriers, intake manifolds, and many more (Holmström et al., 2020; Wang et al., 2025). Hybrid materials that combine both metals and polymer composites have been developed to produce high-quality components that are lighter and cheaper (Raheem and Subbaya, 2023; Wang et al., 2025). Using insert-injection molding, these components can be easily produced;

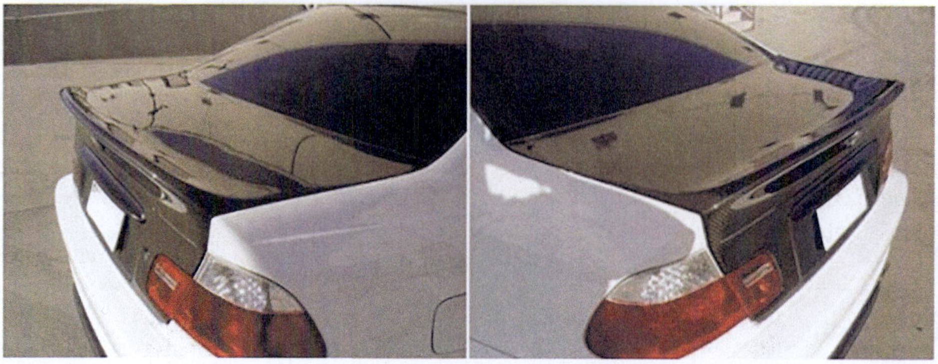

FIGURE 7–8 Trunk lid based on composite product. *Reproduced from Rajak, D.K., Pagar, D.D., Kumar, R., Pruncu, C.I., 2019a. Recent progress of reinforcement materials: a comprehensive overview of composite materials. Journal of Materials Research and Technology 8 (6), 6354–6374. https://doi.org/10.1016/j.jmrt.2019.09.068.*

however, the main issue is the interaction that often limits their application in automotive. Such difficulty can be addressed using different techniques, such as pretreatment and using different joining processes, which are often expensive and time-consuming. With recent works that are dedicated in improving interfacial interaction to realize the enhanced overall properties, hybrid composites made of polymeric materials and different fillers have been produced for the fabrication of automotive components. For instance, Wang et al. (2025) prepared hybrid composites for the fabrication of vehicle components using insert-injection molding process. The resulting composites exhibited strong dimensional stability and mechanical strength due to strong interfacial bonding. The resultant hybrid composites can be used for the production of car components. Fig. 7–8 shows the trunk lid of automotive (Rajak et al., 2019a). This demonstrates the capability of hybrid composite materials (Rajak et al., 2019a).

7.8 Challenges and future direction

The primary challenge of hybrid composites is the compatibility between hydrophobic polymers and inherently hydrophilic natural fibers. Natural fibers also face limitations such as low thermal degradation temperatures and poor chemical resistance, which must be addressed to make them suitable for high-end applications. Ensuring good interfacial bonding with polymers is crucial for better stress transfer, resulting in improved mechanical performance and durability. For synthetic fibers, the major limitations are their high cost and finite resources. However, they offer excellent thermal and mechanical performance. Polymers are also susceptible to fire and exhibit low thermal stability. Hybridization is an essential solution to enhance thermal stability, enabling the resultant materials to meet high-end applications where high-temperature resistance is required. The preparation methods for hybrid composites each have their own limitations in terms of scalability, cost, and the quality of the final products. Therefore there is a trade-off that depends directly on the anticipated applications. In the future, application-specific optimization, from the processing method to the filler, will be the main driver for the manufacturing of hybrid composites. For instance, natural fibers can be used where moisture and thermal stability are not strict requirements, such as in internal components of cars like door panels or seats. With ongoing research focused on understanding the properties of fibers, their interaction with polymers, and other fillers, there is potential to fully realize the capabilities of hybrid high-temperature composites in various high-performance applications.

7.9 Conclusion

More thermal and heat-resistant materials are important for modern applications and therefore, improving hybrid polymer materials is important to ensure that polymers improve their thermal properties and extend their applications to different areas. Hybrid polymer-based high-temperature composite materials represent a significant advancement in material science, offering a combination of lightweight, high strength, and excellent thermal stability. These materials are crucial for various high-end applications, particularly in industries such as aerospace, automotive, and construction. The development and optimization of these composites involve addressing several challenges and exploring future directions to enhance their performance and applicability. The thermal stability of hybrid polymer composites is a critical factor for their use in high-temperature applications. Incorporating fillers such as silicon carbide, boron nitride, and graphene nanoplatelets has been shown to significantly improve the thermal conductivity and heat resistance of these materials. Hybrid composites exhibit superior mechanical performance, namely, flexural strength, impact resistance, and tensile strength. The combination of natural and synthetic fibers, along with appropriate polymer matrices, allows for the customization of these properties to meet the specifications of the anticipated application. Various manufacturing methods, such as hand layup, compression molding, RTM, and pultrusion, are employed to produce hybrid composites. Each technique has its advantages and limitations, and ongoing research aims to optimize these processes for better quality and scalability. The use of natural fibers in hybrid composites offers environmental benefits, such as biodegradability and reduced carbon footprint. However, the integration of synthetic fibers and polymers poses challenges for recycling and disposal. Future research should focus on developing fully biodegradable composites and improving the recyclability of hybrid materials. Tailoring the properties of hybrid composites for specific applications is essential. For instance, in the aerospace industry, these materials are used to create lightweight, durable components that enhance fuel efficiency and performance. In the automotive sector, hybrid composites contribute to the production of lightweight, high-strength parts that improve vehicle efficiency and safety. Continued research and development are necessary to address the current challenges and unlock the full potential of hybrid polymer composites. This includes exploring new material combinations, enhancing interfacial interactions, and developing smart materials with self-healing and shape-memory properties. In conclusion, hybrid polymer-based high-temperature composite materials hold great promise for a wide range of applications. By addressing the existing challenges and focusing on future research directions, these materials can be further optimized to meet the growing demands of modern industries.

AI disclosure

During the preparation of this work the author(s) used Quilbots to improve grammar. After using this tool/service, the author(s) reviewed and edited the content as needed and take(s) full responsibility for the content of the publication.

References

Adhikari, J., Biswas, B., Chabri, S., Bandyopadhyay, N.R., Halder, S., Mitra, B.C., et al., 2018. Mechanical properties of metal oxide dispersed jute fiber reinforced polyester biocomposites. Polymer Composites 39, E101. Available at https://doi.org/10.1002/pc.24410.

Afolabi, O.A., Ndou, N., 2024. Synergy of hybrid fillers for emerging composite and nanocomposite materials—a review. Polymers 16 (13). Available at https://doi.org/10.3390/polym16131907.

Ahmad, D., Parancheerivilakkathil, M.S., Kumar, A., Goswami, M., Ajaj, R.M., Patra, K., et al., 2024. Recent developments of polymer-based skins for morphing wing applications. Polymer Testing 135, 108463. Available at https://doi.org/10.1016/j.polymertesting.2024.108463.

Asim, M., Abdan, K., Jawaid, M., Nasir, M., Dashtizadeh, Z., Ishak, M.R., et al., 2015. A review on pineapple leaves fibre and its composites. International Journal of Polymer Science 2015, 1–16. Available at https://doi.org/10.1155/2015/950567.

Asim, M., Jawaid, M., Saba, N., Ramengmawii, Nasir, M., Sultan, M.T.H., 2017. 1 - Processing of hybrid polymer composites—a review. In: Thakur, V.K., Thakur, M.K., Gupta, R.K. (Eds.), Hybrid Polymer Composite Materials: Processing. Woodhead Publishing, pp. 1–22. Available at 10.1016/b978-0-08-100789-1.00001-0.

Azam, S., Lee, B.K., 2022. Experimental measurement and modeling of film thickness distribution in film insert injection molding process. International Journal of Advanced Manufacturing Technology 120 (1-2), 1347–1357. Available at https://doi.org/10.1007/s00170-022-08886-x.

Balthes, G.E., 2013. Blow-molded composite compositions and methods. *Google Patents.*

Baysal, A., Yayla, P., Turkmen, H.S., Karaca Ugural, B., 2023. Mechanical characterization of hybrid biocomposites reinforced with nonwoven hemp and unidirectional flax fibers. Polymer Composites 44 (6), 3555–3566. Available at https://doi.org/10.1002/pc.27344.

Bhanuprakash, L., Parasuram, S., Varghese, S., 2019. Experimental investigation on graphene oxides coated carbon fibre/epoxy hybrid composites: mechanical and electrical properties. Composites Science and Technology 179, 134–144. Available at https://doi.org/10.1016/j.compscitech.2019.04.034.

Bieniaś, J., Jakubczak, P., Surowska, B., 2017. Properties and characterization of fiber metal laminates. Hybrid Polymer Composite Materials: Properties and Characterisation. Elsevier Inc., Poland, pp. 253–277. Available at http://doi.org/10.1016/B978-0-08-100787-7.00011-1.

Bledzki, A.K., Franciszczak, P., Meljon, A., 2015. High performance hybrid PP and PLA biocomposites reinforced with short man-made cellulose fibres and softwood flour. Composites Part A: Applied Science and Manufacturing 74, 132–139. Available at https://doi.org/10.1016/j.compositesa.2015.03.029.

Bocz, K., Szolnoki, B., Farkas, A., Verret, E., Vadas, D., Decsov, K., et al., 2020. Optimal distribution of phosphorus compounds in multi-layered natural fabric reinforced biocomposites. Express Polymer Letters 14 (7), 606–618. Available at https://doi.org/10.3144/expresspolymlett.2020.50.

Cheewawuttipong, W., Fuoka, D., Tanoue, S., Uematsu, H., Iemoto, Y., 2013. Thermal and mechanical properties of polypropylene/boron nitride composites. Energy Procedia 34, 808–817. Available at https://doi.org/10.1016/j.egypro.2013.06.817.

Corbridge, D.M., Harper, L.T., De Focatiis, D.S.A., Warrior, N.A., 2017. Compression moulding of composites with hybrid fibre architectures. Composites Part A: Applied Science and Manufacturing 95, 87–99. Available at https://doi.org/10.1016/j.compositesa.2016.12.018.

Devendra, K., Rangaswamy, T., 2012. Evaluation of thermal properties of E-Glass/Epoxy composites filled by different filler materials. International Journal Of Computational Engineering Research 2 (5), 1708–1714.

Ding, J., Ye, F., Liu, Q., Yang, C., Gao, Y., Zhang, B., 2019. Co-continuous hollow glass microspheres/epoxy resin syntactic foam prepared by vacuum resin transfer molding. Journal of Reinforced Plastics and Composites 38 (19-20), 896–909. Available at https://doi.org/10.1177/0731684419857173.

Dvir, H., Gottlieb, M., Daren, S., 2003. E. Tartakovsky, optimization of a flame-retarded polypropylene composite. Composites Science and Technology 63 (13), 1865–1875. Available at https://doi.org/10.1016/S0266-3538(03)00170-2.

Dyer, W.E., Kumru, B., 2023. Polymers as aerospace structural components: how to reach sustainability? JMacromolecular Chemistry and Physics 224 (24), 2300186. Available at https://doi.org/10.1002/macp.202300186.

Fu, S.-Y., Lauke, B., Mäder, E., Yue, C.-Y., Hu, X., 2000. Tensile properties of short-glass-fiber- and short-carbon-fiber-reinforced polypropylene composites. Composites Part A: Applied Science and Manufacturing 31 (10), 1117–1125. Available at https://doi.org/10.1016/s1359-835x(00)00068-3.

Gardiner, J., 2015. Fluoropolymers: origin, production, and industrial and commercial applications. Australian Journal of Chemistry 68 (1), 13–22. Available at https://doi.org/10.1071/CH14165.

Gomez-Caturla, J., Lascano, D., Montanes, N., Balart, R., Dominici, F., Puglia, D., et al., 2023. Manufacturing and characterization of highly environmentally-friendly composites with polylactide matrix and mango kernel seed flour. Express Polymer Letters 17 (3), 334–351. Available at https://doi.org/10.3144/expresspolymlett.2023.24.

Gonçalves, A., Ferreira, B., Leite, M., Ribeiro, I., 2023. Environmental and economic sustainability impacts of metal additive manufacturing: a study in the industrial machinery and aeronautical sectors. Sustainable Production and Consumption 42, 292–308. Available at https://doi.org/10.1016/j.spc.2023.10.004.

Hajare, B., Patil, O., Barpande, G.S., Radhakrishnan, S., Kulkarni, M.B., 2022. Tensile, impact and thermal properties of farm-waste based hybrid basalt polymer composites. Cleaner Materials 6, 100157. Available at https://doi.org/10.1016/j.clema.2022.100157.

Holmström, P.H., Hopperstad, O.S., Clausen, A.H., 2020. Anisotropic tensile behaviour of short glass-fibre reinforced polyamide-6. Composites Part C: Open Access 2. Available at https://doi.org/10.1016/j.jcomc.2020.100019.

Indra Reddy, M., Anil Kumar, M., Rama Bhadri Raju, C., 2018. Tensile and flexural properties of jute, pineapple leaf and glass fiber reinforced polymer matrix hybrid composites. Materials Today: Proceedings 5 (1), 458–462. Available at https://doi.org/10.1016/j.matpr.2017.11.105.

Jagadeesh Chandra, B.R., Shivamurthy, B., Kulkarni, S.D., Sathish Kumar, M., 2019. Hybrid polymer composites for EMI shielding application- a review. Materials Research Express 6, 082008. Available at https://doi.org/10.1088/2053-1591/aaff00.

Jiang, J., Zhao, C., Li, G., 2020. Tribological properties of polyether-ether-ketone-based coating undermixed and boundary aviation kerosene-lubrication condition. Coatings 10 (1). Available at https://doi.org/10.3390/coatings10010001.

Kaleni, A., Magagula, S.I., Motloung, M.T., Mochane, M.J., Mokhena, T.C., 2022. Preparation and characterization of coal fly ash reinforced polymer composites: an overview. Express Polymer Letters 16 (7), 735–759. Available at https://doi.org/10.3144/expresspolymlett.2022.54.

Kanitkar, Y.M., Kulkarni, A.P., Wangikar, K.S., 2017. Characterization of glass hybrid composite: a review. Materials Today: Proceedings 4 (9), 9627–9630. Available at https://doi.org/10.1016/j.matpr.2017.06.237.

Krishnakumari, A., Devaraju, A., Saravanan, M., 2018. Evaluation of mechanical properties of hybrid rootfiber reinforced polymer composites. Materials Today: Proceedings 5 (6), 14560–14566. Available at https://doi.org/10.1016/j.matpr.2018.03.046.

Kumar, R., Rai, B., Gahlyan, S., Kumar, G., 2021. A comprehensive review on production, surface modification and characterization of nanocellulose derived from biomass and its commercial applications. Express Polymer Letters 15 (2), 104–120. Available at https://doi.org/10.3144/expresspolymlett.2021.11.

Lee, D., Chen, W.-A., Huang, T.-W., Liu, S.-J., 2013. Factors influencing the warpage in in-mold decoration injection molded composites. International Polymer Processing 28 (2), 221–227. Available at https://doi.org/10.3139/217.2690.

Li, S., Jin, Y., Wang, Z., He, Q., Chen, R., Wang, J., et al., 2019. Preparation and characterisation of nickel-plated carbon fibre/polyether ether ketone composites with high electromagnetic shielding and high thermal conductivity. Colloid and Polymer Science 297 (7-8), 967–977. Available at https://doi.org/10.1007/s00396-019-04522-5.

Li, X., Tabil, L.G., Panigrahi, S., Crerar, W.J., 2006. The influence of fiber content on properties of injection molded flax fiber–HDPE biocomposites. American Society of Agricultural and Biological Engineers, St. Joseph, Michigan. Available at https://doi.org/10.13031/2013.22101.

Mallick, P.K., 1986. Effect of fiber misorientation on the tensile strength of compression molded continuous fiber composites. Polymer Composites 7 (1), 14–18. Available at https://doi.org/10.1002/pc.750070104.

Matumba, K.I., Mokhena, T.C., Ojijo, V., Sadiku, E.R., Ray, S.S., 2024. Morphological characteristics, properties, and applications of polylactide/Poly(ε-caprolactone) blends and their composites—a review. Macromolecular Materials and Engineering 309 (8), 2400056. Available at https://doi.org/10.1002/mame.202400056.

Mochane, M.J., Mokhena, T.C., Mokhothu, T.H., Mtibe, A., Sadiku, E.R., Ray, S.S., et al., 2019. Recent progress on natural fiber hybrid composites for advanced applications: a review. Express Polymer Letters 13 (2), 159–198. Available at https://doi.org/10.3144/expresspolymlett.2019.15.

Mokhena, T.C., Mochane, M.J., 2018. Chapter 11. Transforming fishery waste into chitin and chitin based materials. In: Linganiso, L.Z. (Ed.), "Waste-to-Profit" (W-t-P): Value added Products to Generate Wealth for a Sustainable Economy. Volume 1. Nova Science Publishers, Inc, South Africa, pp. 227–249.

Mokhena, T.C., Jacobs, V., Luyt, A.S., 2015. A review on electrospun bio-based polymers for water treatment. Express Polymer Letters 9 (10), 839–880. Available at https://doi.org/10.3144/expresspolymlett.2015.79.

Mokhena, T.C., Sadiku, E.R., Mochane, M.J., Ray, S.S., 2021. Mechanical properties of fire retardant wood-plastic composites: a review. Express Polymer Letters 15 (8), 744–780. Available at https://doi.org/10.3144/expresspolymlett.2021.61.

Mokhena, T.C., Chabalala, M.B., Mapukata, S., Mtibe, A., Hlekelele, L., Cele, Z., et al., 2024. Electrospun PCL-based materials for health-care applications: an overview. Macromolecular Materials and Engineering 309 (8), 2300388. Available at https://doi.org/10.1002/mame.202300388.

Mokhothu, T.H., Guduri, B.R., Luyt, A.S., 2011. Kenaf fiber-reinforced copolyester biocomposites. Polymer Composites 32 (12), 2001–2009. Available at https://doi.org/10.1002/pc.21233.

Moothoo, J., Bar, M., Ouagne, P., 2021. Mechanical properties of compression moulded aggregate-reinforced thermoplastic composite scrap. Journal of Composites Science 5 (11). Available at https://doi.org/10.3390/jcs5110299.

Mortazavian, S., Fatemi, A., 2015. Tensile and fatigue behaviors of polymers for automotive applications. Materialwissenschaft und Werkstofftechnik 46 (2), 204–213. Available at https://doi.org/10.1002/mawe.201400376.

Mtibe, A., Muniyasamy, S., Mokhena, T.C., Ofosu, O., Ojijo, V., John, M., 2023. Recent insight into the biomedical applications of polybutylene succinate and polybutylene succinate-based materials. Express Polymer Letters 17 (1), 2–28. Available at https://doi.org/10.3144/expresspolymlett.2023.2.

Nguyen, H., Zatar, W., Mutsuyoshi, H., 2017. Mechanical properties of hybrid polymer composite. Hybrid Polymer Composite Materials: Properties and Characterisation. Elsevier Inc., United States, pp. 83–113. Available at 10.1016/B978-0-08-100787-7.00004-4.

Nurul, M.S., Mariatti, M., 2013. Effect of thermal conductive fillers on the properties of polypropylene composites. Journal of Thermoplastic Composite Materials 26 (5), 627–639. Available at https://doi.org/10.1177/0892705711427345.

Ou, X., Lu, X., Chen, S., Lu, Q., 2020. Thermal conductive hybrid polyimide with ultrahigh heat resistance, excellent mechanical properties and low coefficient of thermal expansion. European Polymer Journal 122, 109368. Available at https://doi.org/10.1016/j.eurpolymj.2019.109368.

Pérez-Fonseca, A.A., Reynoso-Llamas, C., Robledo-Ortíz, J.R., Rodrigue, D., Martín del Campo, A.S., 2025. Recycling of biodegradable biobased polymer/agave fiber biocomposites. Polymer Composites 46 (5), 4059–4074.

Rafiee, M., Nitzsche, F., Laliberte, J., Hind, S., Robitaille, F., Labrosse, M.R., 2019. Thermal properties of doubly reinforced fiberglass/epoxy composites with graphene nanoplatelets, graphene oxide and reduced-graphene oxide. Composites Part B: Engineering 164, 1–9. Available at https://doi.org/10.1016/j.compositesb.2018.11.051.

Raheem, A., Subbaya, K.M., 2023. Performance evaluation of hybrid polymer composite materials in marine applications: a review. Materials Today: Proceedings. Available at https://doi.org/10.1016/j.matpr.2023.01.346.

Ravi Raj, V., Vijaya Ramnath, B., 2021. Mechanical, thermal and wear behavior of sic particle strengthening of PMMA-toughened glass-epoxy hybrid composite. Silicon 13, 1925–1932. Available at https://doi.org/10.1007/s12633-020-00580-5.

Rajak, D.K., Pagar, D.D., Menezes, P.L., Linul, E., 2019b. Fiber-reinforced polymer composites: manufacturing, properties, and applications. Polymers 11 (10). Available at https://doi.org/10.3390/polym11101667.

Rajak, D.K., Pagar, D.D., Kumar, R., Pruncu, C.I., 2019a. Recent progress of reinforcement materials: a comprehensive overview of composite materials. Journal of Materials Research and Technology 8 (6), 6354–6374. Available at https://doi.org/10.1016/j.jmrt.2019.09.068.

Rashid, A.B., Haque, M., Islam, S.M.M., Uddin Labib, K.M.R., 2024. Nanotechnology-enhanced fiber-reinforced polymer composites: recent advancements on processing techniques and applications. Heliyon 10 (2), e24692. Available at https://doi.org/10.1016/j.heliyon.2024.e24692.

Riba, J.R., Moreno-Eguilaz, M., Ibrayemov, T., Boizleau, M., 2022. Surface discharges performance of ETFE- and PTFE-insulated wires for aircraft applications. Materials 15 (5). Available at https://doi.org/10.3390/ma15051677.

Seoane-Rivero, R., Germán, L., Santos, F., Gondra, K., 2023. Development of new hybrid composites for high-temperature applications. Polymers 15 (22), 4380. Available at https://doi.org/10.3390/polym15224380.

Seretis, G.V., Kouzilos, G., Manolakos, D.E., Provatidis, C.G., 2017. On the graphene nanoplatelets reinforcement of hand lay-up glass fabric/epoxy laminated composites. Composites Part B: Engineering 118, 26–32. Available at https://doi.org/10.1016/j.compositesb.2017.03.015.

Sethupathi, M., Khumalo, M.V., Skosana, S.J., Muniyasamy, S., 2024. Recent developments of pineapple leaf fiber (palf) utilization in the polymer composites—a review. Separations 11 (8), 245. Available at https://doi.org/10.3390/separations11080245.

Sezer Hicyilmaz, A., Celik Bedeloglu, A., 2021. Applications of polyimide coatings: a review. SN Applied Sciences 3, 363. Available at https://doi.org/10.1007/s42452-021-04362-5.

Shih, C.H., You, J.L., Lee, Y.L., Cheng, A.Y., Chang, C.P., Liu, Y.M., et al., 2022. Design and ballistic performance of hybrid plates manufactured from aramid composites for developing multilayered armor systems. Polymers 14 (22). Available at https://doi.org/10.3390/polym14225026.

Shrivastava, R., Telang, A., Rana, R.S., Purohit, R., 2017. Mechanical properties of coir/ G lass fiber epoxy resin hybrid composite. Materials Today: Proceedings 4 (2), 3477–3483. Available at https://doi.org/10.1016/j.matpr.2017.02.237.

Singh, P., Singari, R.M., Mishra, R.S., Bajwa, G.S., 2022. A review on recent development on polymeric hybrid composite and analysis of their enhanced mechanical performance. Materials Today: Proceedings 56, 3692–3701. Available at https://doi.org/10.1016/j.matpr.2021.12.443.

Sinha, A.K., Bhattacharya, S., Narang, H.K., 2019. Experimental determination and modelling of the mechanical properties of hybrid abaca-reinforced polymer composite using RSM. Polymers and Polymer Composites 27 (9), 597–608. Available at https://doi.org/10.1177/0967391119855843.

Sinha, A.K., Narang, H.K., Bhattacharya, S., 2020. Mechanical properties of hybrid polymer composites: a review. Journal of the Brazilian Society of Mechanical Sciences and Engineering 42 (8). Available at https://doi.org/10.1007/s40430-020-02517-w.

Sivaraj, P., Rajeshkumar, G., 2014. Prediction of mechanical properties of hybrid fiber reinforced polymer composites. International Journal of Engineering Research 3 (1), 21–25. Available at https://doi.org/10.17950/ijer/v3s1/106.

Srinivas, K., Bhagyashekar, M.S., 2015. Thermal conductivity enhancement of epoxy by hybrid particulate fillers of graphite and silicon carbide. Journal of Minerals and Materials Characterization and Engineering 03 (02), 76–84. Available at https://doi.org/10.4236/jmmce.2015.32010.

Steinberg, M.A., 1986. Materials for aerospace. Scientific American 255 (4), 66–72. Available at https://doi.org/10.1038/scientificamerican1086-66.

Todkar, S.S., Patil, S.A., 2019. Review on mechanical properties evaluation of pineapple leaf fibre (PALF) reinforced polymer composites. Composites Part B: Engineering 174, 106927. Available at https://doi.org/10.1016/j.compositesb.2019.106927.

Trzepieciński, T., Najm, S.M., Sbayti, M., Belhadjsalah, H., Szpunar, M., Lemu, H.G., 2021. New advances and future possibilities in forming technology of hybrid metal-polymer composites used in aerospace applications. Journal of Composites Science 5 (8), 217. Available at https://doi.org/10.3390/jcs5080217.

Uma Devi, L., Bhagawan, S.S., Thomas, S., 2012. Polyester composites of short pineapple fiber and glass fiber: tensile and impact properties. Polymer Composites 33 (7), 1064–1070. Available at https://doi.org/10.1002/pc.22217.

Vaggar, G.B., Kamate, S.C., Shetty, R., 2021. A review on thermal properties of hybrid polymer matrix composites. IOP Conference Series: Materials Science and Engineering 1065 (1), 012030. Available at https://doi.org/10.1088/1757-899x/1065/1/012030.

Valente, M., Rossitti, I., Sambucci, M., 2023. Different production processes for thermoplastic composite materials: sustainability versus mechanical properties and processes parameter. Polymers 15 (1). Available at https://doi.org/10.3390/polym15010242.

Van de Perre, D., Serbruyns, L., Coltelli, M.-B., Gigante, V., Aliotta, L., Lazzeri, A., et al., 2024. Tuning biodegradation of poly (lactic acid) (PLA) at mild temperature by blending with poly (butylene succinate-co-adipate) (PBSA) or polycaprolactone (PCL). Materials 17 (22), 5436. Available at https://doi.org/10.3390/ma17225436.

Verma, D., Goh, K.L., 2019. 3 - Natural fiber-reinforced polymer composites: Application in marine environments. In: Verma, D., Fortunati, E., Jain, S., Zhang, X. (Eds.), Biomass, Biopolymer-Based Materials, and Bioenergy: Construction, Biomedical, and other Industrial Applications. Woodhead Publishing, pp. 51–73. Available at 10.1016/B978-0-08-102426-3.00003-5.

Vijaya Ramnath, B., Junaid Kokan, S., Niranjan Raja, R., Sathyanarayanan, R., Elanchezhian, C., Rajendra Prasad, A., et al., 2013. Evaluation of mechanical properties of abaca–jute–glass fibre reinforced epoxy composite. Materials & Design 51, 357–366. Available at https://doi.org/10.1016/j.matdes.2013.03.102.

Vora, R.H., 2010. Designing of molecular architecture, synthesis and properties of the next generation of state-of-the-art high-performance thermoplastic fluoro-poly(ether amide)s, (6F-PEA), fluoro-poly(ether amide-imide)s (6F-PEAI), and their co-polymers. Materials Science and Engineering: B 168 (1), 71–84. Available at https://doi.org/10.1016/j.mseb.2009.10.023.

Wang, J., Li, H., Li, F., Fan, C., Liu, T., Wang, D., 2025. Insert-injection moulding and post-thermal treatment of hybrid continuous and discontinuous glass-fibre-reinforced polyamide composite products. Composites Part A: Applied Science and Manufacturing 188, 108534. Available at https://doi.org/10.1016/j.compositesa.2024.108534.

Wang, Y., Zhang, Y., Liu, B., Zhao, Q., Qi, Y., Wang, Y., et al., 2020. A novel phosphorus-containing lignin-based flame retardant and its application in polyurethane. Composites Communications 21, 100382. Available at https://doi.org/10.1016/j.coco.2020.100382.

Zhang, X., Chen, Y., Hu, J., 2018. Recent advances in the development of aerospace materials. Progress in Aerospace Sciences 97, 22–34. Available at https://doi.org/10.1016/j.paerosci.2018.01.001.

Zikmundová, M., Eliášová, M., 2022. Influence of elevated temperature on the mechanical properties of transparent adhesive glass-glass joints. Challenging Glass Conference Proceedings 8. Available at https://doi.org/10.47982/cgc.8.439.

8

Biopolymer composites for high temperature applications

Stephen Shaibu Ochigbo[1], Emmanuel Rotimi Sadiku[2], Esther Aro[1], Hussaina Salihu[1], Musa Aliyu[1]

[1]DEPARTMENT OF CHEMISTRY, FEDERAL UNIVERSITY OF TECHNOLOGY, MINNA, NIGER STATE, NIGERIA
[2]DEPARTMENT OF CHEMICAL, METALLURGICAL AND MATERIALS ENGINEERING, TSHWANE UNIVERSITY OF TECHNOLOGY, PRETORIA, GAUTENG, SOUTH AFRICA

8.1 Introduction

A composite material, also known as a composition material or simply a composite, is a material that is composed of two or more constituent materials, mixed and bonded on a macroscopic scale. It has physical or chemical properties that differ significantly from one another, which, in combination, can create a material with unique properties. The two constituents of composites are the reinforcement and the matrix. The reinforcing phase or the discontinuous phase provides strength and stiffness. In most cases, the reinforcement is harder, stronger, and stiffer than the matrix, and it is usually a fiber or a particulate. The continuous phase is the matrix, which can be polymer, metal, or ceramic. Polymers have low strength and stiffness, metals have intermediate strength and stiffness, but high ductility, and ceramics have high strength and stiffness, but are brittle. The matrix (continuous phase) performs several critical functions, including maintaining the reinforcements in the desired orientation and spacing and protecting them from abrasion and the environment. Fibers, particles, or flakes, which are termed "reinforcement," embedded in a matrix—polymers, metals, or ceramics—make up the composite materials. In general, the reinforcement enhances the mechanical properties of the matrix, while the matrix holds the reinforcement in place to form the desired shape. In the finished product, each component remains separate and distinct. There are several reasons why the newly created materials might be chosen over their older counterparts. These reasons include the fact that it may be stronger, lighter, or less expensive. Composite materials should possess improved properties, such as mechanical properties, durability, and efficiency, when compared to their constituent materials. When designed properly, the newly combined material exhibits better strength than each individual material (Nagavally, 2016). An example is concrete, which is made up of cement, sand, stones, and water.

The use of composite materials has been around for thousands of years. Bricks produced from mud and straw by ancient technology constitute the earliest example. Adobe bricks, which were used to build stable and long-lasting buildings in ancient Mesopotamia and

High-Temperature Polymeric Materials. DOI: https://doi.org/10.1016/B978-0-443-32792-6.00005-1

Egypt, were made by builders using mud and straw. A classic example is the Biblical story about Israelites who, as "beasts of burden," used bricks made from mortar and straw (Exodus 1: 11, 14; 5: 7–8) to build the pharaoh's treasure cities of Pithom and Raamses in Egypt. Straws have been used as reinforcing agents to create ancient composite items, for example, boats and earthenware. As another Biblical example, Moses' mother used an ark made of bulrushes to preserve his life. In 25 BCE, 10 volumes on architectural documents described several types of mortars and limes as well as concrete construction. The earliest composite bows, created by the Mongols around the year 1200 AD, were composed of bamboo, wood, horns, calf tendon, and silk bound together with natural pine resin. Until the 14th century, these bows were the strongest and most precise weapons available.

In the decades between the years 1870 and 1890, new developments in composite materials began to emerge, with plastics playing a vital role. In 1907, Bakelite was developed as the first synthetic plastic-like composite. In terms of composition, plastic composites are created by the addition of reinforcement to synthetic plastics to provide sufficient stiffness and strength for structural purposes (Keya et al., 2019).

In structural applications, composite materials are often referred to as high-performance systems. For them to perform optimally, however, they require controlled manufacturing conditions. The potential of fiber-reinforced polymers for use as high-tech premium materials for military and electrical applications is also well recognized. Even so, fiber-reinforced plastics are still used primarily for small, reinforced plastic components, such as girders, handrails, stay cables, and part of the bridge deck (Subramanian and Senthilvelan, 2011). In recent years, reinforced plastics have been used in electronics, transportation, and architecture. In transportation-related applications, FRPs, are commonly used in the automotive, aviation, and shipping industries. Several highly specialized applications, such as rocket nozzles, have also been made from carbon fiber-reinforced plastics in the recent past.

Plastic composites provide a greater range of materials with tailor-made properties to match varied applications than ancient traditional composites. Plastic composite materials have a high strength-to-weight ratio and superior fatigue resistance, making them a promising material for aeronautical applications (Banakar et al., 2012). With their usage in aircraft, the weight reductions of the composites result in reduced running costs and greater payloads. With limitless possibilities for combining highly processable polymers with functional fillers, plastic-based composite systems can perform a plethora of functions, thereby expanding their application domains (Shilpa et al., 2017).

8.2 Categories of composites

Generally, composite materials are classified either according to the matrix or reinforcement, as shown in Fig. 8–1. Based on the matrix, there are organic matrix composites (OMCs), metal matrix composites (MMCs), and ceramic matrix composites (CMCs). The term OMC generally refers to a class of composite materials, namely, polymer matrix composites and carbon matrix composites, commonly known as carbon–carbon composites. In MMCs, a metal is intimately combined with another, often a nonmetallic phase, to produce a novel material

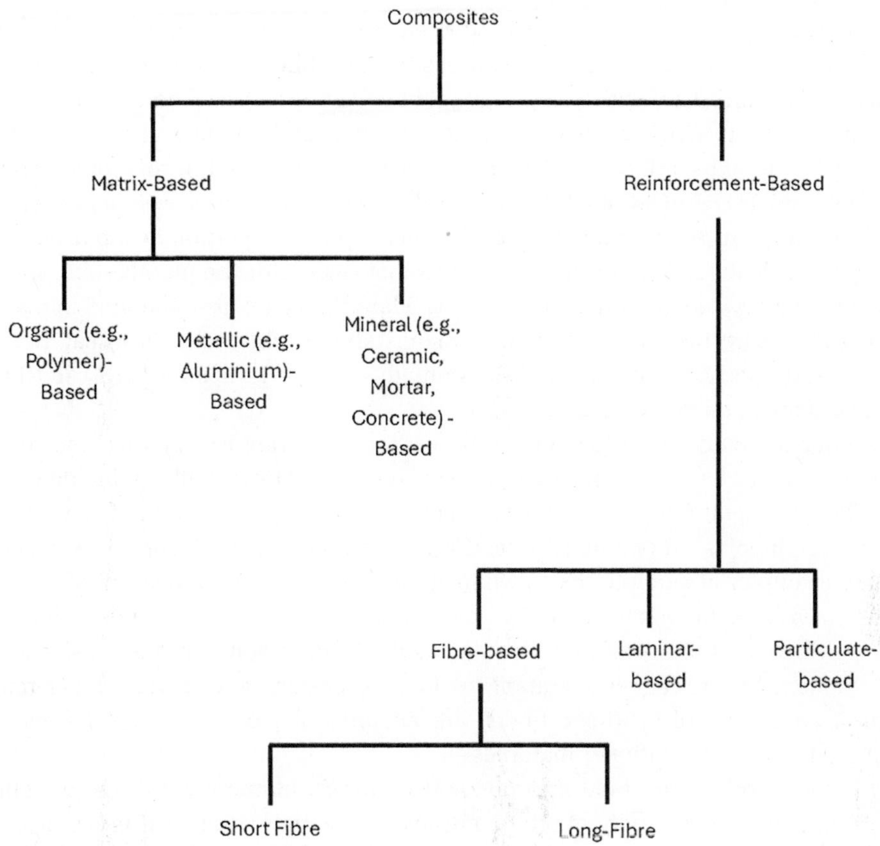

FIGURE 8–1 Different classes of composites. Composite classifications.

having attractive engineering attributes of its own. This class of composites encompasses many different materials that can be distinguished according to their base metal, for example, aluminum, copper, and titanium (Mortensen and Llorca, 2010). CMC is a type of composite that uses ceramic as matrix and fiber, whisker, pellet, or particle as reinforcement. The CMC is also referred to as a multiphase composite. The CMCs have the advantages of high temperature resistance, high hardness, corrosion resistance, and high wear resistance. In addition, the intrinsic brittleness can be improved, and the unexpected destruction can be avoided by introducing reinforcement into the ceramic matrix. The CMCs have been widely used in the high technology domain, the aerospace, national defense, and national economic fields. The most common CMCs are fiber-reinforced CMCs, particle dispersion-strengthened CMCs, in situ growth CMCs, and ceramic nanocomposites (Guo et al., 2011). Ceramic nanocomposites are created by using second-phase dispersion into ceramic matrices with nanosized particles, platelets, whiskers, or fibers (Marimuthu et al., 2023).

In terms of the reinforcement form, there are the fiber-reinforced composites, laminar composites, and particulate composites (Nagavally, 2016). A fiber-reinforced composite is further subdivided into a random fiber-reinforced composite (short fiber-reinforced composite) and a continuous fiber-reinforced composite (long fiber-reinforced composite). Fibers can be naturally sourced or synthesized. Natural fiber can be any raw material that is like hair, derived directly from plants, animals, or minerals. Thousands of natural fibers are found in nature, which are being utilized to create new materials and to improve the properties of composite materials (Faruk et al., 2012; Akil et al., 2011). Plant fibers are mainly cellulosic in nature, found in plant seeds, such as, for example, kapok and cotton; in the leaves, such as sisal and pineapple fibers; and in the bast or skin that surrounds the plant's stem, such as, for example, jute, hemp, kenaf, ramie, and bananas. Plant fibers are also obtained in plant fruits, such as coconut (coir) fiber, as well as in the plant stalks, as obtainable in wheat, rice, barley, maize, rye, and oats. Due to their tensile strength, skin or bast fibers are used in paper, textiles, packaging, and sturdy yarns.

Animal fibers include wool (as derived from sheep and rabbit), silk, fur, and hair (as ox hair, Siberian weasel hair, camel hair, horsehair, etc.). Mineral fibers include asbestos, ceramic fibers or metal fibers. Asbestos is a naturally occurring mineral fiber that resembles serpentine, amphiboles, and anthophyllite. Glass, aluminum oxide, silicon carbide, and boron carbide are examples of ceramic fibers, while aluminum fibers are one group of metal fibers. Conversely, synthetic fibers are made by polymerizing a variety of monomers derived from petroleum, which are then spun into fibers using dry, wet, or solution methods. Examples of suitable monomers are acrylic, aramid, nylon, polyester, and rayon. Fiber-reinforced composites, composed of synthetic fibers, are an integral part of our daily lives, as they have replaced many conventional materials.

These fibers can be processed into nonwoven fabrics, filaments, threads, or ropes, and used as components of a variety of other composite materials. High-end industries, such as automotive, aerospace, and medical science, are increasingly utilizing synthetic fiber-reinforced polymer (FRP) composites. In addition, recent developments in the defense sector have led to the creation of a new high-performance synthetic fiber-reinforced composite with excellent energy accumulation, strength endurance, and specific strength. In today's technological world, retrofitting damaged structures is an essential need and a common technique. For the indentation of retrofitting, fiber-reinforced polymer composites are used. As a result of their beneficial properties, they are used in a wide range of fields. The strength-to-weight ratio of fiber-reinforced polymers is high.

Laminar composites are composed of layers of materials that are held together by matrices. Sandwiched structures fall under this category. Laminar composites are available in as many combinations as there are materials. They are materials composed of layers of materials that have been bonded together. They may consist of several layers of two or more metal materials, alternating or occurring in a determined order, more than once, in as many numbers as needed. Laminar composites are orthotropic, where fibers are oriented in multiple layers and bound together in the matrix. In this composite, each fiber layer shows a two-dimensional orientation (Christian, 2020).

Particulate composites consist of particles that are embedded or distributed in a matrix. Microstructures of metal and ceramic composites, which show particles of one phase strewn in the other, are known as particle-reinforced composites. Reinforcement shapes are square, triangular, and round, but all their sides are approximately equal in dimensions. The size and volume concentration of the dispersoid distinguish it from dispersion-hardened materials. The particles may be flakes or in powder form. Concrete and wood particle boards are examples of this category. Whiskers have a preferred shape but are small both in diameter and length when compared to fibers.

This review focuses on biocomposites, which are composites whose polymer components come from living organisms, such as plants or animals. The polymer component may be reinforcements (cotton, hemp, flax, sisal, jute, and kenaf or recycled wood and paper) or organic matrices, for example, polylactic acid (PLA) and soybean resins. More often than not, the term "polymer" is usually used to refer to its synthetic equivalent, derived from petroleum or other mineral sources. Petroleum-derived commodities are now widely known to create an unsustainable environment. In response to environmental pollution problems associated with conventional polymer composites based on petroleum-derived polymers, biocomposites are increasingly being investigated. In simple terms, biocomposites are those composites obtained from biopolymers. This category of composites is also known as "green composites." One example of a biopolymer composite is the cellulose nanofiber-reinforced starch composite. Cellulose nanofibers, derived from natural sources, such as wood pulp, are incorporated into starch-based materials to improve their mechanical properties, such as strength and stiffness. This composite material has shown promise as a sustainable alternative to traditional plastics in packaging applications. Another example is the chitosan-based composites, where chitosan, a biopolymer derived from chitin found in the exoskeletons of crustaceans, is combined with other materials to create biodegradable and antimicrobial materials. Chitosan composites have been used in a variety of applications, such as wound dressings, food packaging, and water treatment (Díez-Pascual, 2022). However, to better understand biocomposites, biopolymers must first be discussed.

8.3 Composition of biopolymers

A biopolymer is a polymer, either chemically synthesized from a monomer obtained from a biological material or completely naturally biosynthesized in living organisms, such as microorganisms, plants, and animals. The advantages of the biosynthetic methods for biopolymer production include the ability to produce large quantities of biopolymers by using renewable resources, as well as the potential to tailor the properties of the biopolymer by manipulating the genetic makeup of the producing organism. The disadvantages, however, include the need for specialized equipment and expertise, as well as the possibility of variability in the quality and characteristics of the biopolymer, which varies depending on the producing organism and growth conditions.

Biopolymers have attracted increasing attention in recent years due to the environmental consciousness of people. The huge amounts of plastic obtained from fossil sources end up as waste, producing large amounts of carbon dioxide and contributing to global warming, or accumulating in the environment, where they do not degrade, thus negatively impacting ecological sustainability, which is currently a matter of global concern. One of the ways embarked upon to address this concern, besides the regulations by various nations' Environmental Protection Agencies (EPAs), was the decision to establish the United Nations Environmental Programme (UNEP). Established in 1972, with its headquarters in Nairobi, Kenya, for over 50 years, the UNEP has worked with governments, civil society, the private sector, and the UN entities to address humanity's most pressing environmental challenges—from restoring the ozone layer to protecting the world's seas and promoting a green, inclusive economy. The UNEP's work is focused on helping countries transition to low-carbon and resource-efficient economies, strengthening environmental governance and law, safeguarding ecosystems, and providing evidence-based data to inform policy decisions.

Through cutting-edge science, coordination, and advocacy, UNEP supports its 193 Member States to achieve the Sustainable Development Goals and to live in harmony with nature. They have held several conferences to discuss strategies for ensuring environmental sustainability, particularly on how to mitigate the adverse effects of plastic waste from petroleum. These conferences include COP28, referring to the 28th United Nations Climate Change Conference, which was held in Expo City Dubai, Dubai, United Arab Emirates (UAE), from November 30 to December 12, 2023, and COP29, which was held from 11 to 22 November in Baku, Azerbaijan in 2024. Biopolymers appear to be an ideal solution to these environmental problems, and thus a possible alternative to synthetic plastics derived from petroleum.

Although biopolymers currently account for a very tiny portion of the polymer market, it is anticipated that in the future they will likely replace between 30% and 90% of petroleum-based polymers. More recently, developed biopolymers have a greater range of qualities that are almost identical to those of the conventional polymers found on the market. Depending on factors, including cost, availability, moisture absorption, mechanical behavior, degradation stability, and biocompatibility, biopolymers are used in different fields (Christian, 2020). The following subsections provide examples of biopolymers from various sources.

8.3.1 Animal sources

1 Polynucleotides: Polynucleotides (also known as nucleic acids) are key constituents of the animal living cells and viruses. A polynucleotide is a polymer molecule that consists of nucleotides as its monomer units. Nucleotide consists of a sugar molecule (either ribose in ribonucleic acid [RNA] or deoxyribose in deoxyribonucleic acid [DNA]), attached to a phosphate group and a nitrogen-containing base. A single polynucleotide molecule has 14 or more nucleotide monomers in a chain structure. There are two types of polynucleotides that are found in nature. They are RNA and DNA. In DNA, there are two spiral polynucleotide chains that are arranged in a helical fashion, while RNA is a single-stranded molecule. The base in DNA consists of cytosine (C), guanine (G), adenine (A), and thymine, while uracil (U) replaces thymine in RNA. DNA is an organism's basic genetic code.

2 Collagen: Collagen is the primary protein component of animal connective tissues. Twenty-two types of collagens are known. Different polypeptides make up collagen, most of which contain glycine, proline, hydroxyproline, and lysine. A collagen chain's flexibility is determined by its glycine content. Collagen has unique biological properties, and it is enzymatically degradable. It has been extensively investigated for biomedical applications (Nair and Laurencin, 2007). By denaturation and/or physical-chemical degradation of collagen, a high molecular weight polypeptide, called gelatine, is produced (Sperling and Carraher, 1988). Gelatine is also a protein product. It consists of 19 amino acids, and it is water-soluble. Films can be formed easily with gelatine. The mechanical and barrier properties of these films depend on the physical and chemical characteristics of gelatine, especially the amino acid composition and the molecular weight distribution (Vroman and Tighzert, 2009).

3 Hyaluronic acid (HA) and chondroitin sulfate (CS): HA and CS are polysaccharides that are classified under a generic category, called glycosaminoglycans (GAGs), which consist of linear covalently linked disaccharide repeating units. Their disaccharide repeating unit is constituted of an amino sugar (hexosamines, including D-glucosamine and D-galactosamine), and a uronic acid (D-glucuronic acid and L-iduronic acid). They are present in the mammalian tissues as gel-like materials, mainly on the cell surfaces and in the extracellular matrix. HA is a polysaccharide formed of disaccharide repeating units that are comprised of *N*-acetyl-D-glucosamine (GalNAc) and D-glucuronic acid (GlcA) (Lamberg and Stoolmiller, 1974). It is the only GAG that is not sulphated and not bound to proteins (Merry et al., 2017). It is usually comprised of between 100 and 20,000 repeating units and has a molecular weight of between 105 and 108 Da. In the human body, HA is abundant in the intracellular matrix of connective tissues (between 200 and 500 μg g^{-1} in the dermis), the umbilical cord (~4100 μg g^{-1}) and in the fluid of space-filling tissues, such as the synovial fluid (between 1400 and 3600 μg mL^{-1}) and the vitreous humor (between 140 and 500 μg mL^{-1}) (Fakhari and Berkland, 2013). The structure and characteristics of HA, as well as its physicochemical and biological properties, confer valuable features, including biocompatibility, viscoelasticity, lubricity, and immunostimulatory properties. It has been employed in joint injections, ocular surgeries, osteoarthritis treatment, plastic surgeries, and skin treatments, such as major burns and antiaging products (Chong et al., 2005; Kogan et al., 2007). In the biomedical field, HA has been applied in tissue culture scaffolds (Collins and Birkinshaw, 2013). It has been shown to be a potential compound in the development of tailored nanocomposites by combining it with chitosan, for wound and chronic ulcer dressing, due to the antibacterial properties (Abdel-Mohsen et al., 2017, 2013; Abdel-Rahman et al., 2016; Abdelrahman et al., 2020). Furthermore, it has been used as dermal fillers and the treatment of osteoarthritis, vascular diseases, and in cancer progression (Fakhari and Berkland, 2013; Toole, Wight and Tammi, 2002). CS is a GAG formed by repeated disaccharide GalNAc and GlcA (Lamberg and Stoolmiller, 1974). It has a shorter chain than HA as it comprises between 20 and 100 repeating units (Mathews, 1967). It is mainly present in the extracellular matrix of tissues and in plasma membranes (Haylock-Jacobs et al., 2011). It has been highly used in the treatment of osteoarthritis due

to its antiinflammatory action, and its highly negative surface charge is capable of hydrating tissues by absorbing water (Henrotin et al., 2010). It is also used in tissue engineering since the CS hydrogels have been proven to be capable of accelerating wound healing (Gilbert et al., 2004). Therefore safe and pure CS is required for clinical applications.

4 Casein: Casein is a biopolymer, essentially a protein, present in the milk of mammals, in particular, cows. Due to its unique gelling properties, it has been used for many years in the production of a wide variety of fermented foods that fall into two general categories, namely, cheeses and fluid fermented products (Yoghurts). Its applications as biodegradable films, hydrogels, and fibers are also widely reported in the literature.

5 Whey proteins: Usually available as an ingredient in the form of whey powder, whey protein concentrates, and whey protein isolates, whey protein is mainly manufactured from whey, a by-product of the cheese industry. Whey was discovered about 3000 years ago. Apart from being valued as a medicinal agent in the 17th and 18th centuries, whey has primarily been considered a waste by the dairy industry and thus, destined for the "cheapest gutter." Whey began to gain prominence in the late 20th century when regulations against the disposal of untreated whey emerged (Smithers, 2008). The unique characteristics of whey proteins include solubility over a wide range of pHs, favorable nutritional profile with respect to essential amino acids, diverse functionality, and relatively low cost, which make them an ideal ingredient in formulating a wide array of food products (Kilara and Vaghela, 2004).

6 Silk proteins: These are biopolymers, synthesized by silkworms and spiders. These proteins possess distinct characteristics, such as exceptional strength and elasticity. These properties render them valuable for diverse applications, such as textiles, biomedical implants, and regenerative medicine.

8.3.2 Plant sources

1 Wheat gluten: Gluten is a protein found naturally within the wheat seeds. It is a side-stream of the bioethanol industry. The commercially produced wheat gluten contains ~80% proteins on a dry weight basis, while the rest is starch, lipids, mineral fiber, and ash (Cho et al., 2011). Once the water-soluble components and starch are washed off from wheat, what remains is gluten protein. Gluten is considered wheat grain's storage proteins, laid down specifically, to provide an essential supply of amino acids for the developing plant. The storage proteins of the mature wheat may not differ considerably from those of other grains (Shewry and Halford, 2002). However, the distinctive feature that makes wheat unique is the viscoelastic properties of its storage protein. When the grain is milled and mixed with water, what was "merely storage protein," forms a dough with unique rheological properties, capable of retaining gas bubbles, suiting this dough to the wide range of products that have come to be expected from wheat flour. These are the properties that make wheat alone suitable for the preparation of a diversity of food products like breads, noodles, pasta, cookies, cakes, pastries, and many other foods

(Johansson et al., 2013). The viscoelastic properties of wheat gluten have also been exploited to produce biobased films (Cho et al., 2011), foams (Blomfeldt et al., 2012), nanocomposites (Carbonaro et al., 2012), and biocomposites (Dicharry et al., 2006). These have reasonable mechanical and very good gas barrier properties that make them suitable for packaging applications. These unique properties account for wheat being cultivated in such enormous quantities throughout the world. The name given to this unique group of proteins is "gluten," an enigmatic complex of proteins. Although it is sold as a protein, gluten contains more than just protein. As a commodity, dry gluten usually contains ~75% protein, up to 8% moisture, and varying amounts of starch, lipid, and fiber (Day et al., 2006). Wheat gluten is composed of two major protein groups, that is, gliadins and glutenins, which provide wheat gluten with its unique viscoelastic and rheological properties (Hoover et al., 2010). In dough, the gliadins contribute to viscosity and extensibility, while the glutenins provide strength and elasticity. Gliadins are the ethanol-soluble monomeric fraction (molecular weight range of between 30 and 70 kDa) of wheat gluten and can be divided into three subgroups of proteins (D/D, D, and D gliadins), based on their mobility in a gel electrophoresis. The D/D and D-gliadins contain six and eight cysteines, respectively, which are responsible for the intradisulfide cross-links in dough. The D-gliadins lack the amino acid cysteine that is required to form disulfide bonds (Huang et al., 1999). Glutenins are the ethanol nonsoluble polymeric fractions of wheat gluten, with their molecular weights ranging between 30 and 160 kDa; although in an aggregated/cross-linked form, their molecular weights can range from 500 kDa to several million kDa. The glutenins are subdivided into two distinct groups: low molecular weight glutenin subunits (LMW-GS) and high molecular weight glutenin subunits (HMW-GS). The LMW-GS contain similar types of amino acids to D/D gliadins in terms of their sequence and structure (Hoover et al., 2010). In contrast, the HMW-GS are unique in their structure and amino acid composition and are rich in cysteine residues. These characteristics make HMW-GS the main determinant for providing gluten with its unique strength and elasticity in dough (Kuktaite et al., 2012). In dough or in aggregated form, both glutenins and gliadins are cross-linked with inter- and intrachain disulfide bonds and other isopeptide bonds to maintain a three-dimensional structure (Carbonaro et al., 2012).

2 Soy protein: Soy protein is the most popular plant protein source that is isolated from soybeans. Soybeans contain between 35% and 40% protein, of which 90% of this content is storage protein, mostly consisting of D-conglycinin and glycinin. The D-conglycinin contains three subunits with molecular masses of between 150 and 200 kDa, while glycinin is a hexamer with molecular masses of between 300 and 380 kDa. Each Glycinin subunit is composed of acidic (~35 kDa) and basic (~20 kDa) polypeptides, linked together by a disulfide bond. The essential amino acids that are found in soy protein, particularly lysine, are comparable to those found in the best animal products. Despite having a higher lysine content than wheat proteins, soy protein still has a lower lysine content than milk protein, specifically casein (Friedman and Brandon, 2001). Despite this, soy protein is not an ideal source of protein for humans because it is deficient in the amino acid methionine.

Before consumption, cooking soy protein is essential, as raw soy protein is difficult to digest with enzymes.

3 Potato proteins: They are a side-stream of the potato starch industry. After the extraction of starch, potato fruit water (PFW), which contains ~5% solids, is left behind. The 5% solids in PFW mainly consist of potato proteins and other constituents, such as polyphenols, minerals, fiber and reagents, which are used to stop starch browning (Diuk Andrade et al., 2018; Løkra et al., 2009). To recover potato protein from PFW, the pH is dropped between 3.5 and 4.0 to coagulate the protein, and the coagulated mass is further spray-dried at 75°C–100°C to obtain potato protein powder (Domenek et al., 2004). Potato proteins are composed of three major classes of protein: patatin (40%), molecular weight ranging between 40 and 45 kDa, protease inhibitors (~50%), molecular weight ranging between 8 and 25 kDa, and other high molecular weight proteins (~10%), 80 kDa in size (Du et al., 2015; Domenek et al., 2004). Potato proteins also contain several essential and nonessential amino acids. High levels of essential amino acids, such as aspartic acid, glutamic acid, leucine, lysine, phenylalanine, valine, and threonine, have been observed, and the potato proteins are rich in lysine, but low in the sulfur-containing amino acid cysteine (Du et al., 2015). However, the harsh acidic conditions and high temperature used in the potato starch industry during the extraction of potato proteins from PFW limit the activity of the amino acids and induce a high degree of cross-linked protein (Du et al., 2015; Domenek et al., 2004). A few recent studies have reported on the use of commercial potato protein concentrates for making biobased materials (Newson et al., 2015; Du et al., 2015).

4 Pea proteins: They are an important meat-replacing protein sources for human consumption because of their high content of the amino acid lysine, high nutritional value, and low glycaemic index when compared with wheat gluten (Choi and Han, 2001). Recently, fractionation of the pea seed has been aimed at the production of pea protein-rich fractions/isolates that can eventually be used as protein fortifiers in modern day foods (Forssell and Lahtinen, 2002) or to produce edible protein films (Choi and Han, 2001). The pea proteins consist of albumins and globulins, with globulins accounting for more than 80% of the total protein content, which is mainly stored in cotyledons. The pea protein globulins consist of legumin, vicilin, and convicilin, where legumin is the major and vicilin is the second major globulin fraction. Each of the legumin subunits (~60 kDa) consists of disulfide-based acidic and basic polypeptides, with four methionine and between two and seven cysteine amino acids (Gontard et al., 1993). Vicilin is a trimeric protein with a molecular weight of ~150 kDa, and convicilin consists of trimers or tetramers with a molecular weight of ~290 kDa (Choi and Han, 2001). Albumins account for between 13% and 14% of the total protein content, and they are mainly premised as the cytoplasmic proteins that include the sulfur-containing amino acids (Gontard et al., 1993).

5 Zein: It is the major storage protein of corn or maize (*Zea mays* L.) and comprises between 45% and 50% of the protein in corn. It was first identified in 1897, based on its solubility in aqueous alcohol solutions. The zein isolate is not used directly for human consumption due to its negative nitrogen balance and poor water solubility. The protein content of

different corn varieties is between 6% and 12% on a dry basis. About 75% of the protein is contained in the endosperm tissue. The remainder is distributed between the germ and bran. Zein belongs to the characteristic class of proteins, known as prolamins, which occur specifically in cereals (the equivalent of hordein in barley and gliadin in wheat). It was named in 1821 by John Gorham, who first identified it by infusing water in zea, known as "Indian corn," in the United States (Gorham, 1821). Almost all the zein is present in the endosperm, whereas glutelin is distributed between the endosperm and the germ. The potential applications of zein include its usage in fiber, adhesive, coating, ceramic, ink, cosmetic, textile, chewing gum, and biodegradable plastics. These new applications of zein appear to be promising but require the development of low-cost manufacturing methods (Shukla and Cheryan, 2001).

6 Natural rubber: Natural rubber is essentially poly(*cis*-2 methylbut-1, 3-diene) or, in short by nomenclature, *cis*-polyisoprene, which occurs in the form of "plant milk" or latex. The latex is extracted from the rubber trees, particularly the species called Hevea brasiliensis, so-called because, originally, it was indigenous to Brazil. The fresh latex obtained from the trees by incisions through the bark is called "field latex" and it must be immediately ammoniated to prevent coagulation, while in transit, from the plantation to the processing plant. The latex can be converted by creaming, centrifuging, or evaporation processes into a "latex concentrate," which can be stored for as long as desired. Alternatively, field latex may be crumbed into varying grades of solid material based on the curing method used. Natural rubber, either in the crumbed or in latex concentration form, is an exportable product for the manufacture of industrial products, such as tires, shoe soles, and adhesives. The natural rubber chains are susceptible to oxidative degradation due to unsaturation. The polymer is, therefore, vulcanized to obtain cross-linked natural rubber that can be applied to a variety of industrial purposes. The unvulcanized form of natural rubber has a glass transition temperature (T_g) of −70°C and a molecular weight of 3x106 g mol^{-1}; it is biodegradable, but its vulcanized form is not usually classified under the biodegradables. Natural rubber belongs to the specialty or miscellaneous class of polymers, as lignin and shellac, which are not among the distinguished categories of polymers.

7 Starch: Starch is a polysaccharide usually produced in large quantities in plant storage organs such as seeds (e.g., cereals and legumes), tubers (e.g., potato and sweet potato), and roots (e.g., cassava) (Gueguen et al., 1998). Starch is stored in the form of granules, and these granules are extracted during industrial processing. Native starch granules are of different sizes (ranging between 1 and 100 μm) and shapes (oval, round, spherical, elliptical, irregular, etc.), depending on their origin (Hoover et al., 2019). Structurally, it is composed of two types of glucose units, namely, a linear and crystalline fraction called "amylose" (poly-D-1,4-D-glucopyranoside) (between 20% and 30%) and a branched and amorphous fraction known as "amylopectin" (poly-D-1,6-D-glucopyranoside) (of between 70% and 80%) (Gueguen et al., 1998; Gällstedt et al., 2004). A given starch sample's physico-chemical, mechanical, and biodegradability properties are influenced by the relative amounts and molar masses of amylose and amylopectin. For example, a higher amount of amylose leads to higher viscosity and higher gelatinization temperature during processing but improves the strength and stiffness of the materials (Knorr, 1980). Starch is cheap, abundant, adaptable, and relatively easy to process. The use of starch started in the 20th century, when it was added as a filler to synthetic plastics to reduce the cost and enhance their disintegration in nature (Gällstedt et al., 2004; Khatkar et al., 1995). Today, starch is commonly used as a precursor to produce a biobased polymer called PLA. Challenges involved in the utilization of starch range from poor mechanical properties, moisture-sensitiveness, and tendency to retrogradation, which refers to the process of reorganizing intermolecular hydrogen bonds between its molecules at low temperatures that leads to crystallization and loss of plasticity, which is a key requirement for its application as thermoplastic films. Therefore starches are chemically and/or genetically modified to alter their primary structure and the ratio of amylose and amylopectin content, thereby improving the interactions between glucose molecules and thus enhancing the functional performance of the product (Kluger and Alagic, 2004). Biodegradation of starch is achieved via hydrolysis by amylases at the D-1,4 link in amylose and by glucosidases at the D-1,6 link in amylopectin, and the resulting degradation products are nontoxic.

8 Cellulose: Cellulose is a biopolymer of natural origin that is synthesized by plants and some bacteria. The biosynthesis of cellulose involves the assembly of glucose monomers into long chains, which form the structural components of plant cell walls. Cellulose can be utilized in various applications, including paper manufacturing, textiles, and food packaging. Cellulose is the main constituent of wood and vegetable fibers and the main structural component of cell walls. For industrial purposes, it can be sourced from wood, cotton fibers and cotton linters. Chemically, cellulose is a polysaccharide consisting of very long linear macromolecular chains of cellobiose [(1→4)-D-D-glucopyranosyl] repeating unit. It is mostly crystalline, hence infusible and insoluble in all organic solvents. To make it processable, it is usually converted to its derivatives, such as its ethers, esters, and acetals, by reaction of one or more of the hydroxyl groups present in the repeating unit (Yu and Chen, 2009). Biodegradation of cellulose occurs through enzymatic oxidation, with peroxidase secreted by fungi and bacteria, yielding nontoxic products, such as starch (Klemm et al., 2002).

8.3.3 Microorganisms-based sources

1 Chitin and chitosan: Chitin and chitosan are biopolymers that are exudated from the exoskeletons of insects and crustaceans, such as crab shells, shrimp shells, crawfish shells, and the exoskeletons of insects. Chitin is also found in the cell walls of fungi, where it contributes to the stabilization of strength (Yu and Chen, 2009). Chitin and chitosan can be exudated from these sources or synthesized through biosynthesis using microbial cells, such as bacteria and fungi. These biopolymers have various applications, which include drug delivery, wound dressings, and food packaging. Chitin is the second most abundant natural polymer after cellulose. It is a linear copolymer of *N*-acetylglucosamine and *N*-glucosamine with the D-1,4 linkages. Thus it could be regarded as an amino cellulose.

These units are randomly or block-distributed throughout the biopolymer chain, depending on the processing method used to obtain the biopolymer (Vroman and Tighzert, 2009). Chitin is degradable by the action of an enzyme, chitinase. Chitosan is produced via the partial alkaline *N*-deacetylation of chitin. The degree of deacetylation is measured in terms of the ratio of glucosamine to acetyl glucosamine. Crystallinity, surface energy, and the degradation rate of chitosan are affected by this degree, which can range between 30% and 100%, depending on the preparation method. Owing to its rigid and compact crystalline structure and strong intra- and intermolecular hydrogen bonding, chitosan is insoluble in both water and alkaline media, but soluble in dilute acid solutions. Enzymes, such as chitosanase and lysozymes, can degrade chitosan. The usefulness of chitosan stems from its ability to act as a cationic polyelectrolyte, its bioactivity and biocompatibility, its use as a thickening agent in water, and its selective chelation properties. Furthermore, chitosan is easily converted into several useful products, including fibers, films, coatings, and beads (Yu and Chen, 2009).

2 Alginate/Alginic acid: Alginate is an edible heteropolysaccharide that is abundantly available in the brown algae (*Phaeophyceae* sp.), also called brown seaweeds. Alginate is a nonbranched, binary copolymer. It is composed of the D-D-mannuronic acid monomer, linked to the D-L-guluronic acid monomer, through the 1, 4-glycoside linkage. The ratio between the two monomers varies with the sources (Vroman and Tighzert, 2009). Apart from these seaweeds, alginates are produced by the bacteria: *Azotobacter vinelandii* sp. and some *Pseudomonas* sp. strains. Since its discovery by Stanford in 1881 and its ability to form hydrogels, alginate has been widely used as a stabilizer, a thickener, a gelling agent, and an emulsifier (Szekalska et al., 2016). After the patent for alginate extraction from algae was granted in the 1930s, commercially available alginates are mostly isolated from brown algae (*Phaeophyceae* sp.), such as *Laminaria hyperborea, Macrocystis pyrifera,* and *Ascophyllum nodosum*. Some other minor sources are *Laminaria digitata, Laminaria japonica, Ecklonia maxima, Lessonia nigrescens,* and *Sargassum* sp. (Smidsrød and Skjåk-Bræk, 1990; Szekalska et al., 2016). In brown algae (also known as brown seaweed), the presence of alginate in their cell walls provides flexibility and a strong structure to the algae, thereby buffering them against possible injury when exposed to strong seawater waves (Venkatesan et al., 2014). In bacteria, it forms a protective capsule, which aids in biofilm formation (Hay et al., 2013), and it assists in bacterial adherence and colonization (Tipton, 2010). Owing to alginate's gel-forming capability, it is widely used in the food, textile, and paper industries, and to a lesser extent in biomedical applications as biomaterials to promote wound healing and tissue regeneration. Soluble alginic acid is extracted from algae using a solution of 0.1–0.2 N mineral acid. Alginate fibers are commonly made by extruding sodium alginate solution into a calcium chloride bath, thereby producing calcium alginate fibers.

3 PLA: PLA is synthesized from lactic acid produced via starch fermentation from lactic bacteria. PLA is not soluble in water; nevertheless, microorganisms in marine environments can degrade it. PLA is a hard material like acrylic plastic. PLA has the disadvantages of brittleness and poor thermal stability (Vroman and Tighzert, 2009).

4 Poly(β-hydroxyalcanoate)s (PHAs): PHAs are produced by microorganisms like bacteria and algae. Several bacteria can accumulate PHAs as intracellular reserve materials. Some organisms accumulate PHA, comprising between 30% and 80% of their cellular dry weight, in the presence of an abundant source of carbon and under limited nitrogen conditions (Mercan et al., 2002). The general formula of the monomer unit is -[O-CH(R)-CH2-CO]-. These polymers have been utilized in sectors such as packaging, biomedical implants, and tissue engineering. Besides all these, there are carbohydrate polymers, such as xanthan, curdlan, pullulan, and HA, produced by fermenting sugar feedstock with bacteria or fungi (Luengo et al., 2003; Piermaria et al., 2009). Biosynthesis of poly(amino acid)s from microorganisms was also reported (Kunioka, 1997).

8.4 Characteristics of biopolymers

Biopolymers have several important characteristics, such as biodegradability, renewability, biocompatibility, and sustainability, which drive their growing interest. Briefly, each of these characteristics is discussed as follows:

8.4.1 Biodegradability

With minimal exception, most biopolymers are biodegradable, meaning that they may break down into carbon dioxide (CO_2), water (H_2O), methane (CH_4), and inorganic chemicals, through microbiological means (Christian, 2020), hence making them environment-friendly alternatives to the traditional synthetic polymers. The type of polymer, its chemical makeup, and the surrounding environment are only a few of the variables that affect how quickly biopolymers degrade (Mariano et al., 2017). Biodegradation occurs through the action of enzymes and/or chemical deterioration associated with microorganisms. This event occurs in two steps. The first one is the fragmentation of the polymers into lower molecular mass species, by means of either abiotic reactions, that is, oxidation, photodegradation, or hydrolysis, or biotic reactions, that is, degradations by microorganisms. This is followed by bioassimilation of the polymer fragments by microorganisms and their mineralization. The mechanisms and estimation techniques of polymer biodegradation have been reviewed (Lucas et al., 2008). This property reduces waste and pollution.

8.4.2 Renewability

Biopolymers are derived from renewable resources, such as plants, animals, and microbes, thereby making sustainable materials that can be produced without depleting finite resources. This supports the move toward a more sustainable and circular economy.

8.4.3 Biocompatibility

Biopolymers are biocompatible, meaning that they are compatible with living organisms and can be safely used in medical and pharmaceutical applications. This property makes biopolymers ideal for use in drug delivery systems, tissue engineering, and implantable devices.

8.4.4 Sustainability

Biopolymers are considered sustainable materials due to their renewable nature, biodegradability, and low environmental impact when compared to traditional synthetic polymers. Their production typically requires less energy and reduces greenhouse gas emissions.

Unfortunately, biopolymers have certain drawbacks, for example, their hydrophilic nature, which tends to limit their wide range of applications, leading to their low-lasting degradation ratio in damp environments, poor mechanical properties, poor thermal resistance, and limited processing capabilities, resulting in their limited application possibilities (Mohamed et al., 2018). To mitigate these drawbacks, various strategies are often employed during their processing. Biopolymers can be reinforced by fillers and nanofillers to form biocomposites and bionanocomposites that overcome these drawbacks. The choice of fillers is largely determined by the properties and intended applications of composite products. Sometimes, it is necessary to blend a biopolymer with another polymer before applying the reinforcing fillers, depending on the desired properties. This review discusses methods for producing biopolymer composites for high-temperature applications.

8.5 Trends in the applications of biocomposites

A recent survey specifies that, during the period 2016–24, the global market for biocomposites grew at a compound annual growth rate of 11.8%. In 2016 the biocomposite market size was valued at $4.46 billion, with the highest usage emanating from the construction sector (56.0%). The biocomposites market forecast report estimated that the market size was likely to reach $10.89 billion by 2024, from $4.46 billion in 2016. In other words, the fabrication of materials from biobased feedstock is likely to rise from 5% in 2004 to 12% in 2010, ~18% in 2020, and ~25% in 2030 (Gurunathan et al., 2015). Biocomposite materials are not new to mankind. Their application dates to ancient times, namely, the Great Wall of China, built from biomaterials, such as clay bricks, clay, stone, red willow weeds, willow branches, reeds, and sand (121 BCE) (Satyanarayana et al., 2009), Mongolian bows made from animal horns and tendons, wood, or silk (1200 AD). Biocomposites have been largely employed for several secondary applications in the aerospace, automobile, packaging, electronics, and construction (Zini and Scandola, 2011). In recent times, the applications of biocomposites in automotive components have been considerably expanded, particularly by European automotive manufacturers. According to the EURO 6 regulations of the European Union, introduced in 2020, substantial taxes are being levied on the automotive industry that generates 95 g km^{-1} of CO_2 emissions (Hooftman et al., 2018). Weight reduction via the employment of biocomposites supports the European Automobile Manufacturers Association's and the European Commission's goal of decreasing CO_2 emissions. According to the European Directive 2000/53/EC, the European Union End-of-Life Vehicle Directive aims to recycle 95 wt.% of automobiles.

The striking characteristics of biocomposites compared to their conventional counterparts include high specific mechanical properties, thermal insulation, CO_2 neutrality, good damping properties, high health and safety, good fatigue, abrasive, and corrosion resistance, availability, high acoustic insulation, low density, reduced production energy, and lightweight.

Biocomposites are innovative composite materials in which one of the components in the polymer matrix/reinforcement pair is biobased, as shown in Fig. 8–2. The matrices currently used in the development of biocomposites are petroleum-based (thermoplastics or thermosets) or biobased (PLA, polyhydroxy butyrate, starch, etc.) (Faruk et al., 2012). Thermoplastics include polymers, such as polypropylene, polyethylene, polystyrene, and polyvinyl chloride, while polyesters, epoxy, and polyurethane represent examples of thermosets that are used for the manufacturing of natural fiber-reinforced composites.

Attention has been focused on the production, processing, and application of biodegradable composites prepared from biopolymers, such as PLA, chitosan, alginate, and natural rubber, which are among the most promising matrices for green composites in the future. Polylactide (PLA) is currently receiving considerable attention for conventional use, due to its beneficial qualities, which include biodegradability, renewability, and multiple applications, such as packaging materials, production of fibers and composites for technical applications. Furthermore, the PLA's application as a base material for devices, such as suture fibers and scaffolds in tissue engineering, makes the biopolymer highly popular.

Biodegradable composites, created by reinforcing matrices based on natural and/or synthetic biopolymers with different natural fibers, derived from plant and animal sources, are known as "biopolymer composites" (Nagalakshmaiah et al., 2018), or more precisely, termed the "Green biocomposites." If only one out of the polymer/fiber is biodegradable, the composites are termed "partial green." Biocomposites reinforced with high-strength natural fibers represent an ecofriendly and inexpensive alternative to the conventional petroleum-based materials, and they have been increasingly utilized in a wide variety of industrial applications. Examples of such applications are in the development of automotive components, agricultural equipment, cabinets, and consoles of different electrical and electronic equipment, packaging, etc.

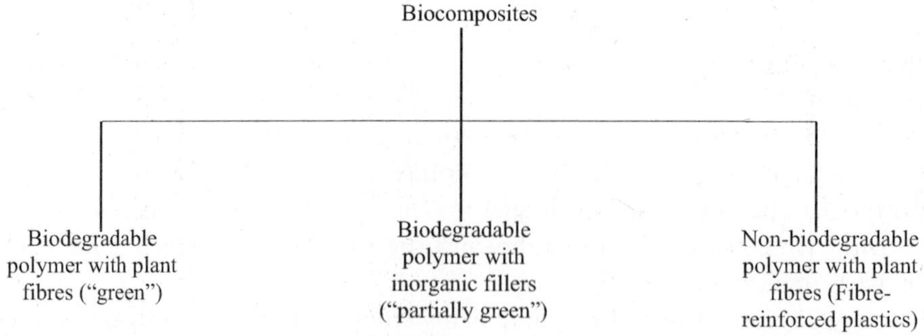

FIGURE 8–2 Categories of biocomposites. Biocomposite categories.

8.6 Creation of biopolymer composites for high-temperature applications

Biocomposite materials are not a problem-free substitute, and they have some drawbacks, such as poor mechanical strength, typically having maximum stiffness and tensile strength ranges of between 1 and 4 GPa and between 20 and 200 MPa, respectively (Christian, 2020); hydrophilicity, resulting in poor moisture resistance as well as a mismatch in hydrophobic polymers; other times fiber/matrix incompatibility; low thermal stability, which limits their high temperature applications; flammability; poor electrical properties; extraction, processing, surface modification, machining, manufacturing and characterization associated challenges; and highly anisotropic properties (Andrew and Dhakal, 2022).

As such, extensive research has been conducted over the past few years to address the aforementioned challenges in producing biocomposites with long-term durability, reliability, serviceability, and sustainable production, while adopting a circular economy approach with biocomposites (Andrew and Dhakal, 2022). This review, however, examines the strategies for enhancing the thermal stability of biocomposites to make them suitable for high-temperature applications. The term "thermal stability" is the resistance to permanent change (degradation) in properties caused solely by heat. This is crucial, since biocomposites may be subjected to high-temperature conditions during manufacturing and in service. The majority of biofibres and matrices tend to be unstable at temperatures higher than 200°C and are not suitable for processing at elevated temperatures (Facca et al., 2007). Typically, the thermal stability of biopolymer composites is assessed by using techniques such as differential scanning calorimetry, thermogravimetric analysis, and dynamic mechanical thermal analysis. As a principle, to employ biocomposites in novel applications, it is vital to identify the processing techniques and properties of fibers, matrices, and their interfaces. Thus the production of biocomposites requires a deep understanding and knowledge of the constituent materials to achieve the specific properties. The synergic combination of features resulting from the constituents and their interfaces determines their performance (Awais et al., 2021; Dicker et al., 2014; Gurunathan et al., 2015; Satyanarayana et al., 2009). Generally, the matrix component of a biocomposite controls its chemical composition, molecular weight, mechanical properties, morphological features, and processing method (George et al., 2020), while the fiber component helps to improve, modify, or overcome certain inherent properties or impart desired properties to the biocomposites. This implies that improving the properties of biocomposites to make them prepared for high-temperature applications prerequisite modification of the fiber component to impart the desired thermal stability property. In order for "green biocomposites" that use biopolymers instead of petroleum-based materials to be more thermally resistant, their fiber components must be modified. When fiber surface modifications are performed, they generally enhance the compatibility between the fibers and the biopolymer, resulting in enhanced thermal stability. Modifications could be physical, such as Corona or Plasma treatments, or chemical, which is preferred in most cases. The various chemical treatments that improve the thermal stability of the fibers, and hence in the subsequent "green biocomposites," can include, but are not limited to, the following strategies, which are highlighted as follows:

8.6.1 Alkaline treatment (mercerization)

Treating natural fibers with an alkaline solution eliminates waxes, hemicellulose, and oils, thus augmenting the roughness of their external surfaces and imparting thermal stability. This heightened roughness fosters improved interlocking between the fibers and the matrix, thereby boosting the overall mechanical properties of the resulting composites. Often, the rate at which the fiber fragments and separates rises when subjected to alkaline treatment. Typically, sodium hydroxide (NaOH) is employed to treat natural fibers, where it effectively removes hemicellulose, waxes, oils, and lignin. Consequently, the fibrils become exposed, leading to a heightened roughness on the fiber surface. This process introduces amorphous regions within the densely packed cellulose lattice, consequently shifting the orientation of the cellulose lattice. The altered lattice structure features a relatively wider gap between cellulose molecules, hence accommodating the infiltration of water molecules into the interstitial spaces. The OH groups, susceptible to NaOH, are thereupon broken down and eliminated from the fiber structure. Consequently, the hydrophilic OH groups are extracted, thereby enhancing moisture resistance. Additionally, oils, waxes, and hemicelluloses are likewise eliminated (Pokharel et al., 2022).

8.6.2 Acetylation treatment

The process of acetylation alters the mechanical and thermal properties of natural fibers, and it involves the treatment of the fibers with acetic anhydride. During treatment, the acetyl groups are incorporated into the cell walls of the fibers, substituting for the OH groups and thereby rendering the fibers hydrophobic in nature (Pokharel et al., 2022).

8.6.3 Benzoylation treatments

The process of benzoylation involves treating fibers with benzoyl chloride to decrease their hydrophilic properties, increase interfacial adhesion, and consequently enhance the strength of the composite material. Prior to benzoylation, an alkali pretreatment is performed to eliminate extractable substances, such as lignin, waxes, and oils covering the fibers, thus exposing hydroxyl (OH) groups on the fiber surface. Subsequently, the fibers undergo benzoyl chloride treatment, during which the OH groups are substituted with benzoyl groups and attach to the cellulose backbone. The peroxide-induced grafting of the biopolymer is bonded to the surface of the fiber. The free radicals, initiated by the peroxide, react with the OH groups that are present in both the fiber and the matrix. Consequently, the peroxide treatment enhances the interface properties between the fiber and the matrix. Additionally, it reduces moisture absorption and boosts thermal stability (Kabir et al., 2012). The application of benzoyl peroxide enhances the bonding interface between the fiber and matrix by eliminating the OH bond. This process includes treating the fibers with a NaOH solution, followed by applying the required amount of benzoyl peroxide in acetone (Madhu et al., 2019).

8.7 Conclusion

Composites are composed of a matrix (a continuous phase) and reinforcement (a dispersed phase), which are different on each side. Wood may be considered a natural example, being made up of lignin matrix dispersed with cellulose fibers. A pristine composite ever made by man contained fibers, notably plant-based fibers, dispersed in mortar, and was used to build Pithom and Raamses for the pharaoh by the Israelites in ancient Egypt. As synthetic plastics emerged in the late 19th century, composites made from these materials became increasingly popular. The ability to tailor properties to suit various applications has led to synthetic plastic composites becoming very popular and ubiquitous. However, since synthetic plastic composites are derived from petroleum, they continue to cause environmental pollution and deplete nonrenewable petroleum resources. Therefore research has been conducted on biocomposites, innovative composite materials obtained from biopolymers dispersed with plant fibers, inorganic fillers, or synthetic plastics. A biocomposite made with a synthetic plastic matrix is not biodegradable in its entirety, but it may be more thermally stable than one made with an organic matrix based on the same biobased reinforcements. In contrast, green biocomposite materials are renewable, biodegradable, and environmentally friendly since they contain both a biodegradable matrix (biopolymer) and a biodegradable reinforcement (plant fiber) but lack the thermal stability needed. A green biocomposite can be made more thermally efficient by modifying its reinforcements. Modification methods commonly used in this situation are mercerization, acetylation, and benzoylation. In view of the growing number of regulations being imposed against petroleum-based (synthetic) polymer composites as a result of their environmental and energy concerns, the future holds promise for biodegradable biocomposites, which can be used for packaging, disposable products, automotive housings for electronics devices, and furniture, among other things.

References

Abdel-Mohsen, A.M., Hrdina, R., Burgert, L., Abdel-Rahman, R.M., Hašová, M., Šmejkalová, D., et al., 2013. Antibacterial activity and cell viability of hyaluronan fiber with silver nanoparticles. Carbohydrate Polymers 92 (2), 1177–1187. Available at https://doi.org/10.1016/j.carbpol.2012.08.098.

Abdel-Mohsen, A.M., Jancar, J., Abdel-Rahman, R.M., Vojtek, L., Hyršl, P., Dušková, M., et al., 2017. A novel in situ silver/hyaluronan bio-nanocomposite fabrics for wound and chronic ulcer dressing: in vitro and in vivo evaluations. International Journal of Pharmaceutics 520 (1-2), 241–253. Available at https://doi.org/10.1016/j.ijpharm.2017.02.003.

Abdelrahman, R.M., Abdel-Mohsen, A.M., Zboncak, M., Frankova, J., Lepcio, P., Kobera, L., et al., 2020. Hyaluronan biofilms reinforced with partially deacetylated chitin nanowhiskers: extraction, fabrication, in-vitro and antibacterial properties of advanced nanocomposites. Carbohydrate Polymers 235, 115951. Available at https://doi.org/10.1016/j.carbpol.2020.115951.

Abdel-Rahman, R.M., Abdel-Mohsen, A.M., Hrdina, R., Burgert, L., Fohlerova, Z., Pavliňák, D., et al., 2016. Wound dressing based on chitosan/hyaluronan/nonwoven fabrics: preparation, characterization and medical applications. International Journal of Biological Macromolecules 89, 725–736. Available at https://doi.org/10.1016/j.ijbiomac.2016.04.087.

Akil, H.M., Omar, M.F., Mazuki, A.A.M., Safiee, S., Ishak, Z.A.M., Abu Bakar, A., 2011. Kenaf fiber reinforced composites: a review. Materials & Design 32 (8-9), 4107–4121. Available at https://doi.org/10.1016/j.matdes.2011.04.008.

Andrew, J.J., Dhakal, H.N., 2022. Sustainable biobased composites for advanced applications: recent trends and future opportunities – a critical review. Composites Part C: Open Access 7, 100220. Available at https://doi.org/10.1016/j.jcomc.2021.100220.

Awais, H., Nawab, Y., Amjad, A., Anjang, A., Md Akil, H., Abidin, M.S.Z., 2021. Environmental benign natural fibre reinforced thermoplastic composites: a review. Composites Part C: Open Access 4, 100082. Available at https://doi.org/10.1016/j.jcomc.2020.100082.

Banakar, P., Shivanand, H.K., Niranjan, N.B., 2012. Mechanical properties of angle, ply laminated composites-a review. International Journal of Pure and Applied Sciences and Technology 9, 127–133.

Blomfeldt, T.O.J., Kuktaite, R., Plivelic, T.S., Rasheed, F., Johansson, E., Hedenqvist, M.S., 2012. Novel freeze-dried foams from glutenin- and gliadin-rich fractions. RSC Advances 2 (16), 6617–6627. Available at https://doi.org/10.1039/c2ra20946b.

Carbonaro, M., Maselli, P., Nucara, A., 2012. Relationship between digestibility and secondary structure of raw and thermally treated legume proteins: a fourier transform infrared (FT-IR) spectroscopic study. Amino Acids 43 (2), 911–921. Available at https://doi.org/10.1007/s00726-011-1151-4.

Cho, S.-W., Gällstedt, M., Johansson, E., Hedenqvist, M.S., 2011. Injection-molded nanocomposites and materials based on wheat gluten. International Journal of Biological Macromolecules 48 (1), 146–152. Available at https://doi.org/10.1016/j.ijbiomac.2010.10.012.

Choi, W.S., Han, J.H., 2001. Physical and mechanical properties of pea-protein-based edible films. Journal of Food Science 66 (2), 319–322. Available at https://doi.org/10.1111/j.1365-2621.2001.tb11339.x.

Chong, B.F., Blank, L.M., Mclaughlin, R., Nielsen, L.K., 2005. Microbial hyaluronic acid production. Applied Microbiology and Biotechnology 66 (4), 341–351. Available at https://doi.org/10.1007/s00253-004-1774-4.

Christian, J., 2020. 5 - Natural fibre-reinforced noncementitious composites (biocomposites). In: Harries, K.A., Sharma, B. (Eds.), Nonconventional and Vernacular Construction Materials: Characterisation, Properties and Applications. Woodhead Publishing, pp. 111–126. Available at https://doi.org/10.1016/B978-0-08-100038-0.00005-6.

Collins, M.N., Birkinshaw, C., 2013. Hyaluronic acid based scaffolds for tissue engineering - a review. Carbohydrate Polymers 92 (2), 1262–1279. Available at https://doi.org/10.1016/j.carbpol.2012.10.028.

Day, L., Augustin, M.A., Batey, I.L., Wrigley, C.W., 2006. Wheat-gluten uses and industry needs. Trends in Food Science & Technology 17 (2), 82–90. Available at https://doi.org/10.1016/j.tifs.2005.10.003.

Dicharry, R.M., Ye, P., Saha, G., Waxman, E., Asandei, A.D., Parnas, R.S., 2006. Wheat gluten-thiolated poly (vinyl alcohol) blends with improved mechanical properties. Biomacromolecules 7 (10), 2837–2844. Available at https://doi.org/10.1021/bm060432n.

Dicker, M.P.M., Duckworth, P.F., Baker, A.B., Francois, G., Hazzard, M.K., Weaver, P.M., 2014. Green composites: a review of material attributes and complementary applications. Composites Part A: Applied Science and Manufacturing 56, 280–289. Available at https://doi.org/10.1016/j.compositesa.2013.10.014.

Díez-Pascual, A.M., 2022. Biopolymer composites: synthesis, properties, and applications. International Journal of Molecular Sciences 23 (4), 2257. Available at https://doi.org/10.3390/ijms23042257.

Diuk Andrade, F., Newson, W.R., Bernardinelli, O.D., Rasheed, F., Cobo, M.F., Plivelic, T.S., et al., 2018. An insight into molecular motions and phase composition of gliadin/glutenin glycerol blends studied by 13C solid-state and 1H time-domain NMR. John Wiley and Sons Inc., Brazil. Journal of Polymer Science, Part B: Polymer Physics 56 (9), 739–750. Available at https://doi.org/10.1002/polb.24586.

Domenek, S., Feuilloley, P., Gratraud, J., Morel, M.H., Guilbert, S., 2004. Biodegradability of wheat gluten based bioplastics. Chemosphere 54 (4), 551–559. Available at https://doi.org/10.1016/S0045-6535(03)00760-4.

Du, Y., Chen, F., Zhang, Y., Rempel, C., Thompson, M.R., Liu, Q., 2015. Potato protein isolate-based biopolymers. Journal of Applied Polymer Science 132 (44). Available at https://doi.org/10.1002/app.42723.

Facca, A.G., Kortschot, M.T., Yan, N., 2007. Predicting the tensile strength of natural fibre reinforced thermoplastics. Composites Science and Technology 67 (11-12), 2454–2466. Available at https://doi.org/10.1016/j.compscitech.2006.12.018.

Fakhari, A., Berkland, C., 2013. Applications and emerging trends of hyaluronic acid in tissue engineering, as a dermal filler and in osteoarthritis treatment. Acta biomaterialia 9 (7), 7081–7092. Available at https://doi.org/10.1016/j.actbio.2013.03.005.

Faruk, O., Bledzki, A.K., Fink, H.P., Sain, M., 2012. Biocomposites reinforced with natural fibers: 2000-2010. Progress in Polymer Science 37 (11), 1552–1596. Available at https://doi.org/10.1016/j.progpolymsci.2012.04.003.

Forssell, P., Lahtinen, R., Lahelin, M., Myllärinen, P., 2002. Oxygen permeability of amylose and amylopectin films. Carbohydrate Polymers 47 (2), 125–129. Available at https://doi.org/10.1016/S0144-8617(01)00175-8.

Friedman, M., Brandon, D.L., 2001. Nutritional and health benefits of soy proteins. Journal of Agricultural and Food Chemistry 49 (3), 1069–1086. Available at https://doi.org/10.1021/jf0009246.

Gällstedt, M., Mattozzi, A., Johansson, E., Hedenqvist, M.S., 2004. Transport and tensile properties of compression-molded wheat gluten films. Biomacromolecules 5 (5), 2020–2028. Available at https://doi.org/10.1021/bm040044q.

George, A., Sanjay, M.R., Srisuk, R., Parameswaranpillai, J., Siengchin, S., 2020. A comprehensive review on chemical properties and applications of biopolymers and their composites. International Journal of Biological Macromolecules 154, 329–338. Available at https://doi.org/10.1016/j.ijbiomac.2020.03.120.

Gilbert, M.E., Kirker, K.R., Gray, S.D., Ward, P.D., Szakacs, J.G., Prestwich, G.D., et al., 2004. Chondroitin sulfate hydrogel and wound healing in rabbit maxillary sinus mucosa. The Laryngoscope 114 (8), 1406–1409. Available at https://doi.org/10.1097/00005537-200408000-00017.

Gontard, N., Guilbert, S., Cuq, J., 1993. Water and glycerol as plasticizers affect mechanical and water vapor barrier properties of an edible wheat gluten film. Journal of Food Science 58 (1), 206–211. Available at https://doi.org/10.1111/j.1365-2621.1993.tb03246.x.

Gorham, J., 1821. Analysis of Indian corn. Quarterly Journal Science, Literature and Arts 2, 206–208.

Gueguen, J., Viroben, G., Noireaux, P., Subirade, M., 1998. Influence of plasticizers and treatments on the properties of films from pea proteins. Industrial Crops and Products 7 (2-3), 149–157. Available at https://doi.org/10.1016/s0926-6690(97)00043-5.

Guo, J., Li, J., Kou, H., 2011. Chapter 19 - Chemical preparation of advanced ceramic materials. In: Xu, R., Pang, W., Huo, Q. (Eds.), Modern Inorganic Synthetic Chemistry. Elsevier, pp. 429–454. Available at 10.1016/B978-0-444-53599-3.10019-8.

Gurunathan, T., Mohanty, S., Nayak, S.K., 2015. A review of the recent developments in biocomposites based on natural fibres and their application perspectives. Composites Part A: Applied Science and Manufacturing 77, 1–25. Available at https://doi.org/10.1016/j.compositesa.2015.06.007.

Hay, I.D., Rehman, Z.U., Moradali, M.F., Wang, Y., Rehm, B.H.A., 2013. Microbial alginate production, modification and its applications. Microbial Biotechnology 6 (6), 637–650. Available at https://doi.org/10.1111/1751-7915.12076.

Haylock-Jacobs, S., Keough, M.B., Lau, L., Yong, V.W., 2011. Chondroitin sulphate proteoglycans: extracellular matrix proteins that regulate immunity of the central nervous system. Autoimmunity reviews 10 (12), 766–772. Available at https://doi.org/10.1016/j.autrev.2011.05.019.

Henrotin, Y., Mathy, M., Sanchez, C., Lambert, C., 2010. Chondroitin sulfate in the treatment of osteoarthritis: from in vitro studies to clinical recommendations. Therapeutic Advances in Musculoskeletal Disease 2 (6), 335–348. Available at https://doi.org/10.1177/1759720x10383076.

Hooftman, N., Messagie, M., Van Mierlo, J., Coosemans, T., 2018. A review of the European passenger car regulations – real driving emissions vs local air quality. Renewable and Sustainable Energy Reviews 86, 1–21. Available at https://doi.org/10.1016/j.rser.2018.01.012.

Hoover, N., Eynard, L., Lavelli, V., Cerletti, P., 2019. Interactions of protein and starch studied through amyloglucosidase action. Cereal Chemistry 74, 846–850.

Hoover, R., Hughes, T., Chung, H.J., Liu, Q., 2010. Composition, molecular structure, properties, and modification of pulse starches: a review. Food Research International 43 (2), 399–413. Available at https://doi.org/10.1016/j.foodres.2009.09.001.

Huang, H.C., Chang, T.C., Jane, J., 1999. Mechanical and physical properties of protein-starch based plastics produced by extrusion and injection molding. Journal of the American Oil Chemists' Society 76 (9), 1101–1108. Available at https://doi.org/10.1007/s11746-999-0210-4.

Johansson, E., Malik, A.H., Hussain, A., Rasheed, F., Newson, W.R., Plivelic, T., et al., 2013. Wheat gluten polymer structures: the impact of genotype, environment, and processing on their functionality in various applications. Cereal Chemistry 90 (4), 367–376. Available at https://doi.org/10.1094/CCHEM-08-12-0105-FI.

Kabir, M.M., Wang, H., Lau, K.T., Cardona, F., 2012. Chemical treatments on plant-based natural fibre reinforced polymer composites: an overview. Composites Part B: Engineering 43 (7), 2883–2892. Available at https://doi.org/10.1016/j.compositesb.2012.04.053.

Keya, K.N., Kona, N.A., Koly, F.A., Maraz, K.M., Islam, M.N., Khan, R.A., 2019. Natural fiber reinforced polymer composites: history, types, advantages, and applications. Materials Engineering Research 1 (2), 69–87. Available at https://doi.org/10.25082/mer.2019.02.006.

Khatkar, B.S., Bell, A.E., Schofield, J.D., 1995. The dynamic rheological properties of glutens and gluten sub-fractions from wheats of good and poor bread making quality. Journal of Cereal Science 22 (1), 29–44. Available at https://doi.org/10.1016/s0733-5210(05)80005-0.

Kilara, A., Vaghela, M.N., 2004. 4 - Whey proteins. In: Yada, R.Y. (Ed.), Proteins in Food Processing. Woodhead Publishing, pp. 72–99. Available at https://doi.org/10.1533/9781855738379.1.72.

Klemm, D., Schmauder, H.P., Heinze, T., 2002. Cellulose. In: De Baets, S., Vandamme, E., Steinbüchel, A. (Eds.), Biopolymers: Volume 6—Polysaccharides II Polysaccharides from Eukaryotes. Wiley-VCH, Weinheim, pp. 275–287.

Kluger, R., Alagic, A., 2004. Chemical cross-linking and protein–protein interactions—a review with illustrative protocols. Bioorganic Chemistry 32 (6), 451–472. Available at https://doi.org/10.1016/j.bioorg.2004.08.002.

Knorr, D., 1980. Effect of recovery methods on yield, quality and functional properties of potato protein concentrates. Journal of Food Science 45 (5), 1183–1186. Available at https://doi.org/10.1111/j.1365-2621.1980.tb06516.x.

Kogan, G., Šoltés, L., Stern, R., Gemeiner, P., 2007. Hyaluronic acid: a natural biopolymer with a broad range of biomedical and industrial applications. Biotechnology Letters 29 (1), 17–25. Available at https://doi.org/10.1007/s10529-006-9219-z.

Kuktaite, R., Plivelic, T.S., Türe, H., Hedenqvist, M.S., Gällstedt, M., Marttila, S., et al., 2012. Changes in the hierarchical protein polymer structure: urea and temperature effects on wheat gluten films. RSC Advances 2 (31), 11908–11914. Available at https://doi.org/10.1039/c2ra21812g.

Kunioka, M., 1997. Biosynthesis and chemical reactions of poly(amino acid)s from microorganisms. Applied Microbiology and Biotechnology 47 (5), 469–475. Available at https://doi.org/10.1007/s002530050958.

Lamberg, S.I., Stoolmiller, A.C., 1974. Glycosaminoglycans. A biochemical and clinical review. Journal of Investigative Dermatology 63 (6), 433–449. Available at https://doi.org/10.1111/1523-1747.ep12680346.

Løkra, S., Schüller, R.B., Egelandsdal, B., Engebretsen, B., Strætkvern, K.O., 2009. Comparison of composition, enzyme activity and selected functional properties of potato proteins isolated from potato juice with two different expanded bed resins. LWT - Food Science and Technology 42 (4), 906–913. Available at https://doi.org/10.1016/j.lwt.2008.11.011.

Lucas, N., Bienaime, C., Belloy, C., Queneudec, M., Silvestre, F., Nava-Saucedo, J.E., 2008. Polymer biodegradation: mechanisms and estimation techniques - a review. Chemosphere 73 (4), 429–442. Available at https://doi.org/10.1016/j.chemosphere.2008.06.064.

Luengo, J.M., García, B., Sandoval, A., Naharro, G., Olivera, E.R., 2003. Bioplastics from microorganisms. Current Opinion in Microbiology 6 (3), 251–260. Available at https://doi.org/10.1016/S1369-5274(03)00040-7.

Madhu, P., Sanjay, M.R., Pradeep, S., Subrahmanya Bhat, K., Yogesha, B., Siengchin, S., 2019. Characterization of cellulosic fibre from *Phoenix pusilla* leaves as potential reinforcement for polymeric composites. Journal of Materials Research and Technology 8 (3), 2597–2604. Available at https://doi.org/10.1016/j.jmrt.2019.03.006.

Mariano, M., Pilate, F., De Oliveira, F.B., Khelifa, F., Dubois, P., Raquez, J.M., et al., 2017. Preparation of cellulose nanocrystal-reinforced poly(lactic acid) nanocomposites through noncovalent modification with PLLA-based surfactants. ACS Omega 2 (6), 2678–2688. Available at https://doi.org/10.1021/acsomega.7b00387.

Marimuthu, S., Malathi, A.C.J., Raghavan, V., Grace, A.N., 2023. 2 - Processing of ceramics. Advanced Ceramics for Energy Storage, Thermoelectrics and Photonics. Elsevier, India, pp. 19–39. Available at 10.1016/B978-0-323-90761-3.00018-8.

Mathews, M.B., 1967. Macromolecular evolution of connective tissue. Biological Reviews of the Cambridge Philosophical Society 42 (4), 499–551. Available at https://doi.org/10.1111/j.1469-185x.1967.tb01528.x.

Mercan, N., Aslim, B., Yüksekdağ, Z.N., Beyatli, Y., 2002. Production of poly-b-hydroxybutyrate (PHB) by some *Rhizobium* bacteria. Turkish Journal of Biology 26 (4), 215–219.

Merry, C.L.R., Lindahl, U., Couchman, J., Esko, J.D., 2017. Chapter 17Proteoglycans and sulfated glycosaminoglycans. In: Varki, A., Cummings, R.D., Esko, J.D. (Eds.), Essentials of Glycobiology [Internet] fourth ed. Cold Spring Harbor Laboratory Press, Cold Spring Harbor (NY). Available at https://www.ncbi.nlm.nih.gov/books/NBK579925/.

Mohamed, S.A.N., Zainudin, E.S., Sapuan, S.M., Azaman, M.D., Arifin, A.M.T., 2018. 1 - Introduction to natural fiber reinforced vinyl ester and vinyl polymer composites. In: Sapuan, S.M., Ismail, H., Zainudin, E.S. (Eds.), Natural Fibre Reinforced Vinyl Ester and Vinyl Polymer Composites: Development, Characterization and Applications. Woodhead Publishing, pp. 1–25. Available at https://doi.org/10.1016/B978-0-08-102160-6.00001-9.

Mortensen, A., Llorca, J., 2010. Metal Matrix Composites. Annual Review of Materials Research 40 (1), 243–270. Available at https://doi.org/10.1146/annurev-matsci-070909-104511.

Nagalakshmaiah, M., Afrin, S., Malladi, R.P., Elkoun, S., Robert, M., Ansari, M.A., et al., 2018. Chapter 9 - Biocomposites: Present trends and challenges for the future. In: Koronis, G., Silva, A. (Eds.), Green Composites for Automotive Applications. Woodhead Publishing, pp. 197–215. Available at https://doi.org/10.1016/B978-0-08-102177-4.00009-4.

Nagavally, R.R., 2016. Composite materials - history, types, fabrication techniques, advantages, and applications. International Journal of Mechanical and Production Engineering 5, 87–92.

Nair, L.S., Laurencin, C.T., 2007. Biodegradable polymers as biomaterials. Progress in Polymer Science 32 (8-9), 762–798. Available at https://doi.org/10.1016/j.progpolymsci.2007.05.017.

Newson, W.R., Rasheed, F., Kuktaite, R., Hedenqvist, M.S., Gällstedt, M., Plivelic, T.S., et al., 2015. Commercial potato protein concentrate as a novel source for thermoformed bio-based plastic films with unusual polymerisation and tensile properties. RSC Advances 5 (41), 32217–32226. Available at https://doi.org/10.1039/c5ra00662g.

Piermaria, J.A., Pinotti, A., Garcia, M.A., Abraham, A.G., 2009. Films based on kefiran, an exopolysaccharide obtained from kefir grain: development and characterization. Food Hydrocolloids 23 (3), 684–690. Available at https://doi.org/10.1016/j.foodhyd.2008.05.003.

Pokharel, A., Falua, K.J., Babaei-Ghazvini, A., Acharya, B., 2022. Biobased polymer composites: a review. Journal of Composites Science 6 (9), 255. Available at https://doi.org/10.3390/jcs6090255.

Satyanarayana, K.G., Arizaga, G.G.C., Wypych, F., 2009. Biodegradable composites based on lignocellulosic fibers—an overview. Progress in Polymer Science 34 (9), 982–1021. Available at https://doi.org/10.1016/j.progpolymsci.2008.12.002.

Shewry, P.R., Halford, N.G., 2002. Cereal seed storage proteins: structures, properties and role in grain utilization. Journal of Experimental Botany 53 (370), 947–958. Available at 10.1093/jexbot/53.370.947.

Shilpa, K.N., Nithin, K.S., Sachhidananda, S., Madhukar, B.S., Siddaramaiah, 2017. Visibly transparent PVA/sodium doped dysprosia (Na2Dy2O4) nano composite films, with high refractive index: an optical study. Journal of Alloys and Compounds 694, 884–891. Available at https://doi.org/10.1016/j.jallcom.2016.10.004.

Shukla, R., Cheryan, M., 2001. Zein: the industrial protein from corn. Industrial Crops and Products 13 (3), 171–192. Available at https://doi.org/10.1016/s0926-6690(00)00064-9.

Smidsrød, O., Skjåk-Bræk, G., 1990. Alginate as immobilization matrix for cells. Trends in Biotechnology 8 (C), 71–78. Available at https://doi.org/10.1016/0167-7799(90)90139-O.

Smithers, G.W., 2008. Whey and whey proteins-From 'gutter-to-gold' International Dairy Journal 18 (7), 695–704. Available at https://doi.org/10.1016/j.idairyj.2008.03.008.

Sperling, L.H., Carraher, C.E., 1988. Polymers from renewable resources. Encyclopedia of Polymer Science and Engineering. John Wiley, New York, pp. 658–690.

Subramanian, C., Senthilvelan, S., 2011. Joint performance of the glass fiber reinforced polypropylene leaf spring. Composite Structures 93 (2), 759–766. Available at https://doi.org/10.1016/j.compstruct.2010.07.015.

Szekalska, M., Puciłowska, A., Szymańska, E., Ciosek, P., Winnicka, K., 2016. Alginate: current use and future perspectives in pharmaceutical and biomedical applications. International Journal of Polymer Science 2016, 1–17. Available at https://doi.org/10.1155/2016/7697031.

Tipton, P.A., 2010. 8.12 - Synthesis of Alginate in Bacteria. In: Liu, H.W., Mander, L. (Eds.), Comprehensive Natural Products II: Chemistry and Biology, Volume 8. Elsevier, Oxford, pp. 423–441. Available at https://doi.org/10.1016/B978-008045382-8.00651-1.

Toole, B.P., Wight, T.N., Tammi, M.I., 2002. Hyaluronan-cell interactions in cancer and vascular disease. Journal of Biological Chemistry 277 (7), 4593–4596. Available at https://doi.org/10.1074/jbc.R100039200.

Venkatesan, J., Nithya, R., Sudha, P.N., Kim, S.K., 2014. Role of alginate in bone tissue engineering. Advances in Food and Nutrition Research 73, 45–57. Available at https://doi.org/10.1016/B978-0-12-800268-1.00004-4.

Vroman, I., Tighzert, L., 2009. Biodegradable polymers. Materials 2 (2), 307–344. Available at https://doi.org/10.3390/ma2020307.

Yu, L., Chen, L., 2009. Polymeric materials from renewable resources. In: Yu, L. (Ed.), Biodegradable Polymer Blends and Composites from Renewable Resources. John Wiley & Sons, Inc., pp. 1–15. Available at https://doi.org/10.1002/9780470391501.ch1.

Zini, E., Scandola, M., 2011. Green composites: an overview. Polymer Composites 32 (12), 1905–1915. Available at https://doi.org/10.1002/pc.21224.

9

Development of high temperature polymers via 3D printing approach: properties and limitation

Kobe Samuel Mojapelo[1], Williams Kehinde Kupolati[1], Everardt Andre Burger[1], Idowu David Ibrahim[2]

[1]DEPARTMENT OF CIVIL ENGINEERING, TSHWANE UNIVERSITY OF TECHNOLOGY, PRETORIA, GAUTENG, SOUTH AFRICA [2]DEPARTMENT OF INDUSTRIAL ENGINEERING, OPERATIONS MANAGEMENT AND MECHANICAL ENGINEERING, VAAL UNIVERSITY OF TECHNOLOGY, VANDERBIJLPARK, SOUTH AFRICA

9.1 Introduction

9.1.1 Overview of high-temperature polymers in advanced applications

With the increased demand for advanced materials that can withstand extreme thermal, chemical, and mechanical environments, high-temperature polymers have gained significant attention through additive manufacturing (AM). Polyaryletherketones (PAEKs), an essential class of thermoplastics, such as polyetheretherketone (PEEK) and polyetherketoneketone (PEKK), are renowned for their exceptional properties in harsh and extreme environments, making them suitable for aerospace, automotive, electronics, and biomedical applications (Zol et al., 2023). AM techniques, such as fused filament fabrication (FFF) and other extrusion-based approaches, have been explored for the processing of these polymers and represent strong potential pathways. Using AM, geometric complexity can be achieved, resulting in desirable mechanical properties, and process parameters such as extrusion temperature and print speed of the AM process can be easily tuned to yield specific properties such as crystallinity and interlayer adhesion (El Magri et al., 2022). Furthermore, advancements such as high-temperature photopolymerization allow for the processing of high-performance photopolymers with elevated glass transition temperatures, which have historically been restricted owing to thermal degradation during AM (Gao et al., 2024). However, other issues persist, such as specifically constrained processing windows, weak interlayer bonding, and thermal stresses during printing. Researchers have addressed these problems through polymer reformulations, process optimization, and reinforcements and composites that provide mechanical strength and heat resistance (Heathman et al., 2024). The rapid development of AM technologies will be accompanied by the ongoing convergence of high-temperature polymer chemistry, broadening the possibilities for robust manufacturing

High-Temperature Polymeric Materials. DOI: https://doi.org/10.1016/B978-0-443-32792-6.00013-0

of high-performance components designed for use under extreme conditions (Weyhrich and Long, 2022).

9.1.2 Relevance of 3D printing additive manufacturing in polymer engineering

AM has transformed polymer engineering, especially in creating high-temperature materials with complex geometries and finely tuned microstructuring, which are not always achievable using conventional methods (Egon et al., 2024). This is particularly beneficial in high-temperature applications where thermal resistance and mechanical integrity are critical. Polymer network architectures are formed by in situ photo- and thermal-curing processes, leading to increased tensile toughness and thermal stability of polymers, hence less prone to thermal reflow during UV-assisted direct ink writing (DIW), resulting in better performance (Chen et al., 2018). Techniques for printing highly flame-retardant and thermal-resistant glass fiber polymers using fused deposition modeling (FDM) require innovations in 3D printing, including adjustments to print speed and layer height to enhance molecular alignment, ultimately creating materials tailored for extreme thermal environments (Prajapati et al., 2021; Battistelli et al., 2024). These principles emphasize the potential of AM in designing interfacial bonding and spatial microstructural features for enhanced performance. A recent branch of AM development involves the conversion of printed structures into ceramics through the pyrolysis of polymeric precursors. This allows for the fabrication of thermally robust components with tunable properties such as porosity and electrical conductivity, which are essential in the energy and aerospace industries (Lacelle et al., 2023). AM has proven to be an enabling platform for the advancement of high-temperature polymer applications, where material innovation and design flexibility converge.

9.2 Fundamentals of high-temperature polymers

9.2.1 Definition and classification

High-temperature polymers are a specialized subset of engineering thermoplastics that meet the elevated temperature requirements of mechanical strength, dimensional stability, and chemical resistance. These materials have rigid aromatic and heterocyclic backbone structures, which provide thermal stability at high temperatures. This includes PEEK, polyetherimide (PEI), polyphenylsulfone (PPSU), polytetrafluoroethylene (PTFE), and polyimide (PI), as well as applications in medical and automotive fields. PEEK and PEI are two high-temperature polymers that are widely used owing to their exceptional resistance and thermomechanical properties. PEEK is a semicrystalline polymer known for its high thermal resistance, which exceeds 250°C, making it suitable for applications in aerospace under various processing parameters (Garcia-Leiner et al., 2021). Derived from PEI, an amorphous, flame-retardant, and optically transparent thermoplastic, this material is recognized for its suitability in FDM-based AM due to its melt-processing capabilities. High-temperature polymers, such as PPSU, PTFE, and PI, are essential for advanced applications (Yilmaz et al., 2022). The oxidative and hydrolytic stability is attributed to the sulfones in PPSU; the strong C–F bonds in PTFE provide chemical inertness and low friction; and PIs offer superior heat resistance and dimensional stability, which have also been explored for capacitive energy storage recently (Dong et al., 2023). The ongoing development of AM and composite formulation technologies has further expanded the range of these polymers to extreme environments (Sarker et al., 2022).

9.2.2 Thermal and mechanical properties of high-temperature polymers

Generally, the development of high-temperature polymers for AM has focused on enhancing their thermal and mechanical performances. These materials exhibit high glass transition (T_g) and melting (T_m) temperatures (T_g>200°C; T_m>450°C), providing dimensional stability and mechanical robustness during prolonged thermal stress (Weyhrich and Long, 2022). All these characteristics render them indispensable in industries such as aerospace, automotive, and electronics, where thermal reliability is a crucial factor. Mechanical integrity is also critical because heat-resistant polymers must withstand dynamic loading without failure. AM, particularly extrusion-based methods such as FFF, subjects the material to rapid thermal cycling, leading to interlayer delamination, residual stresses, and geometric deformations such as warping and shrinkage (Santiago et al., 2021). Although PEEK possesses excellent strength and high crystallinity, its semicrystalline nature results in deformation owing to cooling during printing, which affects the accuracy and performance of the final part.

To overcome these drawbacks, stringent control of the processing parameters is essential. The nozzle temperatures typically range from 350°C to 450°C, which is close to the thermal degradation limit of many high-performance polymers, necessitating careful optimization of flow, adhesion, and degradation (Tu et al., 2023). Specific environments minimize thermal effects, such as temperature differences between layers, whereas others utilize UV-assisted DIW and thermal curing chambers to achieve improved microstructure and binding (Rau et al., 2018). A comprehensive understanding of the thermal transitions, mechanical behavior, and optimal printing conditions is necessary to maximize the potential of high class polymers for extreme applications.

9.2.3 Thermal degradation, glass transition temperature, and crystallinity

Key thermal properties are essential for the development of high-temperature polymers for use in AM, particularly the thermal degradation resistance, glass transition temperature (T_g), and crystallinity. These polymers are typically constructed using heat-stable architectures of the polymer backbone, such as imidazole linkages, aromatic rings, or cyanate groups, to prevent decomposition during preparation. PI systems exhibit remarkable thermal stability owing to their aromatic, chirality-rich backbones, which allow them to maintain structural integrity during thermal cycling (Guzman et al., 2022). The glass transition temperature (T_g) is a crucial factor for determining the working temperature range of a polymer. High T_g values ensure material rigidity and dimensional integrity at relatively high temperatures. Interpenetrating network structures with high T_g availability have been achieved using cyanate ester resins, which provide advantages in processing and in-service performance (Zhou et al., 2020). The ability to tailor T_g through chemical design significantly impacts the success of 3D printing by influencing the flow of the material during deposition and its thermal resistance during end use. Crystallinity influences the mechanical strength and thermal stability of high-temperature polymers; however, it must be carefully controlled during 3D printing (Li et al., 2023). Rapid cooling can lead to anisotropic properties in fabricated products and excessive crystallinity, which is linked to warpage and shrinkage. The cooling rates and thermal management strategies during laser sintering allow for tuning of the crystallization behavior of PEKK, enhancing dimensional accuracy and material performance (Cherri et al., 2022). Therefore it is essential to strike a balance between crystallinity and degradation resistance to achieve high-performance printed parts in aerospace, electronics, and other thermally demanding sectors.

9.3 3D printing technologies for high-temperature polymers

9.3.1 Fused deposition modeling

FDM, also known as FFF, is one of the most common forms of AM for thermoplastic polymers, including high-temperature materials such as PEEK, polyphenylene sulfide, and PEI. This process builds objects by extruding molten filaments through a heated nozzle. The melting time, interlayer adhesion, and mechanical strength are highly dependent on proper control of nozzle temperature, extrusion rate, and ambient heat conditions (Lorenzo, 2025). Consequently, the AM technology of FDM presents underlying thermal challenges when processing high-performance polymers as they demand processing temperatures higher than 300°C, as well as maintaining controlled cooling to avoid degradation, warping, and poor interlayer adhesion. Even small thermal changes can affect the degree of crystallinity, which may alter tensile strength and dimensional stability. Poor thermal and mechanical performance can be enhanced using reinforcing additives, such as carbon nanotubes, and ceramic fillers, such as alumina (Smirnov et al., 2022). Additionally, rheological and calorimetric analyses determine printability while preventing polymer structure degradation (Shah and Jain, 2024). Despite the advantages, FDM faces challenges such as the inability to print high-temperature polymers, the need for significant capital investment in hardware, the anisotropic nature of the final products, and shrinkage. Simulation tools and response surface methodologies have contributed to improvements in parameter tuning alongside material innovations and filament modifications, resulting in enhanced print quality.

9.3.2 Selective laser sintering

Selective laser sintering (SLS) has emerged as a developing AM process for high-temperature polymer materials because of its ability to consolidate powdered feedstock layer-by-layer and construct intricate 3D structures without the need for support structures. The introduction of high-performance polymers, such as PEEK, for use in SLS systems further expands the applications of this process owing to their desirable thermal and mechanical properties (Salehi and Pircheraghi, 2021). These high melting temperatures require elevated processing temperatures, creating significant thermal gradients between layers, which can lead to internal stress, decreased dimensional accuracy, and degradation of the mechanical performance of the part (Shakhmurzova et al., 2020). To obtain the optimal sintering conditions, the laser energy input, powder properties, and scanning parameters must be strictly controlled. The powder characteristics, like morphology, size distribution, and flowability, significantly influence the sintering quality. The poor management of these factors can lead to inadequate layer adhesion and increased porosity (Lei et al., 2019). Additionally, lower resolutions (~20 μm) and the need for in situ thermal management systems introduce further complexity, particularly for high-temperature semicrystalline polymers (Kafle et al., 2021). This improved control over processing has led to research efforts aimed at optimizing the process itself, such as preheating the powder bed and adjusting the laser power and scanning speeds to promote uniform crystallinity and minimize warpage. Nevertheless, this approach still faces a significant challenge of balancing effective sintering with the prevention of thermal degradation. Furthermore, the nonuniform temperature distribution during multilayer builds contributes to anisotropic properties and hinders the reliability of SLS-fabricated parts for load-bearing or high-performance applications (Lupone et al., 2022). Continued improvements in powder formulation, thermal modeling, and laser control would enable the full advantages of SLS for obtaining high-temperature polymers.

9.3.3 Stereolithography for heat-resistant photopolymers

Stereolithography (SLA) is a high-stability AM process known for achieving good dimensional tolerances and the ability to produce dimensionally stable polymer-based components. However, adapting SLA for high-temperature applications poses challenges for resin formulation and process control. The synthesis of suitable photopolymers involves optimizing the chemistry of the monomers to increase the glass transition temperature and thermal stability without sacrificing the low viscosity and high photoreactivity, which are essential for efficient layer-by-layer printing (Basilia et al., 2024). The thermal and mechanical properties have been enhanced with silicone and epoxy functional groups in newer photopolymer systems. The curing rates of epoxy-silicone hybrid systems are higher than those of pure siloxanes and can achieve excellent thermal stability, which is necessary for applications in which mechanical integrity must remain consistent after printing (Zhou et al., 2020). Moreover, high-temperature photopolymerization technology has standardized the curing process, making the postprocessing of printed parts more convenient and promoting rapid manufacturing of various components. These approaches have increased the deployment of SLA for manufacturing thermally resistant parts, despite some remaining issues. These limitations include thermoset chemistry, as higher performance additives can act as glass formers and hinder flow and layer formation, along with a trade-off between resin viscosity and thermal stability (Tu et al., 2020). Furthermore, due to poor light penetration, nonuniform UV curing can lead to inconsistent crosslinking and uncontrolled thermal characteristics (Schlotthauer et al., 2021). Although modified SLA resins are promising, they do not match the thermal stability of traditional thermosets. This performance gap must be narrowed to fully adopt SLA in high-temperature applications, which requires further research into nanofiller-enhanced and dual-cure systems, as well as improved curing kinetics models (Tuteski and Kochov, 2024).

9.3.4 Direct ink writing and extrusion-based approaches

While some methods are already used for part production, DIW and similar extrusion-based 3D printing techniques are potentially very attractive for manufacturing high-temperature polymer components. These methods allow the handling of inks with high viscosity and customized viscoelastic properties, which facilitates the precise deposition of intricate structures without the need for significant heat. DIW offers a viable approach for processing polymers such as PTFE, which has proven challenging using conventional methods, by employing formulation strategies that provide tunable mechanical performance at elevated temperatures (Saadi et al., 2022). One of the significant advantages of DIW is that it enables the exploitation of shear-thinning behavior as well as in situ gelation or cross-linking after deposition to maintain structural stability. Recent approaches utilize sequential thermally motivated and chemotherapeutics-responsive polymeric networks, curing immediately postextrusion and promoting high-dimensional stability along with exceptional feature resolution. Numerous printing parameters, such as nozzle diameter, extrusion pressure, and print speed, must be adjusted along with ink chemistry to ensure support for performance at high temperatures (Li et al., 2025). The challenges exist in achieving an optimal balance between ink extrudability and solidification speed. Rheological mismatches or inadequate curing kinetics can negatively impact the mechanical strength and print resolution, especially when components are exposed to significant thermal loads. Many of these challenges can be addressed through ongoing efforts in polymer formulation and postprocessing, which will be particularly useful in tackling issues related to cross-linking behavior and matrix reinforcement, ensuring reliable performance under high-temperature conditions (Wei et al., 2023).

9.3.5 Hybrid/Additive-subtractive processes

Combining additive and subtractive manufacturing processes is a potential method for enhancing the properties and performance of high-temperature polymer parts. Hybrid approaches have the potential to address common limitations associated with AM, such as staircase effects, anisotropic properties, and surface roughness, by combining the geometric flexibility of AM with the precision of subtractive methods such as milling (Sebbe et al., 2022). This ability to anticipate behavior is especially useful for polymers such as PEEK, which exhibit properties that can be challenging to process owing to their high melting points, the tendency to cool rapidly, and sensitivities to interlayer adhesion (Sommer et al., 2022). The use of processes such as high-speed milling in AM workflows as postprocessing methods has been shown to dramatically improve the surface finish, dimensional accuracy, and mechanical performance, particularly in applications that require thermal and structural reliability

(Zhou et al., 2023). By eliminating or reducing any visibly discernible defects, this process contributes to the overall quality of the part and ensures adherence to high-performance industry standards. This approach enhances the versatility of high-performance polymers in the aerospace, automotive, and energy industries, where factors such as precision, durability, and heat resistance are crucial (Sommer et al., 2022).

9.4 Material preparation and processability

9.4.1 Feedstock requirements

The high viscosity of high-temperature polymers for 3D printing necessitates carefully engineered feedstock that typically forms filaments, powders, and resins. The type of feedstock determines the processability and thermal-mechanical performance of the final part. FDM employs filament-based feedstocks, which are thermoplastic polymers that must be processed under precise conditions to achieve controlled melt viscosity and thermal stability for adequate interlayer adhesion and optimal thermal properties, while preventing warpage during 3D printing (Hua et al., 2023). Improvements in the weld strength and dimensional accuracy of components printed from materials such as PEEK have been observed owing to innovations such as core-shell filament architectures (Wang et al., 2020). Ceramics have gained significant attention for SLS, where feedstocks consist of powder that is selectively fused using a thermal input. Powder flowability and overall sintering quality are critical factors for optimization. A well-defined particle size, morphology, and narrow thermal resistance result in a distributed bonding area between the layers, reducing porosity. While these feedstocks tend to be less common than filaments, powder-based feedstocks designed for high-performance polymers are increasingly becoming the focus of interest as techniques for designing powders have become widespread (Syrlybayev et al., 2021). In contrast, resin-based feedstocks, which migrate to SLA and DLP processes, are chemically engineered for high glass transition temperatures and fast curing. Newer formulations, such as cyanate esters and dual-cure chemicals, have been found to possess higher thermal performance and mechanical stability (Appuhamillage et al., 2019). Rheological behavior, that is, the internal flow properties of the feedstock, is an essential consideration for all feedstock types. Materials must easily flow out of the nozzle and be compatible with curing methods, yet they need to solidify quickly enough to maintain the shape and structure of the components. This trade-off is critical for thermoplastic elastomers to achieve both dimensional stability and functional performance at elevated temperatures (Das et al., 2021). Therefore improving the set parameters through new feedstock formulation and rheological tuning will enable new possibilities for high-temperature 3D printing.

9.4.2 Rheological and thermal considerations for printability

Rheology determines the flow behavior through the printer nozzle and the interlayer adhesion. If the system is used for an extrusion process, such as the FDM approach, it must meet criteria such as complex viscosity, shear-thinning behavior, and elastic modulus, which govern the optimization of the printed path (Mahmoudi et al., 2021). Techniques, such as oscillatory rheometry, previously used to characterize the behavior of phases across temperatures, have been applied to quantify these behaviors, enabling further parameter tuning, such as nozzle temperature and processing parameters, to influence the stability and mechanical properties of the print (Ritzen et al., 2021). Thermal management is crucial for high-temperature polymers. This means that the materials need to maintain their thermal stability while providing controlled flow during printing. Meanwhile, commercial advancements, including dual-stage curing in thermosetting systems and high-T_g photopolymer designs, have opened new applications for 3D printing, even in demanding environments. Moreover, the use of supramolecular polymers with reversible bonding and heat-resistant formulations, such as phthalonitrile-based resins, enhances interlayer bonding and heat dimensional stability. They interlink with bending strength, making thinner parts more durable in the long run and reducing weight (Öblom et al., 2019). However, refining the viscosity for effective extrusion while maintaining the structural integrity remains challenging. Very low viscosity can cause shape deformation, whereas very high viscosity results in poor flowability and nozzle blockage. Fluctuations in extrusion and curing temperatures can also lead to warpage or suboptimal crosslinking, which affects the part's performance. Therefore comprehensive rheological and thermal profiling remains crucial in optimizing both material formulations and processing protocols for high-temperature 3D-printed polymers (Elhattab et al., 2021).

9.4.3 Role of fillers, reinforcements, and nanocomposites

To enhance the printability and the thermal and mechanical properties of high-temperature polymers used in 3D printing, the trend is shifting toward the use of filled and reinforced polymers and nanocomposites. To address the longstanding issues of low strength and heat resistance in photopolymers, fillers such as graphene and metal particles are added to improve the thermal conductivity and mechanical integrity (Kausar et al., 2023). However, these materials must be handled with care and optimized to avoid complications, such as increased viscosity and nozzle clogging, which can negatively impact the AM process. Graphene-based nanocomposites offer multifunctional benefits, including enhanced mechanical properties owing to the 2D characteristics of graphene and improved electrical conductivity, which supports various high-performance industrial and electronic applications (Bilisik and Akter, 2022). In SLA printing, the covalent functionalisation of graphene oxide has resulted in homogeneous dispersion, reduced agglomeration, and enhanced interfacial bonding (Palaganas et al., 2019). Moreover, metal fillers such as nickel can be added to improve the thermal behavior and structural strength, while surface treatment and adhesion at the matrix-filler interface are also essential for their performance (Mackiewicz et al., 2022). Continuous fiber-reinforced and hybrid nanocomposite high-temperature 3D printed parts have gained significant attention, further enhancing their mechanical and thermal performance for aerospace and defense applications (Xiao et al., 2023). Despite these advantages over traditional materials, processing difficulties such as narrow temperature windows and filler-induced rheological alterations continue to pose challenges in achieving reproducible print quality and resolution.

9.4.4 Preprocessing treatments

Material preparation is critical for achieving uniform printing and component performance in high-temperature polymer AM. The PAEK family of polymers, which is susceptible to moisture uptake, requires drying and other preprocessing treatments to an exceptionally high degree (Yap et al., 2023). Moisture can increase melt viscosity and induce porosity or hydrolytic degradation during printing. Therefore residual moisture differences were removed through vacuum or oven drying at temperatures specific to the polymer, ensuring thermal stability and greater homogeneity in the print. This is a significant factor for biomedical, aerospace, and automotive applications (Basgul et al., 2021). Annealing is a postprocessing step aimed at alleviating internal strains that can build up in AM, thereby complementing the drying process. This process allowed for the rearrangement of molecular structure, increased crystallinity, improved interlayer adhesion, and enhanced dimensional stability. Numerous studies have indicated that annealing semicrystalline polymers such as PAEK decreases the interfacial voids and enhances the overall mechanical performance without warping the geometry of the part, making it particularly relevant for high-performance applications (Bolat and Ergene, 2024). Drying and annealing work together to create a controlled and reproducible material environment, leading to consistent printing results. Various preprocessing methods significantly influence the structural, thermal, and mechanical properties of high-temperature polymer components. Therefore incorporating both treatments into the manufacturing workflow enhances the predictability, durability, and performance of 3D-printed components in extreme environments (Basgul et al., 2021).

9.5 Properties of 3D-printed high-temperature polymers

9.5.1 Mechanical properties

The mechanical properties of 3D-printed high-temperature polymers are significantly influenced by polymer chemistry and process parameters. The molecular orientation and strong interlayer adhesion achieved during extrusion enhance the tensile strength. The high degree of molecular alignment during thermal extrusion led to a significant increase in the tensile modulus and tensile strength, approaching their theoretical limits (Ghodbane et al., 2019). Improved filament architecture and warpage control also enhance dimensional precision and interlayer weld strength, directly boosting flexural rigidity. Elhattab et al. (2022) emphasized the importance of nozzle temperature in FDM because it affects the flow behavior and crystallinity of the melt, both of which are determinants of flexural performance. These data underscore the need to control the thermal profiles during deposition to achieve optimized mechanical performance. Impact resistance is influenced by both the design and postprocessing of materials. Additionally, thermally curing them for an extended period and embedding materials within interpenetrating polymer networks can enhance both the transition and toughness (Ge et al., 2021). Two-stage curable thermoset-based strategies can improve tensile and impact properties, although their applications are limited by the cure rate and cost. Overall, the processing parameters, material architecture, and postcuring conditions must be optimized to harness the high-performance mechanical properties of the printed materials under challenging high-temperature conditions.

9.5.2 Thermal performance and heat distortion temperature

Thermal performance is a crucial criterion for determining the suitability of polymers for applications at elevated operating temperatures, particularly in 3D printed materials for high-temperature applications. A significant performance indicator is the heat distortion temperature (HDT), which is the temperature at which the material loses its load-bearing capacity, meaning that it deforms under stress. Furthermore, both a high HDT and a high T_g are vital for maintaining dimensional stability under challenging conditions. Material development has introduced innovations in which thermal endurance takes precedence over mechanical properties, enabling these polymers to remain functional with minimal deformation under prolonged heat exposure (Zaman et al., 2024). Improvements in polymer chemistry have led to the development of promising high-performance AM photopolymers and thermosetting systems with superior thermal stability. A high-temperature photopolymerization process reduces the manufacturing complexity and increases the glass transition temperature, thereby enhancing the HDT and durability of the produced parts. Another notable resin is cyanate ester, which forms interpenetrating polymer networks during postprint thermal treatment. These structures stabilize the thermal resistance through triazine ring formations, where T_g values near 349°C and mechanical strength retention as high as 80% at 200°C are ideal for aerospace and electronics applications (Gao et al., 2024). Commercially available low-cost 3D printing methods, such as FDM, can struggle with thermal properties owing to the limitations in material selection. However, specific processing techniques and material adjustments, such as adding fillers or blending different polymers, can significantly increase HDT. SLS of PEEK–nylon mixtures, followed by thermal postprocessing, enhances thermal and mechanical stability (Chavez et al., 2022). Similarly, low-cost FDM-printed polymers have demonstrated improved thermal properties when the processing conditions are optimized and polymer blends are reinforced (Storck et al., 2022).

9.5.3 Chemical resistance and dielectric behaviour

High-temperature polymers fabricated via 3D printing have recently gained significant attention owing to their superior chemical resistance and tunable dielectric properties, which are desirable for electronic and energy storage applications. PI-based systems prepared using DIW have demonstrated excellent chemical inertness and dielectric behavior even in aggressive chemical and high-temperature environments (Guo et al., 2019). Functionalised nanomaterials, such as graphene derivatives, have played a critical role in enhancing their dielectric properties, which are characterized by high dielectric constants and low energy loss, thereby increasing their appeal in advanced electronics (Zhang et al., 2023). The dielectric performance of these polymers is related to their microstructures, which can be modified by the printing conditions and postprocessing treatments. Thermal history and layer morphology can strongly influence the dielectric constant and breakdown strength, and FFF studies position the printing of dielectrics into mainstream printable materials (Kalaš et al., 2021). Dimensional shrinkage from thermal treatment would stress geometry and layer bonding. Furthermore, it is challenging to achieve a balance between chemical resistance and dielectric properties due to the complex interplay between filler dispersion, interfacial bonding, and polymer chain mobility (Zhang et al., 2023). To achieve this goal of high-temperature 3D-printed dielectric polymers, material formulation and process optimization are currently being investigated for their feasibility in high-performance and harsh environments (Kalaš et al., 2021).

9.5.4 Microstructure, porosity, and anisotropy

Recent studies have shown that hierarchical porosity can be engineered to significantly enhance the thermomechanical properties of 3D-printed high-temperature polymers. Integrating submicrometric porosity from polymerization-induced phase separation with millimeter-scale pores via 3D printing improved the response kinetics of poly(N-isopropyl acrylamide) hydrogels. This principle can be applied to high-temperature systems to improve the heat transfer and stress distribution (Liu et al., 2024). The layer-by-layer deposition mechanism characteristic of 3D printed structures translates directly to mechanical and thermal reliability as well as anisotropic behavior. The alignment of porous ducts improves the mechanical strength and anisotropic functional features, such as thermal conductivity (Yu et al., 2024). Adequately adjusting the printing parameters and postprocessing resulted in enhanced interlayer adhesion and fewer voids in carbon fiber-reinforced PEEK composites, leading to better tribological and mechanical performance (Lv et al., 2024). In addition, design strategies, such as triply periodic minimal surfaces, can control directional mechanical responses even when the same material is used (Vigil et al., 2024). Although controlled porosity can aid in heat dissipation and weight reduction, the presence of irregular voids leads to decreased fatigue resistance and thermal stability. Electrical and mechanical anisotropies can arise in conductive polymers through layer-wise fabrication, which requires strategic mitigation approaches (Tilve-Martinez et al., 2023). Therefore optimizing parameters such as nozzle temperature, print speed, and layer thickness is essential for minimizing defects while balancing anisotropic effects. This allows the use of high-performance printed parts in thermal vitreous environments.

9.6 Challenges and limitations

9.6.1 Warpage, shrinkage, and interlayer adhesion

Although the fabrication of high-temperature polymers through 3D printing has promised new applications, warpage and shrinkage continue to pose significant challenges owing to uneven cooling and steep thermal gradients that lead to residual stresses and dimensional instability. The high melt viscosities of many high-performance polymer compounds exacerbate these issues. Redesigning toolpaths can reduce warpage by more than 40% by counteracting the shrinkage that occurs during the solidification process of the polymer (Das et al., 2021). A gradient structure within a thick section can alleviate the effects of differential cooling and enhance dimensional accuracy (Chen et al., 2024). Another enduring problem in AM is interlayer adhesion, particularly with high-temperature polymers that rapidly cool between layers. Insufficient remelting and poor thermal diffusion during deposition lead to weak interfacial bonding. Postprinting heat treatment effectively enhances the interlayer bonding and mechanical strength of the layers owing to thermal stress relief. A Vitrimer system for heat-induced self-repair addresses cohesion on the basis of a material and is recyclable (Niu et al., 2021). These issues require both material innovation and an accurate process parameter control for resolution. The optimized nozzle temperatures and print speeds enhance layer fusion; however, the fiber-reinforced systems could adjust their strength through interlayer temperature control (Xie et al., 2024). These approaches underscore the importance of synergistic strategies that integrate advanced processing techniques, thermal management, and engineered materials to address the dimensional and structural limitations in high-temperature 3D printed polymers.

9.6.2 Equipment limitations: heated chambers, nozzles, and bed temperatures

To successfully print high-temperature polymers, it is necessary to overcome the limitations of equipment such as heated chambers, extruder nozzles, and heated beds. Uniform heating of the chamber is crucial to minimize thermal gradients that lead to warpage, residual stresses, and nonuniform crystallization of polymers such as PEEK (Rendas et al., 2024). Controlled heating is vital for printed parts in high-performance applications, as slight changes in the chamber temperature result in significant differences in the microstructure and geometric accuracy of printed objects (Tafaoli-Masoule et al., 2024). Metals have much higher melting points than polymers; therefore the extruder nozzle must operate at very high, stable temperatures to melt the polymer without causing thermal degradation. Changes in the nozzle temperature affect the flow characteristics of the polymer and lead to uneven deposition, which degrades the strength and surface quality of the part. Depending on the heating system design, the rheological behavior changes, resulting in rheological instability and the decomposition of sensitive materials during extrusion, necessitating precise nozzle temperature control, real-time monitoring, and high-accuracy sensors, as evidenced by several studies demonstrating heating-induced effects on the surface chemistry of polymers during extrusion (Genest et al., 2023; Lau et al., 2020). Heated beds assist with first-layer adhesion and prevent deformation, particularly for long prints. Uneven bed heating can cause layer lamination and shape distortion, highlighting the importance of aligning the temperatures of the nozzle, bed, and chamber (Xu et al., 2021). These upgrades enhance thermal management and expand the range of high-temperature polymers that can be processed without compromising their mechanical properties.

9.6.3 Postprocessing constraints

Postprocessing methods, such as annealing and machining, are essential for improving the mechanical properties and dimensional stability of 3D-printed high-temperature polymers. Annealing reduces the internal stresses caused by the printing process and the high polymerization temperatures. However, traditional high-temperature polymers have narrow thermal processing windows, as the annealing temperatures must be carefully chosen so as not to underdevelop the mechanical properties or induce thermal degradation in the underlying matrix. Parameters related to annealing, such as temperature, holding time, and heating strategy, need to be optimized, and they are highly material-dependent to achieve a microstructure with desired properties without warping or other permanent changes (Wach et al., 2018). The machining of PAEKs poses challenges for their use owing to the material's intrinsic hardness, thermal resistance, and low ductility. These properties complicate chip formation and increase the risk of microcracking or surface damage if parameters such as cutting speed, tool type, and cooling are not optimal. Localized heat concentration can adversely affect the structural precision and surface finish, necessitating specialized tooling and cooling systems. Consequently, high-temperature polymer machining is a highly complex process that results in higher production costs and lower scalability for intricate geometries (Dizon et al., 2021; Basgul et al., 2021). This requires material-specific strategies for postprocessing and advanced thermal management to strategically leverage the performance potential associated with high-temperature 3D-printed material polymers.

9.6.4 Cost, scalability, and material availability

The use of high-temperature polymers is restricted to 3D printing because of challenges related to cost, scalability, and material availability. As a result, even thermally resistant polymers recently synthesized by Gao et al. (2024) often require complex monomer modifications to increase the glass transition temperature. These changes typically require intricate synthetic pathways and prolonged thermal curing, which increases costs and delays processing, further inhibiting the efficiency of AM. Additionally, the reagents and conditions specific to these polymers do not always support large-scale syntheses. The high cost of materials stems from the low precursor availability and poor formulation scaling behavior, which also hinders industrial adoption. Innovative approaches, such as the in-bath print and cure method, are being implemented to address these limitations, reducing the need for support and postprocessing. This approach holds significant promise for improving the cost efficiency and scalability of the process of printing thermosetting polymers. However, these solutions still depend on specific chemistries that cannot be quickly transferred across diverse 3D printing systems. However, proprietary formulations and processing constraints limit broader applications. To overcome these challenges, advancements are needed in both polymer design and manufacturing methods, considering their performance versus cost and suitability for a wide range of industrial applications.

9.6.5 Environmental and recyclability concerns

Due to complex chemistry and high energy consumption during 3D printing, high-temperature polymers create environmental and recyclability problems. Many of these polymers, particularly thermosets, are highly cross-linked, rendering them nonmoldable and nonrecyclable, leading to material waste. In this context, dynamic covalent network polymers and processable thermosets have been investigated for their ability to undergo repeatable reprocessing through reversible bond exchange pathways without degrading thermal–mechanical performance. The elevated operating temperatures required for 3D printing of high-performance polymers introduce additional sustainability challenges, including higher energy consumption and the emission of volatile organic compounds and particulates. Studies on AM using large formats have highlighted these environmental threats, indicating a need for greater control of emissions and the development of energy-efficient printing systems. Furthermore, high-temperature polymers have limited recycling options compared to common recycling categories like acrylonitrile butadiene styrene or poly (lactic acid), and sustainable manufacturing practices for high-temperature polymers have not been extensively developed. New strategies to achieve recyclability alongside performance depend on integrating biobased feedstock and chemical upcycling. Recycling challenges arise from durable materials due to their strong chemical bonds, such as PAEK (Chen et al., 2024). Many authors have emphasized lifecycle assessment and feedstock innovation, including lactate ester-based materials and recycled polymers, to develop pathways toward circularity while maintaining equivalent print quality and thermal resistance. Ultimately, sustainable 3D printing of high-temperature polymers will require a holistic approach that combines material innovation, lifecycle optimization, and scalable recycling technologies.

9.7 Applications across industries

9.7.1 Aerospace and defense

In recent years, 3D printing has emerged as a vital technology for the development of high-temperature polymers specifically designed for aerospace and defense applications, where extreme thermal and mechanical environments are the norm. These areas require materials with excellent thermal endurance and mechanical reliability, leading to advances in polymer synthesis and printing technologies. Gao et al. (2024) proposed a high-temperature photo-polymerization technique that incorporates high glass transition temperature moieties into the polymer backbone to enhance thermal performance while simplifying the processing. Similarly, Xiao et al. (2023) developed a low-temperature 3D printing method for continuous fiber-reinforced thermosetting composites that can maintain structural integrity under thermal and mechanical loads. These printed materials mainly depend on the mechanical and thermal performance for applications in hypersonic flight and spacecraft reentry, where ablation resistance and dimensional stability are necessary. Wu et al. (2020) have demonstrated the ablation capacity of high-temperature thermoplastics created using additive methods due to their ability to endure extreme heat exposure. These studies illustrate the advancing capabilities of 3D printing technologies to safely produce complex and durable components for operation in demanding aerospace and defense contexts. Despite these advances, challenges remain in synthesizing scalable high-performance polymers and managing the anisotropic behavior caused by layer-by-layer printing. Ongoing issues, such as small-scale production, costly materials, and expensive machinery, have hindered broader adoption. Recent studies have focused on optimizing formulations and processing conditions to make high-performance solutions more suitable for mass production and safety-critical sectors, such as aerospace and defense.

9.7.2 Automotive and motorsports

High-performance polymers created using cutting-edge 3D printing technologies are steadily entering automotive and motorsports applications, where lightweight solutions with good mechanical strength and thermal resistance are essential. Indeed, novel photopolymerization methods have enabled the low-temperature curing of polymers with elevated glass transition temperatures, making engine housings and exhaust system parts serviceable at temperatures exceeding 400°C (Gao et al., 2024; Wang et al., 2021). Furthermore, AM supports complex geometrical designs, reducing part weight while preserving structural integrity. This includes 3D-printed lightweight PI honeycombs with tensile strengths exceeding 500 MPa, which are both thermally stable and mechanically robust (Wang et al., 2021). Nevertheless, challenges remain in the form of complex monomer synthesis, costly processing steps, and printability limitations. There are opportunities for efficiency and performance improvements related to 3D printing within the automotive supply chain, particularly in motorsports, where component precision is a key driver (Delic et al., 2019). However, widespread adoption has been impeded by issues such as interlayer adhesion, variability of the printing process, and the limited availability of low-cost high-temperature polymers. Enhancements in polymer formulations and thermal management to increase scalability and consistency should lead to advancements in broader industrial applications in the future.

9.7.3 Electronics and high-temperature sensors

High-temperature polymers manufactured via 3D printing enable the production of complex geometries for advanced electronic devices and sensors operating in harsh thermal environments. Material innovations have played a crucial role, with studies such as Gao et al. (2024) demonstrating that modified monomers can produce photopolymers with higher glass transition temperatures for use in high-temperature sensors. Likewise, Kausar et al. (2023) developed polymer-graphene ternary nanocomposites utilizing 3D and 4D printing, which exhibited enhanced thermal and electrical properties. These materials support the development of durable sensor components with excellent thermal stability and electrical conductivity. Li et al. (2020) introduced compatible printable copper-based electronics for sensing extreme temperatures, while Hwang et al. (2024) utilized carbon-based thermoelectric materials in soft sensors to facilitate accurate localized thermal mapping, underscoring the importance of material process synergy in sensor performance. Material design, processing strategies, and integration techniques significantly influence the device's functionality. Straubinger et al. (2025) presented molten metal selective soldering as a method for joining components to heat-sensitive 3D-printed elements, thereby enhancing the sensor's resistance to thermal stress. Meanwhile, Hartono et al. (2023) emphasized the importance of process optimization by adjusting the printing parameters to maintain the structural integrity of carbon fiber-reinforced polyamides at high temperatures. Despite these advancements, challenges such as multimaterial compatibility, uniformity of metallization, and synthesis costs for high-energy-density polymers continue to impede progress. It is essential to promote cutting-edge interdisciplinary research to fully leverage 3D printing for high-temperature electronics, which integrates material innovation with precise 3D manufacturing to design sensors suitable for high-performance applications.

9.7.4 Energy

Polymers synthesized via 3D printing at high temperatures are an excellent alternative for energy-related applications, as they are a class of materials with tunable thermal, mechanical, and electrochemical properties. Recent breakthroughs include enhanced ionic conductivity and mechanical integrity in battery systems utilizing 3D printed nanocarbon frameworks and solid polymer electrolytes, resulting in more robust and efficient Li-ion batteries (Gao and Pumera, 2021; Lee et al., 2022). These advancements facilitate optimized internal architectures and improve the electrochemical performance while maintaining the scalability and design freedom associated with AM. They play a vital role in insulation and oil and gas applications due to their solid dielectric performance and long-term thermal stability (Gao et al., 2024; Binyamin et al., 2023), further extending their favorable use in thermal management and extreme stress applications under high temperatures. The oil and gas sector can leverage 3D printing to produce chemically resistant and mechanically stable tools under extreme conditions. The interpenetrating networks of cyanate ester resins show great promise owing to their thermal stability and mechanical properties (Zaman et al., 2024; Caldona et al., 2021). Nonetheless, several issues persist, such as intricate polymerization mechanisms, multistage curing processes, and layer adhesion challenges, which can undermine the reliability of components in extreme environments (Onffroy et al., 2024) and often require additional postproduction treatments, such as annealing or pyrolysis, which increases cost and complexity. However, future studies on the printing parameters and material development are expected to enhance the performance and scalability of high-temperature 3D-printed polymers for use in energy systems.

9.7.5 Biomedical and implantable devices

High-temperature polymers, such as PEEK and shape memory polymers, are increasingly used in biomedical applications owing to their remarkable thermal stability, biocompatibility, and mechanical strength (Fedotkin et al., 2023). PEEK is particularly advantageous for long-term implants because of its chemical inertness and ability to retain structural integrity after manufacturing through processes such as FDM. Furthermore, advanced surface treatments, such as arc evaporation, significantly enhance the bioactivity of PEEK and increase the cell and tissue compatibility, which are essential for implant success (Fedotkin et al., 2023).

Thus far, complex and patient-specific biomedical devices, including shape memory occlusion devices and vascular stents that dynamically respond to physiological conditions, have been developed using 3D and 4D printing technologies (Tamay et al., 2019). Shape memory polymers offer the advantage of postdeployment in vivo adaptability during minimally invasive procedures (Li et al., 2020). These innovations leverage the design freedom of AM to create customized medical solutions that minimize recovery times and improve patient outcomes. However, there are obstacles to the use of high-temperature polymers in medical 3D printing. This includes the need for specialized printers to accommodate the high processing temperatures and viscosities of these materials, which increases the cost and limits the resolution. Achieving dimensional precision and interlayer adhesion in complex shapes remains challenging. Optimizing mechanical and biological performance necessitates postprocessing steps, such as surface modification and thermal treatment. Additionally, research is currently focused on achieving sterility without compromising material properties, and avenues like thermal-assisted 3D printing, which include improved crystallization control and crystallinity uniformity, appear promising (Lin et al., 2019).

9.8 Future directions and emerging trends

3D printing of high-temperature polymers is rapidly evolving owing to innovations in both materials and manufacturing strategies. Emerging trends are now guiding technology toward improvements in performance, sustainability, and functional complexity. One major trend is the use of artificial intelligence (AI) to streamline the 3D printing processes. Real-time monitoring integrated with AI-assisted systems can significantly transform the monitoring of 3D printing environmental parameters, such as print speed, temperature, and curing behavior, providing warnings and real-time corrections. Machine learning models can further enable the prediction and compensation of defects by analyzing the thermal gradients and polymerization kinetics, which is a critical advantage for high-temperature photopolymers with limited processing windows.

At the same time, the development of smart, self-healing materials has overcome service life and durability limitations. High-temperature polymers that exhibit shape-memory and self-repairing functions autonomously reinforce structures by responding to thermal or mechanical damage. These materials are primarily used for components subjected to cyclic stress or changing environments, where long-term operation is essential in critical applications. Another important innovation is the realization of in situ monitoring and adaptive manufacturing. To do so, manufacturers can embed sensors and use thermal imaging or spectroscopy during the printing process to make real-time modifications. This ensures uniform mechanical and thermal properties of the printed parts, which are required for high-performance polymers operating at elevated temperatures. The design envelope of 3D-printed polymers is also expanding with multimaterial and functionally graded printing. Engineers can create spatial variations in stiffness, thermal conductivity, or strength by utilizing materials with complementary properties in a single building. Such an approach allows for the deposition of more native layers over time, facilitating the fabrication of advanced components with optimized performance profiles, such as flexible interfaces within otherwise rigid heat-resistant builds. Sustainability is gaining momentum as circular economy principles are integrated into AM. These principles include recycling technologies, damage-resistant formulations, and energy-efficient processing methods that minimize material loss and carbon dioxide (CO_2) emissions. This shift toward sustainably high-performance materials demonstrates that being environmentally responsible does not require compromising on advanced technical performance. The advancements in high-temperature polymer 3D printing will become accessible in the future through the combination of advanced manufacturing, paired materials, multifunctional design, and sustainability. AM can significantly impact high-temperature applications across multiple sectors, such as aerospace, biomedical, automotive, and energy, by exploring AI-driven control systems, innovative polymer formulations, and green production models.

9.9 Conclusion

The development of high-temperature polymers through 3D printing presents a transformative opportunity for industries that require materials capable of withstanding extreme mechanical and chemical stresses. As this field continues to evolve, advances in polymer formulation, process optimization, and functionalization are unlocking new capabilities. Techniques such as hybrid additive-subtractive manufacturing, functionally graded material design, and AI-driven adaptive printing are addressing long-standing issues of dimensional instability, anisotropy, and interlayer adhesion. Furthermore, the integration of innovative, self-healing polymers and sustainability-focused practices, such as direct recycling and trimer-based networks, is promoting more resilient and circular manufacturing workflows. However, several challenges remain. The high cost of raw materials, limited scalability of specialized processes, equipment constraints, and complex postprocessing requirements continue to hinder widespread adoption. Overcoming these barriers requires interdisciplinary collaboration between materials scientists, mechanical engineers, and computational designers. Emphasis should also be placed on developing universal standards for high-temperature polymer processing and performance benchmarking. The convergence of advanced material science, AM, and digital design is pushing the boundaries of what high-temperature polymers can achieve. With continued research and innovation, these materials are poised to become indispensable in next-generation engineering applications that balance performance, reliability, and sustainability.

AI disclosure

During the preparation of this work, the author(s) used QuillBot and Grammarly to paraphrase and improve the grammar. After using this tool/service, the author(s) reviewed and edited the content as needed and take(s) full responsibility for the content of the publication.

References

Appuhamillage, G.A., Chartrain, N., Meenakshisundaram, V., Feller, K.D., Williams, C.B., Long, T.E., 2019. 110th anniversary: vat photopolymerization-based additive manufacturing: current trends and future directions in materials design. Industrial and Engineering Chemistry Research 58 (33), 15109–15118. Available at https://doi.org/10.1021/acs.iecr.9b02679.

Basgul, C., Spece, H., Sharma, N., Thieringer, F.M., Kurtz, S.M., 2021. Structure, properties, and bioactivity of 3D printed PAEKs for implant applications: a systematic review. Journal of Biomedical Materials Research - Part B Applied Biomaterials 109 (11), 1924–1941. Available at https://doi.org/10.1002/jbm.b.34845.

Basilia, B.A., Boniel, A.J.C., Borilla, G.S.B., de Yro, P.A.N., 2024. 3D printing of urethane dimethacrylate with nanoclay for enhanced curing properties via stereolithography. Advances in Science and Technology 155 (1), 59–64. Available at https://doi.org/10.4028/p-EeP7MJ.

Battistelli, C., Seriani, S., Lughi, V., Slejko, E.A., 2024. Optimizing 3D-printing parameters for enhanced mechanical properties in liquid crystalline polymer components. Polymers for Advanced Technologies 35 (12).

Bilisik, K., Akter, M., 2022. Graphene nanocomposites: a review on processes, properties, and applications. Journal of Industrial Textiles 51 (3_suppl), 3718S. Available at https://doi.org/10.1177/15280837211024252.

Binyamin, I., Grossman, E., Gorodnitsky, M., Kam, D., Magdassi, S., 2023. 3D printing thermally stable high-performance polymers based on a dual curing mechanism. Advanced Functional Materials 33 (24). Available at https://doi.org/10.1002/adfm.202214368.

Bolat, Ç., Ergene, B., 2024. Wear performance of short fiber added polyamide composites produced by additive manufacturing: combined impacts of secondary heat treatment, reinforcement type, and test force. Polymer Composites 45 (8), 6885–6900. Available at https://doi.org/10.1002/pc.28236.

Caldona, E.B., Dizon, J.R.C., Viers, R.A., Garcia, V.J., Smith, Z.J., Advincula, R.C., 2021. Additively manufactured high-performance polymeric materials and their potential use in the oil and gas industry. MRS Communications 11 (6), 701–715. Available at https://doi.org/10.1557/s43579-021-00134-9.

Chaves, L.A., Ibave, P., Hassan, M.S., Hall Sanchez, S.E., Billah, K.M.M., Leyva, A., et al., 2022. Low-temperature selective laser sintering 3D printing of PEEK-Nylon blends: impact of thermal post-processing on mechanical properties and thermal stability. Journal of Applied Polymer Science 139 (23), 1–8.

Chen, K., Kuang, X., Li, V., Kang, G., Qi, H.J., 2018. Fabrication of tough epoxy with shape memory effects by UV-assisted direct-ink write printing. Soft Matter 14 (10), 1879–1886. Available at https://doi.org/10.1039/c7sm02362f.

Chen, Q., Cai, Z., Kuzhandaivel, D., Lin, X., Wang, J., Chen, S., 2024. Three-dimensionally printed ternary composites of polyamide: effect of gradient structure on dimensional stability and mechanical properties. Polymers 16 (19), 2697. Available at https://doi.org/10.3390/polym16192697.

Cherri, A., Iliopoulos, I., Régnier, G., Brulé, B., Lé, G., Tencé-Girault, S., 2022. Thermal and crystallization properties of the alternated tere/Iso PEKK copolymer: importance in high-temperature laser sintering. ACS Applied Polymer Materials 4 (4), 2806–2818. Available at https://doi.org/10.1021/acsapm.2c00096.

Das, A., Gilmer, E.L., Biria, S., Bortner, M.J., 2021. Importance of polymer rheology on material extrusion additive manufacturing: correlating process physics to print properties. ACS Applied Polymer Materials 3 (3), 1218–1249. Available at https://doi.org/10.1021/acsapm.0c01228.

Delic, M., Eyers, D.R., Mikulic, J., 2019. Additive manufacturing: empirical evidence for supply chain integration and performance from the automotive industry. Supply Chain Management: An International Journal 24 (5), 604–621. Available at https://doi.org/10.1108/scm-12-2017-0406.

Di Lorenzo, M.L., Liparoti, S., Cavallo, D., Hashemi, T., Pantani, R., 2025. Fused filament fabrication of poly(l-Lactic Acid): a thermal analysis approach to optimize processing conditions. International Journal of Polymer Science 2025 (1), 9525001. Available at https://doi.org/10.1155/ijps/9525001.

Dizon, J.R.C., Gache, C.C.L., Cascolan, H.M.S., Cancino, L.T., Advincula, R.C., 2021. Post-processing of 3D-printed polymers. Technologies 9 (3). Available at https://doi.org/10.3390/technologies9030061.

Dong, J., Li, L., Qiu, P., Pan, Y., Niu, Y., Sun, L., et al., 2023. Scalable polyimide-organosilicate hybrid films for high-temperature capacitive energy storage. Advanced Materials 35 (20), 1–8.

Egon, A., Bell, C., Shad, R., 2024. Innovations in polymer additive manufacturing: research advancements in polymer materials for additive manufacturing. Preprints. Available at https://doi.org/10.20944/preprints202408.0061.v1.

El Magri, A., Bencaid, S.E., Vanaei, H.R., Vaudreuil, S., 2022. Effects of laser power and hatch orientation on final properties of PA12 parts produced by selective laser sintering. Polymers 14 (17). Available at https://doi.org/10.3390/polym14173674.

Elhattab, K., Bhaduri, S.B., Sikder, P., 2022. Influence of fused deposition modelling nozzle temperature on the rheology and mechanical properties of 3D printed β-tricalcium phosphate (TCP)/Polylactic Acid (PLA) Composite. Polymers 14 (6). Available at https://doi.org/10.3390/polym14061222.

Elhattab, K., Bhaduri, S.B., Lawrence, J.G., Sikder, P., 2021. Fused filament fabrication (Three-Dimensional Printing) of amorphous magnesium phosphate/polylactic acid macroporous biocomposite scaffolds. ACS Applied Bio Materials 4 (4), 3276–3286. Available at https://doi.org/10.1021/acsabm.0c01620.

Fedotkin, A.Y., Akimchenko, I.O., Tran, T.H., Shugurov, A.R., Shesterikov, E.V., Kozelskaya, A.I., et al., 2023. Improvement of the surface properties of polyether ether ketone via arc evaporation for biomedical applications. Materials 16 (8). Available at https://doi.org/10.3390/ma16082990.

Gao, W., Pumera, M., 2021. 3D printed nanocarbon frameworks for Li-Ion battery cathodes. Advanced Functional Materials 31 (11). Available at https://doi.org/10.1002/adfm.202007285.

Gao, W., Duan, K., Jin, Q., Wang, H., Zhou, X., 2024. 3D printing high-performance photopolymers based on high-temperature photopolymerization technology. ACS Applied Polymer Materials 6 (19), 12123–12130. Available at https://doi.org/10.1021/acsapm.4c02240.

Garcia-Leiner, M., Streifel, B., Başgül, C., MacDonald, D.W., Kurtz, S.M., 2021. Characterization of polyaryletherketone (PAEK) filaments and printed parts produced by extrusion-based additive manufacturing. Polymer International 70 (8), 1128–1136. Available at https://doi.org/10.1002/pi.6231.

Ge, Q., Chen, Z., Cheng, J., Zhang, B., Zhang, Y.F., Li, H., et al., 2021. 3D printing of highly stretchable hydrogel with diverse UV curable polymers. Science Advances 7 (2). Available at https://doi.org/10.1126/sciadv.aba4261.

Genest, M., Reynolds, L., Cole, R.G., 2023. High-accuracy infrared thermography measurements of polymer additive manufacturing. Proceedings of SPIE - The International Society for Optical Engineering 12536 (1). Available at https://doi.org/10.1117/12.2663359.

Ghodbane, S.A., Murthy, N.S., Dunn, M.G., Kohn, J., 2019. Achieving molecular orientation in thermally extruded 3D printed objects. Biofabrication 11 (4). Available at https://doi.org/10.1088/1758-5090/ab1d44.

Guo, Y., Xu, J., Yan, C., Chen, Y., Zhang, X., Jia, X., et al., 2019. Direct ink writing of high performance architectured polyimides with low dimensional shrinkage. Advanced Engineering Materials 21 (5). Available at https://doi.org/10.1002/adem.201801314.

Guzman, W., Johnson, I., Wiggins, J.S., 2022. Thermal rearrangement conversion of cross-linked ortho-hydroxy polyimide networks. ACS Applied Polymer Materials 4 (10), 7135–7143. Available at https://doi.org/10.1021/acsapm.2c01031.

Hartono, N.R., Mohd Radzuan, N.A., Khalid, N.N., Sulong, A.B., 2023. 3D printed carbon fibre reinforced polyamides in high temperature. Jurnal Kejuruteraan 35 (6), 1351–1361. Available at https://doi.org/10.17576/jkukm-2023-35(6)-08.

Heathman, N., DeLay, M., Tehrani, M., 2024. Five-axis material extrusion of high-performance structural parts with continuous carbon fiber-reinforced LM-PAEK. Research Square. Available at https://doi.org/10.21203/rs.3.rs-5412244/v1.

Hua, L., Wang, X., Ding, L., Zeng, S., Liu, J., Wu, Z., 2023. Effects of fabrication parameters on the mechanical properties of short basalt-fiber-reinforced thermoplastic composites for fused deposition modeling-based 3D printing. Polymer Composites 44 (6), 3341–3357. Available at https://doi.org/10.1002/pc.27325.

Hwang, S., Jang, D., Kim, H., Kwak, J., Chung, S., 2024. 3D-printed soft temperature sensors based on thermoelectric effects for fast mapping of localized temperature distributions. ACS Applied Materials & Interfaces 16 (19), 25071–25079. Available at https://doi.org/10.1021/acsami.4c04021.

Kafle, A., Luis, E., Silwal, R., Pan, H.M., Shrestha, P.L., Bastola, A.K., 2021. 3D/4D printing of polymers: fused deposition modelling (FDM), selective laser sintering (SLS), and stereolithography (SLA). Polymers 13 (18), 3101. Available at https://doi.org/10.3390/polym13183101.

Kalaš, D., Šíma, K., Kadlec, P., Polanský, R., Soukup, R., Řeboun, J., Hamáček, A., et al., 2021. FFF 3D printing in electronic applications: dielectric and thermal properties of selected polymers. Polymers 13 (21), 3702. Available at https://doi.org/10.3390/polym13213702.

Kausar, A., Ahmad, I., Zhao, T., Aldaghri, O., Eisa, M.H., 2023. Polymer/graphene nanocomposites via 3D and 4D printing—design and technical potential. Processes 11 (3), 868. Available at https://doi.org/10.3390/pr11030868.

Lacelle, T., Sampson, K.L., Yazdani Sarvestani, H., Rahimizadeh, A., Barroeta Robles, J., Mirkhalaf, M., et al., 2023. Additive manufacturing of polymer derived ceramics: materials, methods, and applications. APL Materials 11 (7). Available at https://doi.org/10.1063/5.0151661.

Lau, N.C., Tsai, M.H., Chen, D.W., Chen, C.H., Cheng, K.W., 2020. Preparation and characterization for antibacterial activities of 3D printing polyetheretherketone disks coated with various ratios of ampicillin and vancomycin salts. Applied Sciences (Switzerland) 10 (1). Available at https://doi.org/10.3390/app10010097.

Lee, K., Shang, Y., Bobrin, V.A., Kuchel, R., Kundu, D., Corrigan, N., et al., 2022. 3D printing nanostructured solid polymer electrolytes with high modulus and conductivity. Advanced Materials 34 (42). Available at https://doi.org/10.1002/adma.202204816.

Lei, J., Chen, J., Hong, Y., Zhang, Q., Chen, Q., Tong, J., et al., 2019. The effect of laser sintering on the microstructure, relative density, and cracking of sol-gel-derived silica thin films. Journal of the American Ceramic Society 103, 70–81.

Li, C., Feng, C., Zhang, L., Zhang, L., Wang, L., 2025. Direct ink writing of polymer-based materials—a review. Polymer Engineering & Science 65 (2), 431–454. Available at https://doi.org/10.1002/pen.27038.

Li, S.C., Hao, C.C., Sun, W.F., 2023. Cross-link degree dependence of electronic, water uptakes and thermal-mechanical properties of epoxy resin polymers: molecular simulations. ECS Journal of Solid State Science and Technology 12 (2). Available at https://doi.org/10.1149/2162-8777/acb969.

Li, Z., Khuje, S., Chivate, A., Huang, Y., Hu, Y., An, L., et al., 2020. Printable copper sensor electronics for high temperature. ACS Applied Electronic Materials 2 (7), 1867–1873. Available at https://doi.org/10.1021/acsaelm.0c00358.

Lin, C.T., Lv, J., Li, Y., Zhang, F., Li, J., Liu, Y., et al., 2019. Printed biodegradable and remotely controllable shape memory occlusion devices. Advanced Functional Materials 29 (51), 1–10.

Liu, W., Wang, Z., Serna, J.A., Debastiani, R., Gomez, J.E.U., Lu, L., et al., 2024. Enhancing temperature responsiveness of PNIPAM through 3D-printed hierarchical porosity. Advanced Functional Materials 34 (41). Available at https://doi.org/10.1002/adfm.202403794.

Lupone, F., Padovano, E., Casamento, F., Badini, C., 2022. Process phenomena and material properties in selective laser sintering of polymers: a review. Materials 15 (1). Available at https://doi.org/10.3390/ma15010183.

Lv, X., Yang, S., Pei, X., Zhang, Y., Wang, Q., Wang, T., 2024. Enhanced mechanical and tribological properties of 3D printed short carbon fiber reinforced polyether ether ketone composites. Polymer Composites 45 (10), 8840–8860. Available at https://doi.org/10.1002/pc.28380.

Mackiewicz, E., Wejrzanowski, T., Adamczyk-Cieslak, B., Oliver, G.J., 2022. Polymer–nickel composite filaments for 3D printing of open porous materials. Materials 15 (4), 1360. Available at https://doi.org/10.3390/ma15041360.

Mahmoudi, M., Burlison, S.R., Moreno, S., Minary-Jolandan, M., 2021. Additive-free and support-free 3D printing of thermosetting polymers with isotropic mechanical properties. ACS Applied Materials & Interfaces 13 (4), 5529–5538. Available at https://doi.org/10.1021/acsami.0c19608.

Niu, W., Zhang, Z., Chen, Q., Cao, P.-F., Advincula, R.C., 2021. Highly recyclable, mechanically isotropic and healable 3D-printed elastomers via polyurea vitrimers. ACS Materials Letters 3 (8), 1095–1103. Available at https://doi.org/10.1021/acsmaterialslett.1c00132.

Öblom, H., Zhang, J., Pimparade, M., Speer, I., Preis, M., Repka, M., et al., 2019. 3D-printed isoniazid tablets for the treatment and prevention of tuberculosis—personalized dosing and drug release. AAPS PharmSciTech 20 (2). Available at https://doi.org/10.1208/s12249-018-1233-7.

Onffroy, P.R., Chiovoloni, S., Kuo, H.L., Saccone, M.A., Lu, J.Q., DeSimone, J.M., 2024. Opportunities at the intersection of 3D printed polymers and pyrolysis for the microfabrication of carbon-based energy materials. JACS Au 4 (10), 3706–3726. Available at https://doi.org/10.1021/jacsau.4c00555.

Palaganas, J.O., Palaganas, N.B., Ramos, L.J.I., David, C.P.C., 2019. 3D printing of covalent functionalized graphene oxide nanocomposite via stereolithography. ACS Applied Materials and Interfaces 11 (49), 46034–46043. Available at https://doi.org/10.1021/acsami.9b12071.

Prajapati, A.R., Dave, H.K., Raval, H.K., 2021. An experimental study on mechanical, thermal and flame-retardant properties of 3D-printed glass-fiber-reinforced polymer composites. Journal of Materials Engineering and Performance 30 (7), 5266–5277. Available at https://doi.org/10.1007/s11665-021-05731-2.

Rau, D.A., Herzberger, J., Long, T.E., Williams, C.B., 2018. Ultraviolet-assisted direct ink write to additively manufacture all-aromatic polyimides. ACS Applied Materials and Interfaces 10 (41), 34828–34833. Available at https://doi.org/10.1021/acsami.8b14584.

Rendas, P., Imperadeiro, A., Martins, R.F., Soares, B.A.R., 2024. High-cycle fatigue behaviour of polyetheretherketone (PEEK) produced by additive manufacturing. Polymers 16 (1). Available at https://doi.org/10.3390/polym16010018.

Ritzen, L., Montano, V., Garcia, S.J., 2021. 3D printing of a self-healing thermoplastic polyurethane through FDM: from polymer slab to mechanical assessment. Polymers 13 (2), 305. Available at https://doi.org/10.3390/polym13020305.

Saadi, M.A.S.R., Maguire, A., Pottackal, N.T., Thakur, M.S.H., Ikram, M.M., Hart, A.J., et al., 2022. Direct ink writing: a 3D printing technology for diverse materials. Advanced Materials 34 (28). Available at https://doi.org/10.1002/adma.202108855.

Salehi, A., Pircheraghi, G., 2021. Thermo-oxidative degradation during sintering of polyethylene particles. Journal of Applied Polymer Science 138 (19). Available at https://doi.org/10.1002/app.50373.

Santiago, C.C., Yelamanchi, B., Diosdado De la Peña, J.A., Lamb, J., Roguski, K., Turzyński, F., et al., 2021. Thermoplastic extrusion additive manufacturing of high-performance carbon fiber peek lattices. Crystals 11 (12), 1453. Available at https://doi.org/10.3390/cryst11121453.

Sarker, M.R.H., Glassmeyer, J., Ruble, A., Hamidi, Y., Billah, K.M.M., 2021. Investigation of the Mechanical Properties of High-Temperature Polymer (Polyether Ether Ketone-PEEK) With Material Extrusion Additive Manufacturing. Proceedings of the ASME 2022 International Mechanical Engineering Congress and

Exposition. Volume 2A: Advanced Manufacturing. Columbus, Ohio, USA. October 30–November 3, 2022. V02AT02A019. ASME. Available at https://doi.org/10.1115/IMECE2022-95419.

Schlotthauer, T., Nitsche, J., Middendorf, P., 2021. Evaluation of UV post-curing depth for homogenous cross-linking of stereolithography parts. Rapid Prototyping Journal 27 (10), 1910–1916. Available at https://doi.org/10.1108/rpj-12-2020-0309.

Sebbe, N.P.V., Fernandes, F., Sousa, V.F.C., Silva, F.J.G., 2022. Hybrid manufacturing processes used in the production of complex parts: a comprehensive review. Metals 12 (11), 1874. Available at https://doi.org/10.3390/met12111874.

Shah, A.K., Jain, A., 2024. Physical, mechanical, and thermal characterization of unfilled and carbon fiber filled polyamide (PA) filaments and corresponding fused deposition modelling fabricated samples. Proceedings of the ASME 2024 19th International Manufacturing Science and Engineering Conference. Volume 1: Additive Manufacturing; Advanced Materials Manufacturing; Biomanufacturing; Life Cycle Engineering. Knoxville, Tennessee, USA. June 17–21, 2024. V001T02A001. ASME. Available at https://doi.org/10.1115/MSEC2024-121544.

Shakhmurzova, K., Kurdanova, Z., Baykaziev, A., Zhansitov, A., Khashirova, S., 2020. Key engineering materials. Federation 3d-Printing Methods for Crystalline Polyetheretherketone 869, 466–473. Available at https://doi.org/10.4028/www.scientific.net/KEM.869.466.

Smirnov, A., Seleznev, A., Peretyagin, P., Bentseva, E., Pristinskiy, Y., Kuznetsova, E., et al., 2022. Rheological characterization and printability of polylactide (PLA)-alumina (Al2O3) filaments for fused deposition modeling (FDM). Materials 15 (23), 8399. Available at https://doi.org/10.3390/ma15238399.

Sommer, D., Pape, D., Esen, C., Hellmann, R., 2022. Tool wear and milling characteristics for hybrid additive manufacturing combining laser powder bed fusion and in situ high-speed milling. Materials 15 (3), 1236. Available at https://doi.org/10.3390/ma15031236.

Storck, J.L., Ehrmann, G., Güth, U., Uthoff, J., Homburg, S.V., Blachowicz, T., et al., 2022. Investigation of low-cost FDM-printed polymers for elevated-temperature applications. Polymers 14 (14). Available at https://doi.org/10.3390/polym14142826.

Straubinger, D., Koltay, P., Zengerle, R., Kartmann, S., Shu, Z., 2025. Versatile selective soldering via molten metal printing for heat-sensitive 3D electronics and smart wearables. Advanced Functional Materials 35 (33), 2419653. Available at https://doi.org/10.1002/adfm.202419653.

Syrlybayev, D., Zharylkassyn, B., Seisekulova, A., Akhmetov, M., Perveen, A., Talamona, D., 2021. Optimisation of strength properties of FDM printed parts—a critical review. Polymers 13 (10), 1587. Available at https://doi.org/10.3390/polym13101587.

Tafaoli-Masoule, M., Shakeri, M., Zahedi, S.A., Vaezi, M., 2024. Experimental investigation of process parameters in polyether ether ketone 3D printing. Proceedings of the Institution of Mechanical Engineers, Part E: Journal of Process Mechanical Engineering 238 (1), 267–279. Available at https://doi.org/10.1177/09544089221141554.

Tamay, D.G., Usal, T.D., Alagoz, A.S., Yucel, D., Hasirci, N., Hasirci, V., 2019. 3D and 4D printing of polymers for tissue engineering applications. Frontiers in Bioengineering and Biotechnology 7, 164. Available at https://doi.org/10.3389/fbioe.2019.00164.

Tilve-Martinez, D., Neri, W., Vukadinovic, N., Berton, B., Pénicaud, A., Yuan, J., et al., 2023. Electrical anisotropy and its mitigation in conductive polymers printed by digital light processing. ChemRxiv. Available at https://doi.org/10.26434/chemrxiv-2023-b7jfg.

Tu, J., Makarian, K., Alvarez, N.J., Palmese, G.R., 2020. Formulation of a model resin system for benchmarking processing-property relationships in high-performance photo 3D printing applications. Materials 13 (18). Available at https://doi.org/10.3390/ma13184109.

Tu, R., Kim, H.C., Sodano, H.A., 2023. Additive manufacturing of high-temperature thermoplastic polysulfone with tailored microstructure via precipitation printing. ACS Applied Materials and Interfaces 15 (38), 45270–45280. Available at https://doi.org/10.1021/acsami.3c09048.

Tuteski, O., Kochov, A., 2024. SLA-based injection molding tool inserts: challenges, innovations, and future prospects. Advanced Technologies & Materials 49 (2), 29–36. Available at https://doi.org/10.24867/atm-2024-2-003.

Vigil, J., Lewis, K., Norris, N., Karakoç, A., Becker, T.A., Mari, M., 2024. Design, fabrication, and characterization of 3d-printed multiphase scaffolds based on triply periodic minimal surfaces. Advances in Polymer Technology 2024 (7), 1. Available at https://doi.org/10.1155/2024/4616496.

Wach, R.A., Wolszczak, P., Adamus-Wlodarczyk, A., 2018. Enhancement of mechanical properties of FDM-PLA parts via thermal annealing. Macromolecular Materials and Engineering 303 (9). Available at https://doi.org/10.1002/mame.201800169.

Wang, C., Ma, S., Li, D., Zhao, J., Zhou, H., Wang, D., et al., 2021. 3D printing of lightweight polyimide honeycombs with the high specific strength and temperature resistance. ACS Applied Materials & Interfaces 13 (13), 15690–15700. Available at https://doi.org/10.1021/acsami.1c01992.

Wang, Y., Müller, W.D., Rumjahn, A., Schwitalla, A., 2020. Parameters influencing the outcome of additive manufacturing of tiny medical devices based on PEEK. Materials 13 (2). Available at https://doi.org/10.3390/ma13020466.

Wei, P., Cipriani, C., Hsieh, C.M., Kamani, K., Rogers, S., Pentzer, E., 2023. Go with the flow: rheological requirements for direct ink write printability. Journal of Applied Physics 134 (10). Available at https://doi.org/10.1063/5.0155896.

Weyhrich, C.W., Long, T.E., 2022. Additive manufacturing of high-performance engineering polymers: present and future. Polymer International 71 (5), 532–536. Available at https://doi.org/10.1002/pi.6343.

Wu, H., Kim, S., Fahy, W., Koo, J.H., Kafi, A., Bateman, S., et al., 2020. Ablation performances of additively manufactured high-temperature thermoplastic polymers. AIAA Scitech 2020 Forum 1 (1). Available at https://doi.org/10.2514/6.2020-1125.

Xiao, H., He, Q., Duan, Y., Wang, J., Qi, Y., Ming, Y., et al., 2023. Low-temperature 3D printing and curing process of continuous fiber-reinforced thermosetting polymer composites. Polymer Composites 44 (4), 2322–2330. Available at https://doi.org/10.1002/pc.27247.

Xie, J.-M., Zhou, X., Liu, Y., Zhang, J., Wang, W., 2024. Process parameters optimization for 3d printing of continuous carbon fiber reinforced composite. Matéria (Rio de Janeiro) 29 (3). Available at https://doi.org/10.1590/1517-7076-rmat-2024-0201.

Xu, Q., Shang, Y., Jiang, Z., Wang, Z., Zhou, C., Liu, X., et al., 2021. Effect of molecular weight on mechanical properties and microstructure of 3D printed poly(ether ether ketone). Polymer International 70 (8), 1065–1072. Available at https://doi.org/10.1002/pi.6166.

Yap, T., Heathman, N., Phillips, T., Beaman, J., Tehrani, M., 2023. Additive manufacturing of polyaryletherketone (PAEK) polymers and their composites. Composites Part B: Engineering 266, 111019. Available at https://doi.org/10.1016/j.compositesb.2023.111019.

Yilmaz, M., Yilmaz, N.F., Kalkan, M.F., 2022. Rheology, crystallinity, and mechanical investigation of interlayer adhesion strength by thermal annealing of polyetherimide (PEI/ULTEM 1010) parts produced by 3D printing. Journal of Materials Engineering and Performance 31 (12), 9900–9909. Available at https://doi.org/10.1007/s11665-022-07049-z.

Yu, D., Chi, G., Mao, X., Li, M., Wang, Z., Xing, C., et al., 2024. Volume-Metallization 3D-Printed Polymer Composites. Advanced Materials 36. Available at https://doi.org/10.1002/adma.202403088.

Zaman, S., Favela, S., Herrera, N.E., Gandara, A., Molina, L., Hassan, M.S., et al., 2024. 3D printing of cyanate ester resins with interpenetration networks for enhanced thermal and mechanical properties. Journal of Applied Polymer Science 141 (21), e55423. Available at https://doi.org/10.1002/app.55423.

Zhang, P., Liu, X., Zuo, P., Mi, P., Zhuang, Q., 2023. Amine-functionalized reduced graphene oxide/polyimide nanocomposite as a material with high dielectric constant and thermal resistance. ACS Applied Nano Materials 6 (3), 1696–1706. Available at https://doi.org/10.1021/acsanm.2c04630.

Zhou, H., Cheng, X., Jiang, X., Zheng, G., Zhang, J., Li, Y., et al., 2023. Study of the interlayer bonding strength for combined 3D printing and milling of polyetheretherketone. Journal of Applied Polymer Science 140 (17). Available at https://doi.org/10.1002/app.53773.

Zhou, Z.X., Li, Y., Zhong, J., Luo, Z., Gong, C.R., Zheng, Y.Q., et al., 2020. High-performance cyanate ester resins with interpenetration networks for 3D printing. ACS Applied Materials and Interfaces 12 (34), 38682–38689. Available at https://doi.org/10.1021/acsami.0c10909.

Zol, S.M., Alauddin, M.S., Said, Z., Mohd Ghazali, M.I., Hao-Ern, L., Mohd Farid, D.A., et al., 2023. Description of poly(aryl-ether-ketone) materials (PAEKs), polyetheretherketone (PEEK) and Polyetherketoneketone (PEKK) for application as a dental material: a materials iew. Polymers 15 (9). Available at https://doi.org/10.3390/polym15092170.

PART 2

Material modelling

10

Behavioral modeling of polymer composites at elevated temperature

Deju Zhu[1], Yejun Cao[1], Md. Zillur Rahman[2]

[1]COLLEGE OF CIVIL ENGINEERING, HUNAN UNIVERSITY, CHANGSHA, HUNAN, P.R. CHINA [2]DEPARTMENT OF MECHANICAL ENGINEERING, AHSANULLAH UNIVERSITY OF SCIENCE AND TECHNOLOGY, DHAKA, BANGLADESH

10.1 Introduction

Polymer composites, which are composed of a polymer matrix reinforced with fibers, are renowned for their remarkable strength-to-weight and stiffness-to-weight ratios. These characteristics render them invaluable in a multitude of applications, particularly within the realm of civil engineering, where they are employed in the construction of structures that require materials capable of withstanding significant loads while maintaining a low weight. The inherent advantages of these composites include their durability, resistance to corrosion, and ease of fabrication, which further contribute to their widespread use in various industries (Qureshi et al., 2021; Zhou et al., 2019).

However, these materials are not devoid of vulnerabilities. A critical challenge confronting polymer composites is their susceptibility to high temperatures. Exposure to elevated temperatures can soften the polymer matrix, compromising its load-bearing capacity. Furthermore, thermal degradation of the matrix may ensue, resulting in a loss of structural integrity. Additionally, the interface between the reinforcing fibers and the polymer matrix may undergo changes in properties, including a diminution in adhesion strength. These modifications can exert profound impacts on the composite material's overall mechanical performance, thermal stability, and physical attributes (Bazli and Abolfazli, 2020).

Predicting the mechanical behavior of fiber-reinforced polymer (FRP) materials accurately at high temperatures is an area of intense research, offering substantial promise for practical applications. As our scientific understanding deepens and technological capabilities expand, the complexity of engineering problems continues to grow. This evolution necessitates the development of more precise methods for forecasting the performance of FRP materials under thermal stress. Such advancements are crucial for ensuring the reliability and safety of structures that incorporate these materials, especially in environments where they are exposed to high temperatures (Zhou et al., 2019; Manalo et al., 2017).

Consequently, this study examines the latest research advancements in the mathematical modeling of the high-temperature behavior of polymer composites. The mathematical model and its effectiveness in forecasting the high-temperature behavior of polymer

High-Temperature Polymeric Materials. DOI: https://doi.org/10.1016/B978-0-443-32792-6.00011-7

composites are evaluated. Furthermore, the paper identifies the limitations of current research and highlights future directions for enhancing and expanding these models to better suit engineering applications.

10.2 Mathematical models of fiber-reinforced polymer under high temperatures

During the past two decades, considerable efforts have been expended on modeling the behavior of composites under high-temperature conditions. Understanding the degradation of resins, particularly their rapid deterioration at high temperatures, forms the cornerstone of these endeavors. Thermoset resins, prevalent as matrix materials in composites, undergo a transition known as the glass transition temperature (T_g), leading to a sudden decline in epoxy properties. This decline directly correlates with the overall strength reduction in composites at high temperatures. As depicted in Fig. 10–1, the epoxy resin's property deterioration follows a sigmoid curve, rendering it suitable for assessing mechanical properties like tensile modulus and strength (Qureshi et al., 2021).

The important properties of the sigmoid curve include ultimate strength (σ_u), residual strength (σ_r), and symmetry (T_{sym}, $(\sigma_u+\sigma_r)/2$).

Several notable models include the following:

1. Mahieux et al. (2001) proposed an exponential temperature model that postulates temperature-induced disruption of resin intermolecular bonds, with bond strength varying

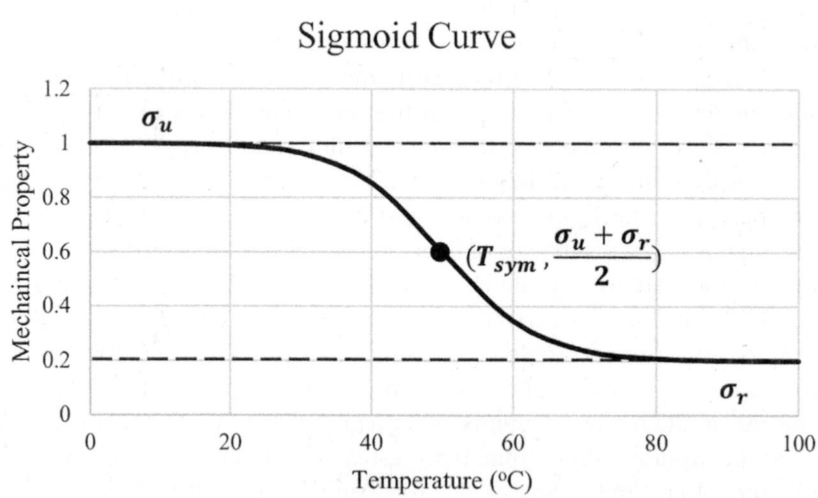

FIGURE 10–1 Sigmoid curve (Qureshi et al., 2021). *From Qureshi, M.F., Sheikh, S.A., Almansour, H., 2021. Temperature effects on the mechanical properties of the GFRP sheets. In: International Conference on Fibre-Reinforced Polymer (FRP) Composites in Civil Engineering. 27 Nov 2021. Springer International Publishing, Cham, pp. 809–822.*

according to the Weibull distribution. However, this model necessitates multiple Weibull exponents for acceptable results, yielding sharp corners in the sigmoid curve's extremities.

2. Bisby (2003) constructed a semiempirical analytical model, utilizing a sigmoid curve to predict high-temperature composite properties, with fitting parameters determined through regression analysis.
3. Gibson et al. (2006) introduced a sigmoid curve model incorporating a residual resin content factor, addressing resin decomposition and stress-dependent factors in fiber/resin-dominated composites.
4. Cao et al. (2011) observed a link between matrix strength decrease and composite strength degradation, developing a hyperbolic tangent function-based model to predict carbon fiber composite tensile strength.
5. The predictions deviated from the actual experimental results because the Bisby model used a large amount of scattered data. In addition, the sigmoid curve is only valid below the epoxy decomposition temperature. After that, the curve fails to capture the second decrease in mechanical properties. Wang et al. (2011) proposed a refined model accounting for resin decomposition and fiber dominance effects, enhancing prediction accuracy.

Table 10–1 summarizes various mathematical models proposed to predict composite mechanical properties at high temperatures.

10.3 Models for the prediction of fiber-reinforced polymer materials at high temperatures and moderate strain rates

Previous research endeavors have introduced diverse models aimed at forecasting the mechanical attributes of FRP materials at high temperatures. Notably, a substantial proportion of these models are rooted in semiempirical curve-fitting methodologies. Furthermore, models tailored to predict the strain rate-dependent variations in FRP materials have been devised, particularly within the moderate strain rate spectrum (Ahmed et al., 2021). The following equations elucidate the intricate relationship between the tensile strength and strain rate in FRP composites, as shown in Table 10–2

Herein, ε denotes the strain rate, while σ_{static} and $\sigma_{dynamic}$ represent the static and dynamic tensile strengths, respectively. E_{static} and $E_{dynamic}$ signify Young's modulus under static and dynamic conditions. Additionally, σ and E are utilized to represent the tensile strength and Young's modulus in a generic sense. The empirical equations incorporate material constants such as φ_σ and β_σ for the tensile strength dependence, φ_E and β_E for the Young's modulus dependence, along with a_σ, b_σ, γ_σ, a_E, b_E, and γ_E, which account for the specific properties and behaviors of the FRP material under investigation. These constants facilitate the accurate prediction of mechanical responses across varying strain rate scenarios.

Table 10–1 Models for predicting composite mechanical properties at elevated temperatures (Qureshi et al., 2021).

References	Models	Parameters
Mahieux et al. (2001)	$P(T) = P_R + (P_U + P_R) \times e^{-\left(\frac{T}{T_o}\right)^m}$	*P(T)*=Mechanical property at temperature *T* P_U=Unrelaxed mechanical property P_R=Relaxed mechanical property T_o=Relaxation temperature (Kelvin) *m*=Weibull exponent
Bisby (2003)	$\frac{f}{f_o} = \left(\frac{1-a}{2}\right)\tanh[-b(T-c)] + \left(\frac{1+a}{2}\right)$	*f*=Mechanical property at temperature *T* *a*=Residual mechanical property f_o=Mechanical property at ambient temperature *b, c*=Fitting parameters using least squares regression analysis
Gibson et al. (2006)	$P(T) = R^n\left[\frac{P_U+P_R}{2} - \frac{P_U-P_R}{2}\tanh\{k(T-T')\}\right]$	*P(T)*=Mechanical property at temperature *T* P_U=Unrelaxed mechanical property P_R=Relaxed mechanical property *k, T'*=Fitting parameters where the curve is nearly symmetrical R^n=Residual resin content
Cao et al. (2011)	$\sigma = \frac{1}{2}(\sigma_0 - \sigma_r)\tanh\left[-\frac{1}{\frac{\Delta T}{2}}\left(T - \left(T_g + \frac{\Delta T}{2}\right)\right)\right] + \frac{1}{2}(\sigma_0 + \sigma_r)$	σ=Tensile strength at temperature *T* σ_0=Tensile strength at room temperature σ_r=Residual tensile strength ΔT=Temperature variation in the glass transition region of the matrix T_g=Glass transition temperature
Wang et al. (2011)	$f_{u,T} = f_{u\text{-normal}}\left[A - \frac{(T-B)^n}{C}\right]$	$f_{u,T}$=Ultimate stress at temperature *T* $f_{u,\ normal}$=Ultimate stress at ambient temperature *A, B, C, n*=Coefficients calibrated for different temperature ranges
Feng et al. (2016)	$P(T) = P_i,\ T \leq Ts$ $P(T) = (P_r - P_i)/(T_g - T_s) \times (T - T_s) + P_i,$ $T_s < T < T_g$ $P(T) = P_r,\ T_g \leq T \leq T_r$	*P(T)*=Mechanical property at temperature *T* P_i=Initial value of mechanical property P_r=Residual value of mechanical property T_s=Temperature corresponding to elastic modulus degradation T_g=Glass transition temperature T_r=Temperature corresponding to residual elastic modulus

Source: From Qureshi, M.F., Sheikh, S.A., Almansour, H., 2021. Temperature effects on the mechanical properties of the GFRP sheets. In: International Conference on Fibre-Reinforced Polymer (FRP) Composites in Civil Engineering. 27 Nov 2021. Springer International Publishing, Cham, pp. 809–822.

Table 10–2 Equations for predicting fiber-reinforced polymer (FRP) materials at high temperatures and moderate strain rates.

Formula content (equation no.)	Function
$\sigma_{dynamic} = \sigma_{static}(\phi_\sigma \log_{10}(\varepsilon) + \beta_\sigma)$ (10.1) $E_{djmamic} = E_{static}(\varphi_E \log_{10}(\varepsilon) + \beta_E)$ (10.2)	Eqs. (10.1) and (10.2) employ logarithmic functions to dissect the influence of strain rate on FRP composites (Al-Zubaidy et al., 2013)
$\sigma_{dynamic} = a_\sigma + b_\sigma(\varepsilon)^{\gamma_\sigma}$ (10.3) $E_{dynamic} = a_E + b_E(\varepsilon)^{\gamma_E}$ (10.4)	Eqs. (10.3) and (10.4) are harnessed to anticipate the dynamic tensile strength and Young's modulus of FRP materials, respectively
$\sigma_{dynamic} = a_\sigma + b_\sigma(\varepsilon)$ (10.5) $E_{dynamic} = a_E + b_E(\varepsilon)$ (10.6)	The sensitivity to strain rate can also be characterized through linear formulations, as exemplified by Eqs. (10.5) and (10.6), thereby providing a more comprehensive understanding of the mechanical behavior under dynamic loading conditions (Ou and Zhu, 2015)

10.4 Comparing the results of each model for predicting fiber-reinforced polymer tensile properties

Drawing inspiration from the model advanced by Gibson et al. (2006), the researchers devised an analogous framework, grounded in their experimental findings (Bazli and Abolfazli, 2020). The generalized model proposed by Gibson et al. is articulated as follows, offering a comprehensive depiction of the mechanical behavior under consideration.

$$P(T) = \frac{P_u + P_R}{2} - \frac{P_u - P_R}{2}\,\mathrm{erf}(k(T - T')) \tag{10.7}$$

$$P(T) = \frac{P_u + P_R}{2} - \frac{P_u - P_R}{2}\,\tanh(k(T - T')) \tag{10.8}$$

In this context, *P(T)* denotes the specific property of the composite at temperature *T*, where P_u and P_R represent the low- and high-temperature properties, respectively. The variation is governed by the distribution constant *k*, while *T′* indicates the identified glass transition temperature at which property transitions occur. Table 10–3 summarizes various mathematical models proposed to predict FRP tensile properties at high temperatures.

Fig. 10–2 illustrates a comparative analysis of the predictions derived from the aforementioned models. Notably, the modulus of elasticity in FRP composites exhibits lesser sensitivity to elevated temperatures compared to their tensile strength.

10.5 New mathematical models for predicting the mechanical properties of fiber-reinforced polymer at elevated temperatures

10.5.1 Predictive modeling of tensile strength

10.5.1.1 Prediction modeling of glass fiber–reinforced polymer tensile strength at high temperatures

The testing was conducted on the glass-fiber-reinforced polymer (GFRP) sheet composites up to a temperature of 250°C, which is slightly below the decomposition temperature of epoxy (Qureshi et al., 2021). This analytical model is best suited to use sigmoid curves. Calibrated to

Table 10–3 Equations for each model predicting fiber-reinforced polymer (FRP) tensile properties.

Reference	Formula content		Function
Yu and Kodur (2014)	$R(T) = 0.56 - 0.44\tan h(0.0052(T - 305))$ $R(E) = 0.51 - 0.49\tan h(0.0035(T - 340))$	(10.9) (10.10)	Forecast the tensile strength and elastic modulus, respectively, of CFRP pultruded strips
Saafi (2002)	$R(T) = 1 - 0.0025\,T\ \ 0 < T \leq 400\ °C$ $R(E) = \begin{cases} 1 & 0 \leq T \leq 100°C \\ 1.25 - 0.025T & 100°C \leq T \leq 300°C \\ 2 - 0.005T & 300°C \leq T \leq 400°C \end{cases}$	(10.11) (10.12)	Predictive tools for estimating the tensile strength and elastic modulus of GFRP bars, respectively, provide valuable insights into the mechanical behavior of these materials
Wang et al. (2011)	$R(T) = \begin{cases} 1 - \frac{(T-22)^{0.9}}{200} & 22°C \leq T \leq 150°C \\ 0.59 - \frac{(T-150)^{0.7}}{490} & 150°C \leq T \leq 420°C \\ 0.48 - \frac{(T-420)^{1.8}}{76000} & 420°C \leq T \leq 706°C \end{cases}$	(10.13)	Anticipate the correlation between temperature and tensile strength in CFRP laminates
Hawileh et al. (2016)	$R(T) = 0.795 - 0.205\tan h(0.075(T - 190.58))$ $R(E) = 0.86 - 0.140\tan h(0.035(T - 163.24))$	(10.14) (10.15)	Forecast the tensile strength and modulus of BFRP sheets, respectively, subsequent to their exposure to elevated temperatures
Jafari et al. (2019)	$R(T) = \begin{cases} 1 & 24°C \leq T \leq 45°C \\ a\left(\frac{1}{T^3}\right) + b((\log(t))^{0.333}) + c & 45°C \leq T \leq 500°C \end{cases}$	(10.16)	Estimate the tensile strength of GFRP laminates, taking into account various sample thicknesses and fiber orientations
Ashrafi et al. (2020)	$R(T) = \mathrm{a}\left(\frac{1}{\mathrm{T}^3}\right) + b\left(\frac{1}{\left(\log\left(\frac{t_1}{6}\right)\right)^{0.5}}\right) - c\left(\frac{1}{(\log(t_2))^{0.5}}\right) + d$ $R(T) = -\mathrm{a}(\mathrm{T}^4) + b\left(\frac{1}{\left(\log\left(\frac{t_1}{6}\right)\right)^{0.5}}\right) - c\left(\frac{1}{(\log(t_2))^{0.5}}\right) + d$	(10.17) (10.18)	Prognosticate the tensile strength and elastic modulus of GFRP laminates, respectively, accounting for this variable

CFRP, Carbon fiber–reinforced polymer; *GFRP*, glass fiber–reinforced polymer.

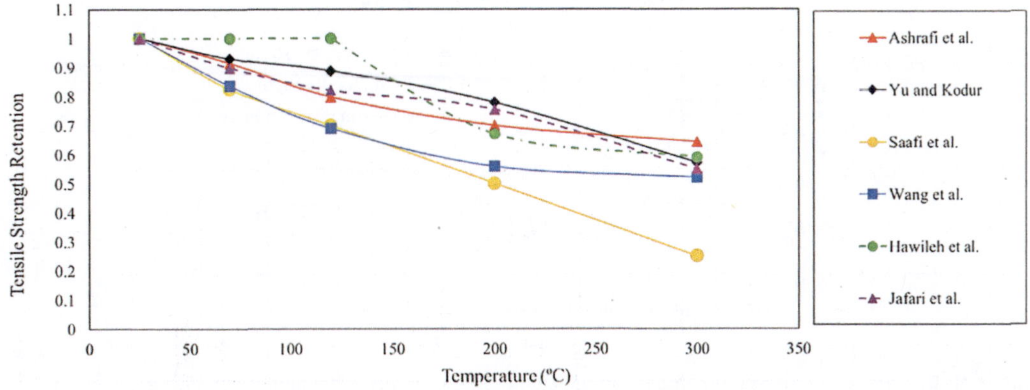

(A)

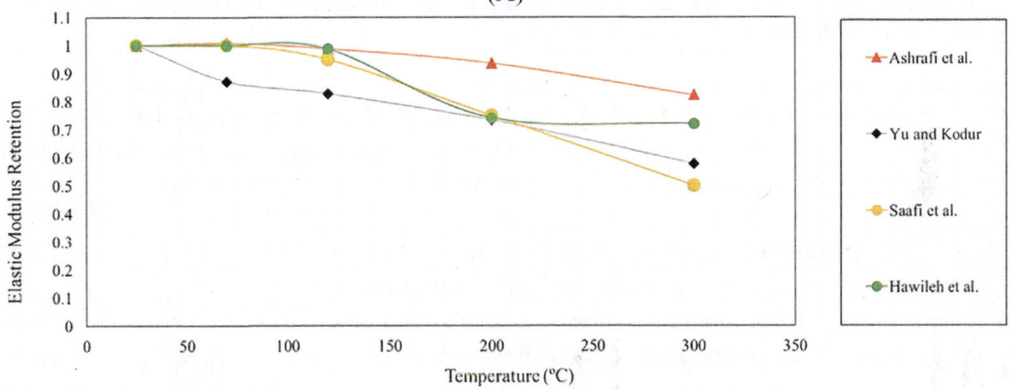

(B)

FIGURE 10–2 Comparison of predicting models: variation in (A) tensile strength retention with temperature and (B) tensile elastic modulus retention with temperature (Bazli and Abolfazli, 2020). *From Bazli, M., Abolfazli, M., 2020. Mechanical properties of fibre reinforced polymers under elevated temperatures: an overview. Polymers 12 (11), 2600. https://doi.org/10.3390/polym12112600.*

the experimental results, the following temperature function model was developed (Qureshi et al., 2021).

$$\sigma = (\sigma_u + \sigma_r)/2 - (\sigma_u - \sigma_r)/2 \times \tanh\left(s \times (T - T_{\text{sym}})\right) \quad (10.19)$$

The equation expresses the tensile strength (σ) of the GFRP at any given temperature (T), with σ_u as the ultimate strength at ambient conditions, σ_r denoting the residual strength post glass transition, and s, T, and T_{sym} signifying the shape constant, the evaluation temperature, and the symmetric temperature of the sigmoid curve, respectively.

This model excels at predicting the mechanical properties of GFRP composites, encompassing tensile strength and modulus, up to the decomposition point of the epoxy. Fig. 10–3 illustrates the

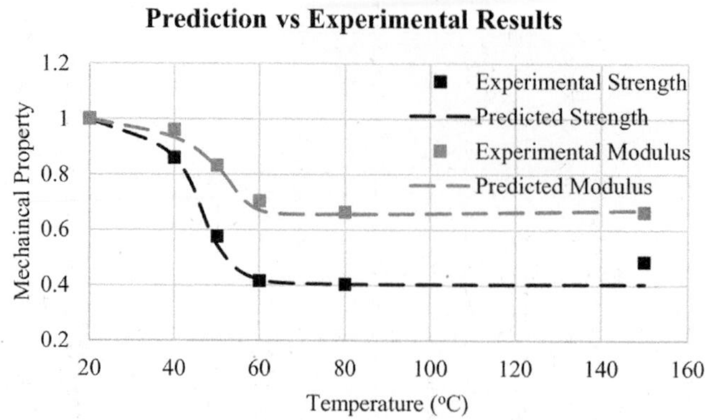

FIGURE 10–3 Comparison of experimental and predicted tensile strength and modulus (Qureshi et al., 2021). *From Qureshi, M.F., Sheikh, S.A., Almansour, H., 2021. Temperature effects on the mechanical properties of the GFRP sheets. In: International Conference on Fibre-Reinforced Polymer (FRP) Composites in Civil Engineering. 27 Nov 2021. Springer International Publishing, Cham, pp. 809–822.*

predictions generated by the model, demonstrating its accuracy in estimating the mechanical behavior of GFRP at elevated temperatures. It is important to emphasize that the applicability of this model is limited to temperatures below the decomposition threshold of the epoxy.

10.5.1.2 Semiempirical model to predict the tensile strength of unidirectional carbon fiber–reinforced polymer composites at high temperatures

A modified semiempirical model was formulated to characterize the temperature-dependent tensile strength of unidirectional carbon fiber-reinforced polymer (CFRP) composites, grounded in the hyperbolic tangent function (y=tanh($-x$)). This foundational equation was adapted to align with the inherent material attributes, resulting in a refined expression that encapsulates all pertinent parameters (a, b, c, and d).

$$\sigma = a\tanh[-b(T - c)] + d \tag{10.20}$$

Integration of these parameters into Eq. (10.20) yielded the comprehensive formulation, Eq. (10.20), which describes the tensile strength of these composites across varying temperatures.

$$\sigma = \frac{1}{2}(\sigma_0 - \sigma_r)\tanh\left[-\frac{1}{\Delta T/2}\left(T - \left(T_g + \frac{\Delta T}{2}\right)\right)\right] + \frac{1}{2}(\sigma_0 + \sigma_r) \tag{10.21}$$

The predictive prowess of (10.21) was validated by juxtaposing its tensile strength forecasts against empirical data, as depicted in Fig. 10–4. This comparison revealed a substantial concordance between the model's projections and the experimentally observed tensile strengths.

10.5.1.3 Improved model for predicting the mechanical properties of carbon fiber–reinforced polymer at high temperatures

Manalo et al. (2017) introduced an advanced exponential function-based model to mimic the thermophysical attributes of CFRP composites at heightened temperatures. This refined approach integrates the multifaceted influences of fiber volume fraction, resin softening, and

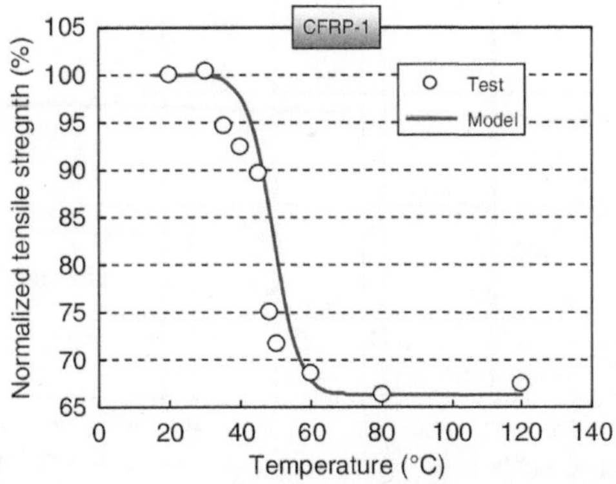

FIGURE 10–4 Comparison of the predicted and experimental tensile strengths of carbon fiber-reinforced polymer (CFRP) sheets (Cao et al., 2011). *From Cao, S., Wang, X., Wu, Z., 2011. Evaluation and prediction of temperature-dependent tensile strength of unidirectional carbon fiber-reinforced polymer composites. Journal of Reinforced Plastics and Composites 30 (9), 799–807. https://doi.org/10.1177/0731684411411002.*

its eventual decomposition, thereby providing a holistic perspective on the mechanical behavior of CFRP composites subjected to elevated temperatures. The model encapsulates the intricate interplay of these factors, enabling more accurate predictions tailored to engineering applications under extreme conditions.

$$\frac{P(T)}{P_0} = (1 - V_f)\exp\left[-k_1\left(\frac{T - T_0}{T_g}\right)^3\right] + V_f \exp\left[-k_2\left(\frac{T - T_0}{T_d}\right)^3\right] \tag{10.22}$$

where k_1 and k_2 represent constants derived through regression analysis of experimental data, whereas V_f signifies the volume fraction of fibers.

The plots in Fig. 10–5 depict the strength degradation profiles attained using various empirical formulations, consistently aligning with the experimental outcomes, thereby attesting to the models' robust fit and predictive capabilities.

10.5.2 Predictive modeling of flexural and shear mechanical properties

10.5.2.1 Empirical modeling of the mechanical properties of glass fiber–reinforced polymer extruded laminates in longitudinal and transverse directions at different temperatures

Based on regression analysis based on experimental and other results, researchers proposed empirical models for predicting the modulus and its corresponding deflection of GFRP sheets under medium and high temperature conditions (Manalo et al., 2017). The following equations were proposed.

Along the pultrusion direction:

$$\begin{aligned}
&\Delta_{L0} = \frac{PL^3}{48E_{L0}I} + \frac{PL}{4G_{L0}A} \\
&E_{L0} = E_{\mathrm{LO-RT}}(-0.0008T + 1.0152)\,\mathrm{whenRT} \le T \le 100°\mathrm{C} \\
&E_{L0} = E_{\mathrm{LO-RT}}(-0.0101T + 1.9184)\,\mathrm{when}100°\mathrm{C} < T \le 150°\mathrm{C} \\
&G_{L0} = G_{\mathrm{LO-RT}}(-0.0066T + 1.0926)\,\mathrm{whenRT} \le T \le 150°\mathrm{C}
\end{aligned} \tag{10.23}$$

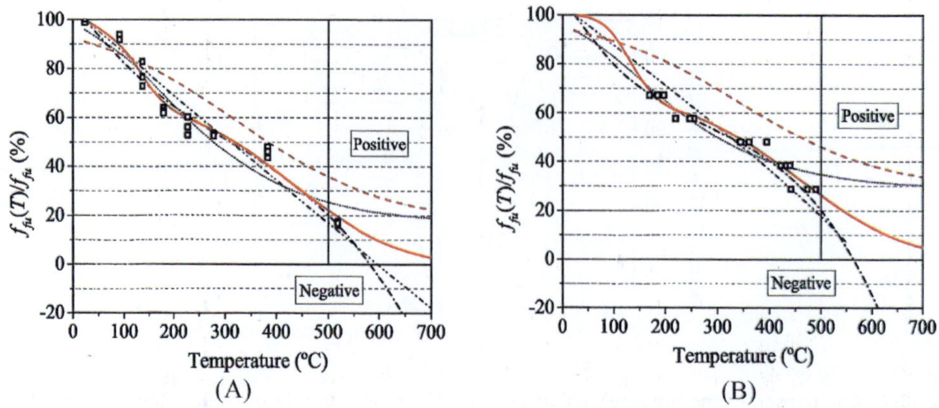

FIGURE 10–5 Nominal strength-temperature modeling curves (Zhou et al., 2019). (A) Steady state tests. (B) Transient state tests. *From Zhou, F., Zhang, J., Song, S., Yang, D., Wang, C., 2019. Effect of temperature on material properties of carbon fiber reinforced polymer (CFRP) tendons: experiments and model assessment. Materials 12 (7), 1025. https://doi.org/10.3390/ma12071025.*

Perpendicular to the pultrusion direction:

$$\begin{aligned} \Delta_{\mathrm{Tr}} &= \frac{PL^3}{48E_{\mathrm{Tr}}I} + \frac{PL}{4G_{\mathrm{Tr}}A} \\ E_{\mathrm{Tr}} &= E_{\mathrm{Tr-RT}}(-0.0053T + 1.1262)\mathrm{whenRT} \le T \le 100^\circ\mathrm{C} \\ E_{\mathrm{Tr}} &= E_{\mathrm{Tr-RT}}(-0.0105T + 1.6064)\mathrm{when}100^\circ\mathrm{C} < T \le 150^\circ\mathrm{C} \\ G_{\mathrm{Tr}} &= G_{\mathrm{Tr-RT}}(-0.0074T + 1.1018)\mathrm{whenRT} \le T \le 150^\circ\mathrm{C} \end{aligned} \tag{10.24}$$

As depicted in Fig. 10–6, the interlaminar shear strength (ILSS) and flexural strength of GFRP laminates exhibit a discernible variation with temperature. By observing a linear decline in both the ILSS and flexural strength across a range from 23°C to 150°C for specimens cut both longitudinally and transversely, predictive equations have been formulated. These equations facilitate the estimation of ILSS and flexural strength at higher temperatures for GFRP laminates.

Along the pultrusion direction:

$$\begin{aligned} \mathrm{ILSS}_{\mathrm{LO}} &= \mathrm{ILSS}_{\mathrm{LO-RT}}(-0.0211T + 1.4766) \ \text{ when } \mathrm{RT} \le T \le 100^\circ\mathrm{C} \\ \sigma_{\mathrm{LO}} &= \sigma_{\mathrm{LO-RT}}(-0.0073T + 1.1484) \ \text{ when } \mathrm{RT} \le T \le 150^\circ\mathrm{C} \end{aligned} \tag{10.25}$$

Perpendicular to the pultrusion direction:

$$\begin{aligned} \mathrm{ILSS}_{\mathrm{Tr}} &= \mathrm{ILSS}_{\mathrm{Tr-RT}}(-0.0109T + 1.2147) \ \text{ when } \mathrm{RT} \le T \le 70^\circ\mathrm{C} \\ \sigma_{\mathrm{Tr}} &= \sigma_{\mathrm{Tr-RT}}(-0.0073T + 1.1484) \ \text{ when } \mathrm{RT} \le T \le 100^\circ\mathrm{C} \end{aligned} \tag{10.26}$$

10.5.2.2 Constitutive deformation model of the mechanical behavior of glass fiber–reinforced polymer composites at high temperatures

The mechanical performance of the modified GEVIPN (glass fiber reinforced epoxy-vinyl ester interpenetrating polymer network) composite at high temperatures can be statistically characterized using the Weibull probability distribution function (Gupta K et al., 2022).

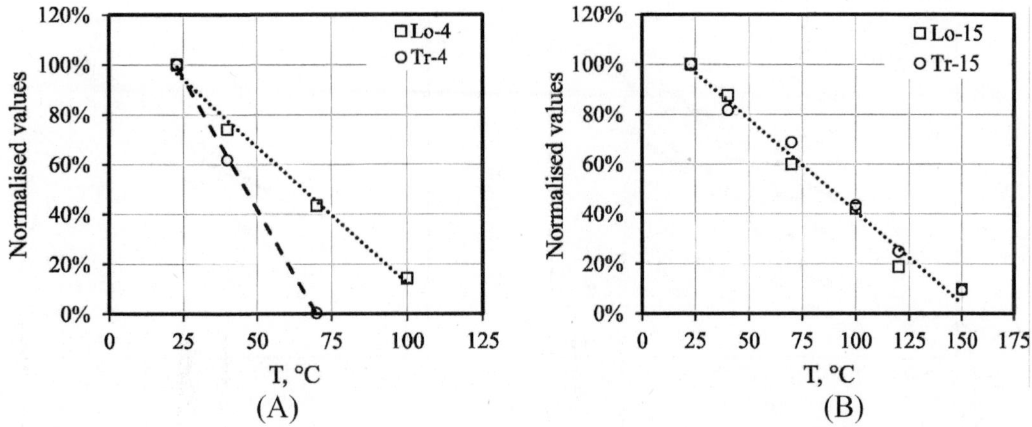

FIGURE 10–6 Interlaminar shear strength (ILSS) and flexural strength as a function of temperature (Manalo et al., 2017). (A) ILSS. (B) Flexural strength. *From Manalo, A., Maranan, G., Sharma, S., Karunasena, W., Bai, Y., 2017. Temperature-sensitive mechanical properties of GFRP composites in longitudinal and transverse directions: a comparative study. Composite Structures 173, 255–267. https://doi.org/10.1016/j.compstruct.2017.04.040.*

The connection between stress (σ) and strain (ε) within this framework is elucidated by Eq. (10.27), facilitating a comprehensive understanding of the material's behavior.

$$\sigma = E\varepsilon \exp\left[-\left(\frac{E\varepsilon}{\sigma_0}\right)^{\beta}\right] \tag{10.27}$$

Fig. 10–7 shows the Weibull linear equations of neat and modified glass fiber composites at different temperatures. Fig. 10–8 displays the shape and proportional parameters obtained from the corresponding fitting lines. At 30°C, 0.1 wt.% nano-Al_2O_3 in the GEVIPN composite shows the largest proportional parameter (Fig. 10–8A). This, like other values, is consistent with the bending strength obtained from bending measurements.

10.5.2.3 Evaluation of the validity of the prediction model for shear properties of glass fiber–reinforced polymer composites under high temperatures

Prior research endeavors (Rosa et al., 2018) have explored diverse modeling approaches to anticipate the mechanical attributes of FRP materials at heightened temperatures. These methodologies encompass (1) empirical mathematical models grounded in curve-fitting techniques applied to experimental data and (2) semiempirical methods. Correia et al. (2013) have critically appraised the veracity of these predictive models, specifically in the context of pultruded GFRP's temperature-dependent tensile, compressive, and 10 degrees off-axis shear strengths, introducing a Gompertz statistical distribution-based model.

$$P(T) = P_r + (P_u - P_r) \times (1 - e^{\mathrm{B}e^{\mathrm{C}\times T}}) \tag{10.28}$$

Furthermore, the pioneering rational modeling framework of Bai and Keller (2008) was originally designed to predict the elastic modulus and was innovatively adapted and extended to cover the strength degradation mechanism of GFRP at different temperatures (Bai and Keller, 2009).

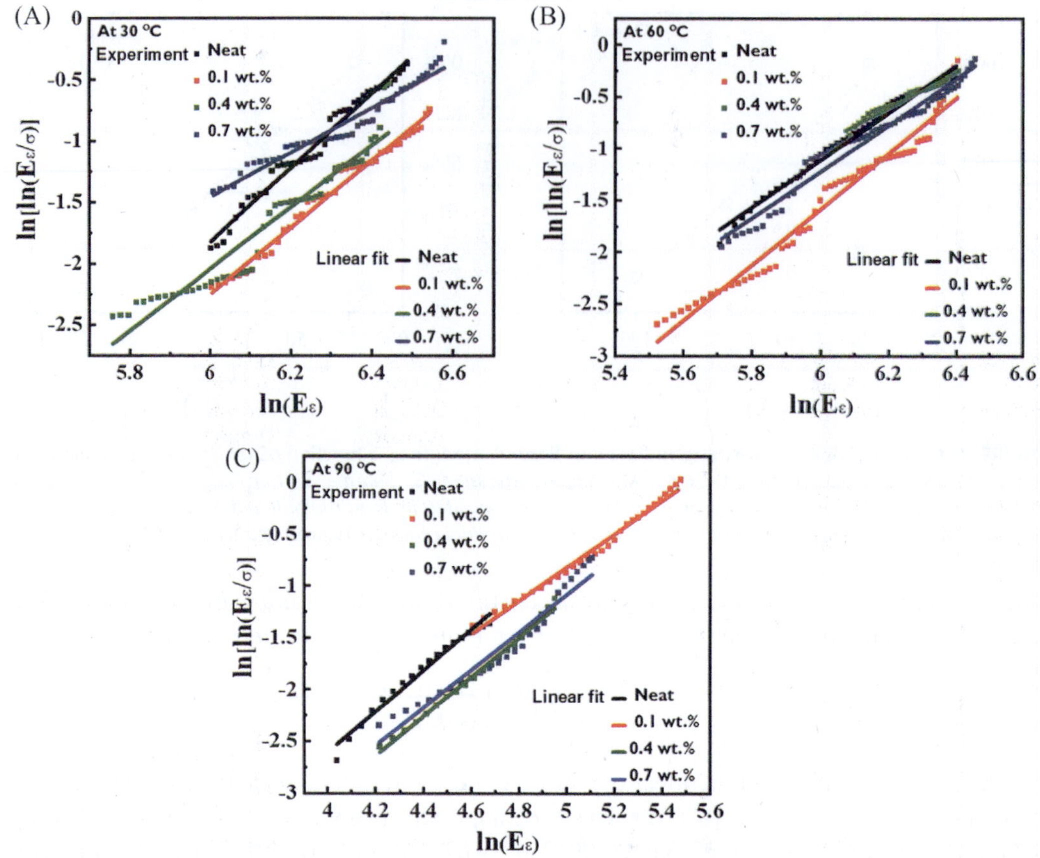

FIGURE 10–7 Weibull fittings of experimental data for neat and modified glass fiber composites at (A) 30°C, (B) 60°C, and (C) 90°C (Gupta K et al., 2022). *From Gupta, K.B.N.V.S.G., Patnaik, S., Ray, B.C., Rai, R.K., Prusty, R.K., 2022. Elevated temperature mechanical behavior of nano Al_2O_3 embedded interpenetrating polymer network/glass fiber composites. Journal of Applied Polymer Science 139 (41), e52991.*

This adaptation recognizes that the strength at the glass transition stage is closely related to the inherent properties of the material before and after this critical stage. Using the mixing law Eq. (10.29) and its inverse Eq. (10.30), two different models are proposed to provide upper and lower bounds for estimating the effective properties of this bimodal material.

$$P(T) = P_g \times [1 - \alpha_g(T)] + P_l \times \alpha_g(T) \times [1 - \alpha_d(T)] + P_d \times \alpha_g(T) \times \alpha_d(T) \tag{10.29}$$

$$\frac{1}{P(T)} = \frac{1 - \alpha_g(T)}{P_g} + \frac{\alpha_g(T) \times [1 - \alpha_d(T)]}{P_l} + \frac{\alpha_g(T) \times \alpha_d(T)}{P_d} \tag{10.30}$$

Table 10–4 consolidates the parameters derived from these assorted models, alongside the absolute mean percentage error (AMPE) metric, which serves as a quantitative assessment of each model's fidelity when compared against experimental benchmarks.

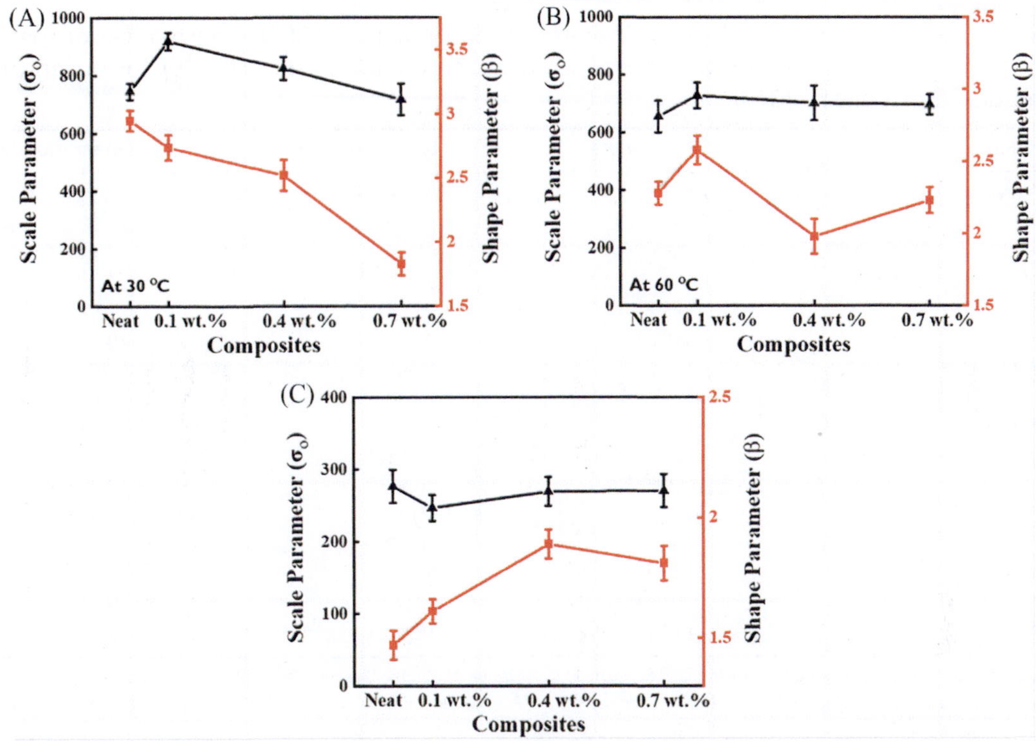

FIGURE 10–8 Variation in the Weibull scale (σ) and shape (β) parameters for neat and modified glass fiber composites at (A) 30°C, (B) 60°C, and (C) 90°C (Gupta K et al., 2022). *From Gupta, K.B.N.V.S.G., Patnaik, S., Ray, B.C., Rai, R.K., Prusty, R.K., 2022. Elevated temperature mechanical behavior of nano Al_2O_3 embedded interpenetrating polymer network/glass fiber composites. Journal of Applied Polymer Science 139 (41), e52991.*

Parameters for different models are shown in Fig. 10–9. The experimental results show that these models can estimate the shear mechanical properties accurately.

10.5.3 Fatigue prediction models: fatigue modeling of a triaxially braided polymer matrix composite at elevated temperatures

A fatigue damage accumulation model, rooted in the principles of damage mechanics, has been constructed utilizing empirical damage observations (Montesano et al., 2013). Elevated temperatures fundamentally modulate the progression of various fatigue damage modalities in braided composites (Mao and Mahadevan, 2002; Wu and Yao, 2010), necessitating the development of models that capture these intricacies. A strategic approach involves fractionating the simulation of material degradation into discrete damage components, such as yarn cracks and interface delamination, which fosters a robust physical alignment between the damage model and the actual degradation patterns manifested in braided materials (Bartley-Cho et al., 1998; Gagel et al., 2006). This approach ensures that the model not only mirrors the experimental observations but also offers a deeper understanding of the fatigue behavior under high-temperature conditions.

Table 10–4 Simulation of glass fiber–reinforced polymer (GFRP) mechanical properties: parameter estimation and absolute mean percentage error (AMPE) for different models (Rosa et al., 2018).

References	Parameters	Shear strengths	Shear moduli
Gibson et al. (2006)	k'	0.0205	0.0195
	$T_{g.mech}$ (°C)	74.86	79.08
	AMPE (%)	8.3	17.9
Mahieux et al. (2001)	m	9	9
	To (K)	365.41	370.35
	AMPE (%)	9.1	20.7
Wang et al. (2011)	A	1.00	1.00
	B	18.00	18.00
	C	29.39	25.94
	n	0.6518	0.6104
	AMPE (%)	10.9	19.3
Correia et al. (2013)	B	−5.10	−4.99
	C	−0.03	−0.03
	AMPE (%)	6.2	14.4
Bai and Keller (2009)	AMPE (%)	52.2	49.2
Bai et al. (2008)	AMPE (%)	15.6	23.9

Source: From Rosa, I.C., Morgado, T., Correia, J.R., Firmo, J.P., Silvestre, N., 2018. Shear behavior of GFRP composite materials at elevated temperature. Journal of Composites for Construction 22 (3). https://doi.org/10.1061/(asce)cc.1943-5614.0000839.

The interfacial crack density of braided yarns and yarns can be estimated by Eqs. (10.31) and (10.32). Eqs. (10.31) and (10.32) can be used to determine the interfacial crack density distribution of braided yarns and yarns for any maximum stress at room or elevated temperatures.

$$\rho_{\mathrm{by}} = a_1 \mathrm{n}^{a_2} \tag{10.31}$$

$$\rho_{\mathrm{int}} = a_3 n^{a_4} \tag{10.32}$$

Analytical cumulative damage models were developed using the observed crack densities at the braided yarn and yarn interface defined by Eqs. (10.31) and (10.32), respectively. Defining the damage model in terms of crack density accurately reflects the actual damage state of the material. The fatigue damage analysis model is derived from the following Eq. (10.33).

$$D = \mathrm{d}_1 + d_2 = \alpha \rho_{by}^{\beta_1} + \alpha \rho_{\mathrm{int}}^{\beta_2} \tag{10.33}$$

In the aforementioned fatigue damage model, the cumulative damage D is conceptualized as the additive decomposition of distinct damage parameters, namely, d_1 and d_2, which individually represent the detrimental effects stemming from yarn cracks and yarn interface delamination, respectively.

It is noteworthy that the versatility of this model extends to the incorporation of additional damage mechanisms, such as matrix cracking (exemplified by d_3, d_4, and so forth), should

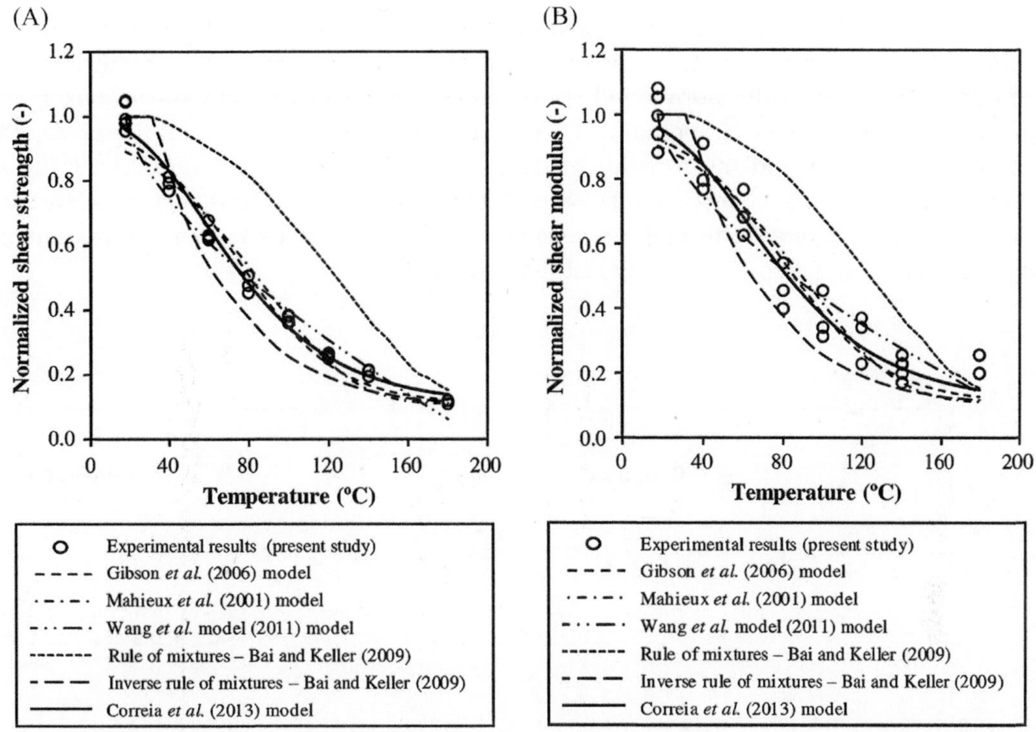

FIGURE 10–9 Experimental results and modeling curves: (A) normalized shear strength (compared to room temperature, τ_{18}=75.7 MPa) versus temperature and (B) normalized shear modulus (compared to room temperature, G_{18}=3.5 GPa) versus temperature (Rosa et al., 2018). *From Rosa, I.C., Morgado, T., Correia, J.R., Firmo, J.P., Silvestre, N., 2018. Shear behavior of GFRP composite materials at elevated temperature. Journal of Composites for Construction 22 (3). https://doi.org/10.1061/(asce)cc.1943-5614.0000839.*

they be deemed pertinent. The coefficient α acts as a scaling factor, remaining unaffected by the applied stress, whereas the material parameters β_1 and β_2 exhibit a functional dependency on the maximum stress, contingent upon the test temperature. Consequently, a fatigue damage model, grounded in Eq. (10.33), has been formulated, enabling the prognostication of both the progressive and consequential damage incurred by braided yarns under any given maximum stress, irrespective of whether it is encountered at ambient or elevated temperatures.

By delineating cumulative damage parameters, it becomes feasible to anticipate the material properties of composites. Leveraging the fundamental tenets of damage mechanics, damage indices can be derived, with particular emphasis on material stiffness. A commonly employed damage index in the realm of scientific literature, denoted as DI, is grounded in standardized stiffness metrics, thereby offering a quantitative assessment of the extent of damage.

$$E(n) = E_o[1 - \alpha(\rho_{\text{by}}^{\beta_1} + \rho_{\text{int}}^{\beta_2})] \tag{10.34}$$

The predictions derived from the analytical model, as formulated in Eq. (10.34), can be validated by comparing them with experimental data. Fig. 10–10 illustrates the predicted and experimental values of the normalized stiffness degradation of the samples at maximum applied stress at ambient temperature. There is a clear correlation between the predicted stiffness degradation and the experimentally measured stiffness degradation at all stress levels. Additionally, the model precisely forecasts the pronounced stiffness degradation at the higher test temperatures, affirming its versatility for application under various maximum applied stresses and across a range of test temperatures.

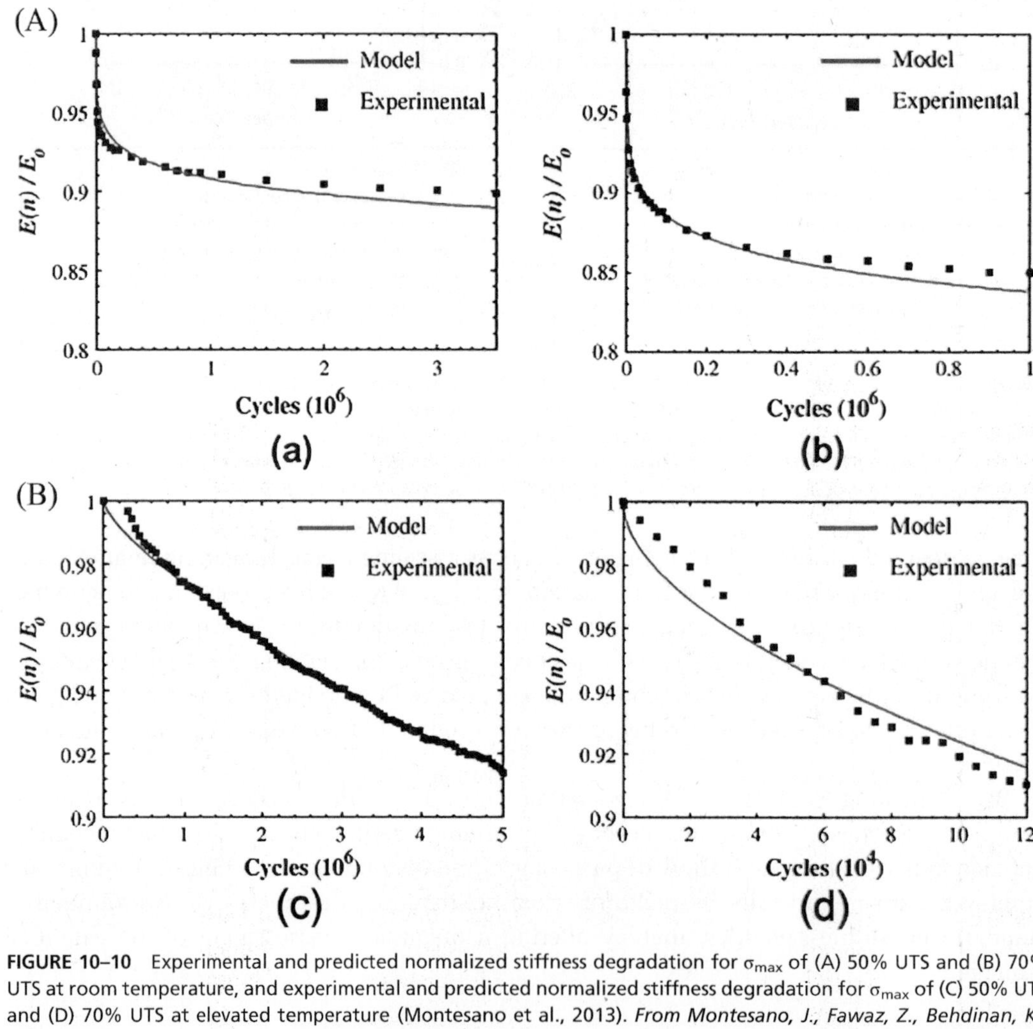

FIGURE 10–10 Experimental and predicted normalized stiffness degradation for σ_{max} of (A) 50% UTS and (B) 70% UTS at room temperature, and experimental and predicted normalized stiffness degradation for σ_{max} of (C) 50% UTS and (D) 70% UTS at elevated temperature (Montesano et al., 2013). *From Montesano, J., Fawaz, Z., Behdinan, K., Poon, C., 2013. Fatigue damage characterization and modeling of a triaxially braided polymer matrix composite at elevated temperatures. Composite Structures 101, 129–137. https://doi.org/10.1016/j.compstruct.2013.01.030.*

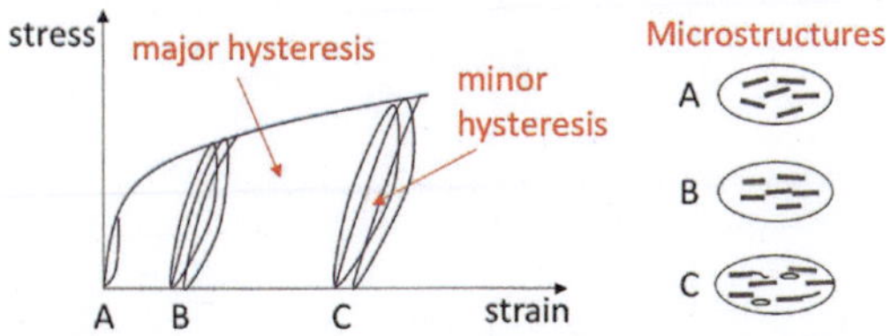

FIGURE 10–11 Macroscopic responses of materials undergoing microstructural changes (Song et al., 2022). *From Song, R., Morak, M., Muliana, A., 2022. A thermo-viscoelastic model of anisotropic polyamide short glass fiber composites. Composite Structures 296, 115850. https://doi.org/10.1016/j.compstruct.2022.115850.*

10.5.4 Creep prediction modeling: thermo-viscoelastic model of anisotropic polyamide short glass fiber composites

Fig. 10–11 illustrates a constitutive material model designed to replicate the macroscopic response of polyamide composites under mechanical stress, specifically tailored for anisotropic viscoelastic materials experiencing microstructural transformations (Song et al., 2022). This model encapsulates the intricate behavior exhibited by these composites, particularly when subjected to mechanical loading, thereby facilitating a comprehensive understanding of their performance under operational conditions.

The formulation of anisotropic, temperature-sensitive viscoelastic materials that undergo microstructural evolution begins with the delineation of Gibbs free energy.

$$G = G(\sigma_V, \sigma_E, T, \alpha) = G_1(\sigma_V, T, \alpha) + G_2(\sigma_E, T, \alpha) \tag{10.35}$$

Here, G_1 represents the Gibbs free energy associated with the viscoelastic component, which possesses the dual capability of energy storage and dissipation. In contrast, G_2 signifies the Gibbs free energy exclusive to the elastic component, responsible solely for energy storage. The viscoelastic and elastic stress responses are denoted by σ_V and σ_E, respectively, with T indicating the ambient temperature. The transformation trajectory from the initial to the ultimate microstructural configuration is parameterized by the scalar α, ranging from 0 to 1. This parameter α serves as a proxy for the volumetric fraction of the nascent (ultimate) microstructural configuration. At $\alpha=0$, the composite's response is solely governed by the initial microstructural configuration, whereas at $\alpha=1$, it is exclusively dictated by the ultimate microstructural state. Consequently, the Gibbs potentials corresponding to the initial and ultimate microstructural configurations are designated as $G(\sigma_V, \sigma_E, T, 0)$ and $G(\sigma_V, \sigma_E, T, 1)$, respectively.

Subsequently, the material's internal energy is formulated by deriving it from the Gibbs free energy framework.

$$u = G + T\eta - \frac{\partial G}{\partial \sigma_V} \cdot \sigma_V - \frac{\partial G}{\partial \sigma_E} \cdot \sigma_E \tag{10.36}$$

$$\xi = \sigma \cdot \varepsilon - u + T\eta \geq 0 \tag{10.37}$$

$$\varepsilon = \varepsilon_E + \varepsilon_V + \varepsilon_P \tag{10.38}$$

$$\varepsilon_E = -\frac{\partial G}{\partial \sigma_E} \tag{10.39}$$

$$\varepsilon_V = \varepsilon_V = \varepsilon_V = -\frac{\partial G}{\partial \sigma_V} \tag{10.40}$$

$$\sigma = \sigma_E = \sigma_V + \sigma_V \tag{10.41}$$

$$\xi|_{T\,\text{fixed}} = (\sigma - \sigma_V)\cdot\dot{\varepsilon}_V + \left(\sigma\cdot\dot{\varepsilon}_P - \frac{\partial G}{\partial \alpha}\dot{\alpha}\right) = \sigma V\cdot\dot{\varepsilon}_V + \left(\sigma\cdot\dot{\varepsilon}_P - \frac{\partial G}{\partial \alpha}\dot{\alpha}\right) = \xi_D + \xi_\alpha \tag{10.42}$$

$$\xi_\alpha = \left(\sigma\cdot\varepsilon_P - \frac{\partial \dot{G}}{\partial \alpha}\dot{\alpha}\right) = \left(\sigma\cdot\frac{\partial \varepsilon_P}{\partial \alpha} - \frac{\partial G}{\partial \alpha}\right)\dot{\alpha} = F_\alpha \dot{\alpha} \tag{10.43}$$

Eqs. (10.35–10.43) describe a model formulation for viscoelastic materials undergoing microstructural changes.

Fig. 10–12 illustrates the predicted axial and lateral creep-recovery behaviors of anisotropic short glass fiber polyamide composites subjected to a 90 degrees load direction. In general, the model provides accurate forecasts of these responses. Constitutive models that account for the anisotropic nature and multiple configurations can satisfactorily anticipate the time- and temperature-sensitive responses observed in glass-polyamide composites.

10.6 Conclusions and future perspectives

This study delves into a comprehensive evaluation of mechanical property models applicable to FRPs, particularly CFRPs and GFRPs, in high-temperature settings.

The research centers on utilizing mathematical frameworks to forecast tensile, shear, fatigue, and creep attributes of CFRP/GFRP at elevated temperatures, alongside validating the accuracy of these predictive models. Based on rigorous experimental data analysis, the models presented in various studies demonstrate their ability to adequately mirror the variations in the strength profiles of FRP composites across a spectrum of test temperatures. Drawbacks in the models, including simplified assumptions about the model and failure to consider complexity factors, are pointed out. Future research can address these issues to continue to improve and extend these models to better meet the needs of engineering practice.

In the future, the application of deep learning algorithms to predict the high-temperature mechanical properties of FRP materials can be explored to improve the models' prediction accuracy and generalization ability. The mechanical behavior of FRP materials at high temperatures can also be further investigated in depth by combining multiscale simulation and experimental validations. The study of high-temperature mechanical property models of FRP materials involves a number of interdisciplinary fields, such as materials science, mechanics, thermodynamics, and so on. In the future, various interdisciplinary fields can be further promoted to improve the accuracy and generalization ability of the models. To sum up, the mechanical property modeling of FRP materials at high temperatures shows a broad research scope and significant application potential. Through the in-depth application of

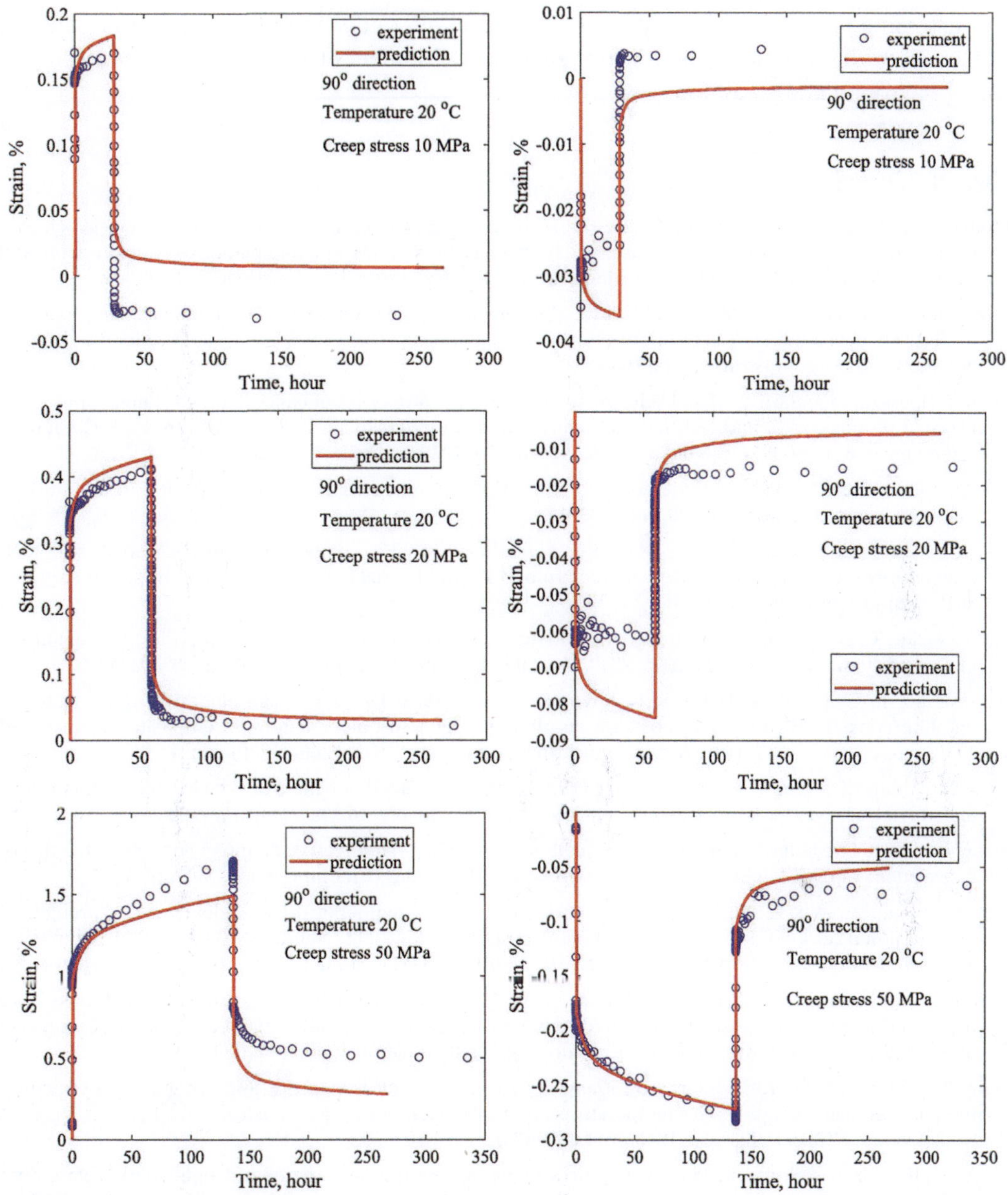

FIGURE 10–12 Creep-recovery response of glass-polyamide composites under loadings at 20°C (Song et al., 2022). *From Song, R., Morak, M., Muliana, A., 2022. A thermo-viscoelastic model of anisotropic polyamide short glass fiber composites. Composite Structures 296, 115850. https://doi.org/10.1016/j.compstruct.2022.115850.*

deep learning algorithms, the combination of multiscale simulation and experimental validation, an interdisciplinary approach, and techniques of continuous exploration and innovation, more accurate and reliable models and tools can be developed for the prediction and assessment of the performance of FRP materials under high-temperature conditions, which will further promote the safe and reliable development of engineering structures under harsh environmental conditions.

References

Ahmed, A., Zillur Rahman, M., Ou, Y., Liu, S., Mobasher, B., Guo, S., et al., 2021. A review on the tensile behavior of fiber-reinforced polymer composites under varying strain rates and temperatures. Construction and Building Materials 294, 123565. Available at https://doi.org/10.1016/j.conbuildmat.2021.123565.

Al-Zubaidy, H., Zhao, X.L., Al-Mahaidi, R., 2013. Mechanical characterisation of the dynamic tensile properties of CFRP sheet and adhesive at medium strain rates. Composite Structures 96, 153–164. Available at https://doi.org/10.1016/j.compstruct.2012.09.032.

Ashrafi, H., Bazli, M., Jafari, A., Ozbakkaloglu, T., 2020. Tensile properties of GFRP laminates after exposure to elevated temperatures: effect of fiber configuration, sample thickness, and time of exposure. Composite Structures 238, 111971. Available at https://doi.org/10.1016/j.compstruct.2020.111971.

Bai, Y., Keller, T. 2008. A kinetic model for predicting stiffness and strength of FRP composites in fire. https://infoscience.epfl.ch/handle/20.500.14299/33017

Bai, Y., Keller, T., Vallée, T., 2008. Modeling of stiffness of FRP composites under elevated and high temperatures. Composites Science and Technology 68 (15-16), 3099–3106. Available at https://doi.org/10.1016/j.compscitech.2008.07.005.

Bai, Y., Keller, T., 2009. Modeling of strength degradation for fiber-reinforced polymer composites in fire. Journal of Composite Materials 43 (21), 2371–2385. Available at https://doi.org/10.1177/0021998309344642.

Bartley-Cho, J., Gyu Lim, S., Hahn, H.T., Shyprykevich, P., 1998. Damage accumulation in quasi-isotropic graphite/epoxy laminates under constant-amplitude fatigue and block loading. Composites Science and Technology 58 (9), 1535–1547. Available at https://doi.org/10.1016/s0266-3538(97)00214-5.

Bazli, M., Abolfazli, M., 2020. Mechanical properties of fibre reinforced polymers under elevated temperatures: an overview. Polymers 12 (11), 2600. Available at https://doi.org/10.3390/polym12112600.

Bisby, L.A. 2003. Fire behaviour of fibre-reinforced polymer (FRP) reinforced or confined concrete. PhD thesis, Department of Civil Engineering, Queen's University, Kingston, Ontario, Canada, 371 pp

Cao, S., Wang, X., Wu, Z., 2011. Evaluation and prediction of temperature-dependent tensile strength of unidirectional carbon fiber-reinforced polymer composites. Journal of Reinforced Plastics and Composites 30 (9), 799–807. Available at https://doi.org/10.1177/0731684411411002.

Correia, J.R., Gomes, M.M., Pires, J.M., Branco, F.A., 2013. Mechanical behaviour of pultruded glass fibre reinforced polymer composites at elevated temperature: experiments and model assessment. Composite Structures 98, 303–313. Available at https://doi.org/10.1016/j.compstruct.2012.10.051.

Feng, P., Wang, J., Tian, Y., Loughery, D., Wang, Y., 2016. Mechanical behavior and design of frp structural members at high and low service temperatures. Journal of Composites for Construction 20 (5). Available at https://doi.org/10.1061/(asce)cc.1943-5614.0000676.

Gagel, A., Fiedler, B., Schulte, K., 2006. On modelling the mechanical degradation of fatigue loaded glass-fibre non-crimp fabric reinforced epoxy laminates. Composites Science and Technology 66 (5), 657–664. Available at https://doi.org/10.1016/j.compscitech.2005.07.037.

Gibson, A.G., Wu, Y.-S., Evans, J.T., Mouritz, A.P., 2006. Laminate theory analysis of composites under load in fire. Journal of Composite Materials 40 (7), 639–658. Available at https://doi.org/10.1177/0021998305055543.

Gupta K, B.N.V.S.G., Patnaik, S., Ray, B.C., Rai, R.K., Prusty, R.K., 2022. Elevated temperature mechanical behavior of nano Al2O3 embedded interpenetrating polymer network/glass fiber composites. Journal of Applied Polymer Science 139 (41). Available at https://doi.org/10.1002/app.52991.

Hawileh, R.A., Abdalla, J.A., Hasan, S.S., Ziyada, M.B., Abu-Obeidah, A., 2016. Models for predicting elastic modulus and tensile strength of carbon, basalt and hybrid carbon-basalt FRP laminates at elevated temperatures. Construction and Building Materials 114, 364–373. Available at https://doi.org/10.1016/j.conbuildmat.2016.03.175.

Jafari, A., Bazli, M., Ashrafi, H., Vatani Oskouei, A., Azhari, S., Zhao, X.L., et al., 2019. Effect of fibers configuration and thickness on tensile behavior of GFRP laminates subjected to elevated temperatures. Construction and Building Materials 202, 189–207. Available at https://doi.org/10.1016/j.conbuildmat.2019.01.003.

Mahieux, C.A., Reifsnider, K.L., Case, S.W., 2001. Property modeling across transition temperatures in PMC's: Part I. Tensile properties. Applied Composite Materials 8 (4), 217–234. Available at https://doi.org/10.1023/A:1011282704357.

Manalo, A., Maranan, G., Sharma, S., Karunasena, W., Bai, Y., 2017. Temperature-sensitive mechanical properties of GFRP composites in longitudinal and transverse directions: a comparative study. Composite Structures 173, 255–267. Available at https://doi.org/10.1016/j.compstruct.2017.04.040.

Mao, H., Mahadevan, S., 2002. Fatigue damage modelling of composite materials. Composite Structures 58 (4), 405–410. Available at https://doi.org/10.1016/s0263-8223(02)00126-5.

Montesano, J., Fawaz, Z., Behdinan, K., Poon, C., 2013. Fatigue damage characterization and modeling of a triaxially braided polymer matrix composite at elevated temperatures. Composite Structures 101, 129–137. Available at https://doi.org/10.1016/j.compstruct.2013.01.030.

Ou, Y., Zhu, D., 2015. Tensile behavior of glass fiber reinforced composite at different strain rates and temperatures. Construction and Building Materials 96, 648–656. Available at https://doi.org/10.1016/j.conbuildmat.2015.08.044.

Qureshi, M.F., Sheikh, S.A., Almansour, H., 2021. Temperature effects on the mechanical properties of the GFRP Sheets. In: Ilki, A., Ispir, M., Inci, P. (Eds.), 10th International Conference on FRP Composites in Civil Engineering. CICE 2021. Lecture Notes in Civil Engineering, vol 198. Springer International Publishing, Cham, pp. 809–822. Available at https://doi.org/10.1007/978-3-030-88166-5_70.

Rosa, I.C., Morgado, T., Correia, J.R., Firmo, J.P., Silvestre, N., 2018. Shear Behavior of GFRP composite materials at elevated temperature. Journal of Composites for Construction 22 (3). Available at https://doi.org/10.1061/(asce)cc.1943-5614.0000839.

Saafi, M., 2002. Effect of fire on FRP reinforced concrete members. Composite Structures 58 (1), 11–20. Available at https://doi.org/10.1016/s0263-8223(02)00045-4.

Song, R., Morak, M., Muliana, A., 2022. A thermo-viscoelastic model of anisotropic polyamide short glass fiber composites. Composite Structures 296, 115850. Available at https://doi.org/10.1016/j.compstruct.2022.115850.

Wang, K., Young, B., Smith, S.T., 2011. Mechanical properties of pultruded carbon fibre-reinforced polymer (CFRP) plates at elevated temperatures. Engineering Structures 33 (7), 2154–2161. Available at https://doi.org/10.1016/j.engstruct.2011.03.006.

Wu, F., Yao, W.X., 2010. A fatigue damage model of composite materials. International Journal of Fatigue 32 (1), 134–138. Available at https://doi.org/10.1016/j.ijfatigue.2009.02.027.

Yu, B., Kodur, V., 2014. Effect of temperature on strength and stiffness properties of near-surface mounted FRP reinforcement. Composites Part B: Engineering 58, 510–517. Available at https://doi.org/10.1016/j.compositesb.2013.10.055.

Zhou, F., Zhang, J., Song, S., Yang, D., Wang, C., 2019. Effect of temperature on material properties of carbon fiber reinforced polymer (CFRP) tendons: experiments and model assessment. Materials 12 (7), 1025. Available at https://doi.org/10.3390/ma12071025.

11

Computational modeling of aging effects of high-temperature composite materials

Oluwaseye Samson Adedoja[1,2], Emmanuel Rotimi Sadiku[1,2], Yskandar Hamam[3]

[1]DEPARTMENT OF CHEMICAL, METALLURGICAL AND MATERIALS ENGINEERING, TSHWANE UNIVERSITY OF TECHNOLOGY, PRETORIA, GAUTENG, SOUTH AFRICA [2]INSTITUTE OF NANOENGINEERING RESEARCH (INER), TSHWANE UNIVERSITY OF TECHNOLOGY, PRETORIA, GAUTENG, SOUTH AFRICA [3]ECOLE SUPERIEURE D'INGENIEURS EN ELECTROTECHNIQUE ET ELECTRONIQUE, 2 BOULEVARD BLAISE PASCAL, NOISY-LE-GRAND, FRANCE

11.1 Introduction

High-temperature composite materials (HTCMs) are essential in several engineering applications like aerospace, energy generation, and others where materials must maintain structural integrity under severe or extreme operating conditions. Computational modeling of aging effects is crucial for predicting service life, optimizing maintenance schedules, and developing more durable materials. There are several types of aging mechanisms that occur in high-temperature materials. These aging mechanisms significantly affect the properties and engineering performance of the composite materials. Examples of these mechanisms are microstructural evolution (such as phase transformation and grain growth), creep deformation, interface degradation between fibers and matrix, fatigue damage accumulation, and thermal oxidation and chemical degradation (Adedoja et al., 2023; Birks et al., 2006; Dong et al., 2024).

The thermal oxidation happens to be one of the primary degradation mechanisms in composite materials (particularly in polymer matrix composites [PMCs]) subjected to elevated temperatures. The oxidation process often begins at the surface and progresses inward, creating an "oxidation layer" with degraded mechanical properties (Pochiraju et al., 2012). The surface oxidation can lead to weight loss, shrinkage, and microcracking, and the rate of oxidation typically follows Arrhenius behavior, increasing exponentially with temperature. The matrix thermal degradation occur because the polymer matrix in composites is particularly susceptible to thermal degradation which is usually through chain scission (breaking of polymer chains) that reduces molecular weight and mechanical properties (Colin and Verdu, 2006), cross-linking that can initially increase stiffness but eventually leads to embrittlement and glass transition temperature (T_g) that may shift due to these chemical changes, affecting performance at elevated temperatures (Yang et al., 2023). Another degradation mechanism is the fiber–matrix interface

High-Temperature Polymeric Materials. DOI: https://doi.org/10.1016/B978-0-443-32792-6.00008-7

degradation. This degradation occurs when the interface between fibers and matrix, which is critical to composite performance, is severely affected by high temperatures. This gives rise to thermal stresses from coefficient of thermal expansion (CTE) mismatches that can weaken the interface (Sethi and Ray, 2015), chemical degradation of sizing agents or coupling agents at elevated temperatures that reduces interfacial strength, and moisture that penetrates to the interface that can cause additional degradation when exposed to high temperatures (Ray, 2006; Suyama et al., 2019)

Physical aging refers to the slow evolution toward thermodynamic equilibrium at temperatures below glass transition temperature. Physical aging causes free volume reduction, which leads to increased density and stiffness but decreased toughness. As a result of physical aging, the creep and stress relaxation rates change over time, affecting long-term dimensional stability. The physical aging effects can be partially or fully reversed by heating above glass transition temperature (T_g). The residual stresses from the manufacturing process can be significantly affected by high temperatures through the occurrence of stress relaxation that can potentially change the dimensions and warpage of the composite. The combination of residual stress and thermal stress leads to the formation of microcracking. The temperature-induced microcracking is a common issue in high-temperature composites due to the cumulative damage through repeated expansion and contraction as a result of thermal cycling. The effect of microcracks is that they provide pathways for environmental ingress, thus accelerating degradation, thereby increasing the possibility of critical crack density as a result of exposure time and temperature. Another form of physical aging is fiber degradation. While most reinforcing fibers are more thermally stable than polymer matrices, they can still degrade. For instance, glass fibers can devitrify at very high temperatures, while carbon fibers may oxidize in air at temperatures above 400°C. Moreover, ceramic fibers undergo phase transformations or grain growth at elevated temperatures (Adedoja et al., 2025; Afzal et al., 2023).

In real scenario applications, high temperature composites usually undergo synergistic effects of multiple aging mechanisms that often operate simultaneously, producing effects greater than the sum of individual mechanisms. For instance, degradation can be accelerated through a thermo-oxidative aging combined with mechanical loading, synergistic interaction of moisture and temperature effect, and radiation combination with thermal effects. To provide a simplified approach to the understanding of how the composite material reacts with various aging mechanisms, several computational modeling approaches have been provided to represent the system. Some of this modeling approaches are multiscale modeling frameworks, physics-based damage models, machine learning approaches, material-specific models, and validation approaches. These approaches are summarized in Fig. 11–1.

11.2 Aging effects of high-temperature composite materials

11.2.1 Thermal aging

High-temperature composites undergo several degradation mechanisms during thermal aging. These degradations include matrix degradation, fiber–matrix interface weakening,

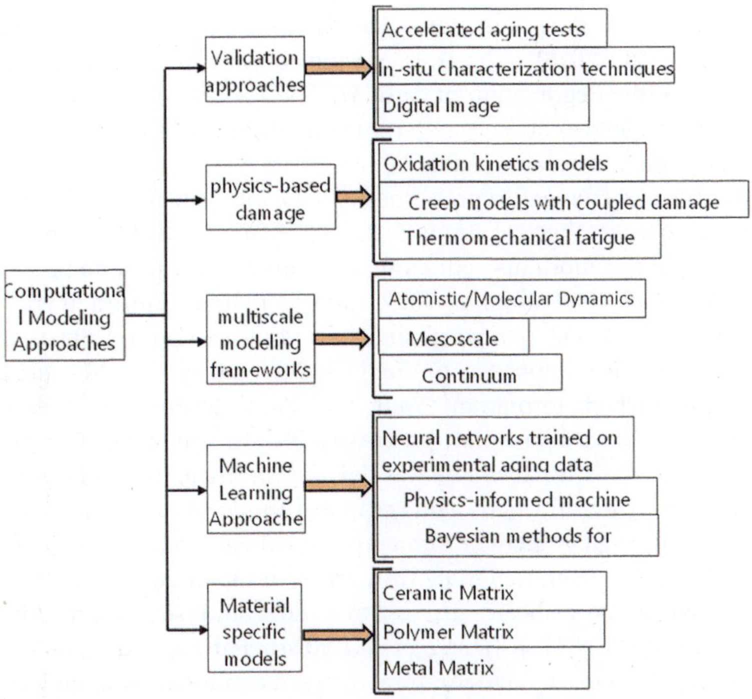

FIGURE 11–1 Computational modeling approaches for modeling high-temperature composites.

microstructural changes, and oxidation effects. In the matrix degradation, the polymer matrices often undergo chain scission, oxidation, and cross-linking. Polyimide matrices typically show a two-stage degradation process with initial weight loss at 450°C–500°C followed by complete decomposition above 550°C (Liu et al., 2020). Also, in fiber–matrix interface weakening, the interface between fibers and matrix is particularly vulnerable. Thermal cycling causes differential expansion rates, thus creating microcracks at interfaces in ceramic matrix composites (CMCs). In addition, crystal growth, phase transformations, and other microstructural changes occur during prolonged high-temperature exposure. For SiC/SiC composites, crystallization of amorphous interlayers occurs after 500 hours at 1200°C (Xu et al., 2023). The effect of oxidation cannot be belittled because oxygen diffusion accelerates degradation. For instance, C/SiC composites form protective silica layers initially, but prolonged exposure leads to deeper oxidation and strength reduction of up to 40% (Cheng et al., 2000).

Further, there are basically four characterization modeling approach or testing methodologies that are utilized in the analysis of thermal aging. These approaches include isothermal aging, thermal cycling, thermogravimetric analysis (TGA), and dynamic mechanical analysis (DMA). In the isothermal aging method, the samples are positioned at constant elevated temperatures for extended periods. Muhamad et al. (2018) presented the use of 1000-hour exposures at 800°C for CMCs to establish lifetime predictions. Thermal cycling involves subjecting the material to repeated heating/cooling cycles (Hu et al., 2021) showed that thermal cycling between room temperature and 1100°C was more damaging than isothermal aging due to accumulated fatigue damage. The TGA measures the weight changes during heating. Chernyak et al. (2016) utilized TGA coupled with mass spectrometry to identify evolving gases during the degradation of carbon/carbon composites (C/C). The DMA tracks viscoelastic property changes. Batista et al. (2013) used the DMA to identify glass transition shifts in polyimide composites after aging.

Thermal aging causes changes in properties of the composite materials. These properties are mechanical properties, thermal properties, and chemical stability. Mechanical properties such as strength and modulus reduction are common in most composites, with 35% strength reduction after 1000 hours at 300°C due to matrix embrittlement. The thermal conductivity and expansion coefficients change in ceramic composites due to thermal aging, thus reducing the ceramic composites' thermal conductivity by 15%–20% after aging, which is attributed to microcrack formation (Yang and Yang, 2020). Moreover, the chemical stability involves surface oxidation that progresses inward over time. The oxidation depth profiles in the composite material can be tracked. For instance, the oxidation depth profiles in C/C composites, when tracked, showed logarithmic progression rates. Mitigation strategies for thermal aging include protective coatings, self-healing composites, and predictive modeling. Moreover, materials with inherent repair capabilities address thermal damage. For instance, borosilicate phases that flow and seal cracks during thermal excursions are utilized in CMCs (Dong et al., 2024). Further, advanced computational approaches are employed to predict long-term aging. A combination of molecular dynamics (MD) with finite element modeling can be applied to predict 10-year performance from 1000-hour test data (Rocha et al., 2024).

11.2.2 Oxidative aging

Oxidative aging is a critical concern for HTCMs. It occurs through several primary mechanisms, depending on the matrix and reinforcement materials. Matrix oxidation occurs via polymer matrices, particularly those based on polyimides, epoxies, and bismaleimides (BMIs), which undergo chain scission and cross-linking when exposed to oxygen at elevated temperatures. This leads to weight loss, shrinkage, and mechanical property degradation (Sayyidmousavi et al., 2015). The fiber–matrix Interface Degradation in oxidative aging occurs when there is oxygen diffusion through microcracks and voids that attack the fiber–matrix interface, thus compromising load transfer efficiency (Pochiraju et al., 2008; Pochiraju et al., 2012). Moreover, the reinforcement oxidation takes place when the carbon fibers and other carbon-based reinforcements are susceptible to direct oxidation at high temperatures, particularly above 400°C, thus resulting in fiber thinning and strength reduction (Jacobson et al., 1999). Moreover, when thermal cycling is combined with oxidation generates internal stresses, leading to microcrack formation, and then oxidation-induced microcracking is formed. This action further accelerates the oxygen ingress in the composite material (Mosallam, 2002).

The effects of oxidative aging on HTCMs include degradation of mechanical properties, reduction in thermal stability, changes in electrical properties, and dimensional instability. Oxidative aging causes 30% decrease in inter-laminar shear strength after prolonged high-temperature oxidative exposure of PMR-15 composites, thus reducing the tensile strength by 40% (Meador, 1990). Moreover, oxidation can reduce the glass transition temperature and thermal decomposition temperature of PMCs (Tsotsis, 1995), and it can also alter electrical conductivity and dielectric properties significantly in composites with multifunctional capabilities. In terms of dimensional instability, differential oxidation often leads to warping, particularly in thick laminates with oxidation gradients (Mosallam, 2002; Bowles et al., 1997).

Further, oxidative aging also affects high-temperature composites in different ways depending on the nature and type of the composites. For instance, CMCs, particularly SiC/SiC systems, form protective SiO_2 scales when exposed to oxidizing environments at high temperatures. However, these protective scales can volatilize in the presence of water vapor, leading to accelerated degradation (Opila, 2003). C/Csare extremely vulnerable to oxidation at temperatures above 400°C. Often times the oxidation can begin at temperatures as low as 350°C in air, with rapid weight loss occurring above 500°C (Wu et al., 2009). Oxidative aging in PMCs also occurs when high-temperature PMCs (HTPMC) based on polyimides show complex oxidation behaviors. Thermo-oxidative stability is often limited by the formation of a surface oxidized layer that can either protect the underlying material or lead to embrittlement (Bowles et al., 2001).

To study the mechanisms of growth of oxidative aging and proffer mitigation strategies, several methods are utilized for their characterization. These methods include TGA that measures the weight loss as a function of temperature and time in oxidizing atmospheres (Issi and Nysten, 1998), DMA that tracks changes in viscoelastic properties during the oxidative aging (Wang et al., 2024), Fourier-transform infrared (FTIR) spectroscopy that is usually utilized in Identifying chemical changes and oxidation products (Dao et al., 2006), and the microscopy techniques such as the scanning electron microscopy (SEM) and optical microscopy that will reveal the physical degradation and oxide layer formation (Pochiraju et al., 2008). In addition, approaches to enhance oxidative resistance are the use of protective coatings, matrix modification, fiber treatment, and self-healing composites. The protective coatings involve the application of environmental barrier coatings (EBCs) and thermal barrier coatings that can significantly reduce oxidation rates. Multilayer coatings based on silicon carbide, silicon nitride, and rare-earth silicates have shown excellent protection on composite materials (Lee, 2000). The matrix modification refers to the Integration of oxidation inhibitors and nanofillers that can enhance oxidative stability. Addition of phosphorus compounds to polyimide matrices has demonstrated improved flame retardancy and reduced oxidation rates. Surface treatments and sizing modifications of fiber interface in composite materials can enhance fiber–matrix adhesion and reduce susceptibility to oxidative attack at interfaces (Jones, 2010). The self-healing composites involve the utilization of Novel self-healing approaches, like the encapsulated healing agents that can release upon microcrack formation, show promise for extending service life (Blaiszik et al., 2010).

Recent modeling approach or work on oxidative aging includes computational modeling which allows the utilization of finite element models that can predict oxidation kinetics and property degradation of high temperature composites (Roy et al., 2008). The nanomaterial reinforcements involve the Incorporation of graphene, carbon nanotubes, and ceramic nanoparticles. The reinforcement has shown promise in enhancing oxidation resistance (Rafiee et al., 2009). The novel preceramic polymers involve the development of preceramic polymers that can convert to ceramic phases during service, providing enhanced oxidation resistance (Colombo et al., 2010).

11.2.3 Hydrothermal aging

Hydrothermal aging refers to the degradation of materials when exposed to both moisture and elevated temperatures over time. For HTCMs, this phenomenon is particularly significant, as it can drastically affect mechanical properties, durability, and service life (Sethi and Ray, 2015; Wang et al., 2021b). The primary mechanisms involved in hydrothermal aging of high temperature composites include matrix hydrolysis, fiber/matrix interface degradation, moisture, physical swelling, microcrack formation, and oxidative degradation (Dao et al., 2010; Wu et al., 2023). The matrix hydrolysis involves the formation of water molecules that can attack polymer matrices through hydrolysis reactions, breaking chemical bonds and reducing molecular weight. Moisture can also weaken the interfacial bond between fibers and matrix materials (Colin and Verdu, 2005; Plagianakos et al., 2020). The absorption of water and the internal stress build-up due to loadings always causes physical swelling, while the differential swelling between components can lead to microcracks that further accelerate moisture ingress. The oxidative degradation usually occurs simultaneously with hydrothermal effects at elevated temperatures (Grammatikos et al., 2016; Hu et al., 2021).

The hydrothermal aging has effects on reinforced polymer, such as Increased void content and microcracking. Some of the effects pointed out by Qin et al. (2024) on the hydrothermal aging of carbon fiber reinforced polyimide composites include; 18% reduction in tensile strength after 1000 hours at 180°C with 85% relative humidity, decreased glass transition temperature, increased void content, and microcracking. Moreover, there are different hydrothermal responses to different composite types. Composite types include CMCs, BMI composites, PMR-15, and other polyimides. According to Tandon et al. (2008), PMR-15 and other polyimides show good hydrothermal resistance but can still experience significant degradation after prolonged exposure, particularly at temperatures above 250°C in humid environments. Suyama et al. (2019) researched that SiC/SiC composites exhibited hydrothermal degradation mainly through oxidation of the interphase material, stress corrosion cracking of fibers, and the formation of silica scale in the presence of moisture at high temperatures. Moreover, BMI composites showed higher moisture absorption rates compared to polyimides and a significant reduction in compressive strength (up to 25%) after hydrothermal aging. They also show accelerated creep behavior in hot/wet conditions (Li et al., 2024a; Zhang et al., 2024).

In addition, methods for evaluating hydrothermal aging include gravimetric analysis to track moisture absorption, DMA to monitor changes in viscoelastic properties, short beam shear and interlaminar fracture toughness tests to evaluate interface degradation, microscopy techniques (SEM, transmission electron microscopy [TEM]) to observe microstructural changes, and FTIR spectroscopy to identify chemical changes in the matrix. Coupled diffusion-reaction models that can predict the long-term behavior of composites under hydrothermal conditions have been developed to incorporate the Fickian and non-Fickian diffusion kinetics, chemical reaction kinetics for degradation processes, temperature-dependent parameters, and mechanical coupling effects (Grace and Altan, 2012; Ray, 2006).

To reduce or mitigate the propagation of hydrothermal aging, several approaches have been developed to enhance hydrothermal resistance. These approaches include surface treatments and coatings (Han and Morscher, 2019; Papakonstantinou and Balaguru, 2007), matrix modification with nanofillers for reduced permeability (Jian, 2021), optimization of fiber sizing and interphase materials (Li et al., 2024a; Zhang et al., 2024), and development of hybrid composites with synergistic moisture resistance (Grace and Altan, 2012; Weitsman, 2012).

11.2.4 Mechanical aging

HTCMs such as CMCs (SiC/SiC, C/SiC, Al_2O_3/Al_2O_3), metal matrix composites (MMCs): Ti-based, Ni-based superalloys, C/C, and PMCs with high-temperature capabilities, experience several aging mechanisms when exposed to mechanical loads over time. Some of these loadings include creep deformation, fatigue damage, microcracking, oxidation-assisted degradation, and fiber–matrix debonding. The creep deformation gives rise to progressive deformation under constant loading. The deformation is significant at elevated temperatures. The fatigue damage is usually caused by the accumulation of microscopic damage from cyclic loading. Moreover, microcracking usually occurs when there is the formation of small cracks in the matrix or at fiber–matrix interfaces. The synergistic effects between mechanical stress and environmental factors give rise to Oxidation-assisted degradation. This causes weakening of the interface between reinforcement and matrix, thereby producing fiber–matrix debonding.

High-temperature composites often experience time-dependent deformation under constant stress, particularly critical in aerospace and energy applications. This creep behavior differs significantly from monolithic materials due to the complex interactions between the matrix and reinforcements. In CMCs, researchers have observed that creep rates are heavily influenced by interfacial sliding between fibers and matrix. Han and Morscher (2019) demonstrated that SiC/SiC composites exhibit primary, secondary, and tertiary creep regimes, with interfacial degradation accelerating creep rates as exposure time increases. The creep behavior of SiC/SiC composites at temperatures up to 1400°C showed that the creep rates are controlled primarily by grain boundary sliding mechanisms in the matrix. In terms of fatigue damage (Talreja and Varna, 2023) established that fatigue damage accumulation in CMCs follows distinctly different patterns from monolithic ceramics, with progressive matrix cracking and subsequent load transfer to fibers. In environmental effects modeling, Liu et al. (2020) presented that oxidation significantly accelerates mechanical degradation in SiC-based composites, creating a complex coupled degradation mechanism.

Moreover, cyclic loading produces cumulative damage via matrix microcracking, fiber/matrix interfacial debonding, fiber fracture, and environmental attack at newly exposed surfaces. Papakonstantinou and Balaguru (2007) presented HTPMCs subjected to tension–tension fatigue, which show progressive stiffness reduction due to matrix crack propagation, with the rate of degradation strongly temperature-dependent. Further, Mechanical loading often interacts synergistically with environmental factors. In C/SiC composites, Santhosh et al. (2021) researched that applied stresses can accelerate oxidation by creating pathways for oxygen ingress along fiber–matrix interfaces, leading to premature failure at temperatures above 500°C.

The factors that influence mechanical aging in high-temperature composites have been found to include temperature Effects and loading parameters. Temperature significantly accelerates aging mechanisms. For instance, in PMCs, exceeding the glass transition temperature dramatically changes viscoelastic properties, while in MMCs, diffusion-controlled processes become dominant above melting temperature. Russell-Stevens et al. (2005) presented that temperature cycling caused more severe degradation than isothermal aging due to differential thermal expansion stresses. The loading parameters affect the nature of mechanical aging by producing load magnitude relative to ultimate strength, increase in stress ratio (R value) in fatigue loading, and holding times at peak stress and increased Loading rate. Yang and Yang (2020) developed lifetime prediction models based on these parameters for oxide/oxide composites, showing that dwell periods at maximum stress were particularly damaging to long-term durability. The techniques that are usually applied for characterization of mechanical aging include nondestructive evaluation (NDE) and microstructural analysis. The modern NDE techniques provide critical insights into aging progression through acoustic emission for tracking damage initiation and growth, digital image correlation (DIC) for strain mapping, and X-ray computed tomography for volumetric damage assessment. Whitlow et al. (2016) demonstrated how in situ acoustic emission monitoring could identify the transition points between damage mechanisms during mechanical aging of SiC/SiC composites. Moreover, microstructural analysis involves microstructural evolution during aging, which is typically characterized through electron microscopy (SEM/TEM) for interface degradation, spectroscopic techniques (Raman, FTIR) for chemical changes, and X-ray diffraction (XRD) for crystallinity and phase transformations. These techniques have been used to establish correlation between microstructural changes and mechanical property degradation in carbon fiber reinforced silicon carbide. To mitigate or reduce the mechanical aging effects on high temperature composites, strategies such as engineered interfaces of the composites with optimal bonding strength, incorporating self-healing matrices in the composite to repair damage during service, and EBCs to limit oxidation effects. Paladugu et al. (2022) developed self-healing CMCs with embedded healing agents that were activated by mechanical damage, extending service life by up to 300% under cyclic loading at 1200°C. Another mitigation strategy is the processing Improvements in manufacturing. These improvements include Controlled fiber architecture for optimized load transfer, tailored porosity distributions, and gradient structures with varying compositions. Liang et al. (2024) demonstrated how controlled porosity gradients in alumina-based composites could redirect crack propagation away from primary load paths, enhancing fatigue resistance.

11.2.5 Radiation aging

Considering the utilization of HTCMs in extreme environments where radiation exposure is significant, such as nuclear reactors, space applications, and fusion energy systems, it is necessary to analyze how radiation affects these materials over time to predict their service life and ensure safety (Noda, 2020). HTCMs exposed to radiation environments experience several degradation mechanisms. This mechanism include atomic displacement damage, ionization effects, transmutation, and gas production. The atomic displacement damage involves the production of energetic particles in the form of neutron and ions that cause atoms to be knocked from their lattice positions, creating vacancies and interstitials. The ionization effect is caused from electromagnetic radiation (gamma rays, X-rays) that can break chemical bonds and create free radicals. Transmutation occurs when neutron capture can transform elements into different isotopes, changing material composition. Moreover, helium and hydrogen generation from nuclear reactions can lead to embrittlement and swelling, thus giving rise to gas production (Katoh et al., 2018).

High-temperature composites, such as PMCs, are particularly susceptible to radiation damage due to the vulnerability of their organic components. The radiation usually causes chain scission, cross-linking, gas evolution, and oxidation acceleration. Koyanagi et al. (2018) presented that gamma radiation exposure of carbon fiber reinforced polymers at doses of 10–100 kGy leads to significant decreases in tensile strength and increases in brittleness. The degradation mechanisms were primarily attributed to polymer matrix decomposition while carbon fibers remained relatively stable. In CMCs, the material exhibits superior radiation resistance compared to PMCs, making them excellent candidates for nuclear applications. However, they still face challenges such as dimensional changes due to point defect accumulation, thermal conductivity reduction, and mechanical property degradation at very high doses. Katoh et al. (2012) conducted extensive studies on SiC/SiC composites, showing that neutron irradiation at temperatures below 1000°C caused volumetric swelling of approximately 1%–2% at doses of 10 dpa (displacements per atom). This swelling plateaued at higher doses, indicating saturation of point defect concentration. Further, MMCs experience radiation effects such as hardening and embrittlement, void formation, precipitate dissolution or coarsening, and interface degradation between matrix and reinforcement. Zinkle and Busby (2009) showed that aluminum matrix composites reinforced with SiC particles exhibited increased yield strength but decreased ductility after neutron irradiation to doses of 1–5 dpa at 250°C–350°C.

The effect of temperature on radiation damage cannot be overemphasized because it plays a critical role in determining radiation damage accumulation and recovery. At very low temperatures ($<0.3\ T_m$, where T_m is melting temperature), defect mobility is limited, resulting in high defect concentration. At intermediate temperatures (0.3–0.5 T_m), some defects anneal out, but significant damage accumulates. Also, at high temperatures ($>0.5\ T_m$), dynamic recovery processes can significantly reduce permanent damage. Koyanagi et al. (2018) presented that SiC composites irradiated at temperatures above 1300°C showed minimal property degradation even at high neutron doses (>10 dpa) due to efficient defect annealing, while the same materials irradiated below 800°C exhibited significant strength reduction.

Sophisticated techniques are utilized to characterize radiation aging. Some of these techniques include TEM, which is utilized for defect analysis in structural analysis, XRD, which is employed for crystallinity changes in the composites, Raman spectroscopy, that are utilized for chemical bonding alterations, and atom probe tomography for elemental segregation. To reduce radiation aging or enhance the resistance of composites to radiation aging, several approaches utilized are development of materials that are nanostructured composites with high interfacial area to trap defects, development of materials with self-healing capabilities that can repair radiation damage, development of materials with gradient structures to manage stress distribution and the development of novel fiber coatings to preserve fiber–matrix interfaces (Thomé et al., 2013).

Radiation aging or exposure of high-temperature composites to radiation typically causes reduced mechanical strength and ductility, dimensional changes (swelling or shrinkage), thermal conductivity degradation, changes in electrical properties, microstructural evolution, and accelerated creep rates. Moreover, different composites respond to radiation in a different pattern. For instance, CMCs such as SiC/SiC show promising radiation resistance up to high doses. The radiation stability depends significantly on processing methods and interfacial properties between fibers and matrix. The C/Cs maintain excellent thermal and mechanical properties at high temperatures but can experience dimensional changes and property degradation under radiation. MMCs with radiation-resistant matrices (like tungsten) can withstand moderate radiation doses while maintaining thermal properties (Dai et al., 2012).

11.3 Mechanisms of aging effects of high-temperature composite materials

11.3.1 Mechanisms of thermal aging

Thermal aging in HTCMs involves several complex degradation mechanisms that affect their performance and durability. Mechanisms responsible for thermal aging of high-temperature composites include oxidation, matrix cracking and microstructural changes, fiber–matrix interface degradation, physical aging and relaxation, and chemical reactions and phase transformations. The oxidation mechanism is one of the primary degradation mechanisms, particularly for carbon-fiber and PMCs at elevated temperatures. In C/Cs, oxygen diffusion and oxidation lead to weight loss and mechanical property degradation and create a chemically altered layer with increased crosslinking, leading to embrittlement in PMCs (Pochiraju and Tandon, 2009). Also, in silicon carbide (SiC) composites, oxidation forms protective silica scales, which usually volatilize at very high temperatures in the presence of water vapor (Opila, 2003).

The matrix cracking and microstructural changes in thermal aging involve the matrix phase of the composite undergoing significant changes during thermal exposure according to the nature of the material of the composites. For instance, CMCs experience microcrack formation due to thermal expansion mismatch between fibers and matrix (DiCarlo and Van Roode, 2006; DiCarlo et al., 2003). Moreover, polymer matrices experience post-cure reactions, leading to additional cross-linking and embrittlement, while the glass-ceramic matrices can undergo devitrification, changing their mechanical properties (Lamon, 2015). In fiber–matrix interface degradation, the fiber–matrix interface is particularly susceptible to thermal aging through; interfacial debonding that occurs due to differential thermal expansion (Naslain, 2004), chemical reactions at fiber–matrix interfaces that creates new phases that can alter load transfer (Morscher et al., 1997), and In SiC/SiC composites where boron nitride interfacial coatings can oxidize, the interface, thus leading to premature failure (More et al., 2000). The physical aging and relaxation mechanism involves structural relaxation toward equilibrium. For instance, in amorphous polymers, there is a free volume reduction that leads to increased stiffness and brittleness (Gates and Grayson, 1999), and also a stress relaxation in fiber reinforcements that can reduce residual stresses initially, but which may later lead to performance degradation over time (Ruggles-Wrenn et al., 2008). The chemical reactions and phase transformations are encountered when there is a long-term high-temperature exposure, which induces various chemical changes. These changes include: formation of interphase regions through the diffusion of elements between the fiber and matrix, crystallization of initially amorphous phases in ceramic composites, and decomposition of thermally unstable phases, particularly in nonoxide ceramics (Padture, 2016).

11.3.2 Mechanisms of oxidative aging

The mechanism of oxidative aging in HTCMs can be categorized based on the nature of the composite material. For instance, PMCs undergo several oxidation mechanisms at elevated temperatures. Some of these mechanisms are chain scission and free radical formation, cross-linking and embrittlement, and thermo-oxidative surface layer formation. The chain scission and free radical formation in PMCs occur when there is thermal energy that breaks covalent bonds in polymer chains, creating highly reactive free radicals. These radicals react with atmospheric oxygen to form peroxyl radicals, initiating oxidation cascades. These chain scission rates increase exponentially with temperature (Pochiraju et al., 2012). Moreover, cross-linking embrittlement causes oxidation that promotes the formation of additional cross-links between polymer chains. This causes increased brittleness, reduced toughness, and decreased strain-to-failure. The cross-linking density correlates with oxygen diffusion rates (Colin and Verdu, 2003). In the thermo-oxidative surface layer formation, the oxidation typically begins at material surfaces where oxygen concentration is highest. This creates distinct oxidized surface layers with different properties than the core material, thus creating the growth kinetics of these layers in polyimide composites (Tandon et al., 2008).

In CMC degradation, CMCs face different oxidation challenges such as fiber–matrix interface, active-to-passive oxidation transitions, and thermal gradient effects. The fiber–matrix interface occurs when there is oxygen diffusion through matrix cracks, which attacks fiber coatings and interfaces, thus degrading load transfer between fibers and matrix. Karadimas et al. (2021) showed that SiC interface coatings experience distinct oxidation mechanisms. In active-to-passive oxidation transitions, when there are temperature variations, silicon-based ceramics (such as SiC, Si_3N_4) experience temperature-dependent oxidation behavior. For instance, in low temperatures, passive oxidation occurs that forms protective SiO_2 scales. Whereas in high temperatures, there is occurrence of active oxidation, which causes volatile SiO formation and material recession (Jin et al., 2018). In thermal gradient effects, there is an occurrence of temperature gradients that accelerate differential oxidation and create internal stresses. Resulting mechanical damage, which further facilitates oxygen ingress. Modeling by thermo mechanically approach can be used to analyze the aging process (Birks et al., 2006; Goulmy et al., 2020).

The mechanism of oxidative aging in MMCs experiences distinct degradation mechanisms such as preferential oxidation at phase boundaries, diffusion-limited oxide growth, oxide spallation, and cyclic oxidation. In the preferential oxidation at phase boundaries, there is a galvanic coupling between reinforcements and matrix, which accelerates oxidation. Ceramic–metal interfaces in composites such as Al–SiC composites often serve as preferential oxidation sites (Chawla, 1998; Paladugu et al., 2022). Also, in diffusion-limited oxide growth, there is a parabolic growth kinetics that typically governs oxide scale formation where there is an ion and electron transport through growing scales, which can be governed by Wagner's theory (Birks et al., 2006). Oxide spallation and cyclic oxidation occur when there is thermal cycling, which causes oxide scale cracking and spallation, thus exposing fresh materials like various high-temperature alloys to further oxidation (Quadakkers et al., 2004).

11.3.3 Mechanisms of hydrothermal aging

Hydrothermal aging is a critical degradation mechanism for HTCMs exposed to moisture and elevated temperatures. The combined effects of moisture absorption and thermal stress can significantly impact mechanical properties, microstructure, and long-term performance of these materials. An analysis of the key mechanisms involved in the hydrothermal aging includes water absorption mechanisms, chemical degradation mechanisms, and physical changes and mechanical degradation. In water absorption mechanism, water molecules infiltrate composite materials through several pathways like; diffusion through the polymer matrix which may be according to the Fickian or non-Fickian behavior depending on the polymer type and temperature (Zhou and Lucas, 1999), Capillary transport along fiber/matrix interfaces where water molecules migrate along microscopic gaps between reinforcement and matrix (Weitsman, 2012), and transport through microcracks and voids in which preexisting defects provide pathways for accelerated moisture ingress (Ray, 2006). In chemical degradation mechanisms, it involves moisture penetration to the composite material through several mechanisms such as hydrolysis reactions where water molecules attack polymer chains, thereby breaking covalent bonds in the resin network (Xiao and Shanahan, 1997), leaching of low molecular weight components in which the water extracts unreacted monomers, plasticizers and other additives (Colin and Verdu, 2006), and through oxidation reactions where moisture can accelerate thermo-oxidative degradation of the composite material at elevated temperatures.

Moreover, the presence of moisture induces various physical changes that impact mechanical performance of composite materials. These physical changes include plasticization, where water molecules reduce intermolecular bonding between polymer chains, decreasing glass transition temperature and compromising high-temperature performance (Hamidi et al., 2005), swelling and residual stresses in which differential swelling between matrix and fibers generates internal stresses that can initiate microcracks (Dao et al., 2006), and fiber/matrix interfacial degradation where water weakens the fiber–matrix bond, particularly in glass fiber composites through hydrolysis of sizing agents (Jones, 2010).

One of the effects of hydrothermal aging is temperature. This occurs through accelerated diffusion kinetics. In this process, higher temperatures increase the rate of moisture diffusion following an Arrhenius relationship (Grace and Altan, 2012). Other mechanisms of temperature effects include synergistic degradation, which combines the thermal and moisture effects that often produce more severe degradation than either of the factors acting separately (Davies et al., 2001), and phase transitions in water where hot water that is near or above boiling point, causes steam that promotes more rapid degradation through pressure effects and accelerated chemical reactions (Wang et al., 2021a). Moreover, it is worthwhile to know that different composites exhibit varying susceptibility to hydrothermal aging. For instance, polyimide composites show excellent hydrothermal stability but experience microcracking at high temperatures in wet environments (Schoeppner et al., 2007). BMI composites are more susceptible to hydrolysis compared to polyimides, with significant temperature depression after moisture exposure (Morgan et al., 1997). Also, in CMCs, the oxide layers and interphase regions are affected by moisture, which potentially accelerates the oxidation at elevated temperatures (Ruggles-Wrenn and Lee, 2016). The C/Cs materials show limited moisture absorption but can experience oxidation acceleration in the presence of moisture at high temperatures (Jacobson et al., 2001; Jacobson et al., 2005).

11.3.4 Mechanisms of mechanical aging

HTCMs experience mechanical aging mechanisms when exposed to elevated temperatures and mechanical loads over extended periods. These mechanisms often work in concert, accelerating degradation and limiting service life. One of the mechanical aging mechanisms is matrix cracking and microcracking. The matrix cracking occurs due to thermal cycling, mechanical stress, and chemical degradation. In PMCs, thermally-induced volumetric changes create internal stresses that lead to microcracking, particularly at the fiber–matrix interface (Russell-Stevens et al., 2005). Also, in CMCs, matrix microcracking becomes significant above critical stress thresholds, creating pathways for oxidation (Zok, 2016). Another aging mechanism is the fiber matrix degradation. The fiber–matrix interface is often the weakest link in high-temperature composites, and as such, degradation mechanisms occur by interfacial debonding due to CTE mismatch, chemical reactions at the interface, particularly in oxidizing environments, diffusion of matrix constituents into the fiber or vice versa, and crack deflection and fiber pullout alterations that affect the toughening mechanisms. Naslain (2004) showed that boron nitride (BN) interphases in SiC/SiC composites degrade through oxidation mechanisms, severely compromising mechanical properties. Considering the oxidation effect mechanism, the oxidation affects both fibers and matrices in high-temperature environments, creating volatile products and solid oxides that change material composition and microstructure. For carbon-fiber composites, fiber oxidation can begin at temperatures as low as 400°C, while SiC fibers form protective silica scales that can be beneficial but also lead to embrittlement (Jacobson et al., 1998; Jacobson et al., 2001). Moreover, creep and stress relaxation mechanisms occur at elevated temperatures because at elevated temperatures, time-dependent deformation becomes significant through the primary, secondary, and tertiary creep stages, leading to progressive deformation, which gives rise to creep cavitation at grain boundaries and interfaces. The mechanism also causes stress redistribution between fibers and matrix, thus affecting load transfer and the residual stress relaxation, altering internal stress states. Creep in SiC/SiC composites can occur at temperatures as low as 1000°C under sustained loading (Ruggles-Wrenn et al., 2011; Ruggles-Wrenn and Lee, 2016).

Moreover, thermal exposure under mechanical loading can cause crystallization of amorphous regions in polymeric matrices, precipitation and coarsening of phases in metal matrices, grain growth in ceramic matrices, loss of volatiles, and crosslinking in certain polymer systems. All these processes usually lead to increased brittleness and reduced fracture toughness. In mechanical aging, fibers undergo several degradation mechanisms through environmentally assisted stress corrosion of glass and ceramic fibers, phase transformations at elevated temperatures, recrystallization that reduces the strength of fine-grained ceramic fibers, and coarsening of nanoscale strengthening features in high-performance fiber. Another study presented that SiC fibers undergo microstructural evolution above 1300°C that significantly compromises their strength properties.

Similarly, mechanical aging of high-temperature composites is sometimes material-specific. For instance, CMCs face distinctive challenges at high temperatures, such as slow crack growth under sustained loading, silica volatilization in water vapor environments above 1200°C, oxidation of nonoxide constituents, and pesting phenomena at intermediate temperatures (700°C–900°C). The unique damage tolerance of CMCs, which depends on controlled matrix cracking and fiber bridging, becomes compromised as these mechanisms advance (Zok and Levi, 2001). In PMCs high-temperature capable PMCs (typically polyimides, BMIs, and phenolics) experience thermo-oxidative degradation, creating a damaged surface layer, chain scission, and cross-linking, affecting glass transition temperature, and physical aging below glass transition, altering free volume. Bowles (1999) showed the progressive weight loss and mechanical property degradation in polyimide composites during isothermal aging. In MMCs, mechanical aging mechanisms occur through interfacial reactions forming brittle intermetallic compounds, CTE mismatch that causes fatigue, accelerated creep due to reinforcement/matrix load transfer, and galvanic corrosion between dissimilar constituents.

11.3.5 Mechanisms of radiation aging

Radiation-induced aging of HTCMs involves complex physicochemical processes that affect structural integrity and performance. Radiation aging mechanisms include matrix degradation, fiber degradation, and fiber-matrix interface degradation. Considering radiation matrix degradation, the radiation exposure causes polymer matrix degradation through several mechanisms, such as chain scission, cross-linking, and free radical formation (Noda, 2020). In the chain scission, high-energy radiation breaks chemical bonds in the polymer backbone, reducing molecular weight and mechanical properties, while the radiation can simultaneously induce additional cross-links between polymer chains, causing embrittlement, which leads to cross-linking (Deng et al., 2024; Sun et al., 2024). The free radical formation involves the generation of reactive species that propagate degradation. According to Hemath et al. (2020), these processes lead to embrittlement, decreased toughness, and reduced strain-to-failure in PMCs exposed to radiation at elevated temperatures.

Considering fiber degradation, different fiber types respond differently to radiation. For instance, carbon fibers show relatively radiation-resistant but can experience crystallinity changes at high doses, while glass fibers are more susceptible to radiation damage through network depolymerization and color center formation. The ceramic fibers show radiation-induced crystallization or amorphization depending on the dose of exposure and temperature. Yan et al. (2025) presented that SiC fibers in high-temperature composites undergo microstructural evolution under neutron irradiation at 600°C–800°C, with significant dimensional changes affecting overall composite stability. In fiber–matrix interface degradation, the fiber–matrix interface is particularly vulnerable to radiation by virtue of the radiation causing debonding through localized chemical changes. This interface degradation often precedes bulk material failure, thereby causing radiation-induced oxidation at interfaces that accelerates debonding and differential swelling between fibers and matrix, which creates interfacial stresses and decreased interfacial shear strength that reduces load transfer efficiency (Katoh et al., 2018). Neutron radiation exposure at temperatures above 300°C significantly degrades the fiber–matrix adhesion in CMCs, leading to premature failure.

The synergistic effects of radiation and temperature are another prominent mechanism of radiation aging because the existence of high-temperature amplifies the temperature-dependent reaction rates that increases the radiation-induced chemistry that causes radiation damage through enhanced chemical reactions (Sudheendra et al., 2023). This occurs at elevated temperatures, thus leading to thermal degradation mechanisms that overlap with radiation effects. There is also an accelerated diffusion of radiation-induced species that occur at elevated temperatures, which leads to thermal annealing competing with radiation damage. Katoh et al. (2018) researched that composites that are exposed to neutron irradiation at temperatures above 800°C exhibit fundamentally different degradation mechanisms than those irradiated at lower temperatures.

The mechanism of radiation aging in high-temperature composites is somewhat material-specific. The radiation aging in PMCs makes the material to experience gas evolution (H_2, CO_2, CH_4) from radiolysis, oxidative degradation enhanced by radiation, decreased glass transition temperature, and loss of mechanical properties primarily through matrix damage. In CMCs, it shows point defect accumulation leading to swelling, formation of lattice defects and voids, amorphization of crystalline regions, transmutation effects resulting in new elements, thermal conductivity degradation through phonon scattering, and crystalline structure changes that affect the mechanical properties. Noda (2020) presented that SiC/

SiC composites experienced a 30%–50% reduction in thermal conductivity after neutron irradiation at doses between 10 and 100 dpa. Also, in MMCs, radiation aging exhibits radiation-induced precipitation and segregation, irradiation creep and swelling, helium and hydrogen embrittlement, and helium bubble formation that affects mechanical properties and changes the interfacial chemistry between matrix and reinforcement. Moreover, the effect of radiation aging is such that different radiation types cause dissimilar damage patterns. For instance, the neutron radiation in composite materials causes displacement damage and transmutation, particularly problematic for nuclear applications, unlike gamma radiation, which primarily affects polymeric components of the composites through ionization and electron radiation that causes surface damage in the composite with limited bulk penetration (Katoh et al., 2012; Sales et al., 2025).

11.4 Computational modeling of various aging effects

11.4.1 Computational modeling of thermal aging

The computational modeling of thermal aging in HTCMs involves complex modeling approaches that combine materials science, thermodynamics, and computational methods. The thermal aging of high-temperature composites is primarily driven by several interrelated mechanisms. These mechanisms include matrix decomposition, fiber–matrix interface degradation, microstructural evolution, and oxidation effects (Colin and Verdu, 2005). The matrix decomposition occurs when the polymer matrices in composites undergo chemical degradation at elevated temperatures, leading to changes in their mechanical properties. Also, at high temperatures, thermal aging occurs in fiber matrix when the bond between fibers and matrix material weakens over time at high temperatures. The microstructural evolution occurs as a result of the formation of microcracks, void growth, and phase transformations that evolve over time. The oxidation effects affect the composite material when there is a surface reaction with the environment that can penetrate into the material over time (Ruggles-Wrenn and Lee, 2016).

Multiscale approaches have become essential in accurately capturing the complex aging phenomena across different length scales. The multiscale modeling approach includes atomistic/MDs simulations, micromechanical models, and macroscale continuum approaches. The MDs simulations are employed to predict chemical degradation pathways in the matrix of the composite material, while the micromechanical models are applied to capture the composite behavior based on constituent properties and microstructure. The macroscale continuum approaches are modeling approaches used to model the overall structural response incorporating aging effects. According to Pochiraju et al. (2012), the integration of these scales provides the most comprehensive predictive capability for long-term aging effects. The reaction-diffusion models, which predict the oxidation exposure level of the material, are also other computational approaches because oxidation is a primary aging mechanism that can be typically modeled using reaction-diffusion equations. The reaction diffusion model has been successfully employed to predict oxidation layer growth in PMR-15 composites (Colin et al., 2001). Moreover, the damage evolution model, such as the continuum damage mechanics, is commonly used to track the material degradation considering the stress state and temperature variations. This allows tracking of progressive deterioration (Talreja, 2014; Talreja and Varna, 2023). The time-temperature superposition approach is another computational approach that relies on accelerated testing and employs the time-temperature superposition principles to predict long-term behavior of the material, considering the shift factor typically following an Arrhenius relationship. Ruggles-Wrenn, Siegert et al. demonstrated this approach for CMCs at temperatures up to 1300°C. Recent advances in computational modeling include phase field methods that are used in capturing microstructural evolution during aging and machine learning approaches that are used by considering data-driven models to predict complex aging behavior based on experimental data (Ruggles-Wrenn et al., 2008). Other recent advances include the digital twin concept. The digital twin concepts use real-time updating of computational models based on in-service data of the material in operation.

11.4.2 Computational modeling of oxidative aging

Oxidative aging in high-temperature composites typically involves oxygen diffusion through the matrix, chemical reactions at fiber or matrix interfaces, formation of oxide scales, changes in mechanical properties due to microstructural evolution. The computational modeling of these processes spans multiple length and time scales, requiring different simulation approaches. At the atomistic scale, density functional theory (DFT) and MD simulations provide insights into oxidation reaction pathways and energetics. Wang et al. (2021a) used DFT to investigate oxygen adsorption and diffusion on SiC surfaces, identifying critical reaction sites and energy barriers relevant to CMC oxidation. It was discovered that oxygen preferentially attacks carbon sites in SiC, creating initial defects that accelerate degradation. Moreover, reactive MD simulation was applied to model the initial stages of oxidation in SiC/SiC composites, capturing the formation of silica layers and their protective effects against further oxidation (He et al., 2024).

Further, mesoscale approaches assist in bridging the gap between atomistic mechanisms and macroscopic behavior. These mesoscale models include phase-field models by He et al. (2024), which were applied to simulate the growth of oxide scales in CMCs, incorporating diffusion kinetics and volumetric expansion effects. The formation of protective oxide layers and its evolvement under different temperature and oxygen partial pressure conditions were presented. Moreover, Zhou et al. (2023) developed a kinetic Monte Carlo approach to model oxygen diffusion through complex microstructures, accounting for grain boundaries and porosity effects on oxidation rates in CMC. Another computational approach is the application of continuum and multiscale approaches. The Continuum models are essential for predicting component-level behavior. Barbero and Cortes (2010) developed a coupled diffusion-reaction model that captures the oxidation weight gain and mechanical property degradation in PMCs. The model accurately predicted the formation of oxidized layers and the resulting changes in composite stiffness. More recently, Meng et al. introduced a

Multiphysics framework that couples oxygen diffusion, chemical reactions, and mechanical damage to predict the service life of C/SiC composites in hypersonic vehicle applications. The model was validated against experimental data from NASA's high-enthalpy facilities (Songhe Meng et al., 2016b).

Another computational approach for modeling oxidative aging is the microstructure-based models. The microstructure plays a crucial role in oxidation behavior. Examples of these microstructure-based models are representative volume element (RVE) and finite element. Xue et al. used RVE approaches to model preferential oxidation at fiber/matrix interfaces in SiC/SiC composites. The computational results obtained from the simulation demonstrated how fiber coatings can significantly delay oxidation-induced degradation (Xue et al., 2025). The finite element models by Zhang (2022) incorporated realistic microstructures from X-ray tomography to predict stress distributions and crack formation during oxidative aging. In recent times, emerging machine learning approaches have been applied in computational modeling of oxidative aging. For instance, Hamza and Chattopadhyay (2024) used deep learning to predict oxidation rates from microstructural features, creating surrogate. Models that are computationally efficient yet preserve the accuracy of physics-based simulations, and Wang et al. (2023) integrated physics-informed neural networks with experimental data to model the time-dependent property degradation in composites exposed to oxidative environments.

11.4.3 Computational modeling of hydrothermal aging

Hydrothermal aging represents a significant challenge for HTCMs used in demanding environments. The mechanisms of hydrothermal aging in high temperature composites include moisture absorption/diffusion, chemical degradation of the matrix, interfacial weakening between fibers and matrix, potential microcracking, and changes in thermo-mechanical properties. Majorly, hydrothermal aging of composites typically involves moisture absorption and diffusion, matrix plasticization, fiber–matrix interface degradation, micro-cracking and void formation, chemical reactions (hydrolysis), and thermal-moisture coupling effects (Mackerle, 1996; Meng et al., 2016a).

The computational approaches used in modeling hydrothermal aging are diffusion-based models, MD simulations, finite element methods, continuum damage mechanics, and machine learning approaches. Fick's laws of diffusion form the foundation of many hydrothermal aging models using the diffusion-based approach. The classic approach uses: Langmuir-type non-Fickian diffusion models (Roy et al., 2000), temperature-dependent diffusivity via Arrhenius relationships, and multiscale approaches linking nano/micro/macro behaviors (Pochiraju et al., 2008) developed a comprehensive model accounting for both reversible and irreversible moisture absorption in PMCs, incorporating non-Fickian behavior. The extended non-Fickian models incorporate temperature-dependent diffusion coefficients, Concentration-dependent diffusion, and Langmuir-type diffusion with bound/unbound water. Afzal et al. (2023) developed a comprehensive multiphysics model coupling thermal, hygroscopic, and mechanical aspects for carbon/epoxy composites, showing good correlation with experimental data for aerospace applications. Moreover, micromechanical models are used to represent the composite at the microstructural level using the RVE approaches, fiber–matrix interface models, and Homogenization techniques. Wang (2019) implemented a 3D micromechanical model with explicit fiber–matrix interfaces to predict property degradation in glass/epoxy composites under hydrothermal aging.

The MD Simulations are employed for atomic-level insights. They include simulating water molecule interactions with polymer chains, predicting chain scission and hydrolysis reactions, and tracking free volume changes and plasticization effects. The MD simulations examine molecular-level interactions such as bond breakage during hydrolysis, and Interface weakening mechanisms. Li et al. (2024b) utilized MD simulations to model water diffusion paths in high-temperature polyimide matrices, revealing preferential water clustering around specific chemical groups. The MD simulations determined the water molecule diffusion mechanisms in epoxy resins at elevated temperatures, revealing preferential pathways and energy barriers. Moreover, the finite element approaches incorporate coupled hygro-thermo-mechanical formulations, residual strength predictions, property degradation as functions of moisture/temperature history, and damage evolution models. Zhou and Lucas (1999) developed finite element models incorporating viscoelastic effects coupled with moisture diffusion to predict structural response under hydrothermal conditions. de Parscau du Plessix et al. (2016) developed a multiscale FEA framework linking microlevel damage to macrolevel property degradation for carbon/BMI composites under hydrothermal cycling. The Continuum Damage Mechanics approach models the material degradation through internal state variables representing degradation mechanisms, evolution equations for damage parameters, and coupling between moisture, temperature, and mechanical loading (Grammatikos et al., 2016). Patel and Case (2000) presented a continuum damage mechanics framework specifically for glass/epoxy composites under hydrothermal aging conditions, accounting for fiber/matrix interfacial weakening. In recent times, the machine learning approach employed in the computational modeling of hydrothermal aging of high-temperature composites includes neural networks that will assist in predicting long-term behavior from short-term test data, Gaussian process models for uncertainty quantification, surrogate models reducing computational costs, and Bayesian methods accounting for uncertainty, and data-driven approaches combining experimental and simulation data. Hamza and Chattopadhyay (2024) demonstrated a deep learning approach to predict the time-dependent mechanical properties of CMCs exposed to hydrothermal environments at temperatures exceeding 800°C. Wang et al. (2024) demonstrated the effectiveness of recurrent neural networks in predicting moisture-induced property degradation in carbon/epoxy composites.

11.4.4 Computational modeling of mechanical aging

Computational modeling of mechanical aging in high-temperature materials is essential for predicting service life, optimizing designs, and developing more durable materials. To effectively model the mechanical aging of HTCMs, several key computational approaches and methodologies that have been considered include multi-scale modeling, constitutive models, and phase field methods. The computational models typically address primary aging mechanisms such as thermally activated degradation (oxidation, chemical reactions), creep deformation under sustained loading, fatigue damage accumulation under cyclic loading, microcracking and void formation, and interface degradation between fibers and matrix. Moreover, advanced constitutive models for high-temperature composites incorporate viscoelastic and viscoplastic material behavior, temperature-dependent material properties, damage evolution laws for progressive failure, coupled thermo-mechanical formulations, and time-temperature superposition. For long-term aging predictions, computational models often employ time-temperature superposition principles to extrapolate short-term test data to service lifetimes. The finite element Implementation typically occurs via custom material subroutines (UMAT/VUMAT) in commercial FE packages, specialized element formulations for composite structures, adaptive meshing techniques for damage progression, and the model validation involves correlating predictions with accelerated aging tests, in situ monitoring data, NDE results, residual strength and stiffness measurements. In recent times, advanced computational techniques involve machine learning integration for improved prediction accuracy, uncertainty quantification in aging simulations, digital twin frameworks for real-time health monitoring, and phase field methods for damage evolution.

The multiscale modeling approaches can be categorized into several models, such as micromechanical models, mesoscale models, and macroscale models. The micromechanical models examine damage evolution at the fiber/matrix interface level, which is critical for high-temperature composites. Matzenmiller et al. (1995) developed a continuum damage mechanics model that accounts for fiber-matrix debonding under cyclic thermal loading. Fish and Yu (2001) proposed a multiscale computational homogenization technique that bridges microscopic damage to macroscopic property degradation in CMCs. The mesoscale models focus on the yarn/tow architecture level, particularly important for textile composites. Ramachandran et al. (2022) developed a mesh-free approach to model oxidation and crack propagation in SiC/SiC composites at the mesoscale. The virtual testing approach simulates progressive damage in textile composites under thermomechanical cycling (McMahon and Prise, 2019). Moreover, the macroscale models address component-level behavior and often incorporate phenomenological elements. Chaboche (2008) unified viscoplastic model has been adapted for high-temperature composites to capture creep-fatigue interactions. Moreover, Kirane et al. (2016) proposed a microplane model to capture the anisotropic damage evolution in composites under complex loading histories.

The key aging mechanisms and modeling techniques in mechanical aging modeling of HTCMs incude creep modeling, fatigue damage accumulation, chemical degradation and oxidation, and thermal cycling effects. In creep modeling, Barbero and De Vivo (2001) developed a computational framework for predicting creep response in PMCs at elevated temperatures. The Norton–Bailey power law has been modified by Xu et al. (2023) to account for anisotropic creep in directionally reinforced composites. Further, a stochastic approach to fatigue damage modeling was proposed by Kawai and Koizumi (2007), incorporating probabilistic aspects of damage initiation. Alabdullah and Ghoniem (2019) developed an energy-based fatigue model that accounts for cycle-dependent property degradation in CMCs.

Also, in chemical degradation and oxidation, oxidation kinetics models coupled with mechanical damage models were proposed by Feng et al. (2021) for SiC/SiC composites in oxidizing environments, while Longbiao (2016) integrated diffusion-reaction equations with mechanical analysis to predict oxidation-induced stress in C/SiC composites. The thermal cycling degradation has been modeled by Zhang et al. (2024) using a coupled thermo-mechanical damage framework. Thermally induced microcracking was simulated by Zhu et al. (2023) using a generalized method of cells approach (Zhang et al., 2024).

Advanced computational methods, which include incorporating machine learning techniques to predict damage evolution, have also gained attention recently. The application of neural networks trained on micromechanical simulation data to accelerate composite aging predictions has been presented by Wang and Sun (2019). Bayesian networks have been employed by Chiachío et al. (2015) to quantify uncertainty in damage progression models. Other advanced computational methods, such as phase field methods and validation approaches, have been applied in the computational modeling of mechanical aging of HTCMs. The phase field approaches have gained popularity for modeling complex crack propagation. De Lorenzis and Gerasimov (2020) extended phase field fracture models to account for environmentally assisted cracking in composites. Kuhn and Müller (2009) developed a phase field model that captures the interaction between oxidation and mechanical damage. DIC combined with in situ mechanical testing has been used to validate computational models (Genin and Birman, 2009). Another validation model is the Neutron diffraction techniques, which have enabled validation of internal stress predictions in models as presented by Scheuerlein et al. (2007).

11.4.5 Computational modeling of radiation aging

Computational modeling of radiation aging in HTCMs involves simulating the complex interactions between radiation damage, thermal effects, and material degradation over time (Luptáková et al., 2024; Singh et al., 2024). Modeling approach employed in radiation aging simulation is the multiscale modeling (Premanand et al., 2025). Several computational approaches categorized under the multiscale modeling includes; atomic-scale, mesoscale, and macroscale. This multiscale modeling is considered most effective computational approach because of its integration of multiple lengths and time scales. The atomic-scale model captures radiation-induced defect formation by virtue of the MD and DFT (Faddoul et al., 2023; Singh et al., 2024). Moreover, considering the mesoscale and macroscale, the finite element analysis, phase field, and kinetic Monte Carlo methods are used to predict the mechanical property and components' lifetimes, and defect evolution and microstructural changes, respectively (Ghaderi et al., 2024; Thomé et al., 2013). The MD simulations offer atomic-level insights into radiation-induced structural changes. Several modeling approaches include interatomic potential-based models, kinetic Monte Carlo simulations, and calculations on density function theory. Comprehensive radiation damage models are achieved using MD simulations for ceramic composites (Zinkle and Snead, 2014), while advanced computational techniques are employed in investigating microstructural evolution under neutron irradiation (Dolezal, 2022). The computational models in the radiation damage mechanisms typically address the displacement cascade dynamics, microstructural restructuring, and radiation-induced property modifications (Thomé et al., 2013), while the integrated modeling strategies under the multiscale approaches are targeted at bridging atomistic and continuum scales, coupling thermal and radiation effects (Zinkle and Snead, 2014).

11.5 Factors influencing the aging of high-temperature composite materials

HTCMs experience several aging mechanisms when exposed to elevated temperatures over extended periods. Several factors that significantly influence the aging behavior of high HTCMs include thermal exposure, oxidation effects, moisture and environmental contaminants, mechanical loading, and fiber/matrix interface stability. Considering thermal exposure, prolonged exposure to elevated temperatures accelerates degradation mechanisms. According to Jian (2021), the effects of this prolonged exposure include matrix softening and subsequent creep deformation, interfacial debonding between fibers and matrix, and microstructural evolution affecting load transfer mechanisms. Muna et al. (2022) presented that thermal cycling induces more severe damage than isothermal aging due to differential thermal expansion between constituent materials. The oxygen penetration represents one of the most significant degradation factors under the oxidation effect. Several key oxidation processes have been identified by Krauklis and Echtermeyer (2018) to include formation of an oxidized layer with altered mechanical properties, progressive weight loss due to volatilization of oxidation products, and shrinkage and cracking of the matrix phase. The oxidation kinetics follow a temperature-dependent Arrhenius relationship, with activation energies typically ranging from 80 to 150 kJ/mol (Mor et al., 2024). The effect of thermal exposure is thermal oxidation, which is often the primary degradation mechanism for PMCs at elevated temperatures. The diffusion of oxygen into the material leads to formation of an oxidized layer with altered mechanical properties, reduction in overall mechanical performance, and weight loss and dimensional changes. Research by Pochiraju et al. (2008) demonstrated that thermal oxidation in PMR-15 composites resulted in a distinct oxidized layer with significantly different properties compared to the core material. The effect of prolonged high-temperature exposure leads to microstructural changes, which cause the formation and growth of microvoids and cracks, crystallinity changes in semicrystalline matrices, and fiber degradation (particularly for organic fibers). Xue et al. (2019) used microscopy techniques to document the evolution of microcracks and voids during thermal aging of carbon/BMI composites.

Moreover, moisture and environmental contaminants are tagged as environmental factors that significantly accelerate degradation of high-temperature composites. According to He et al. (2024), these factors cause effects such as Water molecules disrupting hydrogen bonding in polymer matrices, salt contaminants creation in localized corrosion cells, and UV radiation induction in chain scission in organic components. These effects are particularly pronounced in PMCs but can also affect ceramic matrix systems through grain boundary degradation. The outcome or effects of the matrix degradation are the rapid degradation of the polymer matrix than reinforcing fibers, which leads to chain scission and crosslinking reactions, volatilization of low molecular weight compounds, and embrittlement and microcracking. Polyimide matrices undergo significant molecular restructuring during thermal aging, with competing chain scission and cross-linking mechanisms that affect the mechanical performance (Li et al., 2024a). The environmental factors that play a critical role in aging include humidity, which accelerates degradation through hydrolysis reactions, radiation (UV, ionizing) that can trigger photochemical degradation, and combined effects of temperature and environment that are often synergistic in operation. Flanigan (2022) demonstrated that the combined effect of moisture and elevated temperature significantly accelerated aging compared to dry thermal aging alone.

The mechanical loading is also a major factor contributing to aging of high-temperature composites. Stress states during thermal aging substantially influence degradation rates. Yang et al. (2023) highlighted several stresses that contribute to aging, such as tensile stresses that accelerate the crack formation and propagation, compressive stresses that may induce localized buckling and delamination, and combined thermo-mechanical fatigue that leads to accelerated damage accumulation. Moreover, the fiber/matrix interface stability is a factor that contributes to aging because the interface region often represents the weakest link during aging. Several causes or reactions that contribute to the interface stability effects have been highlighted in literature (Wang et al., 2024). These reactions include chemical reactions at the interface that reduce the bonding strength, differential thermal expansion that creates interfacial shear stresses, and Interdiffusion phenomena that alter local compositions and properties. Moreover, the deterioration of the fiber–matrix interface affects the integrity of the fiber–matrix interface, which is critical for load transfer and overall performance. These effects cause thermal cycling, which can induce interfacial debonding due to mismatch, chemical degradation of sizing or coupling agents, which weakens the interface, and Interfacial microcracks, which serve as pathways for accelerated oxidation. Studies by Zhang et al. (2024) using micromechanical testing demonstrated a significant decrease in interfacial shear strength in carbon/epoxy composites after thermal aging at 200°C. The stress state as a result of mechanical loadings causes aging effects via the Stress-accelerated chemical reactions, Stress-assisted diffusion of oxygen and other reactive species, and creep and stress relaxation phenomena. Colin and Verdu (2003) developed models showing how mechanical stress accelerates oxidation reactions in high-temperature polyimide composites.

11.6 Conclusion

The utilization of HTCMs is at the forefront in the aerospace, energy, and automotive industries, where high demand is characterized by thermal, mechanical, and environmental extremes, and long service times are essential. This results in a rapid decline in mechanical properties, structural integrity, and performance, thereby creating a need for advanced computer modeling methods to predict the long-term behavior of these components. In this chapter, an in-depth discussion of the fundamental aging mechanisms affecting HTCMs, including thermal degradation, oxidation, creep, fatigue, and microstructural changes, has been presented. Computational modeling has been an active area of research, and a variety of methods have been investigated and discussed in the chapter. Each method has its own advantages and limitations. However, Multiscale modeling allows a more integrated framework, linking atomistic and meso- or continuum-scale phenomena to better capture the overall effects of aging. To enable the practical application and aging simulations, numerical implementation strategies, discretization methods, solution schemes, and computational software tools are essential. These approaches enable scientists and engineers to model intricate aging processes, evaluate material behavior at various service configurations, and improve composite architecture for advanced durability.

In addition, experimental validation is critical to the accuracy and credibility of computational models. This is because the results need to be complemented by validation data, including comparative studies, so as to provide further evidence that computational models can help correlate with the observed real-world aging behavior. Nevertheless, bridging the gap between lab-based events and the in-service condition, especially for long-term mass diffusion aging prediction, still presents challenges. Going forward, additional developments in areas such as multiphysics modeling, real-time environmental monitoring, and AI-based predictive analytics will augment the research efforts to model the effect of aging even more accurately.

References

Adedoja, O.S., Sadiku, E.R., Hamam, Y., 2023. An overview of the emerging technologies and composite materials for supercapacitors in energy storage applications. Polymers 15 (10). Available at https://doi.org/10.3390/polym15102272.

Adedoja, O.S., Adekoya, G.J., Sadiku, E.R., Hamam, Y., 2025. Ab initio investigation of graphene-polythiophene nanocomposite as electrode material for mg-ion batteries: a computational perspective. ACS Applied Energy Materials 8 (3), 1427–1436. Available at https://doi.org/10.1021/acsaem.4c01505.

Afzal, A., Bangash, M.K., Hafeez, A., Shaker, K., 2023. Aging effects on the mechanical performance of carbon fiber-reinforced composites. International Journal of Polymer Science 2023. Available at https://doi.org/10.1155/2023/4379307.

Alabdullah, M., Ghoniem, N.M., 2019. Damage mechanics modeling of the non-linear behavior of SiC/SiC ceramic matrix composite fuel cladding. Journal of Nuclear Materials 524, 296–311. Available at https://doi.org/10.1016/j.jnucmat.2019.07.017.

Barbero, E.J., Cortes, D.H., 2010. A mechanistic model for transverse damage initiation, evolution, and stiffness reduction in laminated composites. Composites Part B: Engineering 41 (2), 124–132. Available at https://doi.org/10.1016/j.compositesb.2009.10.001.

Barbero, E.J., De Vivo, L., 2001. Constitutive model for elastic damage in fiber-reinforced PMC laminae. International Journal of Damage Mechanics 10 (1), 73–93. Available at https://doi.org/10.1106/6PQ6-31JW-F69K-74LU.

Batista, N.L., Iha, K., Botelho, E.C., 2013. Evaluation of weather influence on mechanical and viscoelastic properties of polyetherimide/carbon fiber composites. Journal of Reinforced Plastics and Composites 32 (12), 863–874. Available at https://doi.org/10.1177/0731684413482994.

Birks, N., Meier, G.H., Pettit, F.S., 2006. Introduction to the high temperature oxidation of metals, Second Edition. Cambridge University Press, United States, pp. 1–338. Available at http://doi.org/10.1017/CBO9781139163903, http://doi.org/10.1017/CBO9781139163903.

Blaiszik, B.J., Kramer, S.L.B., Olugebefola, S.C., Moore, J.S., Sottos, N.R., White, S.R., 2010. Self-healing polymers and composites. Annual Review of Materials Research 40 (1), 179–211. Available at https://doi.org/10.1146/annurev-matsci-070909-104532.

Bowles, K.J., Roberts, G.D., Kamvouris, J.E., 1997. Long-term isothermal aging effects on carbon fabric-reinforced PMR-15 composites: compression strength. ASTM Special Technical Publication 1302, 175–190. Available at https://doi.org/10.1520/stp11374s.

Bowles, K.J., 1999. Durability of graphite-fiber-reinforced PMR-15 composites aged at elevated temperatures. Journal of Composites Technology and Research 21 (3), 127–132. Available at https://doi.org/10.1520/ctr10612j.

Bowles, K.J., Papadopoulos, D.S., Inghram, L.L., McCorkle, L.S., Klan, O.V., 2001. Longtime durability of PMR-15 matrix polymer at 204, 260, 288, and 316 °C. NASA Technical Memorandum. Available at https://ntrs.nasa.gov/api/citations/20010069579/downloads/20010069579.pdf.

Chaboche, J.L., 2008. A review of some plasticity and viscoplasticity constitutive theories. International Journal of Plasticity 24 (10), 1642–1693. Available at https://doi.org/10.1016/j.ijplas.2008.03.009.

Chawla, K.K., 1998. Metal matrix composites. Composite Materials. Springer Nature America, Inc, New York, NY, pp. 164–211. Available at https://doi.org/10.1007/978-1-4757-2966-5_6.

Cheng, L., Xu, Y., Zhang, L., Yin, X., 2000. Oxidation behavior of three dimensional C/SiC composites in air and combustion gas environments. Carbon 38 (15), 2103–2108. Available at https://doi.org/10.1016/s0008-6223(00)00068-3.

Chernyak, S.A., Ivanov, A.S., Strokova, N.E., Maslakov, K.I., Savilov, S.V., Lunin, V.V., 2016. Mechanism of thermal defunctionalization of oxidized carbon nanotubes. The Journal of Physical Chemistry C 120 (31), 17465–17474. Available at https://doi.org/10.1021/acs.jpcc.6b05178.

Chiachío, J., Chiachío, M., Saxena, A., Sankararaman, S., Rus, G., Goebel, K., 2015. Bayesian model selection and parameter estimation for fatigue damage progression models in composites. International Journal of Fatigue 70, 361–373. Available at https://doi.org/10.1016/j.ijfatigue.2014.08.003.

Colin, X., Marais, C., Verdu, J., 2001. Thermal oxidation kinetics for a poly(bismaleimide). Journal of Applied Polymer Science 82 (14), 3418–3430. Available at https://doi.org/10.1002/app.2203.

Colin, X., Verdu, J., 2003. Thermal ageing and lifetime prediction for organic matrix composites. Plastics, Rubber and Composites 32 (8-9), 349–356. Available at https://doi.org/10.1179/146580103225004117.

Colin, X., Verdu, J., 2005. Strategy for studying thermal oxidation of organic matrix composites. Composites Science and Technology 65 (3-4), 411–419. Available at https://doi.org/10.1016/j.compscitech.2004.09.011.

Colin, X., Verdu, J., 2006. Polymer degradation during processing. Comptes Rendus Chimie 9 (11-12), 1380–1395. Available at https://doi.org/10.1016/j.crci.2006.06.004.

Colombo, P., Mera, G., Riedel, R., Sorarù, G.D., 2010. Polymer-derived ceramics: 40 Years of research and innovation in advanced ceramics. Journal of the American Ceramic Society 93 (7), 1805–1837. Available at https://doi.org/10.1111/j.1551-2916.2010.03876.x.

Dai, Y., Odette, G.R., Yamamoto, T., 2012. 1.06 - The effects of helium in irradiated structural alloys. Comprehensive Nuclear Materials 1-5, 141–193. Available at https://doi.org/10.1016/B978-0-08-056033-5.00006-9.

Dao, B., Hodgkin, J., Krstina, J., Mardel, J., Tian, W., 2006. Accelerated aging versus realistic aging in aerospace composite materials. II. Chemistry of thermal aging in a structural composite. Journal of Applied Polymer Science 102 (4), 3221–3232. Available at https://doi.org/10.1002/app.24573.

Dao, B., Hodgkin, J., Krstina, J., Mardel, J., Tian, W., 2010. Accelerated aging versus realistic aging in aerospace composite materials. V. The effects of hot/wet aging in a structural epoxy composite. Journal of Applied Polymer Science 115 (2), 901–910. Available at https://doi.org/10.1002/app.31029.

Davies, P., MazÉas, F., Casari, P., 2001. Sea water aging of glass reinforced composites: shear behaviour and damage modelling. Journal of Composite Materials 35 (15), 1343–1372. Available at https://doi.org/10.1106/mnbc-81ub-nf5h-p3ml.

De Lorenzis, L., Gerasimov, T., 2020. Numerical implementation of phase-field models of brittle fracture. In: De Lorenzis, L., Düster, A. (Eds.), Modeling in Engineering Using Innovative Numerical Methods for Solids and Fluids. CISM International Centre for Mechanical Sciences, vol 599. Springer Science and Business Media LLC, Cham, pp. 75–101. Available at https://doi.org/10.1007/978-3-030-37518-8_3.

de Parscau du Plessix, B., Jacquemin, F., Lefébure, P., Le Corre, S., 2016. Characterization and modeling of the polymerization-dependent moisture absorption behavior of an epoxy-carbon fiber-reinforced composite material. Journal of Composite Materials 50 (18), 2495–2505. Available at https://doi.org/10.1177/0021998315606510.

Deng, K., Luo, B., Suo, H., Zhang, K., Wang, L., Cheng, H., et al., 2024. Characterization of material degradation mechanism of carbon fiber reinforced epoxy resin composites under ultraviolet radiation and salt-fog synergistic environment. Polymer Composites 45 (1), 805–824. Available at https://doi.org/10.1002/pc.27816.

DiCarlo J.A., Yun H.M., Morscher G.N., Bhatt R.T., 2003. High-Performance SiC/SiC Ceramic Composite Systems Developed for 1315 C (2400 F) Engine Components. Research and Technology.

DiCarlo, J.A., Van Roode, M., 2006. Ceramic composite development for gas turbine engine hot section components. Proceedings of the ASME Turbo Expo 2, 221–231. Available at https://doi.org/10.1115/GT2006-90151.

Dolezal T.D., 2022. A Study on the Early Stages of Degradation of Multi-component Alloy Surfaces in Extreme Environments Using the Multi-cell. Technical Report, Master's Thesis. Available at https://apps.dtic.mil/sti/citations/AD1176048.

Dong, X., Wang, L., Luo, R., Zhang, G., Wang, J., Yang, X., et al., 2024. Thermal protection and dielectric properties of borosilicate coatings for SiCf/SiC composites under high-temperature oxidation. Coatings 14 (11), 1456. Available at https://doi.org/10.3390/coatings14111456.

Faddoul, J., Rahme, P., Guines, D., Leotoing, L., 2023. Thermo-visco mechanical behavior of glass fiber reinforced thermoplastic composite. Journal of Composite Materials 57 (23), 3741–3754. Available at https://doi.org/10.1177/00219983231192147.

Feng, L., Song, Q., He, X., Zhang, Z., Hou, X., Ye, X., et al., 2021. High-performance multifunctional carbon-silicon carbide composites with strengthened reduced graphene oxide. ACS Nano 15 (2), 2880–2892. Available at https://doi.org/10.1021/acsnano.0c08924.

Fish, J., Yu, Q., 2001. Multiscale damage modelling for composite materials: theory and computational framework. International Journal for Numerical Methods in Engineering 52 (1-2), 161–191. Available at https://doi.org/10.1002/nme.276.

Flanigan M.D., 2022. Accelerated aging of adhesively bonded composite joints for use in material screening and selection during development: a review and case study. Theses and Dissertations. 519. Available at https://repository.fit.edu/etd/519.

Gates T.S., Grayson M.A., 1999. On the use of accelerated aging methods for screening high temperature polymeric composite materials. Collection of Technical Papers - AIAA/ASME/ASCE/AHS/ASC Structures, Structural Dynamics and Materials Conference. Available at https://doi.org/10.2514/6.1999-1296.

Genin, G.M., Birman, V., 2009. Micromechanics and structural response of functionally graded, particulate-matrix, fiber-reinforced composites. International Journal of Solids and Structures 46 (10), 2136–2150. Available at https://doi.org/10.1016/j.ijsolstr.2008.08.010.

Ghaderi, A., Ayoub, G., Dargazany, R., 2024. Constitutive behavior and failure prediction of crosslinked polymers exposed to concurrent fatigue and thermal aging: a reduced-order knowledge-driven machine-learned model. Journal of Materials Science 59 (12), 5066–5084. Available at https://doi.org/10.1007/s10853-023-09131-w.

Goulmy, J.P., Caty, O., Rebillat, F., 2020. Characterization of the oxidation of C/C/SiC composites by Xray micro-tomography. Journal of the European Ceramic Society 40 (15), 5120–5131. Available at https://doi.org/10.1016/j.jeurceramsoc.2020.06.042.

Grace, L.R., Altan, M.C., 2012. Characterization of anisotropic moisture absorption in polymeric composites using hindered diffusion model. Composites Part A: Applied Science and Manufacturing 43 (8), 1187–1196. Available at https://doi.org/10.1016/j.compositesa.2012.03.016.

Grammatikos, S.A., Evernden, M., Mitchels, J., Zafari, B., Mottram, J.T., Papanicolaou, G.C., 2016. On the response to hygrothermal aging of pultruded FRPs used in the civil engineering sector. Materials and Design 96, 283–295. Available at https://doi.org/10.1016/j.matdes.2016.02.026.

Hamidi, Y.K., Aktas, L., Altan, M.C., 2005. Effect of packing on void morphology in resin transfer molded E-glass/epoxy composites. Polymer Composites 26 (5), 614–627. Available at https://doi.org/10.1002/pc.20132.

Hamza, M.H., Chattopadhyay, A., 2024. Multi deep learning-based stochastic microstructure reconstruction and high-fidelity micromechanics simulation of time-dependent ceramic matrix composite response. Composite Structures 345. Available at https://doi.org/10.1016/j.compstruct.2024.118360, https://www.sciencedirect.com/science/journal/02638223.

Han, Z., Morscher, G.N., 2019. Monitoring fatigue of SiC/SiC ceramic matrix composites with acoustic emission and electrical resistance. Proceedings of the ASME Turbo Expo 6. Available at https://doi.org/10.1115/GT2019-91330, http://www.asmedl.org/journals/doc/ASMEDL-home/proc/.

He, L., Zhao, J., Yao, G., Tang, J., Cao, Y., Wen, D., 2024. Nano-roughness effect on the interfacial thermal oxidation and surface catalytical recombination characteristics for silicon carbide based materials under non-equilibrium flow. Applied Surface Science 664, 160263. Available at https://doi.org/10.1016/j.apsusc.2024.160263.

Hemath, M., Mavinkere Rangappa, S., Kushvaha, V., Dhakal, H.N., Siengchin, S., 2020. A comprehensive review on mechanical, electromagnetic radiation shielding, and thermal conductivity of fibers/inorganic fillers reinforced hybrid polymer composites. Polymer Composites 41 (10), 3940–3965. Available at https://doi.org/10.1002/pc.25703, http://onlinelibrary.wiley.com/journal/10.1002/(ISSN)1548-0569.

Hu, Q., Zhou, X., Tu, Y., Lu, X., Huang, J., Jiang, J., et al., 2021. High-temperature mechanical properties and oxidation resistance of SiCf/SiC ceramic matrix composites with multi-layer environmental barrier coatings for turbine applications. Ceramics International 47 (21), 30012–30019. Available at https://doi.org/10.1016/j.ceramint.2021.07.176.

Issi, J.-P., Nysten, B., 1998. Carbon Fibers. Carbon Fiberrs. Marcel Dekker, pp. 371–462.

Jacobson, N.S., Fox, D.S., Opila, E.J., 1998. High temperature oxidation of ceramic matrix composites. Pure and Applied Chemistry 70 (2), 493–500. Available at https://doi.org/10.1351/pac199870020493.

Jacobson, N., Farmer, S., Moore, A., Sayir, H., 1999. High-temperature oxidation of boron nitride: I, monolithic boron nitride. Journal of the American Ceramic Society 82 (2), 393–398. Available at https://doi.org/10.1111/j.1551-2916.1999.tb20075.x.

Jacobson, N.S., Opila, E.J., Lee, K.N., 2001. Oxidation and corrosion of ceramics and ceramic matrix composites. Current Opinion in Solid State and Materials Science 5 (4), 301–309. Available at https://doi.org/10.1016/S1359-0286(01)00009-2.

Jacobson, N.S., Opila, E.J., Myers, D.L., Copland, E.H., 2005. Thermodynamics of gas phase species in the Si-O-H system. Journal of Chemical Thermodynamics 37 (10), 1130–1137. Available at https://doi.org/10.1016/j.jct.2005.02.001.

Jian, L., 2021. Effect of sizing agent on interfacial properties of carbon fiber-reinforced PMMA composite. Composites and Advanced Materials 30. Available at https://doi.org/10.1177/2633366x20978657.

Jin, X., Fan, X., Lu, C., Wang, T., 2018. Advances in oxidation and ablation resistance of high and ultra-high temperature ceramics modified or coated carbon/carbon composites. Journal of the European Ceramic Society 38 (1), 1–28. Available at https://doi.org/10.1016/j.jeurceramsoc.2017.08.013.

Jones, F.R., 2010. A review of interphase formation and design in fibre-reinforced composites. Journal of Adhesion Science and Technology 24 (1), 171–202. Available at https://doi.org/10.1163/016942409x12579497420609.

Karadimas, G., Salonitis, K., Georgarakis, K., 2021. Oxide ceramic matrix composite materials for aero-engine applications: literature review. Advances in Transdisciplinary Engineering 15, 153–158. Available at https://doi.org/10.3233/ATDE210029, https://www.iospress.nl/bookserie/advances-in-transdisciplinary-engineering/.

Katoh, Y., Snead, L.L., Szlufarska, I., Weber, W.J., 2012. Radiation effects in SiC for nuclear structural applications. Current Opinion in Solid State and Materials Science 16 (3), 143–152. Available at https://doi.org/10.1016/j.cossms.2012.03.005.

Katoh, Y., Koyanagi, T., McDuffee, J.L., Snead, L.L., Yueh, K., 2018. Dimensional stability and anisotropy of SiC and SiC-based composites in transition swelling regime. Journal of Nuclear Materials 499, 471–479. Available at https://doi.org/10.1016/j.jnucmat.2017.12.009.

Kawai, M., Koizumi, M., 2007. Nonlinear constant fatigue life diagrams for carbon/epoxy laminates at room temperature. Composites Part A: Applied Science and Manufacturing 38 (11), 2342–2353. Available at https://doi.org/10.1016/j.compositesa.2007.01.016.

Kirane, K., Salviato, M., Bažant, Z.P., 2016. Microplane-triad model for elastic and fracturing behavior of woven composites. Journal of Applied Mechanics 83 (4). Available at https://doi.org/10.1115/1.4032275.

Koyanagi, T., Katoh, Y., Nozawa, T., Snead, L.L., Kondo, S., Henager, C.H., et al., 2018. Recent progress in the development of SiC composites for nuclear fusion applications. Journal of Nuclear Materials 511, 544–555. Available at https://doi.org/10.1016/j.jnucmat.2018.06.017.

Krauklis, A.E., Echtermeyer, A.T., 2018. Mechanism of yellowing: carbonyl formation during hygrothermal aging in a common amine epoxy. Polymers 10 (9). Available at https://doi.org/10.3390/polym10091017, http://www.mdpi.com/2073-4360/10/9/1017/pdf.

Kuhn, C., Müller, R., 2009. Phase field simulation of thermomechanical fracture. PAMM 9 (1), 191–192. Available at https://doi.org/10.1002/pamm.200910070.

Lamon, J., 2015. Ceramic Matrix Composites: Materials, Modeling and Technology. Wiley.

Lee, K.N., 2000. Current status of environmental barrier coatings for Si-Based ceramics. Surface and Coatings Technology 133-134, 1–7. Available at https://doi.org/10.1016/s0257-8972(00)00889-6.

Li, C., Feng, C., Zhang, L., Shan, J., Shi, M., 2024a. A review of the effect of hydrothermal aging on the mechanical properties of 2D fiber-reinforced resin matrix composites. Polymers for Advanced Technologies 35 (1). Available at https://doi.org/10.1002/pat.6239.

Li, X., Dong, C., Li, J., Bin, G., Zhou, C., 2024b. Molecular dynamics simulation of aging properties in polymer materials: a review. Polymer Bulletin 1–31.

Liang, C., Gao, X., Fu, L., Mei, H., Cheng, L., Zhang, L., 2024. Pore evolution and mechanical response under locally varying density defects in ceramic matrix composites. Composites Part B: Engineering 279, 111459. Available at https://doi.org/10.1016/j.compositesb.2024.111459.

Liu, Y., Xu, X.Z., Mo, S., Lan, B.W., Zhu, C.Z., Li, C.H., et al., 2020. Long-term Thermo-oxidative degradation modeling of a carbon fiber reinforced polyimide composite: multistep degradation behaviors and kinetics. Chinese Journal of Polymer Science (English Edition) 38 (11), 1202–1213. Available at https://doi.org/10.1007/s10118-020-2425-7, http://www.springerlink.com/content/0256-7679.

Longbiao, L., 2016. Modeling the effect of oxidation on hysteresis loops of carbon fiber-reinforced ceramic-matrix composites under static fatigue at elevated temperature. Journal of the European Ceramic Society 36 (3), 465–480. Available at https://doi.org/10.1016/j.jeurceramsoc.2015.11.005.

Luptáková, N., Svoboda, J., Bártková, D., Weiser, A., Dlouhý, A., 2024. The irradiation effects in ferritic, ferritic–martensitic and austenitic oxide dispersion strengthened alloys: a review. Materials 17 (14), 3409. Available at https://doi.org/10.3390/ma17143409.

Mackerle, J., 1996. Finite element analysis and simulation of welding: a bibliography (1976-1996). Modelling and Simulation in Materials Science and Engineering 4 (5), 501–533. Available at https://doi.org/10.1088/0965-0393/4/5/006.

Matzenmiller, A., Lubliner, J., Taylor, R.L., 1995. A constitutive model for anisotropic damage in fiber-composites. Mechanics of Materials 20 (2), 125–152. Available at https://doi.org/10.1016/0167-6636(94)00053-0.

McMahon, S.J., Prise, K.M., 2019. Mechanistic modelling of radiation responses. Cancers 11 (2). Available at https://doi.org/10.3390/cancers11020205, https://www.mdpi.com/2072-6694/11/2/205/pdf.

Meador, M., 1990. High Temperature Polymer Matrix Composites for Extreme Environments. ASM International, pp. 529–539.

Meng, M., Le, H., Grove, S., Jahir Rizvi, M., 2016a. Moisture effects on the bending fatigue of laminated composites. Composite Structures 154, 49–60. Available at https://doi.org/10.1016/j.compstruct.2016.06.078, www.elsevier.com/inca/publications/store/4/0/5/9/2/8.

Meng, S., Zhou, Y., Xie, W., Yi, F., Du, S., 2016b. Multiphysics Coupled Fluid/thermal/ablation Simulation Of Carbon/carbon Composites. Journal of Spacecraft and Rockets 53 (5), 930–935. Available at https://doi.org/10.2514/1.a33612.

Mor, M., Vinci, A., Sciti, D., 2024. Processing and oxidation resistance at 1650 °C of ZrB2-based UHTCMCs with short fibre gradients. Journal of the European Ceramic Society 44 (7), 4536–4548. Available at https://doi.org/10.1016/j.jeurceramsoc.2024.01.099.

More, K.L., Tortorelli, P.F., Ferber, M.K., Keiser, J.R., 2000. Observations of accelerated silicon carbide recession by oxidation at high water-vapor pressures. Journal of the American Ceramic Society 83 (1), 211–213. Available at https://doi.org/10.1111/j.1151-2916.2000.tb01172.x.

Morgan, R.J., Shin, E.E., Rosenberg, B., Jurek, A., 1997. Characterization of the cure reactions of bismaleimide composite matrices. Polymer 38 (3), 639–646. Available at https://doi.org/10.1016/S0032-3861(96)00542-3, http://www.journals.elsevier.com/polymer/.

Morscher, G.N., Bryant, D.R., Tressler, R.E., 1997. Environmental durability of BN-based interphases (for SiCf/SiC$-$/ composites) in H_2O containing atmospheres at intermediate temperatures. Ceramic Engineering and Science Proceedings 18 (3 A), 525–534.

Mosallam A.S., Composites in construction. Handbook of materials selection. (2002), 1369–1420.

Muhamad, I.I., Zahan, K.A., Pa'e, N., Salehudin, M.H., Khairuddin, N., Marsin, A.M., et al., 2018. Accelerated testing methodology for long-term life prediction of cellulose-based polymeric composite materials. Durability and Life Prediction in Biocomposites, Fibre-Reinforced Composites and Hybrid Composites. Elsevier, Malaysia, pp. 149–171. Available at http://doi.org/10.1016/B978-0-08-102290-0.00008-8, http://www.sciencedirect.com/science/book/9780081022900.

Muna, I.I., Mieloszyk, M., Rimasauskiene, R., Maqsood, N., Rimasauskas, M., 2022. Thermal effects on mechanical strength of additive manufactured cfrp composites at stable and cyclic temperature. Polymers 14 (21), 4680. Available at https://doi.org/10.3390/polym14214680.

Naslain, R., 2004. Design, preparation and properties of non-oxide CMCs for application in engines and nuclear reactors: an overview. Composites Science and Technology 64 (2), 155–170. Available at https://doi.org/10.1016/s0266-3538(03)00230-6.

Noda, T., 2020. Advanced SiC-SiC composites for nuclear application. Handbook of Advanced Ceramics and Composites: Defense, Security, Aerospace and Energy Applications. Springer International Publishing, Japan, pp. 641–666. Available at http://doi.org/10.1007/978-3-030-16347-1_20, https://link.springer.com/book/10.1007/978-3-030-16347-1.

Opila, E.J., 2003. Oxidation and volatilization of silica formers in water vapor. Journal of the American Ceramic Society 86 (8), 1238–1248. Available at https://doi.org/10.1111/j.1151-2916.2003.tb03459.x, http://www.blackwellpublishing.com/aims.asp?ref=0002-7820.

Padture, N.P., 2016. Advanced structural ceramics in aerospace propulsion. Nature Materials 15 (8), 804–809. Available at https://doi.org/10.1038/nmat4687, http://www.nature.com/nmat/.

Paladugu, S.R.M., Sreekanth, P.S.R., Sahu, S.K., Naresh, K., Karthick, S.A., Venkateshwaran, N., et al., 2022. A comprehensive review of self-healing polymer, metal, and ceramic matrix composites and their modeling aspects for aerospace applications. Materials 15 (23). Available at https://doi.org/10.3390/ma15238521, http://www.mdpi.com/journal/materials.

Papakonstantinou, C.G., Balaguru, P.N., 2007. Fatigue behavior of high temperature inorganic matrix composites. Journal of Materials in Civil Engineering 19 (4), 321–328. Available at https://doi.org/10.1061/(ASCE)0899-1561(2007)19:4(321).

Patel, S.R., Case, S.W., 2000. Durability of a graphite/epoxy woven composite under combined hygrothermal conditions. International Journal of Fatigue 22 (9), 809–820. Available at https://doi.org/10.1016/S0142-1123(00)00041-4.

Plagianakos, T.S., Muñoz, K., Saenz-Castillo, D., Mendias, M.M., Jiménez, M., Prentzias, V., 2020. Effect of hot-wet storage aging on mechanical response of a woven thermoplastic composite. Aerospace 7 (2). Available at https://doi.org/10.3390/aerospace7020018, https://res.mdpi.com/d_attachment/aerospace/aerospace-07-00018/article_deploy/aerospace-07-00018-v2.pdf.

Pochiraju, K.V., Tandon, G.P., Schoeppner, G.A., 2008. Evolution of stress and deformations in high-temperature polymer matrix composites during thermo-oxidative aging. Mechanics of Time-Dependent Materials 12 (1), 45–68. Available at https://doi.org/10.1007/s11043-007-9042-5.

Pochiraju, K., Tandon, G.P., 2009. Interaction of oxidation and damage in high temperature polymeric matrix composites. Composites Part A: Applied Science and Manufacturing 40 (12), 1931–1940. Available at https://doi.org/10.1016/j.compositesa.2009.03.010.

Pochiraju K.V., Schoeppner G.A., Tandon G.P., Long-term durability of polymeric matrix composites. Long-Term Durability of Polymeric Matrix Composites. 9781441993083 (2012), 1–677, Available at https://doi.org/10.1007/978-1-4419-9308-3. https://doi.org/10.1007/978-1-4419-9308-3.

Premanand, A., Schimmelpfeng, H., Balle, F., 2025. Comparability of fatigue strength and life estimation of a cf-pekk composite under low and ultrasonic frequencies using time-temperature-based approaches. Fatigue & Fracture of Engineering Materials & Structures. Available at https://doi.org/10.1111/ffe.14608.

Qin, Q., Liu, G.-X., Wu, J.-C., Sun, W.-X., Liao, S., 2024. Influence of 1, 1′-Azobis (cyclohexanezonitrile) on the thermo-oxidative aging performance of diolefin elastomers. Polymers 24 (1).

Quadakkers, W.J., Naumenko, D., Wessel, E., Kochubey, V., Singheiser, L., 2004. Growth rates of alumina scales on Fe–Cr–Al Alloys. Oxidation of Metals 61 (1/2), 17–37. Available at https://doi.org/10.1023/B:OXID.0000016274.78642.ae.

Rafiee, M.A., Rafiee, J., Wang, Z., Song, H., Yu, Z.Z., Koratkar, N., 2009. Enhanced mechanical properties of nanocomposites at low graphene content. ACS Nano 3 (12), 3884–3890. Available at https://doi.org/10.1021/nn9010472, http://pubs.acs.org/doi/pdfplus/10.1021/nn9010472.

Ramachandran, K., Leelavinodhan, S., Antao, C., Copti, A., Mauricio, C., Jyothi, Y.L., et al., 2022. Analysis of failure mechanisms of Oxide - Oxide ceramic matrix composites. Journal of the European Ceramic Society 42 (4), 1626–1634. Available at https://doi.org/10.1016/j.jeurceramsoc.2021.11.020, http://www.journals.elsevier.com/journal-of-the-european-ceramic-society/.

Ray, B.C., 2006. Temperature effect during humid ageing on interfaces of glass and carbon fibers reinforced epoxy composites. Journal of Colloid and Interface Science 298 (1), 111–117. Available at https://doi.org/10.1016/j.jcis.2005.12.023.

Rocha, Gonçalves, Reis, Guedes, R.M., 2024. Mechanisms of component degradation and multi-scale strategies for predicting composite durability: present and future perspectives. Journal of Composites Science 8 (6).

Roy, S., Xu, W.X., Park, S.J., Liechti, K.M., 2000. Anomalous moisture diffusion in viscoelastic polymers: modeling and testing. Journal of Applied Mechanics 67 (2), 391–396. Available at https://doi.org/10.1115/1.1304912.

Roy, S., Singh, S., Schoeppner, G.A., 2008. Modeling of evolving damage in high temperature polymer matrix composites subjected to thermal oxidation. Journal of Materials Science 43 (20), 6651–6660. Available at https://doi.org/10.1007/s10853-008-2691-1.

Ruggles-Wrenn, M.B., Christensen, D.T., Chamberlain, A.L., Lane, J.E., Cook, T.S., 2011. Effect of frequency and environment on fatigue behavior of a CVI SiC/SiC ceramic matrix composite at 1200°C. Composites Science and Technology 71 (2), 190–196. Available at https://doi.org/10.1016/j.compscitech.2010.11.008.

Ruggles-Wrenn, M.B., Siegert, G.T., Baek, S.S., 2008. Creep behavior of Nextel™720/alumina ceramic composite with ±45° fiber orientation at 1200°C. Composites Science and Technology 68 (6), 1588–1595. Available at https://doi.org/10.1016/j.compscitech.2007.07.012.

Ruggles-Wrenn, M.B., Lee, M.D., 2016. Fatigue behavior of an advanced SiC/SiC ceramic composite with a self-healing matrix at 1300°C in air and in steam. Materials Science and Engineering: A 677, 438–445. Available at https://doi.org/10.1016/j.msea.2016.09.076.

Russell-Stevens, M., Todd, R., Papakyriacou, M., 2005. The effect of thermal cycling on the properties of a carbon fibre reinforced magnesium composite. Materials Science and Engineering: A 397 (1-2), 249–256. Available at https://doi.org/10.1016/j.msea.2005.02.025.

Sales, A.S.W., De Queiros Pereira, V., Dias, A.N.C., 2025. Advances in nanomaterials for radiation protection in the aerospace industry: a systematic review. Nanotechnology 36 (10). Available at https://doi.org/10.1088/1361-6528/ada38f, https://iopscience.iop.org/journal/0957-4484.

Santhosh, U., Ahmad, J., Ojard, G., Gowayed, Y., 2021. A synergistic model of stress and oxidation induced damage and failure in silicon carbide-based ceramic matrix composites. Journal of the American Ceramic Society 104 (8), 4163–4182. Available at https://doi.org/10.1111/jace.17819.

Sayyidmousavi, A., Bougherara, H., Falahatgar, S.R., Fawaz, Z., 2015. Thermomechanical viscoelastic response of a unidirectional graphite/polyimide composite at elevated temperatures using a micromechanical approach. Journal of Composite Materials 49 (5), 519–534. Available at https://doi.org/10.1177/0021998314521255, http://jcm.sagepub.com/content/by/year.

Scheuerlein, C., Stuhr, U., Thilly, L., 2007. In situ neutron diffraction under tensile loading of powder-in-tube Cu/Nb3Sn composite wires: effect of reaction heat treatment on texture, internal stress state, and load transfer. Applied Physics Letters 91 (4). Available at https://doi.org/10.1063/1.2766685.

Schoeppner, G.A., Tandon, G.P., Ripberger, E.R., 2007. Anisotropic oxidation and weight loss in PMR-15 composites. Composites Part A: Applied Science and Manufacturing 38 (3), 890–904. Available at https://doi.org/10.1016/j.compositesa.2006.07.006.

Sethi, S., Ray, B.C., 2015. Environmental effects on fibre reinforced polymeric composites: evolving reasons and remarks on interfacial strength and stability. Advances in Colloid and Interface Science 217, 43–67. Available at https://doi.org/10.1016/j.cis.2014.12.005.

Singh, G., Yu, J., Xu, F., Yao, T., Xu, P., 2024. Multiscale modeling of silicon carbide cladding for nuclear applications: thermal performance modeling. Energies 17 (23), 6124. Available at https://doi.org/10.3390/en17236124.

Sudheendra, K., Vinodhini, J., Pitchan, M.K., Ajeesh Mannadiar, G., Bhowmik, S., 2023. Effect of High-Energy Radiations on High Temperature-Resistant Thermoplastic Polymeric Composite for Aviation, Space, and Nuclear Applications. Springer Science and Business Media LLC, pp. 219–238. Available at http://doi.org/10.1007/978-981-19-9048-9_7.

Sun, L., Liu, C., Fang, J., Cui, H., Wang, S., Zhang, X., et al., 2024. Multiscale characterization of damage behavior of glass-ceramic bonded SiC joint under He ion irradiation. Journal of the European Ceramic Society 44 (4), 1946–1958. Available at https://doi.org/10.1016/j.jeurceramsoc.2023.11.058.

Suyama, S., Ukai, M., Akimoto, M., Nishimura, T., Tajima, S., 2019. Hydrothermal corrosion behaviors of constituent materials of SiC/SiC composites for LWR applications. Ceramics 2 (4), 602–611. Available at https://doi.org/10.3390/ceramics2040047.

Talreja, R., 2014. Assessment of the fundamentals of failure theories for composite materials. Composites Science and Technology 105, 190–201. Available at https://doi.org/10.1016/j.compscitech.2014.10.014.

Talreja, R., Varna, J., 2023. Modeling Damage, Fatigue and Failure of Composite Materials: A volume in Woodhead Publishing Series in Composites Science and Engineering, Second Edition. Elsevier, United States, pp. 1–598. Available at http://doi.org/10.1016/C2022-0-00099-4, https://www.sciencedirect.com/book/9780443184895.

Tandon, G.P., Pochiraju, K.V., Schoeppner, G.A., 2008. Thermo-oxidative behavior of high-temperature PMR-15 resin and composites. Materials Science and Engineering: A 498 (1-2), 150–161. Available at https://doi.org/10.1016/j.msea.2007.09.103.

Thomé, L., Debelle, A., Garrido, F., Mylonas, S., Décamps, B., Bachelet, C., et al., 2013. Radiation effects in nuclear materials: role of nuclear and electronic energy losses and their synergy. Nuclear Instruments and Methods in Physics Research Section B: Beam Interactions with Materials and Atoms 307, 43–48. Available at https://doi.org/10.1016/j.nimb.2012.11.077.

Tsotsis, T.K., 1995. Thermo-oxidative aging of composite materials. Journal of Composite Materials 29 (3), 410–422. Available at https://doi.org/10.1177/002199839502900307.

Wang, G., 2019. An efficient analytical homogenization technique for mechanical-hygrothermal responses of unidirectional composites with applications to optimization and multiscale analyses. Chinese Journal of Aeronautics 32 (2), 382–395. Available at https://doi.org/10.1016/j.cja.2018.03.025, http://www.sciencedirect.com/science/journal/10009361.

Wang, K., Sun, W.C., 2019. Meta-modeling game for deriving theory-consistent, microstructure-based traction–separation laws via deep reinforcement learning. Computer Methods in Applied Mechanics and Engineering 346, 216–241. Available at https://doi.org/10.1016/j.cma.2018.11.026.

Wang, R., Qi, J.W., Zhang, X., Wang, Y., Qiu, J., Zhao, Z., et al., 2021. A gradient carbide-coated sialon nanowire for synergistic microwave absorption and toughening enhancement. Chemical Engineering Journal 522, 167533. Available at 10.1016/j.cej.2025.167533.

Wang, Y., Meng, Z., Zhu, W., Wan, B., Han, B., Cai, G., et al., 2021. Hygrothermal aging behavior and aging mechanism of carbon nanofibers/epoxy composites. Construction and Building Materials 294, 123538. Available at https://doi.org/10.1016/j.conbuildmat.2021.123538.

Wang, L., Wang, H., Liang, L., Li, J., Zeng, Z., Liu, Y., 2023. Physics-informed neural networks for transcranial ultrasound wave propagation. Ultrasonics 132, 107026. Available at https://doi.org/10.1016/j.ultras.2023.107026.

Wang, J., Karimi, S., Zeinalzad, P., Zhang, J., Gong, Z., 2024. Using machine learning and experimental study to correlate and predict accelerated aging with natural aging of GFRP composites in hygrothermal conditions. Construction and Building Materials 438, 137264. Available at https://doi.org/10.1016/j.conbuildmat.2024.137264.

Weitsman, Y.J., 2012. Fluid Effects in Polymers and Polymeric Composites Mechanical Engineering Series. Springer US. Available at http://doi.org/10.1007/978-1-4614-1059-1.

Whitlow, T., Jones, E., Przybyla, C., 2016. In-situ damage monitoring of a SiC/SiC ceramic matrix composite using acoustic emission and digital image correlation. Composite Structures 158, 245–251. Available at https://doi.org/10.1016/j.compstruct.2016.09.040.

Wu, H., Yip, K., Tian, F., Xie, Z., Li, C.Z., 2009. Evolution of char structure during the steam gasification of biochars produced from the pyrolysis of various mallee biomass components. Industrial and Engineering Chemistry Research 48 (23), 10431–10438. Available at https://doi.org/10.1021/ie901025d.

Wu, R., Wang, X.Q., Zhao, D., Hou, J.A., Wu, C., Lau, D., et al., 2023. Degradation of fiber/matrix interface under various environmental and loading conditions: insights from molecular simulations. Construction and Building Materials 390. Available at https://doi.org/10.1016/j.conbuildmat.2023.131101.

Xiao, G.Z., Shanahan, M.E.R., 1997. Water absorption and desorption in an epoxy resin with degradation. Journal of Polymer Science Part B: Polymer Physics 35 (16), 2659–2670. doi: 10.1002/(SICI)1099-0488(19971130)35:16<2659::AID-POLB9>3.3.CO;2-4.

Xu, X., Luan, X., Zhang, J., Cao, X., Zhao, D., Cheng, L., et al., 2023. Significant improvement of ultra-high temperature oxidation resistance of C/SiC composites upon matrix modification by SiHfBCN ceramics. Composites Part B: Engineering 253, 110553. Available at https://doi.org/10.1016/j.compositesb.2023.110553.

Xue, L., Fan, W., Wu, F., Zhang, Y., Guo, K., Li, J., et al., 2019. The influence of thermo-oxidative aging on the electromagnetic absorbing properties of 3D quasi-isotropic braided carbon/glass bismaleimide composite. Polymer Degradation and Stability 168, 108941. Available at https://doi.org/10.1016/j.polymdegradstab.2019.108941.

Xue, B., Yu, G., Li, J., Ma, W., Ni, Z., Gao, X., et al., 2025. Multiscale modeling and analysis of electro-mechanical properties for plain woven SiCf/PyC/SiC composite under high-temperature tension. Ceramics International 51 (6), 8000–8011. Available at https://doi.org/10.1016/j.ceramint.2024.12.236.

Yan, Z., Feng, Q., Zheng, J., Guo, X., Zhang, X., Zhang, S., et al., 2025. Effects of low temperature neutron irradiation on the microstructure evolution of Amosic-3 SiCf/SiC composites. Ceramics International 51 (9), 12082–12094. Available at https://doi.org/10.1016/j.ceramint.2025.01.060.

Yang, Z., Yang, J., 2020. Investigation of long-term thermal aging-induced damage in oxide/oxide ceramic matrix composites. Journal of the European Ceramic Society 40 (4), 1549–1556. Available at https://doi.org/10.1016/j.jeurceramsoc.2019.10.052.

Yang, Z., Li, W., Chen, Y., Zeng, W., Chen, W., Cao, X., 2023. Life assessment of thermomechanical fatigue in a woven SiC/SiC ceramic matrix composite with an environmental barrier coating at elevated temperature. International Journal of Fatigue 172, 107584. Available at https://doi.org/10.1016/j.ijfatigue.2023.107584.

Zhang B., 2022. Modeling Oxidation-Induced Degradation and Environment-Induced Damage of Thermal Barrier Coatings. Available at http://dx.doi.org/10.20381/ruor-28046.

Zhang, W., Liu, M., Zou, J., Han, Z., Xiong, Y., Liang, B., et al., 2024. Experimental characterization and numerical prediction for mechanical aging of epoxy and its carbon fiber-reinforced composite under hydrothermal conditions. Polymer Composites 45 (12), 11342–11356. Available at https://doi.org/10.1002/pc.28569.

Zhou, J., Lucas, J.P., 1999. Hygrothermal effects of epoxy resin. Part II: variations of glass transition temperature. Polymer 40 (20), 5513–5522. Available at https://doi.org/10.1016/S0032-3861(98)00791-5.

Zhou, S., Liu, X., Yan, Z., Hara, S., Shikazono, N., Zhong, Z., 2023. Kinetic monte Carlo (KMC) simulation of sintering of nickel oxide-yttria stabilized zirconia composites: model, parameter calibration and validation. Materials & Design 232, 112094. Available at https://doi.org/10.1016/j.matdes.2023.112094.

Zhu, T., Ren, Z., Xu, J., Shen, L., Xiao, C., Zhang, C., et al., 2023. Damage evolution model and failure mechanism of continuous carbon fiber-reinforced thermoplastic resin matrix composite materials. Composites Science and Technology 244, 110300. Available at https://doi.org/10.1016/j.compscitech.2023.110300.

Zinkle, S.J., Busby, J.T., 2009. Structural materials for fission & fusion energy. Materials Today 12 (11), 12–19. Available at https://doi.org/10.1016/S1369-7021(09)70294-9.

Zinkle, S.J., Snead, L.L., 2014. Designing radiation resistance in materials for fusion energy. Annual Review of Materials Research 44 (1), 241–267. Available at https://doi.org/10.1146/annurev-matsci-070813-113627.

Zok, F.W., 2016. Ceramic-matrix composites enable revolutionary gains in turbine engine efficiency. American Ceramic Society Bulletin 95 (5), 22–28.

Zok, F.W., Levi, C.G., 2001. Mechanical properties of porous-matrix ceramic composites. Advanced Engineering Materials 3 (1-2), 15–23. doi: 10.1002/1527-2648(200101)3:1/2<15::AID-ADEM15>3.3.CO;2-1.

12

Numerical modeling and simulation of polymer failure at elevated temperatures

Festus Oluwadare Fameso[1], Smith Adamu Salifu[2], Jacques Snyman[3], Mpho Muloiwa[3], Williams Kehinde Kupolati[3], Julius Musyoka Ndambuki[3], Peter Apata Olubambi[4], Dawood Desai[5]

[1]CENTRE FOR CLIMATE CHANGE, WATER SECURITY AND DISASTER MANAGEMENT, DEPARTMENT OF CIVIL ENGINEERING, TSHWANE UNIVERSITY OF TECHNOLOGY, PRETORIA, GAUTENG, SOUTH AFRICA [2]DEPARTMENT OF CHEMICAL, METALLURGICAL AND MATERIALS ENGINEERING, UNIVERSITY OF JOHANNESBURG, JOHANNESBURG, SOUTH AFRICA [3]DEPARTMENT OF CIVIL ENGINEERING, TSHWANE UNIVERSITY OF TECHNOLOGY, PRETORIA, GAUTENG, SOUTH AFRICA [4]CENTRE FOR NANOENGINEERING AND ADVANCED MATERIALS, UNIVERSITY OF JOHANNESBURG, JOHANNESBURG, SOUTH AFRICA [5]DEPARTMENT OF MECHANICAL AND MECHATRONICS ENGINEERING, TSHWANE UNIVERSITY OF TECHNOLOGY, PRETORIA, GAUTENG, SOUTH AFRICA

12.1 Introduction and background

12.1.1 Importance of polymers in modern applications

In recent years, polymers have become of great interest in research and industry, largely as a result of the comparatively improved properties and value they provide. These make them integral to numerous modern applications due to their versatility, lightweight, ease of fabrication, and cost-effectiveness. Uses range from automotive and aerospace to electronics and healthcare, as well as in critical infrastructure, which are key to disaster management and water security value chains, serving as essential materials in the manufacture of parts, insulation, devices, and electronic modules (Shah et al., 2022; Tuli et al., 2024; Saleh Alghamdi et al., 2021; Choudhury et al., 2022). Given the operations of some of these applications, which may be in high-temperature environments, it is essential to understand the behavior of polymers at elevated temperatures and their effects on their structural integrity (Koumoulos et al., 2019; Yaragatti and Patnaik, 2021). As will be further discussed in this chapter, elevated temperatures can significantly affect the mechanical properties and longevity of polymers, making it a source of concern when addressing issues such as operational reliability and safety (Hamad et al., 2017; Bazli and Abolfazli, 2020; Khotbehsara et al., 2019). Understanding these specific structural behaviors can assist in the management of catastrophic failures, explosions, and other disasters while guaranteeing safety and security in their applications as well as reliability in service over time. This is central to the design,

High-Temperature Polymeric Materials. DOI: https://doi.org/10.1016/B978-0-443-32792-6.00014-2

selection, and optimization of polymer-based materials and components. By conducting thorough material characterization, thermal analysis, and accelerated aging studies, researchers and engineers can gain insights that are key to developing more robust and durable polymer formulations, selecting appropriate materials for specific applications, and implementing effective strategies for preventing or mitigating in-service failures.

12.1.2 Overview of polymer failure mechanisms

Polymers, like every other material, are not immune to failure. Though composed of microstructural molecular long chains, loss of structural functionality does occur when certain characteristic limits are exceeded (Hawkins, 1984; Brazel and Rosen, 2012; Jancar et al., 2010). Hence the loss of load-bearing capacity of a polymer can serve as a good definition of polymer failure. The mechanism of this compromise of their structural integrity is often dictated by the type of force(s) producing it, that is, mechanical, thermal, or concurrent actions of both—often referred to as "thermomechanical." The mechanical characteristics of polymers are highly sensitive to the rate of strain and thermochemical nature of their environment (Mobarakian et al., 2020; Gharibi and Abdolmaleki, 2018). While failure mechanisms in polymers may differ significantly from those in metals and ceramics due to their unique molecular structures, the factors influencing these failures are not too far-fetched. These include the following:

- Physical factors (Visser et al., 2010): Stress profiles, loading conditions, usage profile.
- Environmental factors (Sengodan et al., 2020): Temperature, humidity, exposure to chemicals.
- Material property factors (Pukánszky, 1990): Microstructure, molecular weight, manufacturing process, presence of additives.

According to Greenhalgh (2009), polymer failures generally occur as a result of the inability of the polymer material to withstand applied stresses, resulting first in grain boundary separation, crack initiation, and propagation, and eventual fracture when the fracture strength of the material is exceeded. Determining the cause of polymer failure often requires any, or a combination of, visual, analytical, and wide-ranging testing apparatuses to identify the failure mode and underlying causes, which can be related to material properties, design, processing, or service conditions.

12.2 Theoretical background

12.2.1 Failure mechanics of polymers

The modes and mechanisms of failure of polymers stem from their mechanical properties, which are also defined by parameters used in determining the structural behavior of metals and other materials. The failure modes and mechanisms of failure of polymers stem from their mechanical properties, which are also defined by parameters used in determining the structural behavior of metals and other materials in polymers can be either ductile or brittle, depending on the microstructural characteristics and loading conditions (Pineau et al., 2016). Ductile failures are typically accompanied by significant plastic deformation, while brittle failures are normally accompanied by no elongation or plastic deformation (Failure analysis & testing, 2024).

However, as illustrated in Fig. 12–1, properties such as modulus of elasticity, yield, and tensile strengths characteristic of material response on a stress–strain curve are temperature-dependent and vary with the rate of change of strain and the chemical environment, especially at temperatures above the vicinity of room temperature. At low temperatures, deformation proceeds elastically, but the failure mode exhibited is often of the brittle type. At mid-range temperatures in the vicinity of room temperature, polymer materials exhibit plastic behavior before yielding, after extended deformation, even though their initial deformation may be elastic. However, at elevated temperatures, the stress-bearing capacity of polymers is very much limited, and they produce large unrecoverable strains with the least amount of stress. Thus a polymer may exhibit glass-like characteristics at low temperatures, be rubber-like at temperatures above glass transition temperatures, and be a viscous flowing liquid at elevated temperatures.

However, at elevated temperatures, the stress-bearing capacity of polymers is very much limited, and they produce large unrecoverable strains with the least amount of stress.

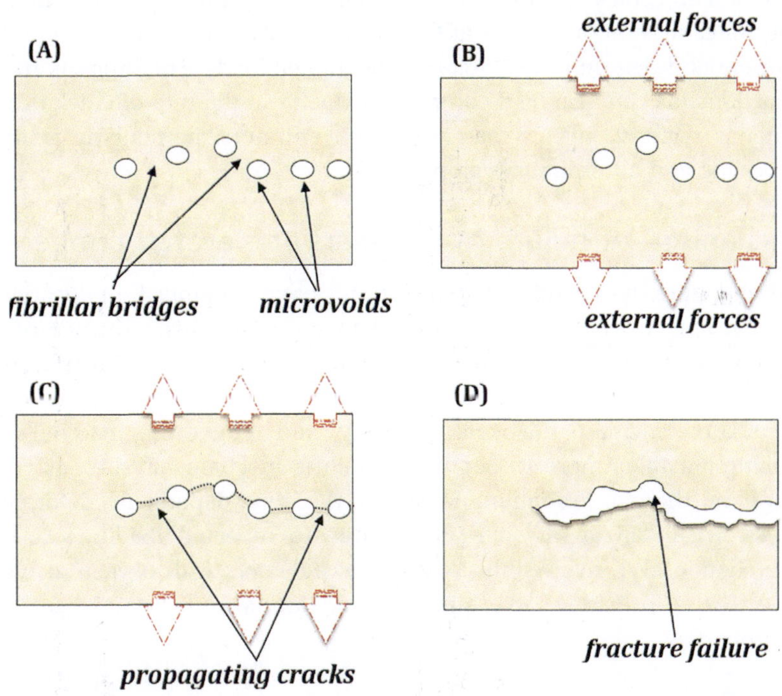

FIGURE 12–1 Schematic drawings of a craze (A) showing microvoids and fibrillar bridges (B) acted upon by external forces (C) showing the initialization and propagation of cracks(D) eventual fracture failure. This figure presents a structural depiction of the failure mechanisms in polymeric materials, which is important for designing and engineering polymer-based components and structures to withstand the expected service loads and environments without premature failure.

At the microstructural level, the separation of grain boundaries plays a crucial role in determining the failure behavior of most polymers (Gutta et al., 2024). This is the region where bond forces acting between adjacent polymer molecules begin to give way to applied external load, which could be thermal or athermal. Recall that polymers are composed of vaguely ordered molecular chain arrangements as well as the presence of impurities, weak Van der Waals forces, and even weaker intermolecular forces, making the boundaries of their intermolecular domains vulnerable to structural compromise. As illustrated in Fig. 12–1, under the influence of the adverse application of thermal loads, these bounding forces begin to give way, resulting in the formation of microvoids, initiation, and subsequent propagation of cracks. As more grain boundaries are separated and more microvoids formed, they begin to coalesce and cross paths with the path of propagating cracks, leading to a more macrostructural manifestation of the failing material and its eventual fracture.

At the macroscopic scale, the fracture behavior of polymers is influenced by the cumulative effect of microstructural defects and grain boundary interactions. Factors such as grain size, strain rate, and the presence of crystal defects can affect the plastic deformation and fracture behavior of polymers. Fine-grained polymers are generally harder and stronger than coarse-grained ones, as they have a greater total grain boundary area to impede dislocation motion and crack propagation. The process of the propagation of the crack is often driven by the concentration of stresses at the crack tips, as well as the ability of the polymer chains to reorient and deform in response to the applied loads. The internal arrangement of the molecular domains can also be a contributing factor to the rate of crack propagation, as nonhomogeneous regions can serve as stress concentrators, particularly at the interfaces between crystalline and amorphous regions.

12.2.2 Mechanisms of failure of polymers at elevated temperatures

Increased temperature is scientifically known to break chemical bonds, with elevated temperatures causing thermal vibrations in polymer chains. This sequence of interactions eventually often leads to the severance of intermolecular bonds within the polymer microstructure. According to Eftekhari and Fatemi (2015), elevated temperatures accelerate thermal degradation, increase molecular mobility, and reduce the material's mechanical strength, making polymers more susceptible to failure mechanisms such as creep, fatigue, impact failure, or thermal oxidation. Creep is the time-dependent deformation under constant stress, while fatigue involves failure due to repeated loading cycles. Thermal-oxidative degradation also involves the reaction of polymers with oxygen at high temperatures, leading to the formation of brittle, oxidized layers and a significant reduction in mechanical properties.

12.2.2.1 Creep failure

Creep is the slow, time-dependent, thermally activated deformation of a material under sustained stress levels (Raghavan and Meshii, 1998). Polymers are particularly susceptible to creep due to their viscoelastic nature. Materials generally creep when exposed to temperatures beginning from 25% of their melting temperatures, even when the loading is below their yield strength. The creeping process proceeds in three distinct stages of slow and continuous deformation of the material over time:

- Primary creep: An initial stage where the rate of deformation decreases over time.
- Secondary creep: Steady-state deformation where the rate of deformation is constant.
- Tertiary creep: Accelerated deformation leading to failure.

12.2.2.2 Fatigue failure

Repeated or cyclic loading can cause fatigue failure in polymers (Katunin, 2012). Even if the applied stress is below the yield strength, repeated cycles can lead to the initiation and growth of cracks, leading to a degradation of the material's mechanical properties and eventual failure once its fracture strength coefficient is exceeded. The number of cycles a polymer can withstand before failing is known as its fatigue life. Factors like the frequency and amplitude of the applied load, the environment, and the inherent properties of the polymer affect fatigue life. Fatigue cracks typically initiate at stress concentrators and grow incrementally with each load cycle until the remaining cross-section can no longer support the applied load. However, according to Crawford and Benham (1974), cycling polymers at high frequencies or stress amplitudes may result in thermal-softening-induced failure due to localized heating, rather than the usual process of incremental initiation, coalescence, and growth of cracks through the material lattice.

12.2.2.3 Impact failure

Polymers can fail under impact loading, where a sudden force is applied to the material, especially when a material is already subjected to high temperatures and stresses. The resistance to impact is influenced by the polymer's toughness, which is a measure of the energy it can absorb before fracturing (Perkins, 1999). Such failures may be brittle in nature or ductile, with resistance to impact significantly reduced in the brittle state. While brittle fracture is generally associated with thermosetting polymers, both ductile and brittle fracture can occur with thermoplastic polymers. However, polymers exhibit a transition from ductile to brittle behavior at lower temperatures, with their impact strength decreasing as temperature increases. Of great importance to polymer applications are the magnitude of their impact strength at ambient temperature, as well as their ductile-to-brittle transition temperature.

12.3 Numerical modeling in polymers

Numerical modeling is a powerful tool for understanding and predicting the failure behavior of polymers subjected to different operating conditions (Starkova et al., 2022). To simulate the failure of polymers at elevated temperatures, various numerical methods are employed, and these methods and the principles behind their implementation and their applications in polymer science are highlighted in subsequent subsections.

12.3.1 Governing equations

The foundation of numerical modeling lies in the governing equations that describe the physical behavior of polymers. These equations encompass the following:

12.3.1.1 Continuum mechanics

Continuum mechanics forms the backbone of numerical modeling approaches for polymer failure, especially at elevated temperatures, as it provides a framework for describing the deformation and failure of materials by considering them as continuous media, even though they are composed of discrete molecules (Bergstrom, 2015). Continuum mechanics takes into account the stress-strain relationships and constitutive models that describe temperature-dependent behaviors, such as viscoelasticity and viscoplasticity (Bergstrom, 2015). The key concepts in continuum mechanics are as follows:

1. Stress-strain relationships
   a. Stress tensor (σ): Represents internal forces within the material per unit area.
   b. Strain tensor (ε): Represents the deformation of the material in response to applied stress.
   c. Constitutive models: Describe the relationship between stress and strain, incorporating temperature-dependent behavior.
2. Constitutive models for polymers
   a. Elasticity: Linear and nonlinear elastic models capture the immediate elastic response of polymers.
   b. Viscoelasticity: Accounts for time-dependent behavior, where the material exhibits both viscous and elastic characteristics. Common models include the Maxwell, Kelvin-Voigt, and Generalized Maxwell models.
   c. Viscoplasticity: Describes irreversible deformation, where the polymer flows like a viscous fluid under high stress. Models like the Perzyna or Bingham models are used.
   d. Thermomechanical coupling: Combines thermal effects with mechanical deformation, critical for polymers operating at elevated temperatures.

12.3.1.2 Heat transfer

Heat transfer plays a crucial role in the numerical modeling of polymer failure, especially at elevated temperatures (Boyard, 2016). Understanding how heat diffuses through polymers and how it affects their mechanical properties is essential for accurate prediction and analysis of failure mechanisms (Boyard, 2016). For example, the heat conduction equation captures the temperature distribution within the polymer. The key concepts in heat transfer are as follows:

1. Heat conduction equations
   a. Fourier's law: Describes the conduction of heat within a material. The heat flux $\boldsymbol{q}$ is proportional to the negative gradient of the temperature field T (Choudhury et al., 2022):

$$\mathbf{q} = -k\nabla T \tag{12.1}$$

   where k is the thermal conductivity of the polymer.

b. Heat equation: Governs the distribution of temperature over time, and it is expressed as follows (Kakaç et al., 2018):

$$\rho c_p \frac{\partial T}{\partial t} = \nabla \cdot (k \nabla T) + Q \tag{12.2}$$

where ρ is the density, C_p is the specific heat capacity, and Q is the internal heat generation per unit volume.

2. Thermal expansion

a. Thermal strain: The strain induced in the material due to temperature changes is given by:

$$\epsilon_{\text{thermal}} = \alpha \Delta T \tag{12.3}$$

where α is the coefficient of thermal expansion, and ΔT is the temperature change. This thermal strain affects the overall stress distribution within the polymer.

II. Temperature-dependent material properties

a. Thermal degradation: High temperatures can cause degradation of polymer chains, altering mechanical properties such as strength, stiffness, and toughness.

b. Viscoelastic and viscoplastic behavior: Temperature significantly influences the time-dependent behavior of polymers, necessitating temperature-dependent constitutive models.

12.3.1.3 Damage mechanics

Damage mechanics is an essential component of numerical modeling approaches for predicting polymer failure (Shahabi and Forouzan, 2017), particularly at elevated temperatures. It provides a framework for describing the initiation and growth of microdamage, such as microcracks, voids, and other defects, which ultimately lead to the macroscopic failure of polymers. The evolution equations for damage variables quantify microcrack formation and growth. Key concepts in damage mechanics are the following:

1. Damage variables

a. Scalar damage variable (***D***): This represents the extent of damage within the material, ranging from 0 (undamaged) to 1 (fully damaged).

b. Tensorial damage variables: In more complex models, damage may be represented by a tensor to account for anisotropic damage effects.

2. Constitutive models with damage

a. Stress–strain relationships: Incorporate damage variables to modify the stress–strain relationship. The effective stress σ' in a damaged material is given by:

$$\sigma' = (1 - D)\sigma \tag{12.4}$$

b. Strain energy density: Damage affects the strain energy density, reducing the material's ability to withstand loads.

Damage evolution laws

c. Damage initiation criteria: Define the conditions under which damage begins to develop, often based on stress or strain thresholds.

d. Damage growth laws: Describe how damage evolves over time or with increasing load, typically using differential equations:

$$\dot{D} = f(\sigma, \epsilon, T) \tag{12.5}$$

where $\dot{D}$ is the rate of damage growth, f is a function of stress and strain, and the absolute temperature is represented by T

II. Coupling with thermal effects

Thermomechanical coupling: Damage mechanics models incorporate thermal effects by recognizing the fact that elevated temperatures can accelerate damage processes such as thermal degradation and viscoelastic or viscoplastic deformation.

12.3.2 Discretization techniques

Discretization techniques are fundamental to numerical modeling as they convert continuous governing equations into a form that can be solved computationally (Sha et al., 1995). In the context of polymer failure modeling at elevated temperatures, discretization allows for the detailed analysis of stress, strain, thermal effects, and damage evolution within polymers. Once the governing equations are established, discretization techniques are employed to convert them into a form suitable for computer simulations, as shown in Fig. 12-2. The common discretization methods, according to Fernandes et al. (2018), Fameso et al. (2021a), Chouet (1986), Cruz-Atienza et al (2007), Benjemaa et al. (2007), include the following.

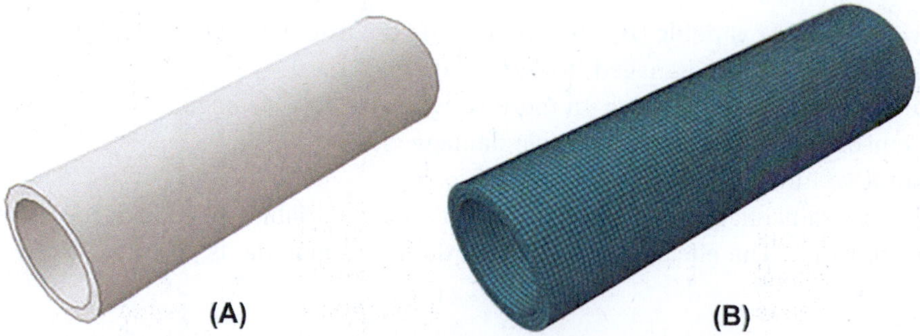

FIGURE 12–2 Developed (A) geometric model and (B) discretized model of a polyvinyl chloride pipe. The figure depicts two stages in the finite element analysis process for a cylindrical pipe: (A) the geometric unmeshed model, and (B) the geometric meshed model.

12.3.2.1 Finite element method

At its core, the finite element method (FEM) of discretization is a spatial discretization method that helps transform complex continuous problems with an infinite number of degrees of freedom into a discrete one with a finite number of unknowns, which can be easily resolved, especially within a computational domain. This technique can be achieved by the methodical implementation of distinct stages, the first of which involves the discretization of the geometry under consideration. This process is often referred to as mesh generation. It is a further subdivision of the initial complex geometry into a finite number of smaller geometries acting as subdomains of the entire geometry. These smaller geometries are often referred to as elements. The elements are smaller units of the entire geometry that can independently represent the properties of the entire geometry. Each element is often defined by a set of "nodes," which serve as the fundamental points of connection between adjacent elements and are locations where primary field variables such as stress, displacement, or temperature are calculated. In the structural analysis of a polymer, for instance, the displacement of the entire domain is described by the displacements computed at these nodal points. The quality of the mesh generation process is paramount to the accuracy, stability, and efficiency of the computation. Highly skewed, elongated, or irregular elements lead to poor convergence and numerical inadequacies. Irregularities in element density and variations also contribute to inaccuracies in the capture of essential variable gradients. Essentially, choosing the right elemental building blocks is critical, as this may be the difference between computational cost and solution accuracy. These building blocks are often characterized by their shapes (triangles, quadrilaterals, hexahedra, and tetrahedra) or order (linear, quadratic, or higher orders). The choice of what building blocks to use depends on the complexity of the problem. For preliminary linear-elastic stress analysis of a polymer part, linear elements suffice. However, in the case of nonlinear phenomena such as hyperelasticity or shock wave propagation, quadratic elements are often desirable to capture the complex strain fields accurately, since quadratic elements have additional nodes at the midpoints of their edges, enabling them to use quadratic polynomials for approximation. Also critical in the finite element discretization technique is the discretization of the fundamental mathematical formulations that describe the physical phenomena under investigation into simple algebraic equations, which hold at every point in the continuous domain that can be solved for the nodal field variable values. The most common method of doing this is by applying the Galerkin method of weighted residuals. A complex partial differential equation is resolved into a system of ordinary differential equations and subsequently assembled into a global matrix of equations by carefully adding their initial and final conditions based on the global node numbering. Thus the FEM of discretization provides a systematic framework for converting the complex physical system under investigation into solvable algebraic equations within the computational domain, allowing simulation and prediction of the response of the system to different physical conditions. This is a foundational step that enables computational analyses to serve as a virtual laboratory for the investigation, evaluation, and design of polymer-based components and processes.

12.3.2.2 Finite difference method

The finite difference method (FDM) is a particularly powerful technique for domain discretization, creating simple, regular domains that form the basis for most computational fluid dynamics (CFDs) and transient heat transfer simulations relevant to polymer processing. It involves the superimposition of a structured grid over the computational domain, and replacing the continuous derivatives of the partial differential equations that represent the governing physics of the phenomena under evaluation with algebraic expressions derived from Taylor series expansions. This directly transforms the differential equations into a system of algebraic equations that can be solved iteratively or directly. As in the case of the FEM, the continuous polymer domain under finite difference discretization is often replaced with a discrete set of points from which solutions will be calculated. The method relies essentially on a structured grid defined by the Cartesian coordinates x, y for two-dimensional geometries and x, y, z for three-dimensional geometries. These Cartesian grid points are often referred to as "nodes" or "grid points," as illustrated in Fig. 12–3, with each point identified by indices i, j, k corresponding to the relevant Cartesian coordinate. The distance between corresponding grid points Δx, Δy, or Δz, is referred to as the grid spacing. Consequent to the geometric discretization, the core of the FDM lies in using the Taylor series to approximate derivatives and the grid points. For a field variable $f(x)$ at grid point i, the value at its neighboring points $i+1$ and $i-1$ will be determined as follows:

Forward expansion:

$$f(i+1) = f(i) + (df(x)/dx)_i \Delta x + (d^2 f(x)/dx^2)_i ((\Delta x)^2/2)! + \cdots \quad (12.6)$$

Backward expansion:

$$f(i-1) = f(i) - (df(x)/dx)_i \Delta x + (d^2 f(x)/dx^2)_i ((\Delta x)^2/2)! - \cdots \quad (12.7)$$

From these series expansions, other fundamental finite difference approximations can be derived. These different formulas are applied to all internal grid points, converting the governing partial differential equations into algebraic equations where the unknowns are the nodal field variables at every grid point.

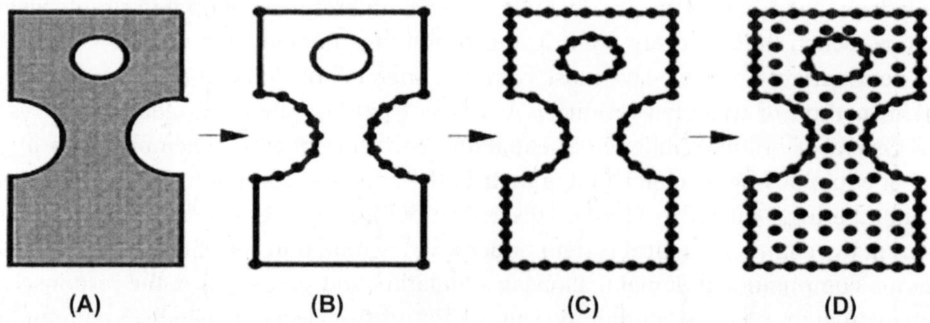

FIGURE 12–3 Sequence for meshless mode generation: (A) Computer Aided design (CAD) model, (B) boundary segment, (C) nodes on boundary segments, and (D) nodes inside the domain (Fameso et al., 2021a). The sequence of shapes from (A) to (D) demonstrates the progression of the mesh generation, starting with a simple initial shape and gradually increasing the complexity and resolution of the mesh through successive refinement steps.

The choice of how to discretize in time is critical for transient variables, that is, time-dependent or subject to change with a change in time. These include scenarios such as the transient heat conduction during the cooling of a polymer melt. Such problems require specific schemes with profound implications for stability and accuracy. For a rapid, high-resolution simulation of a fast injection molding filling phase, an explicit scheme with small, natural time steps is desirable. The advantage is in its computational simplicity and straightforward calculation for each grid point as an independent evaluation. However, the explicit scheme may only be stable at certain time steps smaller than a threshold dictated by the Courant–Friedrichs–Lewy condition. For a slow postfilling cooling phase of the same part, an implicit scheme where the spatial derivatives are evaluated at the future, unknown time level ($t + \Delta t$) is desirable. Meaning the solution at that time level depends on itself, leading to a system of equations that must be solved simultaneously. This allows for much larger time steps and faster simulations, which is highly beneficial for processes with slow transients. However, assembling and solving a system of algebraic equations at each time step may turn out to be more computationally intensive and complex. Another scheme worth mentioning is the Crank–Nicolson scheme, which is more of a hybrid and highly popular approach that balances the merits and demerits of the explicit and implicit methods. It applies an average of the spatial derivatives of the current time (t) and future time ($t + \Delta t$). This generally provides superior accuracy for a given time step, allowing for large time steps without loss of stability.

12.3.2.3 Finite volume method

The finite volume method is another discretization technique particularly adopted in CFDs. It stems from its fundamental property of enforcing local conservation of quantities such as mass, momentum, and energy. This makes it exceptionally robust and physically sound for simulating transport phenomena involved in processes such as the flow of polymer melts, heat transfer during processing, and species diffusion in polymer blends. While the FEM is often preferred for solid mechanics, the finite volume method works best for fluid-related problems. Its application principles revolve around the discretization of the integral form of the governing equations over small control volumes surrounding each node in a computational domain. This ensures that the difference between what goes into the control volume and what leaves it is exactly equal to what is stored or generated within it. The first step in a finite volume discretization is to subdivide the complex domain into a finite number of small nonoverlapping domains called "control volumes." A grid structured as regular rectangular, triangular, or tetrahedral geometry is first defined to define the faces of the control volumes. Each control volume will then have a node located at its geometric center. This is referred to as the "cell-centered scheme." Alternatively, the nodes may be located at the vertices of the grid, and the control volumes constructed around them. This method is referred to as the "vertex-centered scheme." The corresponding governing integral equations, in the form of conservation equations, are thereafter applied to each control volume in the domain and continuously approximated using discrete values at the cell nodes and interpolations at the cell faces. The collective enforcement of conservation over an assembly of a system of algebraic equations across all control volumes guarantees that the governing laws hold not just for the entire domain, but also locally for every control volume. This prevents monphysical mass imbalances that can lead to solver divergence, especially in complex polymer flow simulations with recirculation or sharp fronts.

At the heart of the finite volume method are the calculation of the fluxes of the conserved quantities across the control volumes. These fluxes are the linchpin that connects one control volume to its adjacent neighbors. The stability and accuracy of a finite volume discretization are hinged on this. The primary flux types are the convective and diffusive fluxes F_c and F_D, which define transport due to bulk motion and variable gradients in the fluid, respectively. To compute the flux at a face, interpolation is performed on the conservation equations from the nodes to the face. The choice of interpolation scheme in this case is critical. Convective fluxes can be calculated by any of linear interpolation or central differences, upstream interpolation where the value of the quantity at the face is assumed to be equal to its value in the upstream cell, or by modern high-resolution schemes designed to be highly accurate in smooth regions while remaining stable and sharp in regions with high gradients, such as the flow front of a filling mold. Diffusive fluxes, on the other hand, can simply be calculated directly by approximations of the gradient at a face, leading to a central difference interpolation. The meticulous calculation of fluxes at control volume surfaces makes the finite volume method unique and powerful, making it the preeminent discretization technique for simulating polymer processing operations involving fluid flow and heat transfer.

12.3.2.4 Meshless methods

While the finite elements and final volume methods have been the cornerstones of computational analyses of engineering problems, their reliance on predefined meshes can be a limitation for problems involving extreme deformations, moving boundaries, or complex discontinuities like crack propagation. Mesh-free methods help to overcome these constraints by fundamentally rethinking the spatial discretization of the domain. They iteratively discard predefined boundaries and approximate solutions entirely based on arbitrary sets of nodes scattered within the domain and its boundaries, offering powerful and flexible alternatives for computing challenging processes and failure mechanisms. Discretization in meshless methods begins not with elements but with a cloud of nodes that do not require elemental connectivity. Shape functions are defined in a flexible manner that allows representation of complex geometries without the burden of mesh generation. Two of the most influential meshless methods are the smoothed-particle hydrodynamics (SPH) and the element-free Galerkin (EFG) methods. SPHs is a collocation-based method that represents a material as a collection of discrete particles that carry material properties and discretizes governing equations directly at the nodes (particles). The approximation for any field variable at a particle is computed by a weighted averaging (smoothing) over the values at the neighboring particles within a defined range, known as the smoothing length. This method is essentially well-suited for simulating complex fluid-like processes such as splashing and merging of polymer melt in injection molding and extrusion without mesh tangling and locking issues that plague finite element analyses. The EFG method, on the other hand, is a more rigorous descendant of early meshless ideas, designed primarily for solid mechanics. Although a

Galerkin-based method, which makes it conceptually close to the finite elements method, its core idea is in its use of the moving least squares technique to construct its high-order continuity shape functions. Consequently, essential boundary conditions are not as really enforced as in FEMs. While being meshless for approximation, it requires a background cell structure for the numerical integration of the weak form, avoiding the issues of element distortion in the primary approximation. Thus meshless methods present a shift in paradigms in computational discretization by decoupling the approximation of the solution from a fixed mesh topology. While challenges such as the computational cost of shape function construction and complex imposition of boundary conditions remain, their potential for solving previously intractable problems ensures their continued importance in the computational world.

12.4 Numerical modeling approaches

Numerical modeling approaches are critical for understanding and predicting the behavior of polymers under elevated temperatures. These approaches involve various techniques that simulate the thermomechanical response and failure mechanisms of polymers. Below are the primary numerical modeling approaches used in polymer failure modeling.

12.4.1 Finite element analysis

In finite element analysis (FEA), the principles and techniques adopted involve the discretization (dividing) of the polymer domain into smaller finite elements connected at nodes, such that the governing equations can be used to solve over these elements (Mackerle, 1996). In FEA, the element type can be 1D, 2D, or 3D, as depicted in Fig. 12–4, with higher order elements providing more accuracy. Constitutive material models such as viscoelastic, viscoplastic, and thermomechanical are incorporated to simulate high-temperature behavior.

Implementation of FEA

- Meshing: The quality and density of the mesh significantly influence the accuracy of the simulation. Fine meshes are used in regions with high gradients.
- Boundary conditions: Appropriate thermal and mechanical boundary conditions are applied to reflect real-life scenarios.
- Solvers: Numerical solvers, such as Newton–Raphson for nonlinear problems, are used to solve the system of equations.

Applications of FEA

- Thermomechanical behavior: FEA is extensively used to study stress, strain, and temperature distributions.
- Crack initiation and growth: Predicting where and how cracks initiate and propagate.
- Component design: Optimizing polymer components for high-temperature applications.

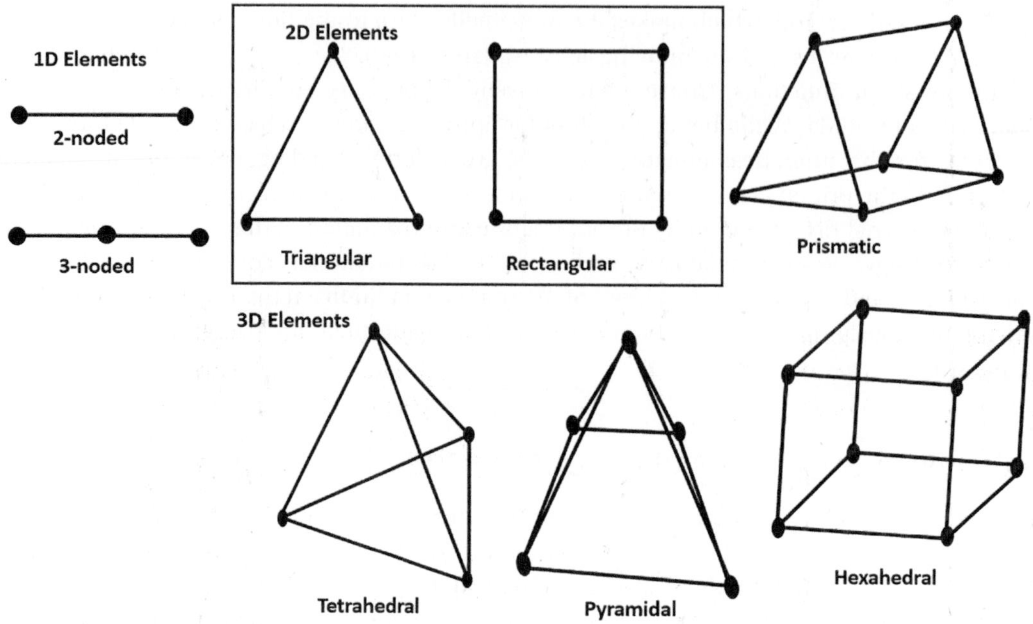

FIGURE 12–4 Different mesh element types (Rathinam, 2015). The figure illustrates various types of mesh elements used in computational modeling and simulation. These elements are fundamental building blocks for representing and discretizing geometric shapes and structures. *From Rathinam A., 2015. Design and development of UWE-4: integration of electric propulsion units, structural analysis, and orbital heating analysis.*

12.4.2 Molecular dynamics simulations

Molecular dynamics (MD) simulations model the behavior of polymers at the atomic scale or a microscopic perspective by solving Newton's equations of motion for atoms (Dewapriya and Miller, 2021; Tam et al., 2017). With MD simulation, accurate force fields such as AMBER and CHARMM are essential for predicting interactions between atoms.

Implementation of MD

- Initial configuration: A realistic initial configuration is necessary, often derived from experimental data.
- Simulation parameters: Environmental factors such as temperature and pressure are set to mimic real operating conditions.
- Time step and duration: Small time steps are used to capture atomic motions, and the duration must be sufficient to observe significant events.

Applications of MD

- Thermal degradation: MD makes it possible to understand mechanisms such as chain scission at the molecular level.
- Mechanical properties: MD makes it possible to study the effect of temperature on polymer mechanical properties.

12.4.3 Finite volume modeling

In finite volume modeling (FVM), the domain is divided into control volumes, and fluxes at their surfaces are calculated to ensure conservation laws are satisfied (Moukalled et al., 2016). FVM is particularly suited for solving fluid flow and heat transfer equations.

Implementation of FVM

- Discretization: The domain is divided into control volumes.
- Flux calculation: Ensures mass, momentum, and energy conservation.
- Solution algorithms: Algorithms like SIMPLE (Semiimplicit Method for Pressure-Linked Equations) are used.

Applications of FVM

- Heat transfer: Simulating heat transfer within polymers at elevated temperatures to predict thermal failure.

12.4.4 Multiscale modeling

Multiscale modeling (MM) combines different scales of analysis (atomistic to macroscopic) to capture the comprehensive behavior of materials (Li et al., 2013). In MM, methods like sequential, concurrent, and hierarchical coupling are used.

Implementation of MM:

- Representative volume elements: Used for mesoscale modeling to capture microstructure.
- Bridging models: Continuum models link with molecular models to provide boundary conditions.

Applications of MM:

- Polymer composites: Studying the interaction between polymer matrix and reinforcements across scales.

12.4.5 Cohesive zone modeling

Cohesive zone modeling (CZM) defines the relationship between traction and separation of crack faces, and it is effective in the simulation of crack growth in polymers (Chandra, 2002).

Implementation CZM

- Element insertion: Cohesive elements are inserted along potential crack paths within the FEM framework.
- Damage evolution: Monitored and elements are progressively degraded to simulate crack growth.

Applications of CZM

- Adhesive failure: It is used to model adhesive failure and delamination in polymer composites.
- Crack growth: It is used to determine crack growth under thermal and mechanical loads.

12.4.6 Phase-field modeling

Phase-field modeling (PFM) simulates the evolution of microstructures without explicitly tracking interfaces (Brighenti et al., 2021). This is achieved by using a continuous field to represent the state of the material.

Implementation of PFM

- Order parameter evolution: PFM is governed by partial differential equations derived from a free energy functional.
- Coupling with mechanics: The phase field of PFM is coupled with mechanical fields.

Applications of PFM

- Crack patterns: It is used in studying complex crack patterns in polymers.
- Crack initiation and propagation: Particularly useful for high-temperature conditions.

12.5 Constitutive models for polymers

In polymers, constitutive models are essential for accurately describing their mechanical behavior under different loading conditions, temperatures, and rates (Heymans, 2003). These models help engineers and researchers understand and predict how polymers deform, flow, and fail in various applications. Here are some commonly used constitutive models for polymers:

12.5.1 Linear elastic model

A linear elastic constitutive model for polymers is a simplified representation of the mechanical behavior of polymers under small deformations. In this model, stress is assumed to be directly proportional to strain, and the behavior displayed is similar to Hooke's law in linear elasticity (Du et al., 2020). While polymers can exhibit complex viscoelastic and nonlinear behavior, the linear elastic model serves as a useful approximation for many engineering applications, especially when the material is subjected to low strains and relatively short durations. In the linear elastic model, polymers are assumed to undergo small deformations. This implies that the strains are proportional to the applied stresses within the elastic limit of the material, where the deformation is reversible and does not result in permanent changes in shape or size. Furthermore, the model assumes isotropic material properties, meaning that material properties such as Young's modulus and Poisson's ratio are independent of direction. This assumption simplifies the modeling process, thus making it suitable for many polymer materials, especially those that are amorphous or lightly crosslinked.

In the linear elastic regime, the stress–strain relationship is described by Hooke's law:

$$\sigma = E \cdot \epsilon \tag{12.8}$$

where

σ is the stress (in Pa or MPa), E is Young's modulus or modulus of elasticity (in Pa or MPa), and ϵ is the strain (dimensionless).

For uniaxial loading along a single direction, the relationship between stress and strain simplifies to:

$$\sigma = E \cdot \epsilon \tag{12.9}$$

The linear elastic model is commonly used in the structural analysis of polymer components under small loads, such as in FEA simulations. It is also used during design optimization to facilitate the prediction of material response and deformation behavior of polymer-based products during the design phase. According to Mahieux and Reifsnider (2001), the linear elastic model does not accurately capture the behavior of polymers under large deformations, time-dependent loading, or in the presence of stress concentrations or material defects. In such cases, more complex constitutive models like hyperelastic or viscoelastic models may be required.

12.5.2 Nonlinear elastic models

Nonlinear elastic models are advanced constitutive models used to describe the mechanical behavior of polymers under large deformations, where the stress–strain relationship deviates significantly from linear elasticity. These models are essential for capturing the complex viscoelastic and nonlinear behavior exhibited by many polymer materials (Mirkhalaf and Vadizadeh, 2024). One of the common nonlinear elastic constitutive models used for polymers is the following:

12.5.2.1 Hyperelastic models

Hyperelastic models describe materials that exhibit nonlinear elastic behavior without plastic deformation (Okeke et al., 2017). These models are suitable for highly deformable materials like elastomers and certain thermoplastic polymers. Some commonly used hyperelastic models include the following:

- Neo-Hookean model: The Neo-Hookean model is a simple yet effective hyperelastic model for polymers (Stoltz et al., 2024). It assumes that the strain energy density (W) depends only on the deformation gradient tensor ($\boldsymbol{F}$). The strain energy density function for the Neo-Hookean model is given by:

$$W = \frac{\mu}{2}(I_1 - 3) - \mu \ln J + \frac{\lambda}{2}(\ln J)^2 \tag{12.10}$$

where

μ is the shear modulus, λ is the Lame's first parameter, I_1 is the first invariant of the deformation tensor $\boldsymbol{F}$, and J is the determinant of the deformation tensor $\boldsymbol{F}$.

The stress tensor (σ) can be obtained by differentiating the strain energy density function with respect to the deformation gradient tensor.

$$\sigma = 2\frac{\partial W}{\partial F} \tag{12.11}$$

- Mooney–Rivlin model: The Mooney–Rivlin model is a more general Neo–Hookean hyperelastic model that includes additional parameters to capture the material behavior more accurately (Kim et al., 2012). The strain energy density function for the Mooney–Rivlin model is given by:

$$W = C_{10}(I_1 - 3) + C_{01}(I_2 - 3) \tag{12.12}$$

where

C_{10} and C_{01} are material constants, and I_1 and I_2 are the first and second invariants of the deformation tensor $\boldsymbol{F}$, respectively.

The stress tensor (σ) can be obtained by differentiating the strain energy density function with respect to the deformation gradient tensor:

$$\sigma = 2\frac{\partial W}{\partial F} \tag{12.13}$$

These equations provide a framework for modeling the nonlinear elastic behavior of polymers under various loading conditions, and the material constants (μ, λ for Neo–Hookean and C_{10}, C_{01} for Mooney–Rivlin) can be determined experimentally or through other means such as FEA.

12.5.2.2 Ogden model

The Ogden model is a nonlinear elastic model that can be extended to describe elastic–plastic behavior as well. This model is commonly used to describe the nonlinear elastic behavior of rubber-like materials. However, it can also be adapted to capture elastic–plastic behavior by incorporating yield criteria and flow rules (El Yaagoubi et al., 2017). The strain energy density function (W) in the Ogden model is expressed as a polynomial function of the principal stretches (λ_1, λ_2, λ_3)

$$W = \sum_{i=1}^{N} \frac{\mu_i}{\alpha_i}[(\lambda_1^{\alpha_i} + \lambda_2^{\alpha_i} + \lambda_3^{\alpha_i}) - 3] \tag{12.14}$$

where

μ_i are material parameters related to the shear modulus, α_i are material parameters related to the nonlinear behavior of the material, and N is the number of terms in the series.

This strain energy density function captures the nonlinear elastic behavior of the material by considering multiple terms with different powers of the principal stretches.

To extend the Ogden model in such a way that it is capable of describing the elastic–plastic behavior, the yield criterion and a flow rule need to be incorporated.

- Yield criterion: A yield criterion, such as the von Mises yield criterion, can be used to determine when plastic deformation begins. Once the stress state exceeds the yield stress predicted by the criterion, plastic deformation initiates.
- Flow rule: The flow rule describes how the material deforms plastically under applied stress. For example, the von Mises flow rule relates the rate of plastic strain to the deviatoric stress.

By combining the Ogden model for nonlinear elasticity with appropriate yield criteria and flow rules, it is possible to develop a constitutive model that accurately captures the elastic-plastic behavior of materials (Jebur et al., 2023). The material parameters in the Ogden model can be determined through fitting experimental data or other calibration methods, allowing the model to be tailored to specific materials and loading conditions.

12.5.3 Viscoelastic models

Viscoelastic models account for both elastic and viscous behavior in polymers, making them suitable for materials that exhibit time-dependent responses (Amirkhizi et al., 2006). Some common viscoelastic models include the following:

12.5.3.1 Maxwell model

The Maxwell model consists of a spring (representing elastic behavior) in series with a dashpot (representing viscous behavior), which shows a linear relationship between stress and strain. This model is effective in modeling short-term viscoelastic behavior in polymers, but it does not capture long-term behavior or stress relaxation accurately (Lan et al., 2018). The expression for the constitutive equation for the Maxwell model is as follows:

$$\sigma(t) = E\varepsilon(t) + \eta\frac{d\varepsilon}{dt} \tag{12.15}$$

where

$\sigma(t)$ is the stress at time t, $\varepsilon(t)$ is the strain at time t, E is the elastic modulus (Young's modulus), representing the stiffness of the material, and η is the viscosity coefficient, representing the resistance to deformation over time.

This equation describes how the stress at time t is influenced by both the instantaneous elastic response ($E\varepsilon(t)$) and the time-dependent viscous response ($\eta\frac{d\varepsilon}{dt}$).

12.5.3.2 Kelvin–Voigt model

The Kelvin–Voigt model is also similar to the Maxwell model because it consists of a spring and a dashpot, but in a parallel configuration, and this represents both the instantaneous and time-dependent responses. This model is useful for modeling creep and stress relaxation in polymers (Marais and Villoutreix, 1998). However, the model oversimplifies the behavior of real polymers, especially at high temperatures or under complex loading conditions. The expression for the constitutive equation for the Kelvin–Voigt model is given as follows:

$$\sigma(t) = E\varepsilon(t) + \eta\varepsilon'(t) \tag{12.16}$$

where

$\sigma(t)$ is the stress at time, t, $\varepsilon(t)$ is the strain at time t, $\varepsilon'(t)$ is the strain rate (the time derivative of strain). E is the elastic modulus (Young's modulus), and η is the viscosity coefficient.

In this model, the stress at time, t depends on both the instantaneous elastic response ($E\varepsilon(t)$) and the viscous response proportional to the strain rate ($\eta\varepsilon'(t)$).

12.5.4 Rate-dependent models

Rate-dependent constitutive models for polymers are essential for capturing the time-dependent behavior of these materials, particularly when subjected to loading rates that vary over time (Siviour and Jordan, 2016). Polymers often exhibit significant rate sensitivity due to their viscoelastic nature. This implies that their mechanical response can vary depending on the rate at which loads are applied. Some of the commonly used rate-dependent constitutive models for polymers are as follows:

12.5.4.1 Quasilinear viscoelastic models

Quasilinear viscoelastic (QLV) models are commonly used to describe the time-dependent behavior of polymers, particularly their viscoelastic response under varying loading conditions. These models are based on the idea of superposition, where the instantaneous elastic response is combined with a time-dependent (viscoelastic) response (Zhi et al., 2017). QLV models are particularly useful for capturing both the elastic and viscous behavior of polymers over a wide range of loading rates. In other words, the rate-dependent constitutive QLV model considers the viscoelastic response of polymers as a combination of an elastic component and a time-dependent component that varies with the loading rate. This model is especially useful for capturing the rate sensitivity of polymers, where the material response changes depending on the rate at which the load is applied (Suchocki, 2013). The constitutive equation for the rate-dependent QLV model is such that the total stress (σ) at time t is expressed as the sum of the instantaneous elastic stress (σ_e) and the time-dependent viscoelastic stress (σ_v)

$$\sigma(t) = \sigma_e(t) + \sigma_v(t) \tag{12.17}$$

and the instantaneous elastic stress (σ_e), which represents the material's immediate response to an applied load, is described using Hooke's law as follows:

$$\sigma_e(t) = E\varepsilon(t) \tag{12.18}$$

where

E is the elastic modulus (Young's modulus) of the material, and $\varepsilon(t)$ is the strain at time t.

The time-dependent viscoelastic stress (σ_v), which accounts for the material's viscoelastic behavior, is often expressed using a Prony series or a similar mathematical representation. This series (Prony series) decomposes the viscoelastic stress into a sum of exponential functions with different relaxation times, as shown in the following equation:

$$\sigma_v(t) = \sum_{i=1}^{N} A_i\left(1 - e^{-t/\tau_i}\right) \tag{12.19}$$

where

A_i are coefficients representing the magnitudes of the relaxation modes, τ_i is the relaxation time associated with each mode, and N is the number of relaxation modes. The coefficients (A_i) and relaxation times (τ_i) in the Prony series are determined experimentally or through fitting to viscoelastic data obtained from rheological tests such as creep, relaxation, or dynamic mechanical analysis (DMA).

12.5.4.2 Modified power law models

The modified power law model is a commonly used empirical model to describe the time-dependent mechanical behavior of polymers, especially in the context of creep and stress relaxation (Ihueze et al., 2023). This model is an extension of the well-known power law model, incorporating additional parameters to improve its accuracy in capturing the viscoelastic response of polymers. The model describes the relationship between stress and time-dependent strain (creep or relaxation) using a power law function, and it accounts for both the instantaneous elastic response and the time-dependent viscous response of polymers. The creep function ($\varepsilon(t)$) of the modified power law model describes the time-dependent deformation of a material under constant applied stress, and can be expressed as follows:

$$\varepsilon(t) = \frac{\sigma_0}{E}\left(\frac{t}{t_0}\right)^n \tag{12.20}$$

where

$\varepsilon(t)$ is the strain at time t, σ_o is the applied stress, E is the elastic modulus (Young's modulus) of the material, t_o is a reference time constant and, n is the power law exponent, representing the rate sensitivity of the material.

On the other hand, the stress relaxation function ($\sigma(t)$) of the model describes the decrease in stress over time when a constant strain is applied to the material, and it can be expressed as follows:

$$\sigma(t) = \sigma_0\left(\frac{t}{t_0}\right)^{-m} \tag{12.21}$$

where

$\sigma(t)$ is the stress at time t, σ_o is the initial stress, t_o is a reference time constant, and m is an empirical parameter that accounts for the time dependence of stress relaxation.

The parameters E, n, and m in the modified power law model are typically determined from experimental data obtained through creep and stress relaxation tests that involve applying constant stress or strain to the material and measuring the resulting deformation or stress relaxation over time. After this, curve-fitting techniques are employed to determine the best fit values of the model parameters. This model finds widespread applications in polymer engineering, materials science, and biomechanics due to its simplicity and effectiveness in characterizing the viscoelastic behavior of polymers under various loading conditions.

12.5.4.3 Three-parameter model

The three-parameter model, also known as the Kelvin–Voigt model with three parameters, is a viscoelastic model commonly used to describe the time-dependent behavior of polymers. It is an extension of the Kelvin–Voigt model, which consists of a spring (representing elastic behavior) and a dashpot (representing viscous behavior) in parallel configuration. The addition of a third parameter in the three-parameter model allows for a better representation of the viscoelastic response of materials, especially in capturing creep and stress relaxation behaviors

(Mortazavian and Fatemi, 2017). In the model, the total stress (σ) at time t is expressed as the sum of the elastic stress (σ_e) and the viscoelastic stress (σ_v) as shown in the following equation:

$$\sigma(t) = \sigma_e(t) + \sigma_v(t) \tag{12.22}$$

The elastic stress (σ_e) represents the immediate response of the material to an applied load and is described using Hooke's law as follows:

$$\sigma_e(t) = E\varepsilon(t) \tag{12.23}$$

where

E is the elastic modulus (Young's modulus) of the material, and $\varepsilon(t)$ is the strain at time t.

On the other hand, the viscoelastic stress (σ_v) accounts for the time-dependent response of the material and is described using a Kelvin–Voigt element with three parameters:

$$\sigma_v(t) = \eta\dot{\varepsilon}(t) + \frac{E}{\tau}\int_0^t \exp\left(-\frac{t-t'}{\tau}\right)\dot{\varepsilon}(t')dt' \tag{12.24}$$

where

η is the viscosity coefficient, $\dot{\varepsilon}(t)$ is the strain rate at time t, and τ is the relaxation time constant.

The parameters E, n, and τ in the three-parameter model are typically determined from experimental data obtained through rheological tests such as creep, stress relaxation, or DMA. These parameters characterize the viscoelastic properties of the material that are crucial for accurately predicting its behavior under different loading conditions. The three-parameter model finds applications in various fields, including polymer engineering, materials science, and biomechanics, as it provides a comprehensive framework for describing the viscoelastic behavior of polymers, thus making it possible to be used by engineers and researchers to predict the response of materials to different loading conditions and also to design polymer-based products with desired mechanical properties.

12.5.4.4 Arrhenius-type models

Arrhenius-type models are used to describe the temperature-dependent behavior of polymers, particularly in relation to their viscoelastic properties (White and Lipson, 2018). These models are based on the Arrhenius equation, which relates the rate of a chemical reaction to temperature. In the context of polymers, Arrhenius-type models are often employed to characterize the effect of temperature on parameters such as viscosity, relaxation time, or modulus. The Arrhenius equation describes the temperature dependence of a reaction rate (k) as follows:

$$k = A\exp\left(-\frac{E_a}{RT}\right) \tag{12.25}$$

where

A is the preexponential factor or frequency factor, E_a is the activation energy, R is the gas constant, and T is the temperature in Kelvin.

In polymers, Arrhenius-type models are often used to describe the temperature dependence of viscoelastic parameters such as viscosity (η), relaxation time (τ), or modulus (E). These models typically assume that the parameter of interest follows an Arrhenius-type relationship with temperature.

Arrhenius-type models find applications in polymer processing, materials engineering, and product design, where understanding the temperature-dependent behavior of polymers is crucial. These models allow engineers and researchers to predict how polymer properties will change with temperature, thus enabling the optimization of processing conditions and the development of materials with tailored performance characteristics.

12.5.5 Plasticity models

According to Vogler et al. (2013), plasticity models for polymers describe the material behavior when subjected to stresses that cause irreversible deformations. Unlike metals, polymers exhibit more complex behaviors due to their viscoelastic and viscoplastic nature. Some of the common plasticity models for polymers, along with their equations, are described in the following sections.

12.5.5.1 Yield criteria for polymers

Similar to metals, polymers have a yield point where they transition from elastic to plastic deformation. Common yield criteria used for polymers include

- Von Mises yield criterion: The Von Mises yield criterion is commonly used for modeling the yield behavior of polymers subjected to multiaxial stress states. This criterion simplifies the complex yielding behavior of polymers but often fails to capture the strain-rate effects accurately.

$$\sigma_e = \sqrt{\frac{3}{2}S:S} \tag{12.26}$$

  where $\mathbf{S}$ is the deviatoric stress tensor. Using this criterion, yielding occurs when σ_e reaches a critical value σ_y.
- Tresca yield criterion: The Tresca yield criterion is a classical theory used in materials science to predict the onset of plastic deformation in ductile materials (Hawkins, 1984). Although it is often applied to metals, it can also be used for polymers, especially those that exhibit ductile behavior. The Tresca criterion is based on the maximum shear stress in the material and can be expressed as follows:

$$\sigma_{\max} - \sigma_{\min} = \sigma_y \tag{12.27}$$

where $\sigma_{\max}$ and $\sigma_{\min}$ are the maximum and minimum principal stresses, respectively.

12.5.5.2 Constitutive plasticity models

Unlike purely elastic materials that return to their original shape upon unloading, polymers exhibit a permanent set or deformation. This inelastic behavior is a controlled yielding process that allows the material to absorb energy and redistribute stress. Accurately capturing this sequence is crucial for the virtual prototyping and design of polymer products.

These necessitate the use of constitutive plasticity models, which are mathematical frameworks built upon key components such as yield criterion, flow rule, and hardening/softening rule. Some of these models include the following:

- Elastic perfectly plastic model: This model assumes that the material behaves elastically up to the yield point and then deforms plastically at a constant stress. The expression for the model is given as follows:

$$\sigma = E\varepsilon \text{ for } \sigma \leq \sigma_y \tag{12.28}$$

$$\sigma = \sigma_y \text{ for } \sigma > \sigma_y \tag{12.29}$$

  where E is the elastic modulus and σ_y is the yield stress.
- Isotropic hardening model: In this model, the yield stress increases with plastic strain, representing strain hardening.

$$\sigma = E\varepsilon_e + K\varepsilon_p^n \tag{12.30}$$

  where
  ε_e is the elastic strain, ε_p is the plastic strain, K and n are material constants.
- Kinematic hardening model: This model accounts for the Bauschinger effect, where the yield surface translates in the stress space.

$$\alpha = C\varepsilon_p \tag{12.31}$$

where α is the back stress, and C is a constant defining the movement of the yield surface.

12.5.5.3 Viscoplastic models

Classical plasticity theory often operates on the assumption that material response to loading is rate-independent. This implies that the stress–strain curve for a material is identical whether it is pulled apart in a fraction of a second or over the course of an hour. For metals, this can be a reasonable first approximation, but for polymers, it is a fundamental misrepresentation of their physics of deformation. The mechanical behavior of polymers is intrinsically time-dependent, manifesting in several ways, such as in strain-rate sensitivity, creep, and stress relaxation. Classical rate-independent plasticity models are incapable of capturing these behaviors; hence overcoming them requires a more advanced class of material models, such as viscoplastic models. This class of plasticity models is more suited for polymers as they incorporate time-dependent plastic deformation. They include the following:

- Perzyna viscoplastic model: This model combines rate-dependent plastic flow with yield criteria.

$$\dot{\varepsilon}_p = \frac{1}{\eta}\left\langle \frac{f(\sigma) - \sigma_y}{\sigma_y} \right\rangle^n \tag{12.32}$$

  where
  $\dot{\varepsilon}_p$ is the plastic strain rate, η is the viscosity parameter<.> is the Macaulay bracket, $f(\sigma)$ is the yield function, and n is a strain rate sensitivity parameter.

- Bodner–Partom viscoplastic model: This model characterizes both the immediate elastic response and time-dependent plastic behavior.

$$\dot{\varepsilon} = \dot{\varepsilon}_e + \dot{\varepsilon}_p \tag{12.33}$$

The plastic strain rate is given by:

$$\dot{\varepsilon}_p = A\left(\frac{\sigma}{\sigma_0}\right)^n \exp\left(-\frac{Q}{RT}\right) \tag{12.34}$$

where

A is a preexponential factor, σ_o is the reference stress, Q is the activation energy, R is the gas constant, T is the temperature, and n is a strain rate sensitivity exponent.

12.6 Industrial applications

The integration of numerical simulation applications in industrial processes provides valuable insights into polymer failure when manufactured components are used across varying ranges of temperatures. This, by extension, has been instrumental in advancing our understanding of the mechanical behavior, processing, and performance of polymers in various industrial and scientific fields (Müzel et al., 2020; Daniyan et al., 2021; Daniyan et al., 2022). This is crucial in ensuring the reliability and performance of these polymer-based components even before they are produced for use in the various industrial sectors where they are used, such as in automotive components, aerospace structures, electronic housings, and other areas of application where polymers can be exposed to high temperatures. These numerical analysis techniques allow product developers to reproduce within a computationally visualizable domain the complex physical and chemical processes that occur within polymers. Consequently, potential failure modes and optimizable design areas can be identified for improved performance and reliability.

12.6.1 Fields of application

Advancements in technology, as revealed by Fameso et al. (2022b), Fameso et al. (2021b), Fameso et al. (2022a), have opened vistas of opportunities for deeper and more broad-based investigations of physical processes and applications. Thus the integration of numerical modeling and simulation in industrial applications will continue to grow, contributing to the development of innovative and resilient polymer materials and products. Some of the applications of FEM to polymer materials in the industry, as revealed by Eftekhari and Fatemi (2015), Solahuddin and Yahaya (2023), Teotia and Soni (2018), include the following:

12.6.1.1 Mechanical behavior modeling

FEM enables the modeling of both linear and nonlinear mechanical behavior of polymers, especially at elevated temperatures. The linear analyses are suitable for small deformations and elastic behavior, while nonlinear analyses are essential for capturing large deformations, viscoelasticity, plasticity, and creep. To model the nonlinear behavior in polymers, various constitutive models have been developed to describe the mechanical behavior of polymers. These include linear elastic, hyperelastic (e.g., Mooney–Rivlin, Ogden models), viscoelastic (e.g., Maxwell, Kelvin–Voigt models), and viscoplastic models.

12.6.1.2 Material properties prediction

FEM can be used in conjunction with experimental techniques to characterize the material properties of polymers. Tensile, compressive, shear, and flexural tests can be simulated to obtain stress–strain curves. If the FEM model is calibrated, it can predict material properties of polymers such as modulus, strength, toughness, fatigue life, and thermal expansion coefficients under different loading and environmental conditions.

12.6.1.3 Processing simulation

FEM is widely used to simulate the injection molding process for polymer parts, and the method helps to optimize mold design, minimize defects (e.g., warpage, sink marks), and predict residual stresses and part distortion. Similar to injection molding, FEM can simulate extrusion and blow molding processes. This makes it possible to optimize die design, control wall thickness distribution, and predict residual stresses.

12.6.1.4 Failure analysis

FEM facilitates the analysis of crack initiation, propagation, and arrest in polymer materials. Fracture mechanics parameters such as stress intensity factors and critical crack tip opening displacements can be computed to assess fracture resistance. Also, fatigue failure of polymers due to cyclic loading can be simulated using FEM, and the stress-life ($S - N$) and strain-life ($\varepsilon - N$) fatigue curves can be predicted and used to estimate the fatigue life of the component.

12.6.1.5 Coupled phenomena

FEM allows for the simulation of the coupled thermomechanical behavior of polymers. This is important for applications involving temperature changes, such as automotive components, electronic devices, and consumer products. In applications such as seals, gaskets, and membranes where polymers interact with fluids, FEM can simulate fluid–structure interaction to predict fluid flow patterns, pressure distribution, and deformation of polymer structures.

12.6.1.6 Design optimization

FEM enables the efficient exploration of design space by performing parametric studies that help in the optimization of polymer component geometry, material selection, and processing parameters to meet performance requirements while minimizing cost and weight. Also, advanced FEM techniques such as topology and shape optimization can be employed to automatically generate optimal designs with improved structural efficiency and reduced material usage.

12.6.1.7 Multiscale modeling

FEM can be coupled with MM techniques to investigate the relationship between polymer microstructure and macroscopic properties. This aids in understanding structure-property relationships and designing materials with tailored properties.

12.6.2 Benefits of numerical simulations of polymer failure in industry

12.6.2.1 Improved reliability

By predicting and analyzing the behavior of polymers under elevated temperatures, manufacturers can ensure the reliability and performance of their products. The long-term durability and service life of polymer components are essential to the enhancement of their value. Ensuring they can withstand high temperatures and other environmental stresses before they are produced and commissioned is one way manufacturers can guarantee their investments in product development and innovation.

12.6.2.2 Enhanced safety

By predicting and analyzing the behavior of polymers under elevated temperatures, manufacturers can, in addition to ensuring the safety of their products and that of users, identify the root cause(s) of polymer component failures while in service. This enables manufacturers to implement corrective measures in design, material selection, or usage instruction. The robustness that simulations have over experimental testing helps in computationally subjecting the material to extensive testing using a wide range of operating parameters and, as such, provides a valuable understanding of the appropriate polymer materials with the ability to withstand high-temperature conditions without degrading.

12.6.2.3 Reduced costs

Experimental and real-life scaled models of products used in performance evaluation and reliability testing of manufactured components can be bulky and costly. The wastage associated with conducting material testing with actual manufactured products can be of great proportion. The use of computational models to replace real-life models in the visualization, testing, and analysis of manufactured polymer-based products helps in identifying potential in-service failure modes and mechanisms. Manufacturers can reduce the costs associated with product development and testing.

12.6.2.4 Increased efficiency

Numerical modeling and simulation enable manufacturers to optimize the design and performance of their products, leading to increased efficiency and reduced waste. By having the predictive edge of the thermal and mechanical behavior of polymer components, designs can be optimized for better performance and durability.

12.6.3 Case studies in numerical simulations of polymer failure in industry

12.6.3.1 Case study 1: automotive under-the-hood components

In the automotive industry, under-the-hood components made of polymers are exposed to high temperatures generated by the engine. As shown in Fig. 12-5, studies, such as Andrearczyk et al. (2021), Mancini et al. (2017), Sangamesh et al. (2022), Zaman et al. (2024), have been conducted using FEA to evaluate the thermal and mechanical performance of polymer-based components in automotive applications. Such FEA models incorporated

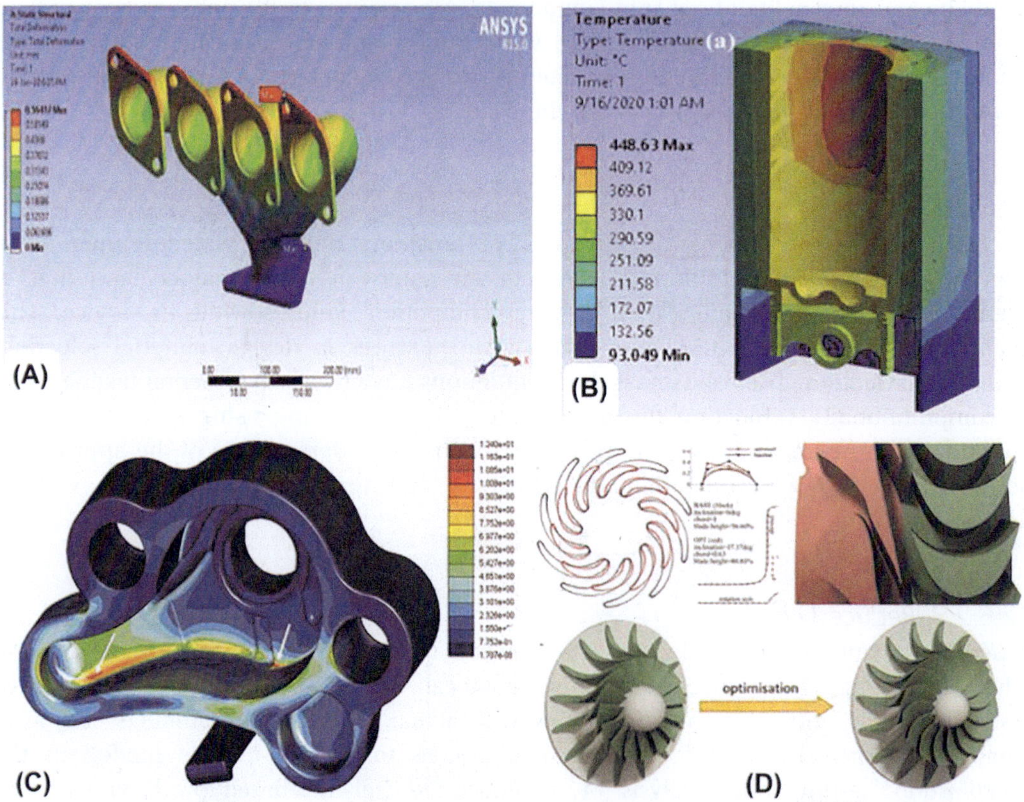

FIGURE 12–5 **Finite element prognosis of the thermal and mechanical performance of some automotive engine parts: (A) an exhaust manifold (Sangamesh et al., 2022), (B) a cylinder (Zaman et al., 2024), (C) a pump casing (Mancini et al., 2017), and (D) an impeller (Andrearczyk et al., 2021).** The images in the figure collectively demonstrate the capabilities and applicability of numerical and computational simulations in performing comprehensive analyses of automotive components, including structural, thermal, and fluid dynamics evaluations, as well as design optimization. *Figure modified from Sangamesh, R., Twinkle, R., Chiniwar, D.S., Vishwanatha, H.M., Sondar, P., Hiremath, S., 2022. Modelling of single and multi-port manifolds and studying the influence of structural and thermal behaviour on exhaust manifolds used in automotive applications. International Journal on Interactive Design and Manufacturing (IJIDeM), 1–10 and Andrearczyk et al. (2021)). These are open access articles distributed under the terms of the Creative Commons CC BY license, which permits unrestricted use, distribution, and reproduction in any medium, provided the original work is properly cited.*

temperature-dependent material properties and simulated the thermal cycles experienced by the component. The results identified potential failure points due to thermal expansion and stress concentration. Design modifications were suggested, including the use of heat-resistant polymer grades and improved cooling strategies. These have led to the development of more robust components with enhanced thermal and mechanical performances.

12.6.3.2 Case study 2: thermal protection systems

Researchers, such as Feng et al. (2022), Bregado et al. (2023), Talapatra and Datta (2023), Lou et al. (2020), Tafrishi et al. (2022), utilized MD simulation to study the thermal degradation of polymer-based thermal protection systems (TPSs) used in different engineering applications. The MD simulation, as pictured in Fig. 12–6 provides insights into the atomic-level mechanisms of thermal degradation, such as bond breaking and formation of free radicals. The results were used to develop new TPS materials with improved thermal stability. Additionally, FEA was employed to evaluate the macroscopic behavior of the TPS under reentry conditions. The combination of MD and FEA simulations resulted in a comprehensive understanding of the TPS performance, leading to the development of more reliable TPSs.

12.6.3.3 Case study 3: heat flow in components' housing design

In the electronics industry, CFDs has been used by , Ferro et al. (2023), Hildebrand et al. (2022) to analyze heat dissipation in polymer housing for the enhanced performance of electronic gadgets and appliances such as computers, electronic displays, and actuators. According to Visser et al., 2010, CFDs produce quantitative predictions of flow behaviors and provide more insight into the flow behaviors that are particularly costly or impossible to represent by experiments. The CFD models simulated the flow of lubricants such as air or oil and the temperature distribution within the lubricant housing. The results are provided in the visualizations in Fig. 12–7 indicating areas with excessive heat buildup, which could lead to thermal degradation of the polymer. Design changes were implemented, including the addition of ventilation slots and the use of a polymer material with higher thermal conductivity. The improved design was validated through further simulations, resulting in components that effectively managed heat dissipation and maintained the integrity of the polymer.

12.6.3.4 Case study 4: construction of piping systems

Manufacturers and industries have continually faced issues with polymer piping systems used for high-temperature distribution. FEA has been employed by Letchumanan et al. (2021), Khademi-Zahedi (2019), and Shi et al. (2021) to study the thermal expansion and stress distribution in the piping system under operational conditions. The FEA model incorporated temperature-dependent material properties and simulated the thermal cycles experienced by the piping. The analysis identified areas prone to thermal expansion and potential failure, as shown in Fig. 12–8. Design modifications, such as the use of expansion joints and the selection of a more heat-resistant polymer, therefore become implementable, improving piping systems and demonstrating enhanced performance and durability in real-world applications.

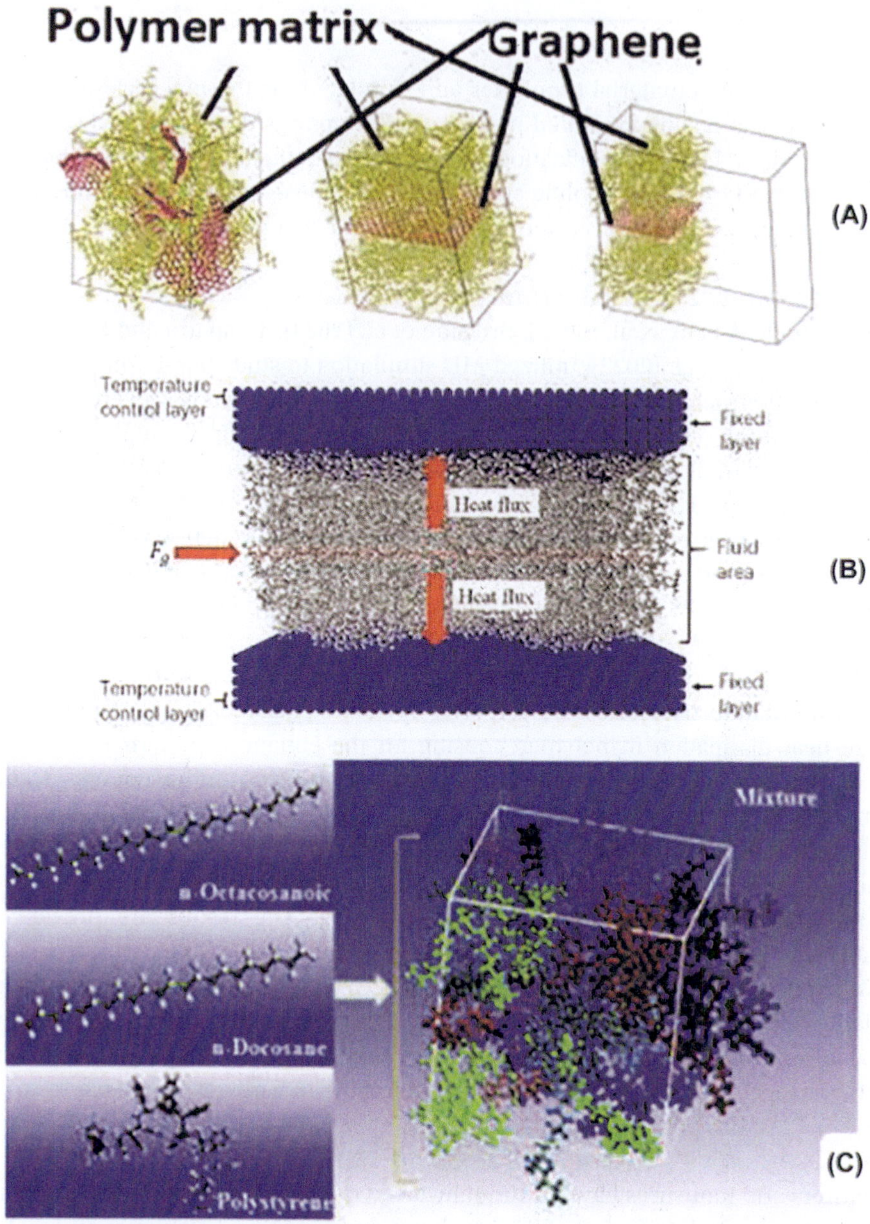

FIGURE 12–6 Molecular dynamics–based investigation of effects of elevated thermal and mechanical forces on intermolecular behavior of polymer matrices, such as (A) graphene-reinforced polymer nanocomposites (Talapatra and Datta, 2023), (B) wall slip heat flow in layers of PMMA (Lou et al., 2020), and (C) energy storage in a phase-change polymer material (Tafrishi et al., 2022). The figure appears to show a molecular dynamics–based investigation focused on understanding how the introduction of graphene and the application of elevated thermal and mechanical forces can influence the intermolecular behavior and structural organization within the polymer matrix. *Modified from Lou, Y., Wu, G., Feng, Y., 2020. Wall slip behaviour of polymers based on molecular dynamics at the micro/nanoscale and its effect on interface thermal resistance. Polymers 12(10), 2182. https://doi.org/10.3390/polym12102182 and Tafrishi, H., Sadeghzadeh, S., Ahmadi, R., 2022. Molecular dynamics simulations of phase change materials for thermal energy storage: a review. RSC Advances 12(23), 14776–14807. https://doi.org/10.1039/d2ra02183h, published as open access articles distributed under the terms and conditions of the Creative Commons Attribution (CC BY) license and Creative Commons Attribution-Non Commercial 3.0 Unported Licence, respectively.*

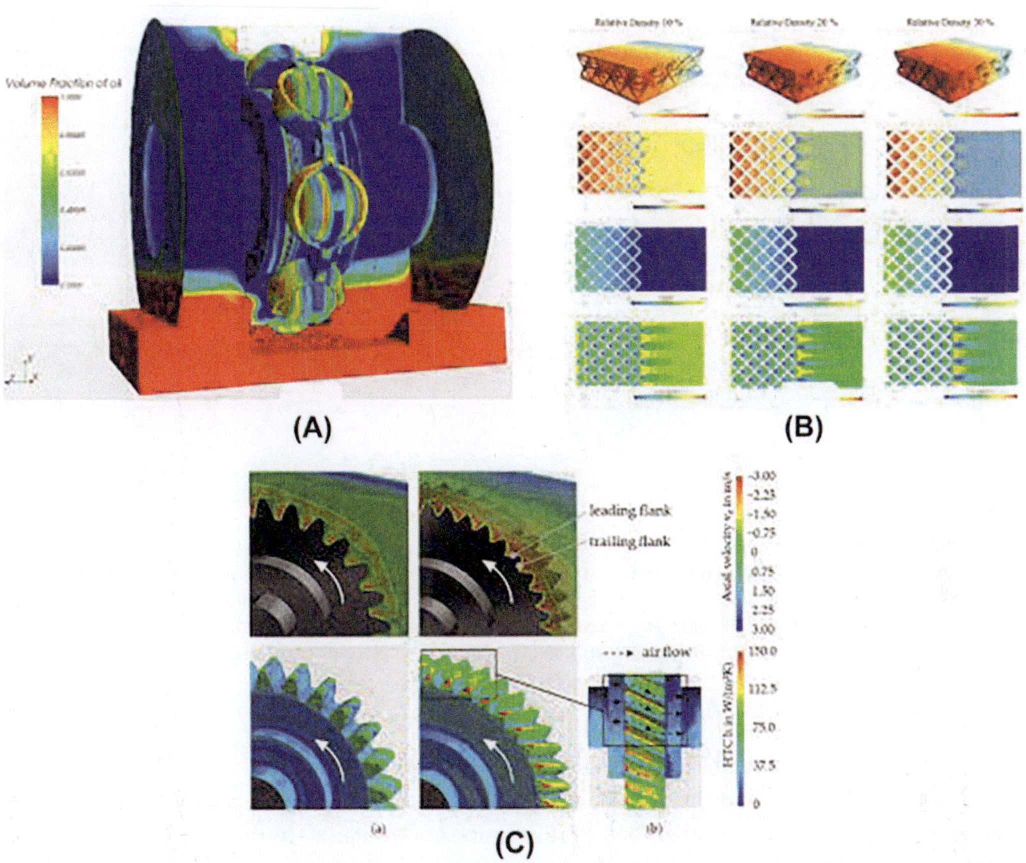

FIGURE 12–7 Computational fluid dynamic investigation of heat distribution and dissipation in components housing, such as in (A) moving bearing casing, (B) lattice panel of aircraft antiicing system (Ferro et al., 2023), and (C) dry-lubricated CD-ROM port gear (Hildebrand et al., 2022). The figure provides a visualized understanding of the profiles of thermal generation, stress distribution, and airflow pattern in the mechanical component, thus helping to optimize lubrication and improve the performance and lifespan, reduce wear, and improve efficiency. *Modified from Hildebrand, L., Dangl, F., Paschold, C., Lohner, T., Stahl, K., 2022. CFD analysis on the heat dissipation of a dry-lubricated gear stage. Applied Sciences 12(20), 10386. https://doi.org/10.3390/app122010386. The article is an open access article distributed under the terms and conditions of the Creative Commons Attribution (CC BY) license (https://creativecommons.org/licenses/by/4.0/).*

12.6.3.5 Case study 5: medical device sterilization

Medical procedure researchers and manufacturers have used FEA to study the effects of sterilization processes on polymer-based medical components. (Campaner et al., 2021; Burkhardt et al., 2024; Sanjari et al., 2024) applied FEA modeling to simulate the thermal and mechanical behavior of the polymer-based components and instruments during various high-temperature utilization cycles. The analyses illustrated in Fig. 12–9 revealed potential deformation and degradation patterns of the different polymer materials under the action of corresponding external thermal loads. Design changes where needed, such as the use of higher temperature-

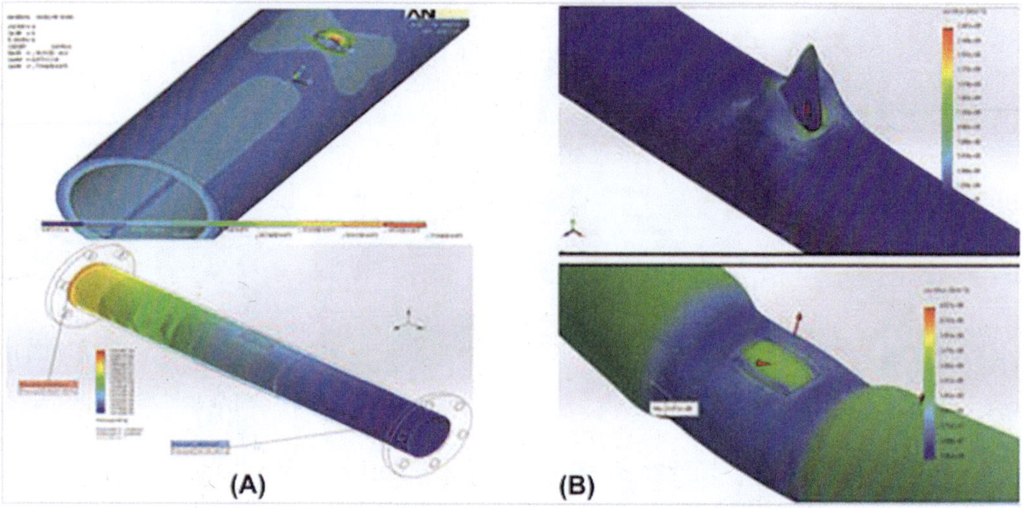

FIGURE 12–8 Finite element analysis investigation of thermal and mechanical stress distribution in (A) a polyethylene gas pipe and (B) a high-density polyethylene piping system under operational conditions. The figure presents a set of computational fluid dynamic simulation visualizations of pipeflow, showing the profiles of stresses generated during the process. The color maps, streamlines, and other visual cues help engineers and researchers understand the underlying fluid dynamics, pressure distributions, and thermal characteristics of the system.

resistant polymers and optimized geometries, were suggested. The improved design and manufacturing procedures of these components demonstrated better resistance to high-thermal gradient processes and maintained their performance and integrity. In the case of Frizziero et al. (2021), the analyses proved that the sterilization procedures do not compromise the reliability of the part, nor the safety of the procedure. Thus confirming that prototypes made with a similar process could be safely used in actual surgery practices.

12.6.4 Challenges associated with industrial applications of numerical modeling in polymer engineering

Polymers exhibit complex and often nonlinear behavior when subjected to elevated temperatures. While numerical modeling and its integration into modern engineering applications have advanced problem-solving, analysis, design, production, and use of polymers in sophisticated engineering systems, the need for higher fidelity and capacity to crack more complex phenomena related to material behavior and performance continues to introduce new challenges that require more profound and painstaking research and development efforts from the scientific community. Rapid advancements in computational power and algorithms notwithstanding, the increasingly diverse applications of polymers (evolving from everyday plastics to high-performance engineering polymers), as well as the limitations of current modeling techniques, have presented modeling accuracies of atypically complex microstructural interactions and behaviors under elevated thermal conditions.

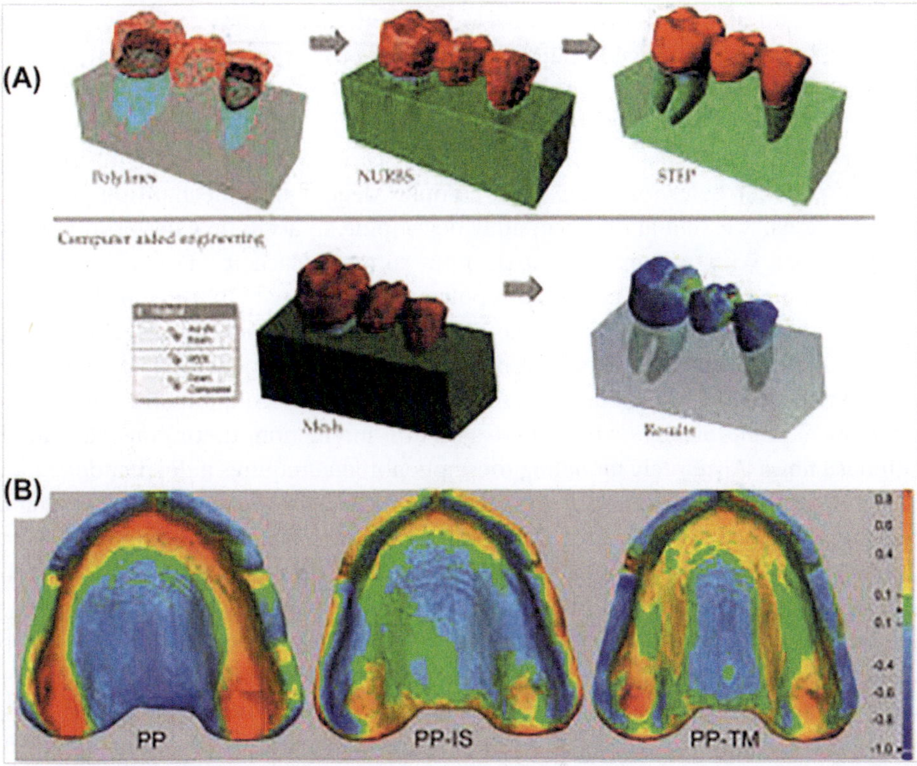

FIGURE 12–9 Numerical simulation of heat sterilization effects on dental implants made from (A) polymeric finite difference method–optimized posterior fixed partial dentures. (Campaner et al., 2021) and (B) optimized polyolefins and biocopolyesters (Burkhardt et al., 2024). The figure provides a visualization of the effects of heat sterilization on polymer-based dental implants. The numerical simulation allows the analysis of the thermal profiles, stress distributions, or other relevant properties of the implant design during the sterilization process. *Modified from Campaner, L.M., Silveira, M.P.M., de Andrade, G.S., Borges, A.L.S., Bottino, M.A., Dal Piva, A.M.D.O., et al., 2021. Influence of polymeric restorative materials on the stress distribution in posterior fixed partial dentures: 3D finite element analysis. Polymers 13(5), 1–11. Available from: https://www.mdpi.com/2073-4360/13/5/758/pdf. https://doi.org/10.3390/polym13050758 and Burkhardt, F., Schirmeister, C.G., Wesemann, C., Baur, L., Vach, K., Nutini, M., et al., 2024. Dimensional accuracy and simulation-based optimization of polyolefins and biocopolyesters for extrusion-based additive manufacturing and steam sterilization. Journal of the Mechanical Behavior of Biomedical Materials 153, 106507. https://doi.org/10.1016/j.jmbbm.2024.106507.*

The complexity of polymer behavior is evident in its time-dependent strain, meaning its deformation is influenced by both the rate and duration of applied loads, thus undergoing significant changes in properties with temperature variations, phase transitions, and crystallinity changes. While numerical modeling can be a good way of analyzing engineering problems, numerical models are as good as how accurately the governing physics have been captured in their algorithm and how closely they have described the real-life phenomena they are built to replicate. This motivates the need for more sophistication in computational resources and methods for predicting and analyzing the behavior of polymers under elevated temperature conditions, enabling the development of more robust, reliable, and durable polymer-based materials in a wide range of industrial products.

12.6.4.1 Constitutive modeling

A complete definition of the material models that capture all aspects of the physics and chemistry of polymer behavior, such as its complex degradation mechanisms or multiaxial loading conditions, is essential to successfully developing an accurately performing computational replication. With numerical models now commonplace in research and industry, several material models describing different polymer behaviors are in use, mostly empirical in nature. There still exist challenges in developing robust multiphysics models, either as holistic stand-alone or as a coupled amalgamation of multivariate empirical models across multiple time and spatial scales. Engineering systems often involve the coupling of multiple physical and mechanical phenomena, such as fluid–structure interaction, thermomechanical effects, or multiphase flows. Accurately modeling these phenomena requires a deep understanding of the material's thermomechanical behavior and often necessitates the incorporation of multiple physics-based models, increasing the complexity and computational demand. In turn, modeling these multiphysics problems requires the development of efficient coupling algorithms that can handle the interactions between different physical processes while maintaining numerical stability and accuracy. This requires the careful formulation of the governing equations, the identification of relevant coupling mechanisms, and the appropriate choice of discretization methods for each physical domain. The challenges of achieving precision in constitutive modeling stem from the challenge of accurately characterizing the material properties, including thermal, mechanical, and chemical properties, identification of the respective physical governing principles, selecting the right coupling approach, and the numerical computational resources required.

12.6.4.2 Domain discretization

Analysis methods selected for numerical simulations often contribute significantly to the accuracy and computational stability of the model. Analysis types could range from implicit to explicit solutions carried out in finite domains of elements or volumes, often referred to as meshes. While large meshes result in lower computational resolution, small-sized meshes generate high resolution and hence consume relatively higher amounts of computational resources. Thus generating high-quality computational meshes, particularly for complex geometric domains, remains a significant challenge. While traditional low-order discretization schemes, such as linear finite elements or first-order finite volumes, are widely used, there is an increasing demand for higher order methods that can provide improved accuracy and convergence rates. Adaptive mesh refinement techniques, which dynamically self-adjust based on the solution, are crucial for capturing localized phenomena, but their development and implementation pose their own set of challenges. These constitute hindrances to the successful implementation of numerical simulations of polymer behaviors at elevated temperatures, given the context of nonlinearity and multiphysics conditions in which polymer failures occur.

12.6.4.3 Computational resources

High-fidelity simulations require significant computational resources, posing challenges for extensive or detailed analyses. The complexity of the mechanics of failure in polymer engineering, ranging from nonlinearities and multiphysics couplings to the need for robust and scalable solvers, requires numerical algorithms that can only be executed by high-end computing resources. For example, nonlinear problems are usually solved by approximate methods such as the Newton–Raphson numerical solution technique. This technique requires multiple steps of iterations to obtain a converged solution, which can be computationally demanding and sensitive to initial conditions. For highly nonlinear polymer failure problems with multiple-phase interactions, access to robust solvers remains a challenge. While cloud computing and other forms of parallel and HPC capabilities have been developed in recent times, the growing demand for these facilities has created systemic queues and overloads, affecting accessibility as well as utilization optimality. Where available, the sheer volume of raw data generated by high-fidelity numerical simulations often exceeds the available memory capacity of individual computing nodes. The performance of multicore processors, graphics processing units (GPUs), and distributed computing clusters, particularly for parallel file systems and distributed storage solutions, can be greatly hampered by storage deficiencies, increasing the likelihood of run-time crashes, software failures, and processing errors and hence has a significant impact on the overall computational efficiency of the computing systems. Furthermore, the availability of such resources and the expertise to utilize them effectively has been a limiting factor for many researchers and engineers.

12.6.4.4 Uncertainty quantification

Numerical models inherently involve uncertainties, stemming from factors such as material properties, boundary conditions, or model assumptions. Just like anything computational, the outcomes of computational simulations follow the rules of "garbage-in-garbage-out," meaning the computer will only produce results whose quality is as good as that of its inputs. Discrepancies between model predictions and experimental results can arise due to simplifications and assumptions made in the models, uncertainties in material properties, and limitations in experimental techniques. Hence deviations of input parameters, definition of physical and constitutive models, initial or boundary conditions, and analysis steps compromise the validity and accuracy of the analysis results. Verifying the correctness of multiphysics models and validating them against experimental data or high-fidelity simulations is crucial for ensuring the reliability of the numerical results. This process is particularly challenging for complex engineering systems, as it requires the identification of appropriate validation metrics, the quantification of uncertainties in both the numerical model and the experimental data, and the development of rigorous verification and validation procedures. Statistical imbalances in model parameters and numerical errors often lead to shear or volumetric locking, a situation that tends to underpredict response to loading, giving an impression of an incompressible material. Multiple input parameters and spatial coordinates sometimes increase the solution orders of numerical problems, thus requiring additional computation in preprocessing analysis to identify the most influential input parameters and guide model simplification or experimental design.

12.6.5 Emerging trends and future directions of numerical modeling in polymer engineering

Advances in computational technology and sophistication in modeling tools have fostered the development of methodologies that hold promise in overcoming many of the current challenges of numerical modeling and simulation of polymer behavior and failure at elevated temperatures. MM, machine learning (ML), and artificial intelligence (AI) are powerful tools that now offer new opportunities for more accurate and efficient simulations. These technologies can help identify patterns in complex data, optimize material properties, and predict behaviors that are difficult to capture with traditional methods.

For example, AI algorithms are being developed for adaptive mesh refinement, which dynamically adjusts the mesh based on the solution. This can be particularly useful in specialized problems involving complex geometries, moving or deforming boundaries, nonlinearities, treatment of material interfaces, and multiphysics coupling. ML algorithms are also advancing in capability to handle traditional numerical computation methods, such as the Newton–Raphson or continuum solvers. This is particularly essential for modeling problems requiring the synchronization of different time scales and the design of efficient data exchange mechanisms between the coupled models to simulate the interactions between different physical processes while maintaining numerical stability and accuracy. ML and AI can enhance material model development, optimization processes, and predictive capabilities through data-driven constitutive models, surrogate modeling techniques, and physics-informed neural networks that can capture the underlying physics more effectively than traditional modeling approaches, thereby reducing the need for extensive experimental data.

With these surges in the scale and complexity of tools required for more in-depth, robust, and accurate computational representation and analysis of complexly interacting polymer behaviors at elevated thermal conditions, there is also a corresponding desire for HPC machines that can handle these tasks efficiently and timeously. More modern HPC architectures, such as multicore processors, GPUs, and distributed computing clusters, are being developed and rolled out to prospective users year in and year out. Systems that provide adequate computational resources, balance analysis loads, optimize data and memory space utilization, and communicate information in real-time are evolving to make innovative approaches and the integration of diverse numerical techniques a possibility. The growing importance of digital twins and the need for real-time or near-real-time simulations have driven the development of new numerical modeling approaches that can provide fast and accurate predictions while still maintaining the fidelity of the underlying physical and mechanical models. This includes the use of reduced-order modeling, adaptive mesh refinement, and efficient solver algorithms that can adapt to the changing conditions and requirements of the simulation.

Current and future research directions should focus on developing more comprehensive constitutive models and validating them with experimental data. Experimental validation remains crucial for ensuring the accuracy and reliability of numerical models, bridging the gap between simulations and real-world behavior. Verifying the correctness of multiphysics models and validating them against experimental data or high-fidelity simulations is crucial for ensuring the reliability of the numerical results. This process is particularly essential, as it requires the identification of appropriate validation metrics, the quantification of uncertainties in both the numerical model and the experimental data, and the development of rigorous verification and validation procedures. Integrating uncertainty quantification techniques into the numerical modeling process provides more reliable and robust predictions. The increasing focus on sustainable and energy-efficient engineering solutions is also driving research and development of numerical modeling techniques that can accurately capture the environmental impact and energy performance of engineering systems. This includes the integration of life-cycle assessment, optimization of energy efficiency, and the modeling of renewable energy technologies and energy storage systems.

12.7 Conclusion

This chapter explored the theoretical background, numerical modeling approaches, simulation methodologies, case studies, and practical applications. The critical importance of numerical modeling and simulation in understanding and predicting polymer failure at elevated temperatures was discussed. Fast becoming an indispensable tool in the field of engineering, the modeling of physical and mechanical problems in polymer technology and applications now enables the analysis and simulation of complex physical and mechanical problems. Through the detailed examination of constitutive modeling, domain discretization, and the necessity for advanced computational resources, it is evident that sophisticated numerical methods are essential for accurately replicating the complex behaviors of polymers under high-temperature conditions.

While the polymer material may exhibit a defined response to rising temperature based on the defined constitutive laws, predicting when and how polymer materials will fail under thermal stress is vital for understanding the failure criteria for polymers and preventing catastrophic failures in practical applications. At elevated temperatures, polymers may fail through mechanisms such as thermal degradation, chain scission, or stress-induced cracking. This chapter has discussed failure mechanisms, which include creep, fatigue, and thermal-oxidative degradation. Failure criteria often involve critical stress or strain values that, when exceeded, indicate imminent failure. These failure criteria provide a comprehensive framework for assessing the long-term reliability and durability of polymer components in high-temperature environments. By integrating these failure criteria into numerical models, simulations can provide early warnings of potential failures, allowing for preventive measures to be taken.

Despite the advancements in simulation technologies and the benefits numerical modeling brings to polymer engineering, challenges remain in achieving highly accurate and predictive models for polymer failure at elevated temperatures. This chapter has also explored the current challenges and emerging trends in the field of numerical modeling. As the demand and application of high-performance polymers continue to grow, ongoing advancements in computational power and algorithmic sophistication will be pivotal in addressing the identified challenges and unlocking new possibilities for polymer engineering. Future research would likely need to focus on developing more integrated and robust multiphysics models, enhancing computational efficiency, and ensuring the practical applicability of simulation results to real-world scenarios. The chapter has thus demonstrated the burgeoning role of numerical modeling and simulation in understanding and predicting polymer failure at elevated temperatures. By leveraging advanced simulation methodologies, detailed constitutive models, and accurate failure criteria, researchers can develop reliable predictions of polymer behavior, leading to improved material performance and safety.

References

Amirkhizi, A.V., Isaacs, J., McGee, J., Nemat-Nasser, S., 2006. An experimentally-based viscoelastic constitutive model for polyurea, including pressure and temperature effects. Philosophical Magazine 86 (36), 5847–5866. Available at https://doi.org/10.1080/14786430600833198.

Andrearczyk, A., Konieczny, B., Sokołowski, J., 2021. Additively manufactured parts made of a polymer material used for the experimental verification of a component of a high-speed machine with an optimised geometry—preliminary research. Polymers 13 (1), 137. Available at https://doi.org/10.3390/polym13010137.

Bazli, M., Abolfazli, M., 2020. Mechanical properties of fibre reinforced polymers under elevated temperatures: an overview. Polymers 12 (11), 2600. Available at https://doi.org/10.3390/polym12112600.

Benjemaa, M., Glinsky-Olivier, N., Cruz-Atienza, V.M., Virieux, J., Piperno, S., 2007. Dynamic non-planar crack rupture by a finite volume method. Geophysical Journal International 171 (1), 271–285. Available at https://doi.org/10.1111/j.1365-246X.2006.03500.x.

Bergstrom, J.S., 2015. Mechanics of Solid Polymers: Theory and Computational Modeling. William Andrew.

Boyard, N., 2016. Heat Transfer in Polymer Composite Materials: Forming Processes. Wiley. Available at http://doi.org/10.1002/9781119116288.

Brazel, C.S., Rosen, S.L., 2012. Fundamental Principles of Polymeric Materials. John Wiley & Sons.

Bregado, J.L., Souto, F., Secchi, A.R., Tavares, F.W., Calado, V., 2023. Molecular dynamics study of thermophysical and mechanical properties in hydrated lignin with compositions close to softwood. ACS Sustainable Chemistry and Engineering 11 (1), 238–255. Available at https://doi.org/10.1021/acssuschemeng.2c05264, http://pubs.acs.org/journal/ascecg.

Brighenti, R., Rabczuk, T., Zhuang, X., 2021. Phase field approach for simulating failure of viscoelastic elastomers. European Journal of Mechanics - A/Solids 85, 104092. Available at https://doi.org/10.1016/j.euromechsol.2020.104092.

Burkhardt, F., Schirmeister, C.G., Wesemann, C., Baur, L., Vach, K., Nutini, M., et al., 2024. Dimensional accuracy and simulation-based optimization of polyolefins and biocopolyesters for extrusion-based additive manufacturing and steam sterilization. Journal of the Mechanical Behavior of Biomedical Materials 153, 106507. Available at https://doi.org/10.1016/j.jmbbm.2024.106507.

Campaner, L.M., Silveira, M.P.M., de Andrade, G.S., Borges, A.L.S., Bottino, M.A., Dal Piva, A.M.D.O., et al., 2021. Influence of polymeric restorative materials on the stress distribution in posterior fixed partial dentures: 3D finite element analysis. Polymers 13 (5), 1–11. Available at https://doi.org/10.3390/polym13050758, https://www.mdpi.com/2073-4360/13/5/758/pdf.

Chandra, N., 2002. Evaluation of interfacial fracture toughness using cohesive zone model. Composites Part A: Applied Science and Manufacturing 33 (10), 1433–1447. Available at https://doi.org/10.1016/s1359-835x(02)00173-2.

Choudhury, M., Bindra, H.S., Singh, K., Singh, A.K., Nayak, R., 2022. Antimicrobial polymeric composites in consumer goods and healthcare sector: a healthier way to prevent infection. Polymers for Advanced Technologies 33 (7), 1997–2024. Available at https://doi.org/10.1002/pat.5660, http://onlinelibrary.wiley.com/journal/10.1002/(ISSN)1099-1581.

Chouet, B., 1986. Dynamics of a fluid-driven crack in three dimensions by the finite difference method. Journal of Geophysical Research: Solid Earth 91 (B14), 13967–13992. Available at https://doi.org/10.1029/jb091ib14p13967.

Crawford, R.J., Benham, P.P., 1974. Cyclic stress fatigue and thermal softening failure of a thermoplastic. Journal of Materials Science 9 (1), 18–28. Available at https://doi.org/10.1007/BF00554752.

Cruz-Atienza, V.M., Virieux, J., Aochi, H., 2007. 3D finite-difference dynamic-rupture modeling along nonplanar faults. Geophysics 72 (5), SM123. Available at https://doi.org/10.1190/1.2766756, http://library.seg.org/loi/gpysa7.

Daniyan, I., Mpofu, K., Muvunzi, R., Fameso, F., Ramatsetse, B., 2021. Model design and finite element analysis of a traction link of a railcar. Procedia CIRP 100, 37–42. Available at https://doi.org/10.1016/j.procir.2021.05.006.

Daniyan, I., Mpofu, K., Fameso, F., Tlhabadira, I., Phokobye, S., 2022. Computer-aided modelling and experimental evaluation of the pocket milling operation for alloy tool steel (AISI D3). The International Journal of Advanced Manufacturing Technology 122, 4453–4466. Available at https://doi.org/10.1007/s00170-022-09979-3.

Daniyan, I., Ale, F., Fameso, F., Mrausi, S., Ndambuki, J., 2024. Computer aided simulation and experimental investigation of the machinability of Al 6065 T6 during milling operation. The International Journal of Advanced Manufacturing Technology 133, 589–607. Available at https://doi.org/10.1007/s00170-024-13772-9.

Dewapriya, M.A.N., Miller, R.E., 2021. Molecular dynamics simulations of shock propagation and spallation in amorphous polymers. Journal of Applied Mechanics, Transactions ASME 88 (10). Available at https://doi.org/10.1115/1.4051238, https://asmedigitalcollection.asme.org/appliedmechanics.

Du, S., Mullins, M., Hamdi, M., Sue, H.-J., 2020. Quantitative modeling of scratch behavior of amorphous polymers at elevated temperatures. Polymer 197, 122504. Available at https://doi.org/10.1016/j.polymer.2020.122504.

Eftekhari, M., Fatemi, A., 2015. Tensile, creep and fatigue behaviours of short fibre reinforced polymer composites at elevated temperatures: a literature survey. Fatigue & Fracture of Engineering Materials & Structures 38 (12), 1395–1418. Available at https://doi.org/10.1111/ffe.12363.

El Yaagoubi, M., Juhre, D., Meier, J., Alshuth, T., Giese, U., 2017. Prediction of energy release rate in crack opening mode (mode I) for filled and unfilled elastomers using the Ogden model. Engineering Fracture Mechanics 182, 74–85. Available at https://doi.org/10.1016/j.engfracmech.2017.07.017.

Fameso, F., Desai, D., Kok, S., Armfield, D., Newby, M., Glaser, D., 2021a. Material Surface Roughness Minimization in Laser Shock Peening of Power Generation Turbine Blades. International Conference on Electrical, Computer, and Energy Technologies, ICECET 2021 https://doi.org/10.1109/ICECET52533.2021.9698545, http://ieeexplore.ieee.org/xpl/mostRecentIssue.jsp?punumber=9698246.

Fameso, F., Desai, D., Kok, S., Newby, M., Glaser, D., 2021b. Coupled explicit-damping simulation of laser shock peening on x12Cr steam turbine blades. Journal of Physics: Conference Series 1780 (1), 012002. Available at https://doi.org/10.1088/1742-6596/1780/1/012002.

Fameso, F., Desai, D., Kok, S., Armfield, D., Newby, M., 2022a. Residual stress enhancement by laser shock treatment in chromium-alloyed steam turbine blades. Materials 15 (16), 5682. Available at https://doi.org/10.3390/ma15165682.

Fameso, F., Ramakokovhu, U., Desai, D., Salifu, S., 2022b. Effects of manufacturing-induced operating-conditions amplified stresses on fatigue-life prediction of micro gas turbine blades – a simulation-based study. Journal of Physics: Conference Series 2224 (1), 012043. Available at https://doi.org/10.1088/1742-6596/2224/1/012043.

Ferro, C.G., Pietrangelo, F., Maggiore, P., 2023. Heat exchange performance evaluation inside a lattice panel using CFD analysis for an innovative aerospace anti-icing system. Aerospace Science and Technology 141, 108565. Available at https://doi.org/10.1016/j.ast.2023.108565.

Feng, C.P., Wei, F., Sun, K.Y., Wang, Y., Lan, H.B., Shang, H.J., et al., 2022. Emerging flexible thermally conductive films: mechanism, fabrication, application. Nano-Micro Letters 14 (1). Available at https://doi.org/10.1007/s40820-022-00868-8, http://www.springer.com/engineering/journal/40820.

Fernandes, C.M.C.G., Rocha, D.M.P., Martins, R.C., Magalhães, L., Seabra, J.H.O., 2018. Finite element method model to predict bulk and flash temperatures on polymer gears. Tribology International 120, 255–268. Available at https://doi.org/10.1016/j.triboint.2017.12.027, www.elsevier.com/inca/publications/store/3/0/4/7/4.

Frizziero, L., Santi, G.M., Leon-Cardenas, C., Ferretti, P., Sali, M., Gianese, F., et al., 2021. Heat sterilization effects on polymeric, fdm-optimized orthopedic cutting guide for surgical procedures. Journal of Functional Biomaterials 12 (4), 63. Available at https://doi.org/10.3390/jfb12040063.

Gharibi, H., Abdolmaleki, A., 2018. Thermo-chemical modification of a natural biomembrane to induce mucoadhesion, pH sensitivity and anisotropic mechanical properties. Journal of the Mechanical Behavior of Biomedical Materials 87, 50–58. Available at https://doi.org/10.1016/j.jmbbm.2018.07.014.

Greenhalgh, E.S., 2009. Failure Analysis and Fractography of Polymer Composites Failure Analysis and Fractography of Polymer Composites. Elsevier Ltd., United Kingdom, pp. 1–595. Available at http://doi.org/10.1533/9781845696818, http://www.sciencedirect.com/science/book/9781845692179.

Gutta, B., Huilgol, P., Perugu, C.S., Kumar, G., Reddy, S.T., Toth, L.S., et al., 2024. A polymer-based metallurgical route to produce aluminum metal-matrix composite with high strength and ductility. Materials 17 (1). Available at https://doi.org/10.3390/ma17010084, http://www.mdpi.com/journal/materials.

Hamad, R.J.A., Megat Johari, M.A., Haddad, R.H., 2017. Mechanical properties and bond characteristics of different fiber reinforced polymer rebars at elevated temperatures. Construction and Building Materials 142, 521–535. Available at https://doi.org/10.1016/j.conbuildmat.2017.03.113.

Hawkins, W.L., 1984. Polymer Degradation. Springer Science and Business Media LLC, pp. 3–34. Available at http://doi.org/10.1007/978-3-642-69376-2_2.

Heymans, N., 2003. Constitutive equations for polymer viscoelasticity derived from hierarchical models in cases of failure of time–temperature superposition. Signal Processing 83 (11), 2345–2357. Available at https://doi.org/10.1016/s0165-1684(03)00187-7.

Hildebrand, L., Dangl, F., Paschold, C., Lohner, T., Stahl, K., 2022. CFD analysis on the heat dissipation of a dry-lubricated gear stage. Applied Sciences 12 (20), 10386. Available at https://doi.org/10.3390/app122010386.

Jancar, J., Douglas, J.F., Starr, F.W., Kumar, S.K., Cassagnau, P., Lesser, A.J., et al., 2010. Current issues in research on structure–property relationships in polymer nanocomposites. Polymer 51 (15), 3321–3343. Available at https://doi.org/10.1016/j.polymer.2010.04.074.

Ihueze, C.C., Okafor, C.E., Onwurah, U.O., Obuka, S.N., Kingsley-omoyibo, Q.A., 2023. Modelling creep responses of plantain fibre reinforced HDPE (PFRHDPE) for elevated temperature applications. Advanced Industrial and Engineering Polymer Research 6 (1), 49–61. Available at https://doi.org/10.1016/j.aiepr.2022.06.001, www.keaipublishing.com/en/journals/advanced-industrial-and-engineering-polymer-research/.

Jebur, Q.H., Jweeg, M.J., Al-Waily, M., 2023. Ogden model for characterising and simulation of PPHR Rubber under different strain rates. Australian Journal of Mechanical Engineering 21 (3), 911–925. Available at https://doi.org/10.1080/14484846.2021.1918375, http://www.tandfonline.com/loi/tmec20#.VhTePLfos5g.

Kakaç, S., Yener, Y., Naveira-Cotta, C.P., 2018. Heat Conduction. CRC press.

Katunin, A., 2012. Thermal fatigue of polymeric composites under repeated loading. Journal of Reinforced Plastics and Composites 31 (15), 1037–1044. Available at https://doi.org/10.1177/0731684412452679.

Khademi-Zahedi, R., 2019. Application of the finite element method for evaluating the stress distribution in buried damaged polyethylene gas pipes. Underground Space 4 (1), 59–71. Available at https://doi.org/10.1016/j.undsp.2018.05.002.

Kim, B., Lee, S.B., Lee, J., Cho, S., Park, H., Yeom, S., et al., 2012. A comparison among Neo-Hookean model, Mooney-Rivlin model, and Ogden model for Chloroprene rubber. International Journal of Precision Engineering and Manufacturing 13 (5), 759–764. Available at https://doi.org/10.1007/s12541-012-0099-y, http://www.springerlink.com/content/m1h6t0831225t4k4/fulltext.pdf.

Khotbehsara, M.M., Manalo, A., Aravinthan, T., Reddy, K.R., Ferdous, W., Wong, H., et al., 2019. Effect of elevated in-service temperature on the mechanical properties and microstructure of particulate-filled epoxy polymers. Polymer Degradation and Stability 170, 108994. Available at https://doi.org/10.1016/j.polymdegradstab.2019.108994.

Koumoulos, E., Trompeta, A.-F., Santos, R.-M., Martins, M., Santos, C., Iglesias, V., et al., 2019. Research and development in carbon fibers and advanced high-performance composites supply chain in Europe: a roadmap for challenges and the industrial uptake. Journal of Composites Science 3 (3), 86. Available at https://doi.org/10.3390/jcs3030086.

Lan, P., Zhang, Y., Dai, W., Polycarpou, A.A., 2018. A phenomenological elevated temperature friction model for viscoelastic polymer coatings based on nanoindentation. Tribology International 119, 299–307. Available at https://doi.org/10.1016/j.triboint.2017.11.009, www.elsevier.com/inca/publications/store/3/0/4/7/4.

Letchumanan, S.M., Tajul Arifin, A.M., Taib, I., Rahim, M.Z., Nor Salim, N.A., 2021. Simulating the optimization of carbon fiber reinforced polymer as a wrapping structure on piping system using solidworks. Journal of Failure Analysis and Prevention 21 (6), 2038–2063. Available at https://doi.org/10.1007/s11668-021-01287-4, https://www.springer.com/journal/11668.

Li, Y., Abberton, B.C., Kröger, M., Liu, W.K., 2013. Challenges in multiscale modeling of polymer dynamics. Polymers 5 (2), 751–832. Available at https://doi.org/10.3390/polym5020751, http://www.mdpi.com/2073-4360/5/2/751/pdf?y=1.

Lou, Y., Wu, G., Feng, Y., 2020. Wall slip behaviour of polymers based on molecular dynamics at the micro/nanoscale and its effect on interface thermal resistance. Polymers 12 (10), 2182. Available at https://doi.org/10.3390/polym12102182.

Mackerle, J., 1996. Finite element analysis and simulation of polymers-an addendum: a bibliography. Modelling and Simulation in Materials Science and Engineering 11 (2).

Mahieux, C.A., Reifsnider, K.L., 2001. Property modeling across transition temperatures in polymers: a robust stiffness–temperature model. Polymer 42 (7), 3281–3291. Available at https://doi.org/10.1016/s0032-3861(00)00614-5.

Mancini, S.D., Santos Neto, A.D.O., Cioffi, M.O.H., Bianchi, E.C., 2017. Replacement of metallic parts for polymer composite materials in motorcycle oil pumps. Journal of Reinforced Plastics and Composites 36 (2), 149–160. Available at https://doi.org/10.1177/0731684416673727, http://jrp.sagepub.com/content/by/year.

Marais, C., Villoutreix, G., 1998. Analysis and modeling of the creep behavior of the thermostable PMR-15 polyimide. Journal of Applied Polymer Science 69 (10), 1983–1991. Available at https://doi.org/10.1002/(sici)1097-4628(19980906)69:10<1983::aid-app10>3.0.co;2-u.

Mirkhalaf, M., Vadizadeh, R., 2024. Micro-mechanical modeling of semi-crystalline polymers: a review. International Journal of Solids and Structures 290, 112691. Available at https://doi.org/10.1016/j.ijsolstr.2024.112691.

Mobarakian, M., Safarabadi, M., Farahani, M., 2020. Developing a thermomechanical and thermochemical model for investigating the cooling rate effects on the distortion of unsymmetrical viscoelastic polymeric composite laminates. Polymer Testing 87, 106503. Available at https://doi.org/10.1016/j.polymertesting.2020.106503.

Mortazavian, S., Fatemi, A., 2017. Tensile behavior and modeling of short fiber-reinforced polymer composites including temperature and strain rate effects. Journal of Thermoplastic Composite Materials 30 (10), 1414–1437. Available at https://doi.org/10.1177/0892705716632863.

Moukalled, F., Mangani, L., Darwish, M., 2016. The finite volume method in computational fluid dynamics: an advanced introduction with openFOAM® and Matlab. Fluid Mechanics and its Applications 113. Available at http://doi.org/10.1007/978-3-319-16874-6.

Müzel, S.D., Bonhin, E.P., Guimarães, N.M., Guidi, E.S., 2020. Application of the finite element method in the analysis of composite materials: a review. Polymers 12 (4). Available at https://doi.org/10.3390/POLYM12040818, https://www.mdpi.com/2073-4360/12/4/818.

Ogunnowo, E., Ogu, E.L.E.M.E.L.E., Egbumokei, P., Dienagha, I. and Digitemie, W., 2024. Conceptual Model for Failure Analysis and Prevention in Critical Infrastructure Using Advanced Nondestructive Testing. Iconic Research and Engineering Journals, 7 (10), 2456–8880.

Okeke, C.P., Thite, A.N., Durodola, J.F., Greenrod, M.T., 2017. Hyperelastic polymer material models for robust fatigue performance of automotive LED lamps. Procedia Structural Integrity 5, 600–607. Available at https://doi.org/10.1016/j.prostr.2017.07.022.

Perkins, W.G., 1999. Polymer toughness and impact resistance. Polymer Engineering and Science 39 (12), 2445–2460. Available at https://doi.org/10.1002/pen.11632.

Pineau, A., Amine Benzerga, A., Pardoen, T., 2016. Failure of metals III: fracture and fatigue of nanostructured metallic materials. Acta Materialia 107, 508–544. Available at https://doi.org/10.1016/j.actamat.2015.07.049.

Pukánszky, B., 1990. Influence of interface interaction on the ultimate tensile properties of polymer composites. Composites 21 (3), 255–262. Available at https://doi.org/10.1016/0010-4361(90)90240-w.

Raghavan, J., Meshii, M., 1998. Creep of polymer composites. Composites Science and Technology 57 (12), 1673–1688. Available at https://doi.org/10.1016/s0266-3538(97)00104-8.

Rathinam A., Design and development of UWE-4: integration of electric propulsion units, structural analysis and orbital heating analysis. (2015).

Sanjari, S., Saraeian, P., Haghighi, S.E., Alinia-ziazi, A., 2024. Modeling degradation behavior of biodegradable polymers for medical devices: A comparative review of phenomenological and stochastic approaches. Journal of Polymers and the Environment. Available at https://doi.org/10.1007/s10924-024-03285-7, https://www.springer.com/journal/10924.

Saleh Alghamdi, S., John, S., Roy Choudhury, N., Dutta, N.K., 2021. Additive manufacturing of polymer materials: progress, promise and challenges. Polymers 13 (5), 753. Available at https://doi.org/10.3390/polym13050753.

Sangamesh, R., Twinkle, R., Chiniwar, Vishwanatha, H.M., Sondar, Hiremath, S., 2022. Modelling of single and multi-port manifolds and studying the influence of structural and thermal behaviour on exhaust manifolds used in automotive applications. International Journal on Interactive Design and Manufacturingf (IJIDeM) 1–10.

Sengodan, G.A., Allegri, G., Hallett, S.R., 2020. Simulation of progressive failure in laminated composites under variable environmental conditions. Materials & Design 196, 109082. Available at https://doi.org/10.1016/j.matdes.2020.109082.

Sha, Y., Hui, C.Y., Ruina, A., Kramer, E.J., 1995. Continuum and discrete modeling of craze failure at a crack tip in a glassy polymer. Macromolecules 28 (7), 2450–2459. Available at https://doi.org/10.1021/ma00111a044.

Shah, V., Bhaliya, J., Patel, G.M., Deshmukh, K., 2022. Advances in polymeric nanocomposites for automotive applications: a review. Polymers for Advanced Technologies 33 (10), 3023–3048. Available at https://doi.org/10.1002/pat.5771, http://onlinelibrary.wiley.com/journal/10.1002/(ISSN)1099-1581.

Shahabi, E., Forouzan, M.R., 2017. A damage mechanics based failure criterion for fiber reinforced polymers. Composites Science and Technology 140, 23–29. Available at https://doi.org/10.1016/j.compscitech.2016.12.023, http://www.journals.elsevier.com/composites-science-and-technology/.

Shi, J., Hu, A., Yu, F., Cui, Y., Yang, R., Zheng, J., 2021. Finite element analysis of high-density polyethylene pipe in pipe gallery of nuclear power plants. Nuclear Engineering and Technology 53 (3), 1004–1012. Available at https://doi.org/10.1016/j.net.2020.08.019.

Siviour, C.R., Jordan, J.L., 2016. High strain rate mechanics of polymers: a review. Journal of Dynamic Behavior of Materials 2 (1), 15–32. Available at https://doi.org/10.1007/s40870-016-0052-8, http://www.springer.com/materials/special+types/journal/40870.

Solahuddin, B.A., Yahaya, F.M., 2023. A state-of-the-art review on experimental investigation and finite element analysis on structural behaviour of fibre reinforced polymer reinforced concrete beams. Heliyon 9 (3), e14225. Available at https://doi.org/10.1016/j.heliyon.2023.e14225.

Starkova, O., Gagani, A.I., Karl, C.W., Rocha, I.B.C.M., Burlakovs, J., Krauklis, A.E., 2022. Modelling of environmental ageing of polymers and polymer composites—durability prediction methods. Polymers 14 (5). Available at https://doi.org/10.3390/polym14050907, https://www.mdpi.com/2073-4360/14/5/907/pdf.

Stoltz, B., Lindvall, M., Kroon, M., 2024. A modified neo-Hookean model for semi-crystalline thermoplastics assessed by relaxation and zero-stress creep tests of recycled polypropylene and polyoxymethylene. Mechanics of Time-Dependent Materials 28 (1), 43–63. Available at https://doi.org/10.1007/s11043-023-09631-x.

Suchocki, C., 2013. A quasi-linear viscoelastic rheological model for thermoplastics and resins. Journal of Theoretical and Applied Mechanics 51 (1), 117–129. Available at http://www.ptmts.org.pl/2013-1-suchocki.pdf.

Tafrishi, H., Sadeghzadeh, S., Ahmadi, R., 2022. Molecular dynamics simulations of phase change materials for thermal energy storage: a review. RSC Advances 12 (23), 14776–14807. Available at https://doi.org/10.1039/d2ra02183h.

Talapatra, A., Datta, D., 2023. A review of the mechanical, thermal and tribological properties of graphene reinforced polymer nanocomposites: a molecular dynamics simulations methods. Polymer Bulletin 80 (3), 2299–2328. Available at https://doi.org/10.1007/s00289-022-04216-z.

Tam, L.H., Zhou, A., Yu, Z., Qiu, Q., Lau, D., 2017. Understanding the effect of temperature on the interfacial behavior of CFRP-wood composite via molecular dynamics simulations. Composites Part B: Engineering 109, 227–237. Available at https://doi.org/10.1016/j.compositesb.2016.10.030, https://www.journals.elsevier.com/composites-part-b-engineering.

Teotia, M., Soni, R.K., 2018. Applications of finite element modelling in failure analysis of laminated glass composites: a review. Engineering Failure Analysis 94, 412–437. Available at https://doi.org/10.1016/j.engfailanal.2018.08.016.

Tuli, N.T., Khatun, S., Rashid, A.B., 2024. Unlocking the future of precision manufacturing: a comprehensive exploration of 3D printing with fiber-reinforced composites in aerospace, automotive, medical, and consumer industries. Heliyon 10 (5). Available at https://doi.org/10.1016/j.heliyon.2024.e27328, https://www.sciencedirect.com/science/journal/24058440.

Visser, H.A., Bor, T.C., Wolters, M., Wismans, J.G.F., Govaert, L.E., 2010. Lifetime assessment of load-bearing polymer glasses: the influence of physical ageing. Macromolecular Materials and Engineering 295 (12), 1066–1081. Available at https://doi.org/10.1002/mame.201000174.

Vogler, M., Rolfes, R., Camanho, P.P., 2013. Modeling the inelastic deformation and fracture of polymer composites – Part I: plasticity model. Mechanics of Materials 59, 50–64. Available at https://doi.org/10.1016/j.mechmat.2012.12.002.

White, R.P., Lipson, J.E.G., 2018. Pressure-dependent dynamics of polymer melts from arrhenius to non-arrhenius: the cooperative free volume rate equation tested against simulation data. Macromolecules 51 (13), 4896–4909. Available at https://doi.org/10.1021/acs.macromol.8b00591, http://pubs.acs.org/journal/mamobx.

Yaragatti, N., Patnaik, A., 2021. A review on additive manufacturing of polymers composites. Materials Today: Proceedings 44, 4150–4157. Available at https://doi.org/10.1016/j.matpr.2020.10.490.

Zaman, M.F., Bukhari, S.M.A., Husnain, N., Abbas, Z., Ayyub, A., Najmi, M.F., et al., 2024. Effect of zirconium carbide coating on thermal behavior of heavy duty V12 diesel engine using finite element method. Engineering Research Express 6 (2), 025549. Available at https://doi.org/10.1088/2631-8695/ad4f51.

Zhi, Y., Muliana, A., Rajagopal, K.R., 2017. Quasi-linear viscoelastic modeling of light-activated shape memory polymers. Journal of Intelligent Material Systems and Structures 28 (18), 2500–2515. Available at https://doi.org/10.1177/1045389x17689936.

PART 3

Application

Application of high-temperature polymers in the automotive industry

Yakubu Azeh[1], Musah Monday[1], Abel Inobeme[2], Elijah Yanda Shaba[3], Umar Muhammad Badeggi[1], Mohammed Zago Dagaci[1], John Tsado Mathew[1]

[1]DEPARTMENT OF CHEMISTRY, IBRAHIM BADAMASI BABANGIDA UNIVERSITY, LAPAI, NIGERIA [2]DEPARTMENT OF CHEMISTRY, EDO STATE UNIVERSITY, UZAIRUE, EDO, NIGERIA [3]DEPARTMENT OF CHEMISTRY, FEDERAL UNIVERSITY OF TECHNOLOGY MINNA, MINNA, NIGERIA

13.1 Introduction

In the realm of automotive engineering, the pursuit of performance, efficiency, and sustainability drives constant innovation. As the automotive industry evolves to meet increasingly stringent regulations and consumer demands, the role of materials becomes pivotal in shaping the design, functionality, and environmental impact of vehicles. Among the diverse array of materials utilized in modern automobiles, high-temperature polymers have emerged as crucial components, offering a unique blend of properties that address the challenges faced by automotive engineers (Jagani et al., 2024).

High-temperature polymers, often referred to as thermosets or engineering plastics, are a class of materials known for their exceptional heat resistance, mechanical strength, and chemical stability. Unlike conventional plastics, which may deform or degrade under elevated temperatures, high-temperature polymers retain their integrity and performance characteristics even in extreme environments, making them ideal candidates for demanding automotive applications. The automotive industry's shift toward high-temperature polymers can be attributed to several factors. First, the quest for lightweight materials has become paramount in the pursuit of fuel efficiency and emissions reduction. High-temperature polymers offer a favorable strength-to-weight ratio, allowing engineers to achieve weight savings without compromising structural integrity or performance. Additionally, the versatility of these polymers facilitates intricate and complex designs, enabling innovative solutions to engineering challenges while optimizing space and functionality within vehicles (Oladele et al., 2023).

Furthermore, the stringent regulations governing emissions and fuel economy drive the need for components that can withstand the harsh conditions encountered in modern automotive systems. From under-the-hood applications to interior and exterior components, high-temperature polymers play a crucial role in enhancing durability, reliability, and overall performance. The application of high-temperature polymers in the automotive industry spans a wide range of components and systems. Engine components such as gaskets, seals, and

High-Temperature Polymeric Materials. DOI: https://doi.org/10.1016/B978-0-443-32792-6.00017-8

intake manifolds benefit from the chemical resistance and thermal stability of these materials, ensuring optimal performance under the extreme temperatures and pressures present in combustion engines. Similarly, under-the-hood applications, including cooling system components and fuel system parts, rely on high-temperature polymers to withstand the rigors of engine operation while maintaining efficiency and reliability (Ravi et al., 2023).

Beyond the confines of the engine bay, high-temperature polymers find utility in interior and exterior applications, contributing to both esthetics and functionality. From engine bay insulation to exterior trims and panels, these materials offer versatility in design and performance, allowing automakers to achieve desired esthetics while meeting stringent performance requirements. The integration of high-temperature polymers in automotive engineering represents a paradigm shift in material selection and design philosophy. By leveraging the unique properties of these materials, engineers can push the boundaries of performance, efficiency, and sustainability, driving innovation across the automotive industry (Vieyra et al., 2022).

This study explores the various applications of high-temperature polymers in the automotive sector, highlighting their key properties, advantages, challenges, and future prospects. Through a comprehensive analysis of real-world examples, case studies, and emerging trends. The study aims to provide insights into the transformative impact of high-temperature polymers on automotive engineering and their role in shaping the vehicles of the future (Li et al., 2023).

13.2 Overview of high-temperature polymers

High-temperature polymers, also known as high-performance polymers, represent a class of materials engineered to withstand extreme heat and harsh environmental conditions while maintaining their structural integrity and mechanical properties. These polymers find applications in various industries ranging from aerospace and automotive to electronics and industrial manufacturing (Table 13–1).

13.3 Importance of materials in automotive engineering

In automotive engineering, materials selection profoundly impacts vehicle design, performance, safety, and sustainability. Engineers meticulously choose materials to strike the optimal balance between performance, cost, and environmental impact. High-strength steel, aluminum, and composite materials are often favored for their lightweight properties, enhancing fuel efficiency and agility while maintaining structural integrity and safety standards. Additionally, advancements in materials science, such as carbon fiber-reinforced polymers and innovative alloys, offer further opportunities to improve vehicle performance and sustainability. By prioritizing materials with superior strength-to-weight ratios and recyclability, automotive engineers contribute to the ongoing evolution of ecofriendly and efficient transportation solutions. An overview of the importance of materials in automotive engineering is highlighted in Fig. 13–1.

Table 13–1 Overview of high-temperature polymers.

Polymer type	Characteristics	Applications	Processing methods	Challenges and future trends	References
Polyimides	Exceptional thermal stability, mechanical strength, and chemical resistance	Aerospace components, electronics, and insulation	Compression molding and injection molding	Development of sustainable variants and cost	Dallaev et al. (2022)
Polyetheretherketone	Outstanding mechanical properties, chemical resistance, and thermal stability up to 250°C	Automotive, aerospace, and medical implants	Injection molding and extrusion	Cost and limited processability	Korycki et al. (2023)
Polyetherimide	High strength, rigidity, and excellent electrical properties at elevated temperatures	Electrical connectors and automotive parts	Injection molding and extrusion	Environmental concerns and recycling	Caldona et al. (2021)
Polyphenylene sulfide	Excellent chemical resistance, flame retardancy, and thermal stability up to 260°C	Automotive parts and electrical components	Injection molding and extrusion	Environmental impact and material costs	Caldona et al. (2021)
Fluoropolymers (e.g., PTFE, FEP, PFA)	Outstanding chemical resistance and low friction properties	Seals, gaskets, linings, and chemical processing	Compression molding and extrusion	Cost and specialized processing equipment	Lohmann et al. (2020)

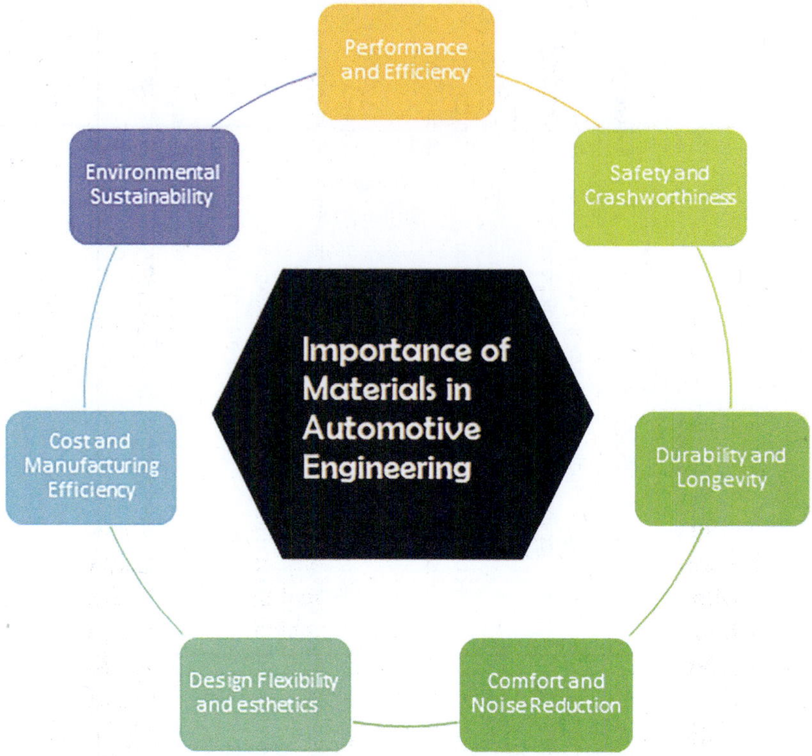

FIGURE 13–1 Overview of the importance of materials in automotive engineering.

Performance and efficiency: The selection of materials significantly influences vehicle performance and efficiency. Utilizing lightweight materials like aluminum, high-strength steel, and carbon fiber composites helps reduce vehicle weight, enhancing fuel efficiency, acceleration, and handling. Lighter vehicles require less energy to operate, resulting in lower emissions and reduced environmental impact. This trend toward lightweighting is a key strategy in automotive engineering to meet stringent fuel economy standards and address environmental concerns. Furthermore, lightweight materials can also improve overall vehicle dynamics, contributing to better stability, agility, and responsiveness on the road. By leveraging advanced materials and manufacturing techniques, automotive engineers continuously strive to optimize the balance between performance, safety, and sustainability in modern vehicles, driving innovation toward a more efficient and ecofriendly automotive industry (Mohammadi et al., 2023).

Safety and crashworthiness: Materials are pivotal in guaranteeing the safety and crashworthiness of vehicles. Advanced materials like high-strength steels, aluminum alloys, and composites are employed to engineer vehicle structures capable of efficiently absorbing and dissipating crash energy, thereby safeguarding occupants during collisions. The continual advancement of materials and structural designs has played a vital role in elevating vehicle safety standards over time. By incorporating innovative materials and manufacturing techniques, automotive engineers strive to enhance crashworthiness, mitigate injury risks, and ensure passenger safety, thereby reinforcing consumer confidence in vehicle safety features and regulatory compliance. Moreover, materials such as polymers and composites offer advantages like corrosion resistance, allowing for longer vehicle lifespans and reducing maintenance costs, further enhancing the overall value proposition for consumers. In essence, materials not only underpin vehicle safety but also contribute to the durability and reliability of automobiles, fostering trust and satisfaction among users (Wang et al., 2024).

Durability and longevity: Automotive materials face a variety of environmental challenges, including temperature fluctuations, moisture, road salts, and corrosive agents. To combat these issues, corrosion-resistant materials such as galvanized steel, stainless steel, and specialized coatings are employed. These materials enhance the durability and longevity of vehicle components by offering protection against rust and corrosion. By incorporating corrosion-resistant materials, automakers can reduce maintenance expenses and extend the service life of vehicles, ultimately enhancing reliability and customer satisfaction. Moreover, ongoing advancements in material science drive innovation in corrosion protection technologies, leading to the creation of more resilient and long-lasting automotive materials capable of enduring harsh environmental conditions for extended periods. By continually improving corrosion resistance, automotive engineers ensure that vehicles remain functional and esthetically pleasing throughout their lifespan, bolstering their appeal to consumers and reinforcing brand reputation for quality and durability (Ramezani et al., 2023).

Comfort and noise reduction: Materials play a crucial role in enhancing comfort and minimizing NVH (noise, vibration, and harshness) levels in vehicles. Manufacturers integrate sound-absorbing materials, acoustic laminates, and vibration-damping technologies into vehicle interiors to reduce noise levels, ensuring a quieter and more pleasant cabin environment for passengers. Additionally, materials with thermal insulation properties are utilized to regulate interior cabin temperature, enhancing overall comfort for occupants. By incorporating these materials and technologies, automotive engineers aim to create a more serene and comfortable driving experience, thereby improving customer satisfaction and reinforcing the vehicle's perceived quality. Moreover, advancements in material science continue to drive innovation in NVH reduction techniques, leading to further improvements in cabin comfort and refinement (Fidan et al., 2023).

Design flexibility and esthetics: Advancements in material science have revolutionized automotive design, granting designers unprecedented flexibility. Materials like carbon fiber composites, glass-reinforced plastics, and polymers empower the creation of intricate shapes and aerodynamic profiles, boosting both performance and esthetics. Furthermore, these materials can be customized to achieve specific visual attributes, bolstering brand identity and consumer allure. This flexibility enables automakers to push the boundaries of design innovation, creating vehicles that not only excel in performance but also resonate with the desires and preferences of consumers. As material science continues to advance, automotive designers have a vast palette of materials and techniques at their disposal, promising even more striking and innovative vehicle designs in the future (Diniță et al., 2024).

Environmental sustainability: As environmental concerns and regulations mount, automotive materials become pivotal in achieving sustainability objectives. Lightweight materials reduce fuel consumption and greenhouse gas emissions during vehicle use, while recyclable materials and biobased polymers facilitate end-of-life recyclability, curbing environmental impact. The automotive sector is incessantly innovating to foster ecofriendly materials and manufacturing procedures, aiming to shrink its carbon footprint. By embracing sustainable materials and processes, automakers not only meet regulatory mandates but also appeal to environmentally conscious consumers, enhancing brand reputation and market competitiveness. This commitment to sustainability underscores the industry's proactive approach to mitigating environmental harm and advancing a more sustainable future for mobility (Onat et al., 2023).

Cost and manufacturing efficiency: Material selection exerts a profound impact on manufacturing costs and production efficiency within the automotive sector. Advanced materials like carbon fiber composites, although offering exceptional performance benefits, frequently entail elevated manufacturing expenses and intricate processes. Consequently, automakers face the challenge of striking a delicate balance between material costs and performance specifications to maintain competitiveness in the market. While innovative materials can confer distinct advantages in terms of weight reduction, durability, and fuel efficiency, their adoption must be justified by their economic viability and scalability. Therefore prudent material selection involves rigorous evaluation of cost-effectiveness, manufacturability, and overall performance, ensuring that the chosen materials align with both technical requirements and financial constraints, thereby optimizing production efficiency and sustaining profitability in the automotive industry (Mohammadi et al., 2023).

Materials play a multifaceted role in automotive engineering, influencing vehicle performance, safety, comfort, design, sustainability, and cost. Technological advancements in material science enable automotive manufacturers to leverage novel materials and manufacturing techniques to meet evolving consumer demands while addressing environmental challenges and regulatory requirements. For instance, lightweight materials like aluminum and composites enhance fuel efficiency and agility, while advanced high-strength steels ensure structural integrity and safety. Additionally, innovative materials and design approaches improve comfort through noise and vibration reduction. Sustainable materials and manufacturing processes reduce environmental impact and contribute to meeting emissions standards. Cost-effective materials help manufacturers maintain competitiveness while delivering quality products. As the automotive industry continues to evolve, materials will remain at the forefront of innovation, driving improvements across various aspects of vehicle design, production, and performance to meet the demands of consumers and regulatory agencies alike (Mathew et al., 2024).

13.4 High-temperature polymers: characteristics and properties

High-temperature polymers are a class of materials designed to withstand extreme heat and harsh environmental conditions while maintaining their structural integrity and mechanical properties. These polymers exhibit exceptional thermal stability, often withstanding temperatures exceeding 150°C (300°F) without significant degradation. They possess high mechanical strength, offering excellent resistance to deformation and mechanical stress even at elevated temperatures. Chemical resistance is another key characteristic of high-temperature polymers, making them suitable for applications where exposure to corrosive substances or harsh chemicals is common. These polymers typically exhibit resistance to solvents, acids, bases, and other aggressive chemicals. Common high-temperature polymers include polyimides, polyetheretherketone (PEEK), polyetherimide (PEI), polyphenylene sulfide (PPS), and various fluoropolymers such as polytetrafluoroethylene (PTFE), fluorinated ethylene propylene (FEP) and perfluoroalkoxy alkane (PFA). Each type of polymer has its own specific thermal stability, mechanical properties, and chemical resistance, making them suitable for a wide range of applications across industries such as aerospace, automotive, electronics, and chemical processing (Table 13–2). In addition, high-temperature polymers offer exceptional thermal stability, mechanical strength, and chemical resistance, making them indispensable materials in environments where traditional polymers would degrade or fail under extreme conditions (Chattopadhyay et al., 2023; Barra et al., 2023).

13.5 Challenges in automotive engineering

Automotive engineering grapples with numerous challenges influenced by evolving consumer demands, technological advancements, regulatory requirements, and environmental concerns. These challenges encompass various facets of vehicle design, manufacturing, performance, safety, and sustainability. Meeting consumer expectations for features such as connectivity, autonomy, and efficiency necessitates continual innovation in vehicle design and technology. Stricter regulations regarding emissions, safety standards, and fuel efficiency push automakers to develop cleaner, safer, and more sustainable vehicles while ensuring compliance. Additionally, advancements in materials, electronics, and propulsion systems present both opportunities and challenges in enhancing vehicle performance and reliability. Moreover, addressing environmental concerns requires reducing the carbon footprint of vehicles through the adoption of alternative fuels, electrification, and lightweight materials. Successfully navigating these challenges demands collaboration, innovation, and adaptation within the automotive industry to deliver vehicles that meet the needs of consumers, regulatory mandates, and the planet (Inobeme et al., 2024). Some main challenges in automotive engineering are as follows:

Electrification and alternative powertrains: The automotive industry is shifting toward electrification and alternative powertrains to curb greenhouse gas emissions and lessen dependence on fossil fuels. Developing cost-effective electric vehicles (EVs), hybrid EVs, and fuel cell vehicles demands progress in battery technology, charging infrastructure, range, and performance to address consumer apprehensions about range anxiety, charging duration, and vehicle affordability. Improving battery energy density and durability enhances EV range and lifespan, while expanding charging infrastructure alleviates concerns about access to charging

Table 13–2 Outlining the characteristics and properties of high-temperature polymers.

Polymer type	Thermal stability	Mechanical strength	Chemical resistance	Applications	References
Polyimides	Exceptional, up to 500°C	High	Excellent	Aerospace, electronics, and insulation	Yun et al. (2023)
Polyetheretherketone	Up to 250°C	Very high	Excellent	Automotive, aerospace, and medical implants	Primc (2022)
Polyetherimide	Up to 180°C	High	Excellent	Electrical connectors and automotive	Caldona et al. (2021)
Polyphenylene sulfide	Up to 260°C	High	Excellent	Automotive and electrical components	Seki et al. (2022)
Fluoropolymers (e.g., PTFE, FEP, and PFA)	Varies, generally high	Varies, generally good	Excellent	Seals, gaskets, linings, and chemical processing	Lohmann et al. (2020)

stations. Furthermore, advancements in electric drivetrain efficiency and performance bridge the gap between electric and conventional vehicles. As these technologies mature and costs decrease, the automotive industry anticipates wider adoption of electrified vehicles, contributing to a greener and more sustainable transportation landscape while meeting regulatory mandates for reduced emissions and fostering energy independence (Rashid and Pagone, 2023). Polyaniline is known for its electrical conductivity, which can be further enhanced by the addition of nanoclay (Yakubu et al., 2023). This property is beneficial for applications in EVs, where components such as sensors, actuators, and energy storage systems require superconductive materials that can withstand high temperatures.

Autonomous driving and advanced driver assistance systems (ADAS): The development and integration of autonomous driving technologies and ADAS present formidable engineering challenges. Attaining higher levels of automation, such as Level 4 and Level 5 autonomy, demands the seamless incorporation of intricate sensor arrays, artificial intelligence, machine learning algorithms, and rigorous safety protocols. These technologies must reliably interpret real-time data from various sensors, including lidar, radar, cameras, and GPS, to make split-second decisions in complex and dynamic driving environments. Additionally, ensuring fail-safe mechanisms and redundancy systems is critical to guaranteeing safe operation, especially in emergencies or unexpected scenarios. As engineers strive to overcome these challenges, advancements in autonomous driving technologies hold the promise of enhancing road safety, reducing accidents, and revolutionizing the future of transportation (Ignatious et al., 2023).

Safety and crashworthiness: Enhancing vehicle safety and crashworthiness stands as a paramount challenge in automotive engineering. Advancements in materials, structural design, active safety systems, and crash avoidance technologies are crucial to reducing the frequency and severity of accidents while mitigating injury risks for both vehicle occupants and pedestrians. Innovative materials like high-strength steel and composites bolster structural integrity, while active safety systems such as automatic emergency braking and lane-keeping assist help prevent collisions. Furthermore, advancements in crash avoidance technologies, including adaptive cruise control and blind-spot monitoring, enhance driver awareness and response capabilities. By continuously refining these technologies and integrating them seamlessly into vehicle design, automotive engineers aim to create safer transportation systems that prioritize the protection of all road users, ultimately working toward the goal of minimizing accidents and saving lives (Li et al., 2023).

Lightweighting and material innovation: Lightweighting plays a pivotal role in enhancing fuel efficiency, reducing emissions, and improving performance in vehicles. Nonetheless, achieving weight reduction while maintaining safety, durability, and cost-effectiveness presents substantial challenges. Automotive engineers are thus focusing on developing innovative materials, manufacturing processes, and joining techniques to fabricate lightweight yet robust vehicle structures. Advanced materials such as high-strength steel, aluminum alloys, and carbon fiber composites offer the desired weight savings without compromising structural integrity. Additionally, novel manufacturing methods like additive manufacturing and tailored forming techniques enable the production of intricate lightweight components with high strength-to-weight ratios. Furthermore, innovative joining techniques such as adhesive bonding and laser welding ensure seamless integration of lightweight materials into vehicle assemblies. By addressing these challenges through continuous research and development efforts, automotive engineers strive to realize the benefits of lightweighting while upholding safety, durability, and cost-effectiveness standards in modern vehicles (Bao et al., 2023).

Environmental regulations and sustainability: Meeting stringent emissions standards and environmental regulations while ensuring sustainable vehicle manufacturing and operation poses an ongoing challenge for the automotive industry. Automakers are compelled to invest in research and development of ecofriendly materials, recyclable components, and energy-efficient technologies to minimize the environmental footprint of vehicles across their lifecycle. This involves the adoption of sustainable materials like biobased plastics, recycled metals, and natural fibers to reduce resource consumption and emissions during production. Additionally, implementing energy-efficient manufacturing processes, such as lean manufacturing and renewable energy utilization, helps mitigate greenhouse gas emissions at production facilities. Furthermore, advancements in vehicle technology, including electrification, hydrogen fuel cells, and lightweighting, contribute to reducing carbon emissions during vehicle operation. By embracing these strategies and fostering innovation in sustainable practices, the automotive industry aims to achieve environmental sustainability while meeting regulatory mandates and addressing societal concerns about climate change (Vieyra et al., 2022).

Connected vehicles and cybersecurity: The rise of connected vehicles and in-vehicle connectivity brings about significant cybersecurity risks, such as hacking, data breaches, and unauthorized access to vehicle systems. Automotive engineers face the critical task of prioritizing cybersecurity measures to mitigate these risks effectively. This entails implementing robust encryption protocols to secure communication channels between vehicle components and external networks. Additionally, the development of intrusion detection systems is crucial for promptly identifying and thwarting potential cyber threats. By integrating cybersecurity features directly into vehicle electronics and software, engineers aim to safeguard vehicle systems from malicious attacks while ensuring user privacy is protected. Furthermore, ongoing research and collaboration within the automotive industry and with cybersecurity experts are essential to staying ahead of emerging threats and continuously enhancing the security posture of connected vehicles. Ultimately, prioritizing cybersecurity measures is vital for ensuring the safety, reliability, and trustworthiness of modern automotive technology (Awaad et al., 2023).

Global supply chain disruptions: The COVID-19 pandemic underscored vulnerabilities in the global automotive supply chain, revealing disruptions in raw material supply, component shortages, and production delays. Automotive engineers confront the challenge of constructing resilient supply chains by diversifying sourcing strategies and mitigating risks associated with geopolitical tensions, natural disasters, and unforeseen disruptions. Implementing robust risk management protocols involves identifying alternative suppliers, establishing contingency plans, and enhancing communication and collaboration across the supply chain network. Furthermore, embracing digital technologies like artificial intelligence and predictive analytics aids in predicting and mitigating potential disruptions before they occur. By adopting these proactive measures, automotive engineers strive to enhance the flexibility, agility, and resilience of the automotive supply chain, enabling faster recovery from disruptions and ensuring the uninterrupted flow of materials and components critical for vehicle production (Raj et al., 2022).

Consumer expectations and user experience: Addressing evolving consumer expectations for comfort, convenience, connectivity, and customization remains an ongoing challenge for automotive engineers. Developing intuitive user interfaces, seamless integration of infotainment systems, and personalized driving experiences necessitates close collaboration between engineering, design, and human factors experts. By leveraging insights from user research and market trends, engineers can design vehicles that resonate with modern consumers. Intuitive controls, voice-activated commands, and user-friendly interfaces enhance the overall driving experience and facilitate safe interaction with vehicle systems. Moreover, seamless connectivity features, such as smartphone integration and Wi-Fi hotspots, provide passengers with on-the-go entertainment and productivity options. Additionally, offering customization options for interior features, seating configurations, and ADAS enables consumers to tailor vehicles to their specific preferences and needs. By prioritizing consumer-centric design principles and embracing emerging technologies, automotive engineers strive to deliver vehicles that exceed expectations and meet the evolving demands of today's discerning consumers (Diraco et al., 2023).

Regulatory compliance and standards harmonization: Navigating complex and evolving regulatory requirements across diverse regions and markets poses a significant challenge for automotive manufacturers. Harmonizing safety, emissions, and performance standards while accommodating regional variations and market preferences demands coordination among regulatory agencies, industry stakeholders, and international organizations. Engineers must remain vigilant to stay abreast of changing regulations and standards across different jurisdictions, ensuring compliance throughout the vehicle development and production process. Moreover, fostering collaboration and dialog between government bodies, industry associations, and technical experts facilitates the development of unified approaches to regulatory compliance. By aligning regulatory frameworks and adopting common standards, automotive manufacturers can streamline compliance efforts, reduce regulatory burden, and enhance market access. Additionally, proactive engagement with regulators enables manufacturers to provide input on proposed regulations and influence policy decisions, fostering a regulatory environment that supports innovation, safety, and environmental sustainability in the automotive industry (Llopis-Albert et al., 2021).

In addition, automotive engineering confronts a myriad of challenges ranging from technological innovation and regulatory compliance to environmental sustainability and consumer expectations. Overcoming these challenges demands collaboration, innovation, and adaptability across the automotive industry to create safer, more efficient, and more sustainable vehicles for the future (Inobeme et al., 2023).

13.6 Applications of high-temperature polymers in automotive

High-temperature polymers find numerous applications in the automotive industry due to their ability to withstand extreme temperatures and harsh operating conditions. Polyimides are commonly used in engine components such as seals, gaskets, and insulators, where they offer high thermal stability and chemical resistance. PEEK is utilized in fuel system components like fuel lines and connectors due to its exceptional chemical resistance and lightweight properties. PEI finds applications in electrical systems for connectors and insulating components, benefiting from its high strength and electrical properties. PPS is employed in transmission components such as gears and pump components, thanks to its excellent thermal stability and durability under high temperatures and mechanical stress. Fluoropolymers like PTFE, FEP, and PFA are used in sealing applications throughout the vehicle, including engine seals, gaskets, and O-rings, owing to their superior chemical resistance and low friction properties. Overall, high-temperature polymers play a crucial role in enhancing the performance, reliability, and longevity of automotive components operating in demanding environments (Yun et al., 2023) (Table 13–3).

13.7 Advantages of high-temperature polymers in automotive applications

High-temperature polymers offer several advantages in automotive applications, contributing to improved performance, durability, and efficiency of vehicle components. Here is a detailed overview of the advantages of high-temperature polymers in automotive applications:

Thermal stability: High-temperature polymers are specifically engineered to endure the elevated temperatures experienced in automotive environments, particularly in under-the-hood components and exhaust systems. These polymers possess remarkable thermal stability, maintaining their mechanical properties and dimensional integrity even when exposed to temperatures surpassing 150°C (300°F). This exceptional property guarantees the reliability and longevity of automotive components operating in high-temperature conditions, minimizing the risk of degradation, warping, or failure. By incorporating high-temperature polymers into critical automotive parts, such as engine components, fuel systems, and exhaust components, manufacturers enhance the performance, durability, and safety of vehicles, ensuring they can withstand the rigorous demands of operating in harsh and demanding environments. Additionally, the utilization of high-temperature polymers contributes to overall weight reduction and fuel efficiency improvements, further enhancing the sustainability and efficiency of automotive systems (Mathew et al., 2024).

Chemical resistance: Automotive systems encounter a multitude of chemicals, fuels, lubricants, and environmental contaminants that can deteriorate conventional materials over time. High-temperature polymers provide outstanding chemical resistance, enduring exposure to acids, bases, solvents, oils, and automotive fluids without undergoing significant degradation or

Table 13–3 Applications of high-temperature polymers in automotive.

Polymer type	Application	Description	Benefits	Examples	References
Polyimides	Engine components	Used in components exposed to high temperatures, such as seals, gaskets, and insulators	High thermal stability and chemical resistance	Turbocharger seals, gaskets, and insulation	
Polyetheretherketone	Fuel systems	Utilized in fuel system components due to its resistance to chemicals and high temperatures	Excellent chemical resistance and lightweight	Fuel lines, connectors, and pump components	Dua et al. (2021)
Polyetherimide	Electrical connectors	Used in electrical systems for connectors and insulating components due to its electrical properties and heat resistance	High strength and electrical properties	Electrical connectors and sensor housings	Alqahtani and Bertola (2023)
Polyphenylene sulfide	Transmission components	Applied to transmission components for its resistance to heat, chemicals, and wear	Excellent thermal stability and durability	Transmission gears and pump components.	Gao et al. (2022)
Fluoropolymers (e.g., PTFE, FEP, PFA)	Seals and gaskets	Employed in sealing applications due to their chemical resistance and low friction properties	Excellent chemical resistance and low friction	Engine seals, gaskets, and O-rings	

corrosion. This property significantly enhances the durability and reliability of automotive components, minimizing the necessity for frequent maintenance and replacement due to chemical exposure. By utilizing high-temperature polymers in critical automotive parts such as seals, gaskets, fuel system components, and electrical connectors, manufacturers can ensure prolonged performance and longevity of vehicles even in harsh operating conditions. Furthermore, the exceptional chemical resistance of these polymers contributes to overall safety and environmental protection by reducing the risk of leaks, spills, and contamination, thereby enhancing the sustainability and reliability of automotive systems (Afonso et al., 2023).

Mechanical strength: In addition to their thermal and chemical resistance, high-temperature polymers also exhibit remarkable mechanical strength, allowing them to withstand mechanical stress, vibration, and impact prevalent in automotive applications. These polymers retain their structural integrity and dimensional stability even under dynamic loading conditions, making them ideal for use in critical automotive components such as engine mounts, transmission gears, and suspension bushings. By incorporating high-temperature polymers into these components, manufacturers enhance the durability and performance of vehicles, ensuring reliable operation over extended periods. Furthermore, the exceptional mechanical strength of these polymers helps minimize wear and tear, reduce maintenance requirements, and extend the service life of automotive systems. Overall, the combination of thermal and chemical resistance with high mechanical strength makes high-temperature polymers indispensable materials for enhancing the robustness and reliability of automotive components in demanding operating environments (Kong et al., 2023).

Weight reduction: High-temperature polymers provide a lightweight alternative to traditional metal components in automotive applications. Materials like PEEK, polyamide-imide (PAI), and PPS offer exceptional strength and durability while being significantly lighter than metals. By replacing metal parts with these lightweight polymers, automakers can achieve substantial weight savings without sacrificing performance or durability. Lighter vehicles consume less fuel, leading to improved fuel efficiency, reduced emissions, and enhanced overall performance. Additionally, the use of high-temperature polymers contributes to better handling, acceleration, and agility due to reduced vehicle mass. Moreover, lightweight polymers offer design flexibility, allowing engineers to optimize vehicle aerodynamics and packaging. Overall, the adoption of high-temperature polymers in automotive manufacturing facilitates the development of more fuel-efficient and ecofriendly vehicles while maintaining, or even improving, performance characteristics (Alshammari et al., 2021).

Design flexibility: High-temperature polymers offer automotive designers enhanced flexibility in crafting intricate shapes, complex geometries, and integrated features that may be challenging to achieve with conventional materials. Techniques such as injection molding, compression molding, and additive manufacturing facilitate the production of custom-designed components with precise tolerances, thereby reducing the requirement for secondary machining operations. This flexibility in design empowers engineers to optimize automotive systems for performance, aerodynamics, and space utilization. Furthermore, it aids in reducing manufacturing complexity and costs by streamlining production processes and minimizing material waste. By leveraging the capabilities of high-temperature polymers, automotive designers can innovate more freely, creating vehicles that not only meet stringent performance and safety standards but also push the boundaries of design and functionality. Ultimately, this versatility contributes to the development of more efficient, esthetically pleasing, and technologically advanced automobiles (Czepiel et al., 2023).

Noise and vibration damping: High-temperature polymers exhibit inherent damping properties that effectively attenuate NVH levels within automotive interiors. Components constructed from polymers like polyurethane, polyethylene, and acrylonitrile butadiene styrene possess the ability to absorb and dampen vibrations, thereby diminishing cabin noise and enhancing passenger comfort during vehicle operation. Furthermore, polymer-based insulation materials can be strategically employed to mitigate engine and road noise transmitted to the vehicle cabin, further enhancing overall acoustic comfort for occupants. By leveraging these damping properties of high-temperature polymers, automotive designers can effectively reduce NVH levels, ensuring a quieter and more pleasant driving experience for passengers. This contributes to the overall quality perception of the vehicle and enhances the satisfaction of occupants, aligning with consumer expectations for comfort and refinement in modern automobiles (Gosar et al., 2023).

Corrosion resistance: Unlike metal components, high-temperature polymers inherently resist corrosion and rust, even in aggressive environments exposed to moisture, salt, and road chemicals. This corrosion resistance significantly prolongs the service life of automotive components, especially in regions with harsh climates or where road salt is prevalent during winter months. By utilizing corrosion-resistant polymers in exterior and under-the-hood components, automakers can effectively mitigate the risk of corrosion-related failures and reduce maintenance costs over the vehicle's lifecycle. Additionally, the use of these polymers enhances the overall durability and reliability of automotive systems, contributing to improved customer satisfaction and brand reputation. Overall, the corrosion resistance of high-temperature polymers offers a compelling advantage in automotive applications, providing long-term protection against environmental factors and ensuring the longevity of critical vehicle components (Sawyer-Beaulieu et al., 2022).

Cost savings: High-temperature polymers provide cost-effective solutions for automotive applications when compared to traditional materials like metals and ceramics. While the initial material costs of high-temperature polymers may be higher, they offer long-term cost savings. These polymers contribute to reduced weight, which enhances fuel efficiency and reduces operational costs over the vehicle's lifespan. Moreover, their inherent durability leads to extended component lifespan and decreased maintenance requirements, further reducing overall ownership costs. Additionally, high-temperature polymers allow for the production of complex shapes and integrated designs in a single manufacturing step. This capability reduces production costs and lead times associated with assembling multiple components, thereby enhancing the cost-effectiveness of using these polymers in automotive applications. Overall, the combination of reduced operational costs, extended lifespan, and streamlined manufacturing processes makes high-temperature polymers a financially prudent choice for automotive engineers seeking cost-effective solutions without compromising performance or reliability (Medri et al., 2023).

Environmental sustainability: High-temperature polymers play a significant role in promoting environmental sustainability within automotive applications. By facilitating lightweighting, these polymers contribute to reduced fuel consumption and lower emissions. Vehicles constructed with lightweight materials, including high-temperature polymers, require less energy to operate, resulting in decreased greenhouse gas emissions and reduced reliance on fossil fuels. This leads to improved air quality and lower carbon footprints, aligning with environmental goals. Moreover, many high-temperature polymers are recyclable or can be processed into secondary materials. This supports circular economy principles by minimizing waste and promoting resource efficiency in automotive manufacturing processes. By incorporating recyclable materials like high-temperature polymers, automakers can reduce the environmental impact of vehicle production and contribute to a more sustainable future. Overall, the adoption of high-temperature polymers in automotive applications fosters environmental stewardship and supports efforts to mitigate climate change and reduce waste (Vieyra et al., 2022).

In general, high-temperature polymers offer a multitude of advantages in automotive applications, including thermal stability, chemical resistance, mechanical strength, weight reduction, design flexibility, noise and vibration damping, corrosion resistance, cost savings, and environmental sustainability. These properties make high-temperature polymers indispensable materials in modern automotive engineering, driving innovation, efficiency, and performance in vehicle design and manufacturing (Valente et al., 2023).

13.8 Real-world examples of high-temperature polymer usage in automotive

High-temperature polymers have revolutionized automotive engineering, offering a versatile and durable solution for a wide range of applications. Here are some real-world examples showcasing the extensive usage of high-temperature polymers in the automotive industry.

Engine components: High-temperature polymers such as polyimides are frequently utilized in critical engine components subjected to extreme heat and corrosive conditions. Polyimide-based gaskets, seals, and insulators play integral roles in turbochargers, exhaust systems, and engine compartments, ensuring consistent and dependable performance even under elevated temperatures. These polymers offer exceptional thermal stability and chemical resistance, making them well-suited for use in harsh automotive environments. In turbocharger applications, polyimide-based gaskets and seals effectively withstand the intense heat generated by exhaust gases, maintaining proper sealing and preventing leaks. Additionally, polyimide insulators provide thermal insulation and electrical insulation properties, enhancing the efficiency and safety of engine components. Overall, the utilization of high-temperature polymers like polyimides in critical engine components significantly contributes to the reliability, durability, and longevity of automotive systems operating under extreme heat and corrosive conditions (Yun et al., 2023).

Fuel system components: PEEK finds extensive use in fuel system components owing to its outstanding chemical resistance and thermal stability. Fuel lines, connectors, and fuel pump components crafted from PEEK effectively endure exposure to diverse automotive fuels and additives without experiencing degradation, thereby ensuring the integrity and safety of the fuel delivery system. PEEK's exceptional resistance to chemicals, including hydrocarbons and common fuel additives, safeguards against fuel leakage and contamination, critical for maintaining optimal engine performance and emissions control. Additionally, PEEK's high thermal stability enables it to withstand the elevated temperatures encountered within the engine compartment, ensuring long-term reliability and durability of fuel system components. By utilizing PEEK in fuel system applications, automakers enhance the robustness and efficiency of vehicles while ensuring compliance with stringent safety and emissions regulations. Overall, PEEK's unique combination of properties makes it an ideal material choice for critical fuel system components in automotive applications (Dos Santos et al., 2021).

Electrical systems: High-temperature polymers like PEI are favored for use in electrical connectors and insulating components within automotive electrical systems. PEI exhibits high strength, dimensional stability, and excellent resistance to heat and chemicals, rendering it suitable for applications requiring dependable electrical connections. Its robust properties ensure reliable performance in challenging automotive environments, where exposure to elevated temperatures, vibration, and harsh chemicals may occur. PEI's exceptional thermal stability allows it to maintain its electrical properties over a wide temperature range, ensuring consistent performance under varying operating conditions. Additionally, its resistance to chemicals and environmental factors further enhances its suitability for automotive electrical applications. By utilizing PEI in electrical connectors and insulating components, automotive manufacturers ensure the reliability and longevity of electrical systems, contributing to the overall performance and safety of vehicles (Tengsuthiwat et al., 2020).

Transmission components: Transmission components demand materials with exceptional thermal stability and mechanical strength. High-temperature polymers such as PPS find application in transmission gears, valve bodies, and pump components owing to their outstanding thermal resistance, wear resistance, and dimensional stability under high temperatures and mechanical stress. PPS exhibits superior thermal stability, retaining its properties even when exposed to elevated temperatures encountered in transmission systems. Additionally, its high mechanical strength ensures reliable performance under the demanding conditions of transmission operation, including torque and load fluctuations. PPS's excellent wear resistance minimizes friction and extends the lifespan of transmission components, contributing to enhanced durability and reliability. By utilizing PPS in transmission applications, automotive manufacturers can achieve optimal performance, efficiency, and longevity of transmission systems while meeting stringent performance and durability requirements. Overall, the exceptional properties of PPS make it a preferred material choice for critical transmission components in automotive applications (Rajak et al., 2021).

Exhaust systems: Advanced high-temperature polymers are utilized in various exhaust system components, including exhaust manifolds, catalytic converters, and exhaust hangers. Compounds like PAI and PEI offer exceptional thermal stability, corrosion resistance, and dimensional stability, thus prolonging the service life of exhaust components and aiding in emissions reduction efforts. PAI and PEI compounds are well-suited for withstanding the high temperatures generated within exhaust systems, ensuring structural integrity and preventing degradation over time. Additionally, their corrosion resistance makes them ideal for exposure to exhaust gases and environmental contaminants, further enhancing durability. Furthermore, the dimensional stability of these polymers maintains proper alignment and functionality of exhaust system components, contributing to optimal performance and emissions control. Overall, the use of advanced high-temperature polymers in exhaust systems promotes longevity, reliability, and environmental sustainability in automotive applications (Yun et al., 2023).

Under-the-hood components: High-temperature polymers find applications in various under-the-hood components like intake manifolds, throttle bodies, and valve covers. Glass-reinforced polymers and PPS compounds offer thermal stability, dimensional accuracy, and chemical resistance, making them suitable for components exposed to high temperatures and mechanical stress (Hong and Cho, 2023).

Interior components: High-temperature polymers are favored in automotive interiors due to their esthetic appeal, durability, and heat resistance. PPS compounds, for instance, find application in interior trim panels, dashboard components, and seatbelt retractors. These polymers offer exceptional heat resistance, ensuring that interior components maintain their structural integrity and appearance even when exposed to high temperatures. Moreover, PPS compounds provide excellent impact strength, enhancing the durability of interior components and increasing passenger safety. Additionally, their dimensional stability ensures that interior trim panels and dashboard components maintain their shape and fitment over time, contributing to a premium feel and appearance within the vehicle cabin. By utilizing high-temperature polymers like PPS in automotive interiors, manufacturers achieve a balance of functionality, esthetics, and durability, enhancing the overall quality and appeal of the vehicle's interior environment (Dydek et al., 2021).

Exterior components: High-temperature polymers are favored in automotive interiors due to their esthetic appeal, durability, and heat resistance. PPS compounds, for instance, find application in interior trim panels, dashboard components, and seatbelt retractors. These polymers offer exceptional heat resistance, ensuring that interior components maintain their structural integrity and appearance even when exposed to high temperatures. Moreover, PPS compounds provide excellent impact strength, enhancing the durability of interior components and increasing passenger safety. Additionally, their dimensional stability ensures that interior trim panels and dashboard components maintain their shape and fitment over time, contributing to a premium feel and appearance within the vehicle cabin. By utilizing high-temperature polymers like PPS in automotive interiors, manufacturers achieve a balance of functionality, esthetics, and durability, enhancing the overall quality and appeal of the vehicle's interior environment (Yun et al., 2023).

Lighting systems: High-temperature polymers play a crucial role in automotive lighting systems due to their thermal stability and optical clarity. Lens materials like polycarbonate blends and PPS compounds are employed to withstand the high temperatures generated by automotive lighting sources. These polymers ensure clarity, durability, and reliability of headlights, taillights, and turn signals. Polycarbonate blends and PPS compounds offer exceptional thermal stability, allowing them to maintain their optical clarity and structural integrity even when exposed to the intense heat produced by lighting elements. This ensures consistent light output and visibility, essential for safe driving conditions. Moreover, these high-temperature polymers provide durability, resisting scratches, impacts, and environmental factors, thereby prolonging the lifespan of automotive lighting components. By utilizing high-temperature polymers in automotive lighting systems, manufacturers enhance the performance, reliability, and esthetics of vehicle lighting, contributing to improved safety and driving experience for vehicle operators and other road users (Hong and Cho, 2023).

Braking systems: High-temperature polymers play a crucial role in braking systems, particularly in components like brake pads and caliper seals. Phenolic resins and aramid-reinforced polymers are utilized for their high-temperature resistance and excellent frictional properties, ensuring efficient braking performance and durability under extreme operating conditions. Phenolic resins provide exceptional thermal stability, allowing brake pads to withstand the high temperatures generated during braking without compromising performance. Additionally, the frictional properties of these polymers contribute to consistent and reliable braking performance, enhancing vehicle safety. Similarly, aramid-reinforced polymers offer high strength and heat resistance, making them suitable for use in caliper seals. These polymers maintain dimensional stability and sealing integrity, ensuring effective operation of the braking system over a wide range of temperatures and environmental conditions. By incorporating high-temperature polymers in braking system components, manufacturers enhance the reliability, performance, and safety of vehicles, providing drivers with confidence and peace of mind during braking maneuvers (Kalel et al., 2022).

Intake systems: High-temperature polymers play a critical role in intake systems, encompassing components like intake manifolds and throttle bodies, where they are required to withstand the high temperatures generated by the engine. Polymers such as PEI and PPS are favored for their exceptional thermal stability, chemical resistance, and dimensional stability, ensuring reliable performance in demanding engine environments. PEI and PPS exhibit remarkable thermal stability, enabling them to withstand the elevated temperatures encountered within intake systems without experiencing degradation or loss of mechanical properties. Additionally, their resistance to chemicals and environmental factors ensures long-term durability and reliability. By utilizing high-temperature polymers like PEI and PPS in intake system components, automotive manufacturers enhance the performance, efficiency, and longevity of vehicles, ensuring consistent operation and optimal engine performance in diverse operating conditions. Overall, the use of these polymers contributes to the reliability and durability of automotive intake systems, thereby enhancing the overall quality and longevity of vehicles.

Heating, ventilation, and air conditioning (HVAC) systems: High-temperature polymers play a vital role in HVAC systems due to their capacity to withstand high temperatures and harsh environmental conditions. Polymers such as PEEK and PPS are commonly utilized in HVAC components, including actuators, valves, and ductwork, ensuring reliable operation and thermal performance. PEEK and PPS offer exceptional thermal stability, allowing them to maintain their structural integrity and mechanical properties even when exposed to elevated temperatures within HVAC systems. Additionally, their resistance to chemicals and environmental factors ensures long-term durability and reliability in harsh operating environments. By incorporating high-temperature polymers like PEEK and PPS in HVAC components, automotive manufacturers enhance the performance, efficiency, and longevity of HVAC systems, ensuring consistent heating, cooling, and ventilation capabilities for vehicle occupants. Overall, the use of these polymers contributes to the reliability and durability of automotive HVAC systems, thereby improving the overall comfort and functionality of vehicles (Rajak et al., 2021).

13.9 Conclusion

The application of high-temperature polymers in the automotive industry offers a transformative solution to address key challenges and enhance vehicle performance, durability, and sustainability. From engine components to fuel systems, electrical connectors, and transmission parts, high-temperature polymers provide thermal stability, chemical resistance, mechanical strength, and weight reduction benefits that enable automakers to innovate and optimize automotive systems. By leveraging the unique properties of high-temperature polymers, automotive engineers can design lighter, more fuel-efficient vehicles with reduced emissions and improved reliability. These polymers also facilitate design flexibility, allowing for the creation of complex shapes and integrated features that enhance functionality and efficiency. Furthermore, the cost-effectiveness and environmental sustainability of high-temperature polymers make them an attractive choice for automotive manufacturers striving to meet regulatory requirements, consumer demands, and industry trends. As the automotive industry continues to evolve toward electrification, autonomy, and sustainability, high-temperature polymers will play an increasingly significant role in shaping the future of automotive engineering, driving innovation, and advancing toward a more efficient and environment-friendly transportation ecosystem.

13.10 Future prospects of high-temperature polymers in automobile

The use of high-temperature polymers in the automobile industry shows great promise due to their unique properties and potential applications for the fabrication of various automobile components. However, the use of traditional polymers comes with a lot of challenges. Thus there is a need to look inward into the use of additive manufacturing by employing the use of nanoclay/polymer technology. The use of polyaniline/nanoclay composites as high-temperature polymers in the automobile industry shows great promise due to their unique properties and potential applications. Research and development in the area of polyaniline/nanoclay composites is likely to lead to further advancements and applications of this innovative material in the automotive sector due to its enhanced thermal stability, which is crucial for components exposed to elevated temperatures, such as engine parts, exhaust systems, and under-the-hood components; mechanical strength where tensile strength, toughness, and impact resistance are significant for structural components in vehicles that require high strength and durability; electrical conductivity where components such as sensors, actuators, and energy storage systems require conductive materials that can withstand high temperatures; corrosion resistance, which is a robust property of polyaniline/nanoclay composites, making them suitable for use in automotive components exposed to harsh environmental conditions, such as salt spray, moisture, and chemicals. This can help extend the lifespan of vehicle parts and reduce maintenance costs; and lightweighting, which can contribute to lightweighting efforts in the automotive industry, as these materials are typically lighter than metals and can help reduce the overall weight of vehicles. This can lead to improved fuel efficiency and performance. Overall, the future prospects of high-temperature polymers in the automotive industry are promising, driven by their unique combination of thermal stability, mechanical strength, electrical conductivity, corrosion resistance, lightweighting benefits, and environmental sustainability.

References

Afonso, I.S., Nobrega, G., Lima, R., Gomes, J.R., Ribeiro, J.E., 2023. Conventional and recent advances of vegetable oils as metalworking fluids (MWFs): a review. Lubricants 11 (4). Available at https://doi.org/10.3390/lubricants11040160, www.mdpi.com/journal/lubricants.

Alqahtani, A.A., Bertola, V., 2023. Polymer and composite materials in two-phase passive thermal management systems: a review. Materials 16 (3), 893. Available at https://doi.org/10.3390/ma16030893.

Alshammari, B.A., Alsuhybani, M.S., Almushaikeh, A.M., Alotaibi, B.M., Alenad, A.M., Alqahtani, N.B., et al., 2021. Comprehensive review of the properties and modifications of carbon fiber-reinforced thermoplastic composites. Polymers 13 (15). Available at https://doi.org/10.3390/polym13152474, https://www.mdpi.com/2073-4360/13/15/2474/pdf.

Awaad, T.A., El-Kharashi, M.W., Taher, M., Tawfik, A., 2023. Detecting cyber attacks in-vehicle diagnostics using an intelligent multistage framework. Sensors 23 (18). Available at https://doi.org/10.3390/s23187941, http://www.mdpi.com/journal/sensors.

Bao, Z.J., Yang, H.Y., Dong, B.X., Chang, F., Li, C.D., Jiang, Y., et al., 2023. Development trend in composition optimization, microstructure manipulation, and strengthening methods of die steels under lightweight and integrated die casting. Materials 16 (18). Available at https://doi.org/10.3390/ma16186235, http://www.mdpi.com/journal/materials.

Barra, G., Guadagno, L., Raimondo, M., Santonicola, M.G., Toto, E., Vecchio Ciprioti, S., 2023. A comprehensive review on the thermal stability assessment of polymers and composites for aeronautics and space applications. Polymers 15 (18), 3786. Available at https://doi.org/10.3390/polym15183786.

Caldona, E.B., Dizon, J.R.C., Viers, R.A., Garcia, V.J., Smith, Z.J., Advincula, R.C., 2021. Additively manufactured high-performance polymeric materials and their potential use in the oil and gas industry. MRS Communications 11 (6), 701–715. Available at https://doi.org/10.1557/s43579-021-00134-9, https://www.springer.com/journal/43579.

Chattopadhyay, J., Pathak, T.S., Santos, D.M.F., 2023. Applications of polymer electrolytes in lithium-ion batteries: a review. Polymers 15 (19). Available at https://doi.org/10.3390/polym15193907, http://www.mdpi.com/journal/polymers.

Czepiel, M., Bańkosz, M., Sobczak-Kupiec, A., 2023. Advanced injection molding methods: review. Materials 16 (17), 5802. Available at https://doi.org/10.3390/ma16175802.

Dallaev, R., Pisarenko, T., Sobola, D., Orudzhev, F., Ramazanov, S., Trčka, T., 2022. Brief review of PVDF properties and applications potential. Polymers 14 (22), 4793. Available at https://doi.org/10.3390/polym14224793.

Diniță, A., Ripeanu, R.G., Ilincă, C.N., Cursaru, D., Matei, D., Naim, R.I., et al., 2024. Advancements in fiber-reinforced polymer composites: a comprehensive analysis. Polymers 16 (1). Available at https://doi.org/10.3390/polym16010002, http://www.mdpi.com/journal/polymers.

Diraco, G., Rescio, G., Caroppo, A., Manni, A., Leone, A., 2023. Human action recognition in smart living services and applications: context awareness, data availability, personalization, and privacy. Sensors 23 (13), 6040. Available at https://doi.org/10.3390/s23136040.

Dos Santos, F.S.F., Vieira, M., da Silva, H.N., Tomás, H., Fook, M.V.L., 2021. Surface bioactivation of polyether ether ketone (Peek) by sulfuric acid and piranha solution: influence of the modification route in capacity for inducing cell growth. Biomolecules 11 (9). Available at https://doi.org/10.3390/biom11091260, https://www.mdpi.com/2218-273X/11/9/1260/pdf.

Dua, R., Rashad, Z., Spears, J., Dunn, G., Maxwell, M., 2021. Applications of 3D-printed PEEK via fused filament fabrication: a systematic review. Polymers 13 (22), 4046. Available at https://doi.org/10.3390/polym13224046.

Dydek, K., Latko-Durałek, P., Sulowska, A., Kubiś, M., Demski, S., Kozera, P., et al., 2021. Effect of processing temperature and the content of carbon nanotubes on the properties of nanocomposites based on polyphenylene sulfide. Polymers 13 (21), 3816. Available at https://doi.org/10.3390/polym13213816.

Fidan, I., Huseynov, O., Ali, M.A., Alkunte, S., Rajeshirke, M., Gupta, A., et al., 2023. Recent inventions in additive manufacturing: holistic review. Inventions 8 (4), 103. Available at https://doi.org/10.3390/inventions8040103.

Gao, Y., Zhou, X., Zhang, M., Lyu, L., Li, Z., 2022. Polyphenylene sulfide-based membranes: recent progress and future perspectives. Membranes 12 (10), 924. Available at https://doi.org/10.3390/membranes12100924.

Gosar, A., Emri, I., Klemenc, J., Nagode, M., Oman, S., 2023. On the vibration-damping properties of the prestressed polyurethane granular material. Polymers 15 (5), 1299. Available at https://doi.org/10.3390/polym15051299.

Hong, S.H., Cho, B.G., 2023. Effect of cooling temperature on crystalline behavior of polyphenylene sulfide/glass fiber composites. Polymers 15 (15). Available at https://doi.org/10.3390/polym15153179, http://www.mdpi.com/journal/polymers.

Ignatious, H.A., El-Sayed, H., Khan, M.A., Mokhtar, B.M., 2023. Analyzing factors influencing situation awareness in autonomous vehicles—a survey. Sensors 23 (8). Available at https://doi.org/10.3390/s23084075, http://www.mdpi.com/journal/sensors.

Inobeme, A., Adetunji, C.O., Mathew, J.T., Okonkwo, S., Bamigboye, M.O., Ajai, A.I., et al., 2023. Advanced Nanotechnology for the Degradation of Persistent Organic Pollutants Microbial Degradation and Detoxification of Pollutants. De Gruyter, Nigeria, pp. 51–71. Available at https://www.degruyter.com/document/doi/10.1515/9783110743623/html, https://doi.org/10.1515/9783110743623-003.

Inobeme, A., Mathew, J.T., Adetunji, C.O., Agbugui, M.O., Osarenren, E.J., Chinedu, O., et al., 2024. Biocomposites with Graphene Derivatives Advances in Biocomposites and their Applications. Elsevier, Nigeria, pp. 149–166. Available at https://www.sciencedirect.com/book/9780443190742, 10.1016/B978-0-443-19074-2.00005-8.

Jagani, S., Marsillac, E., Hong, P., 2024. The electric vehicle supply chain ecosystem: changing roles of automotive suppliers. Sustainability 16 (4), 1570. Available at https://doi.org/10.3390/su16041570.

Kalel, N., Bhatt, B., Darpe, A., Bijwe, J., 2022. Suppression of brake noise and vibration using aramid and zylon fibers: experimental and numerical study. ACS Omega 7 (25), 21946–21960. Available at https://doi.org/10.1021/acsomega.2c02313.

Kong, U., Mohammad Rawi, N.F., Tay, G.S., 2023. The potential applications of reinforced bioplastics in various industries: a review. Polymers 15 (10). Available at https://doi.org/10.3390/polym15102399, http://www.mdpi.com/journal/polymers.

Korycki, A., Carassus, F., Tramis, O., Garnier, C., Djilali, T., Chabert, F., 2023. Polyaryletherketone based blends: a review. Polymers 15 (19), 3943. Available at https://doi.org/10.3390/polym15193943.

Li, M., Zhang, S., Zhang, X., Qiu, M., Liu, Z., He, S., 2023. Optimization study of driver crash injuries considering the body NVH performance. Applied Sciences 13 (22), 12199. Available at https://doi.org/10.3390/app132212199.

Llopis-Albert, C., Rubio, F., Valero, F., 2021. Impact of digital transformation on the automotive industry. Technological Forecasting and Social Change 162, 120343. Available at https://doi.org/10.1016/j.techfore.2020.120343.

Lohmann, R., Cousins, I.T., Dewitt, J.C., Glüge, J., Goldenman, G., Herzke, D., et al., 2020. Are fluoropolymers really of low concern for human and environmental health and separate from other PFAS? Environmental Science and Technology 54 (20), 12820–12828. Available at https://doi.org/10.1021/acs.est.0c03244, http://pubs.acs.org/journal/esthag.

Mathew, J.T., Inobeme, A., Abdullahi, A., Bini, E.M., Shaba, E.Y., Musah, M., et al., 2024. Properties and Characteristics of Nanocoatings for Medicinal Applications Sustainable Approach to Protective Nanocoatings. IGI Global, Nigeria, pp. 203–227. Available at https://www.igi-global.com/book/sustainable-approach-protective-nanocoatings/335038, https://doi.org/10.4018/9798369331361.ch007.

Medri, V., Natali Murri, A., Papa, E., Mingazzini, C., Scafè, M., Landi, E., 2023. Ultra-high-temperature ceramic-doped inorganic polymers for thermo-structural fiber-reinforced composites. Materials 16 (20), 6649. Available at https://doi.org/10.3390/ma16206649.

Mohammadi, H., Ahmad, Z., Mazlan, S.A., Faizal Johari, M.A., Siebert, G., Petrů, M., et al., 2023. Lightweight glass fiber-reinforced polymer composite for automotive bumper applications: a review. Polymers 15 (1). Available at https://doi.org/10.3390/polym15010193, http://www.mdpi.com/journal/polymers.

Oladele, I.O., Okoro, C.J., Taiwo, A.S., Onuh, L.N., Agbeboh, N.I., Balogun, O.P., et al., 2023. Modern trends in recycling waste thermoplastics and their prospective applications: a review. Journal of Composites Science 7 (5). Available at https://doi.org/10.3390/jcs7050198, www.mdpi.com/journal/jcs.

Onat, N.C., Mandouri, J., Kucukvar, M., Sen, B., Abbasi, S.A., Alhajyaseen, W., et al., 2023. Rebound effects undermine carbon footprint reduction potential of autonomous electric vehicles. Nature Communications 14 (1). Available at https://doi.org/10.1038/s41467-023-41992-2, https://www.nature.com/ncomms/.

Primc, G., 2022. Strategies for improved wettability of polyetheretherketone (peek) polymers by non-equilibrium plasma treatment. Polymers 14 (23), 5319. Available at https://doi.org/10.3390/polym14235319.

Raj, A., Mukherjee, A.A., de Sousa Jabbour, A.B.L., Srivastava, S.K., 2022. Supply chain management during and post-COVID-19 pandemic: mitigation strategies and practical lessons learned. Journal of Business Research 142, 1125–1139. Available at https://doi.org/10.1016/j.jbusres.2022.01.037, http://www.elsevier.com/locate/jbusres.

Rajak, D.K., Wagh, P.H., Linul, E., 2021. Manufacturing technologies of carbon/glass fiber-reinforced polymer composites and their properties: a review. Polymers 13 (21). Available at https://doi.org/10.3390/polym13213721, https://www.mdpi.com/2073-4360/13/21/3721/pdf.

Ramezani, M., Mohd Ripin, Z., Pasang, T., Jiang, C.P., 2023. Surface engineering of metals: techniques, characterizations and applications. Metals 13 (7). Available at https://doi.org/10.3390/met13071299, http://www.mdpi.com/journal/metals.

Rashid, S., Pagone, E., 2023. Cradle-to-grave lifecycle environmental assessment of hybrid electric vehicles. Sustainability 15 (14), 11027. Available at https://doi.org/10.3390/su151411027.

Ravi, S.S., Osipov, S., Turner, J.W.G., 2023. Impact of modern vehicular technologies and emission regulations on improving global air quality. Atmosphere 14 (7). Available at https://doi.org/10.3390/atmos14071164, www.mdpi.com/journal/atmosphere.

Sawyer-Beaulieu, S., Tam, E., Hussein, A., 2022. Measuring corrosion on vehicles, in real-time, using digital imaging and analysis techniques. Materials 15 (9), 3053. Available at https://doi.org/10.3390/ma15093053.

Seki, Y., Kizilkan, E., Leşkeri, B.M., Sarikanat, M., Altay, L., Isbilir, A., 2022. Comparison of the thermal and mechanical properties of poly(phenylene sulfide) and poly(phenylene sulfide)-syndiotactic polystyrene-based thermal conductive composites. ACS Omega 7 (49), 45518–45526. Available at https://doi.org/10.1021/acsomega.2c06152, pubs.acs.org/journal/acsodf.

Tengsuthiwat, J., Sanjay, M.R., Siengchin, S., Pruncu, C.I., 2020. 3D-MID technology for surface modification of polymer-based composites: a comprehensive review. Polymers 12 (6). Available at https://doi.org/10.3390/polym12061408, https://res.mdpi.com/d_attachment/polymers/polymers-12-01408/article_deploy/polymers-12-01408-v2.pdf.

Valente, M., Rossitti, I., Sambucci, M., 2023. Different production processes for thermoplastic composite materials: sustainability versus mechanical properties and processes parameter. Polymers 15 (1), 242. Available at https://doi.org/10.3390/polym15010242.

Vieyra, H., Molina-Romero, J.M., Calderón-Nájera, J.de D., Santana-Díaz, A., 2022. Engineering, recyclable, and biodegradable plastics in the automotive industry: a review. Polymers 14 (16), 3412. Available at https://doi.org/10.3390/polym14163412.

Wang, Y., Ding, L., Lin, J., Qiu, X., Wu, C., Liu, C., et al., 2024. Recent developments in polyurea research for enhanced impact penetration resistance and blast mitigation. Polymers 16 (3), 440. Available at https://doi.org/10.3390/polym16030440.

Yakubu, A., Monday, M., Mathew, J.T., Muhammad, B.U., Gambo, U., Tauheed, A., et al., 2023. In Situ Synthesis of Polyaniline Nanohybrid and Formulation of Polyaniline/carboxymethylcellulose/ethylene Glycol Nanocomposite: Study of Its Conducting and Antibacterial Properties. Research Square, Nigeria. Available at https://www.researchsquare.com/browse, 10.21203/rs.3.rs-3334019/v1.

Yun, S., Sheng, X., Wang, S., Miao, X., Shi, X., Zhao, Y., et al., 2023. Preparation and properties of high-temperature-resistant, lightweight, flexible polyimide foams with different diamine structures. Polymers 15 (12), 2609. Available at https://doi.org/10.3390/polym15122609.

14

Application of high-temperature polymers in the aviation industry

Anthony Chidi Ezika[1], Emmanuel Ezeani[2], Victor Okpechi[2], Henry Oyeoka[2], Izuchukwu Madu[2], Martin Ibenta[2], Shadrack Ugwu[2]

[1]CHEMICAL, METALLURGICAL, AND MATERIAL ENGINEERING DEPARTMENT, FACULTY OF ENGINEERING AND BUILT ENVIRONMENT, TSHWANE UNIVERSITY OF TECHNOLOGY, PRETORIA, GAUTENG, SOUTH AFRICA

[2]DEPARTMENT OF POLYMER ENGINEERING, FACULTY OF ENGINEERING, NNAMDI AZIKIWE UNIVERSITY, AWKA, NIGERIA

14.1 Introduction

Polymer materials have significantly influenced the aviation sector. In the early days of aviation, the first flights were constructed by Orville and Wilbur Wright, using readily available materials. Natural materials, including spruce wood and muslin, were utilized as the primary surface covering of the Wright Flyer. These materials facilitate the regulation of heavier-than-air flight. In the mid-1900s, as the aircraft industry continued to develop and demand for higher wing loading and larger aircraft increased, aircraft manufacturing shifted toward using stronger and more rigid materials such as metals (Mat Yazik and Ismail, 2025). The use of fabric and wood in the early stages of aircraft construction paved the way for the development of aircraft based on metal in the mid-20th century. These metals, particularly aluminum alloys, enhanced the durability and efficacy of aircraft.

However, the desire to push boundaries is what drove the investigation of polymers. The transition from natural materials to metals and, subsequently, to polymers indicates a change that is motivated by the pursuit of improved aircraft capabilities. The demand for larger and more efficient aircraft as aviation technology advanced facilitated the dominance of metal materials, such as aluminum alloys, which are used as primary materials for aerospace structures, and eventually titanium and vanadium fiber. This helped establish iconic milestones, including the Apollo mission in 1969. New and innovative materials to enhance aircraft performance are consistently sought after by aircraft engineers and designers. This has resulted in the use of polymer materials. Due to their intriguing properties, polymer materials are a viable alternative to metals in the fabrication of aircraft structures to enhance their agility, payload, longevity, and maneuverability. As fuel consumption is directly influenced by a decrease in takeoff and landing weight, utilizing a lightweight aircraft results in substantial savings in terms of environmental and economic expenses. The utilization of polymer materials in aviation has experienced a significant increase in recent decades (Pierce and Falzon, 2017).

High-Temperature Polymeric Materials. DOI: https://doi.org/10.1016/B978-0-443-32792-6.00016-6

The most intriguing and promising polymeric materials are elastomers, plastics, and fibers. In contemporary mechanical engineering, structural polymer materials are being utilized more frequently, particularly when no other material is capable of meeting the ever-increasing demands of new technology. Fundamental structural materials such as wood and metal have presently been significantly replaced by polymers and materials derived from them. The vast array of polymers and fillers, the methods of their modification, and the infinite variability of the composite compositions based on them render the potential of polymeric materials to be exceedingly broad. There has been a substantial amount of effort dedicated to the development and design of polymers that are more capable, resilient, and durable within the last few decades. Numerous novel polymer families have been reported, each with unique properties that have the potential to be applied in challenging sectors such as aerospace, energy, defense, electronics, or automotive. These polymers demonstrate exceptionally elevated service temperatures and resistance to environmental stress (Mittal, 2011a). The terms "high performance" and "advanced engineering" have been used to describe polymers that have considerably improved properties in comparison to conventional ones (Parker, 2000). It is currently used in a broad sense and frequently denotes polymers that possess improved properties, such as thermal stability, dimensional stability, mechanical performance, environmental stability, chemical resistance, gas barrier, and electrical properties (Mittal, 2011b).

Numerous high-performance polymers with exceptionally low flammability have been discovered over the years. The development of high-performance polymers significantly facilitated the aerospace industry's accelerated expansion in the 1960s. An increasing demand for novel, lightweight polymers with exceptional thermal and mechanical properties has been observed for their potential applications in space environments. Various new polymer classes were developed that met these criteria; however, only a small number of them were extensively adopted due to their high manufacturing costs or poor processability (Lau, 2014). Although sufficient data is gathered regarding the properties of high-performance polymers, the physical and chemical mechanisms that regulate the behavior of these materials remain largely unknown and exclusive. Ramgobin et al. (2019) conducted an exhaustive examination of the thermal properties and fire behavior of numerous high-performance and high-temperature polymers. Cone calorimetry, UL-94, microscale combustion calorimetry, and the limiting oxygen index are among the most frequently used methods for fire characterization. Due to the highly stable aromatic or inorganic groups, many high-performance polymers are difficult to ignite or are self-extinguishing.

14.2 Importance of high-temperature polymers in the aviation industry

In recent years, there has been a growing interest in high-temperature polymers for use in extreme environments. Heat shield materials for thermal protection systems (TPS) are a significant application. Due to the detrimental heat fluxes encountered during high-speed flight through the Earth's atmosphere, TPS materials are intended to safeguard structural components. They are frequently employed in the aerospace sector and are required in various applications, such as hypersonic vehicles, missile launchers, rocket propulsion devices, and space reentry vehicles (Wu and Koo, 2022; Laub and Venkatapathy, 2004). There are numerous methods to prevent heat accumulation through thermal protection. Although active cooling systems, such as direct-fuel-coolant systems, are currently under development, they frequently introduce additional weight, complexity, and cost in comparison to traditional ablative designs (Koo, 2016). As a result, ablative materials are of increasing interest as novel high-temperature polymers. Ablatives are composites that are reinforced with fibers or fillers and are used as sacrificial materials. Pyrolysis of the polymer matrix is frequently induced by exposure to high-heat flux conditions, resulting in the formation of a protective char layer that retards the progression of deterioration. This sacrificial process safeguards the space vehicle from direct exposure to high-heat fluxes, enabling reliable operation in extreme heat environments. To qualify and certify a TPS material for flights, it must endure heat fluxes spanning from 30 to 30,000 W cm^{-2} (Natali et al., 2016).

Subliming or noncharring TPS materials and charring TPS materials are the two categories of polymeric ablative TPS materials classification. Ceramics or metals, including tungsten or rhenium, can be employed to fabricate nonablative TPS (Sutton and Biblarz, 2011). They are frequently employed in low-altitude environments that are susceptible to atomic erosion. However, the primary heat transfer mechanism of insulation for nonablative TPS materials is reradiation. The most prevalent category of ablatives is charring TPS materials or charring ablators. They are typically employed in a diverse array of hyperthermal environments, such as reentry conditions, and are capable of withstanding high heat fluxes. The initial heat transmission mode is pure conduction when a charring TPS material is exposed to a nonoxidizing, high-heat flux environment. When the polymeric material reaches a sufficiently high temperature, the heated materials undergo pyrolysis, which involves the random scission of polymer chains. Volatile small molecules and a solid carbonaceous char are produced during this endothermic decomposition process. Carbon is an optimal choice for high-temperature environments due to its high vaporization energy and high heat capacity. Char yield is a critical metric for evaluating the ultimate performance of charring ablators. Other high-temperature polymers have been developed, but phenolic resin remains the most frequently employed charring polymer for TPS.

Intumescent ablators are a particular form of charring ablators that are frequently constructed from thermoplastics. The polymer undergoes pyrolysis, resulting in the release of volatile products into the gas phase. These products obstruct the convective heat transfer from the environment to the material surface. The expanded porous char layer will function as an insulation stratum, offering supplementary protection to the virgin material beneath. The aerodynamic performance of the space vehicle would be adversely affected by the shape deformation that occurs during pyrolysis, which is why intumescent ablators are not recommended for reentry (Wu and Koo, 2022). The thermal protection of vehicles from atmospheric reentry, the protection of rocket nozzles, and the protection of land-based structures from high-heat environments are all examples of the diverse applications that ablatives can provide. Typically, ablatives are used as a thermal protection material and are frequently sacrificed to protect vehicles or equipment from extreme heat (Rallini and Natali, 2019). Furthermore, these materials may function as structural components, thereby modifying the properties of the surface to which they are affixed by removing thermal energy, thereby altering the heat and mass transfer process between the original material and heat source (Wu and Koo, 2022).

14.3 High-temperature polymers in aircraft components

There is an ongoing discussion and scholarly input over sustainability, material design innovation, and the incorporation of cutting-edge technologies like additive manufacturing, the use of high-performance specialty materials, and so on. The aerospace industry is rapidly tilting toward greener aircraft technologies, environmental friendliness, and cost-effectiveness (Lohawala and Wen, 2024). In its early stage of development, the aeronautical industry relied heavily on aluminum-based alloys (Al metal matrix composites) for aircraft fabrication (Garg et al., 2019). This was due to their special properties such as appreciable specific density, resilience to impact, ability to resist corrosion, and high-temperature resistance. However, over the years, polymers and polymer composites have found application in aircraft components, gradually replacing the traditional aluminum alloys, iron, steel, and other metals. Polymers are preferred to metals in aircraft applications because they are lightweight materials that can lower operational expenses by improving mechanical properties, energy efficiency, flight range, and payload (Parveez et al., 2022). Researchers like Rajak et al. (2019) opined that carbon fiber-reinforced plastics had been used more frequently in aircraft frames and engine components possessing a minimum yield strength of 550 MPa and a density of one-fifth of steel and three-fifths of alloys based on aluminum. According to Mat Yazik and Ismail (2025), the applications of polymer composites in aircraft are versatile as they are used in aircraft engine parts, suction manifolds, pump casings, impellers, etc. These developments call for the creation of multifunctional polymer composites that improve environmental sustainability, safety, and efficiency in aeronautical applications. Table 14–1 shows the various polymer matrices with their fillers for application in the aviation industry.

Apart from various polymer composites that have found applications in aerospace technology, there are polymers with characteristics of high-temperature resistant properties and mechanical strength used in aircraft technology. However, researchers like Das et al. (2020) pointed out that due to their high zero-shear viscosity, processing these high-temperature polymers is often difficult because of their viscosity, which prevents coalescence. The high-temperature polymers can be classified as thermoplastics (such as polyether–ether ketone [PEEK], polymer amides, and polyphenylene sulfide [PPS]), thermosetting (such as polyimide [PI], phenol–formaldehyde, and polyesters), and charring composites/polymers designed for TPS (such as epoxy/phenolic silicon fiber composites and phenolic carbon ablator) (Wu and Koo, 2022; Hassan et al., 2022). According to Xu et al. (2022), hypersonic reentry vehicles must withstand intense aerodynamic heating; therefore a special class of

Table 14–1 Polymer composites for aviation applications.

Polymer matrix	Reinforcements	Aircraft applications	References
Epoxy	Fiberglass	Secondary frame fuselage components, shroud panels, fans, duct fairings, spacers, and seals	Parveez et al. (2022)
Polyester	Fiberglass	Secondary structure on aircraft, such as fairings, radomes, and wing tips	Kalaiyarasan and Ramesh (2018)
Epoxy	Aramid	Aircraft wing and fuselage structure	Nagaraj Goud et al. (2022)
Epoxy	Graphite fiber	Fuselage panels, interior cabin, empennage structures	Bello (2021)
Epoxy	Carbon nanotubes	Radomes, coatings, and interior panel	Aina et al. (2024)
Silicon	Carbon (graphene, carbon black, carbon nanotubes)	Radomes, fuselage skin, upper and lower wing skins, and wing stringers	Siengchin (2023)
Polylactic acid	Glass fiber	Seats, cargo bay liners, wall partitioning, overhead compartments, etc.	Shanmugam et al. (2021)
Polyether–ether ketone	Carbon fibers	Aircraft structural applications	Rinaldi et al. (2023)

materials known as charring composites is often employed to offer the most effective thermal protection layer for their TPS.

Generally, these special polymers and composites are used in aerodynamic engineering in applications requiring low-weight, high-performance, and heat-resistant applications, such as aircraft wings and fuselages, and in ablative TPS for heat shielding in harsh environments (Wu and Koo, 2022). Moreover, high-temperature polymers and composites can be fabricated by techniques such as polymer infiltration pyrolysis, reactive melt infiltration, hot pressing, reactive bonding, and chemical vapor infiltration (Kumar et al., 2024a; Arun et al., 2020).

Some high-temperature polymers are outlined in Fig. 14–1. Moreover, Fig. 14–2 shows high-temperature polymers concerning their various performances during use in high-temperature applications.

The study of polymers and composites in TPS applications is ongoing, and many test methods are being developed to produce safer and more efficient reentry vehicles. The following section discusses the specific applications of high-temperature polymers in harsh environments, such as in aircraft engine components.

14.3.1 Engine components

As outlined earlier, high-temperature polymers possess special characteristics that enable them to withstand extremely high temperatures while retaining their firmness and functional properties. Typically, at the forefront of this class of polymers are the aromatic

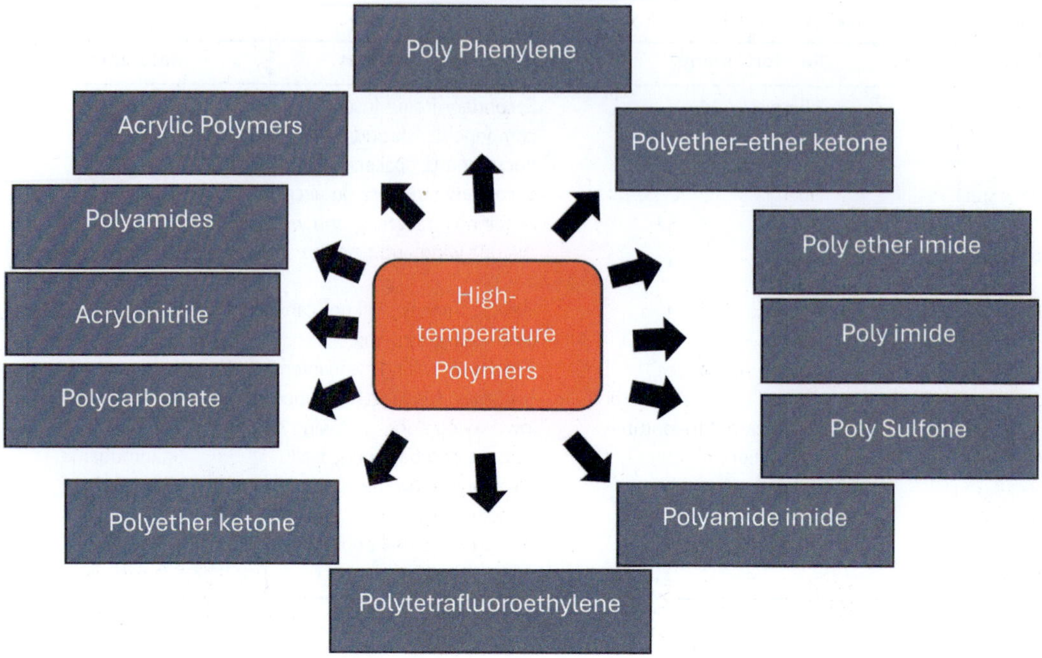

FIGURE 14–1 Outline of high-temperature polymers. Various types of high-temperature polymers.

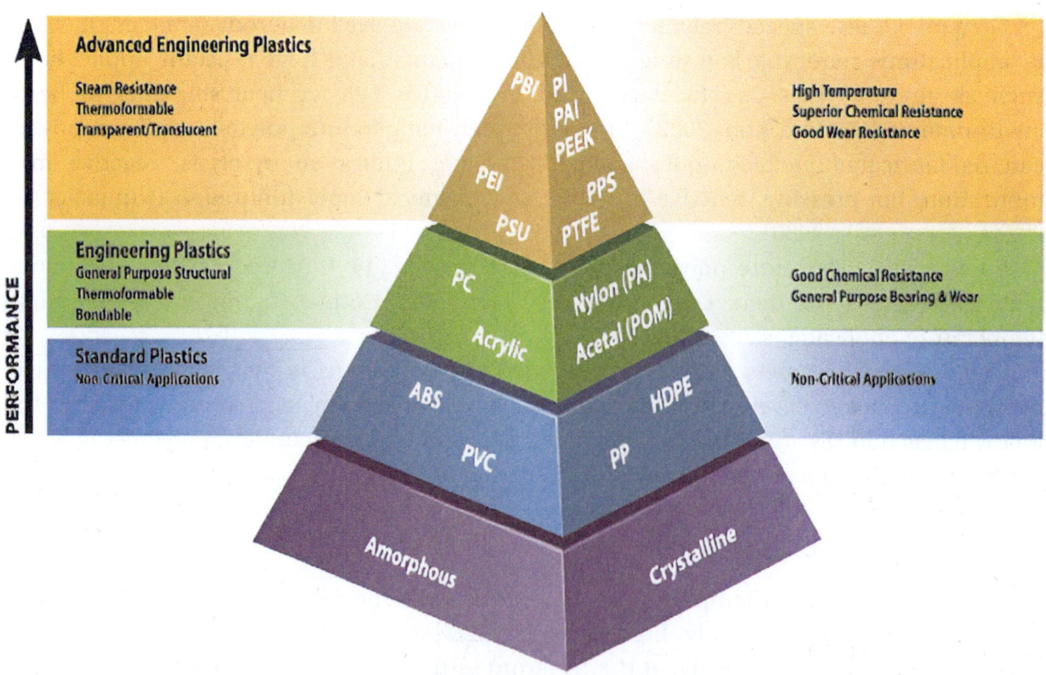

FIGURE 14–2 Performance of high-temperature polymers (Das et al., 2020). The figure showcases the performance of high-temperature polymers.

polymers, characterized by strong molecular backbones that offer remarkable heat stability. The operating conditions of aircraft engines and parts around engines are defined by high temperatures. Therefore materials used as aircraft engine components should be stable at the elevated conditions of the aircraft engine. According to Mat Yazik and Ismail (2025), composites based on PEEK can withstand temperatures as high as 200°C (392°F) without losing their qualities, and they are utilized in parts like thrust reversers, heat shields, and engine parts. Moreover, PI polymers possess exceptional fire resistance and high-temperature stability up to 300°C (572°F), especially the polymerization of monomer reactant PIs (Bystry-King and King, 2020). These PIs find applications as compressor cases for advanced turbine engines. High-temperature polymers offer outstanding thermal insulation, shielding delicate aircraft components from extremely high and low temperatures.

14.3.1.1 Turbine blades

In the development of composites for turbine blade applications, materials such as ceramic matrix composites (CMCs) have been an attractive option (Tayade, 2024). This was because when compared to conventional metals and metallic alloys, CMCs, especially those based on silicon/carbon (Si/C/Si/C) composites, offer better high-temperature capability, decreased weights, and fewer cooling requirements due to conduction, as with the case of metals and alloys (Hussain and Immanuel, 2021). A study by Leicht et al. (2023) showed that CMC turbine rotors like those from silicon nitride can increase thrust and power output by a factor of 1.44, coupled with a better inlet temperature. Similarly, according to Tayade and Lynch (2025), CMC turbine demonstrated a 3.4% gain in thermal efficiency because of improved turbine inlet temperatures. Additionally, using CMC as a turbine component leads to lower density and centrifugal loads, which increase the overall fuel efficiency of the turbine (Tayade, 2024; Leicht et al., 2023).

Recently, high-temperature polymers have begun to gain relevance in the study of materials for turbines for better performance. This class of polymers can endure high temperatures, show outstanding robustness in harsh conditions, and be chemically and thermally stable. For instance, Hernandez-Carrillo et al. (2019) designed a 1000-W microturbine generator for a microorganic Rankine cycle (mORC) using polymers such as polyethylene as the nozzle body and an impeller composed of the PEEK. Researcher like Zirak et al. (2022) pointed out that systems that have the potential to employ polymers or polymer composites in place of their rotating components include refrigeration, mORC, and heating, ventilation, and air conditioning systems. Examples of polymers and composites that have found applications in rotor fabrications include PEEK, polylactic acid, acrylonitrile butadiene styrene (ABS), PPS, nylon, and composites such as epoxy resin reinforced with up to 60% carbon fiber, and PEEK-GF30 (Zirak et al., 2022; Hernandez-Carrillo et al., 2017; Upadhyay et al., 2020).

Grilli et al. (2021) opined that to effectively optimize resources used in cooling systems of turbines and minimize heat transmission, high-temperature polymers in the form of special coatings are employed. Moreover, such polymers have been shown to significantly reduce weight and improve flexibility when integrated as components of engines (Zirak et al., 2022). Polymers generally have a high ease of processability compared to other engineering materials, which makes them more economical and cost-effective.

14.3.1.2 Combustion chambers

Polymers and their composites are used in internal combustion chambers of aircraft as ablative materials. These polymers serve as a TPS that shields aircraft engines from the intense heat and pressure of atmospheric entry. According to Fahy et al. (2021), materials for such applications function by vaporizing or degrading over time to form a layer of gas that shields and diverts heat away from the aircraft. A study by Guo et al. (2024b) showed that polymer composites based on zirconia oxide (ZrO_2) and silicon are suitable for ablation applications in spacecraft because they can lower ablation rates up to 30.6% and can form a thick protective layer of char when they degrade. A similar study by Guo et al. (2024a) showed a novel use of boron trioxide (B_2O_3) in an ablative system for solid rocket motors (SRM) and TPS. Their study showed that B_2O_3 improved the performance of the char layer and offered superior ablation resistance when added to polymer matrices and also formed a liquid coating on the surface. Further studies showed that B_2O_3 can be used to preserve a char layer from thermal oxidation at low temperatures. According to Radyjowski (2021), B_2O_3 has boiling and melting temperatures of 1860°C and 450°C, respectively, and it can melt to form a liquid film when the char layer temperature rises above 450°C thereby reducing oxidative consumption of char layers and improving vapor deposition reaction at elevated temperatures to produce more pyrolytic carbon.

Other important polymers that are used as ablative material in TPS and SRM include Kevlar and silica-filled ethylene propylene diene monomer, carbon nanotubes (CNT) fillers in their polymer matrix, such as phthalonitrile resins, etc. (Guo et al., 2024a; Ma et al., 2023; Deng et al., 2022).

14.3.1.3 Seals and gaskets

Yang et al. (2018) noted that there are two broad types of gaskets, namely, the traditional elastic gaskets and the hydrophilic composite elastic gaskets. The traditional elastic gaskets, like EPDM and chloroprene, rely mainly on elastic properties to bring about a sealing effect, while the hydrophilic composite types like water-swelling rubber and water-swelling polyurethane combine both elastic qualities as well as their capacity to absorb water to bring about sealing effects (Yang et al., 2018). According to (Yang et al. (2018), polymer materials for gaskets and sealing must have exceptional thermal stability and high sealing performance. Wu et al. (2025) stated that polymers such as polytetrafluoroethylene and fluorosilicone rubber had outstanding dimensional stability in high-temperature applications, while fluorine rubber had comparatively low dimensional stability but greater thermal decomposition temperatures. The seals for aircraft engine applications are mostly nonmetallic materials (polymers), and they play a crucial role in preserving the engine by preventing oil leakage. Shabbir et al. (2019) opined that if an aircraft engine seal fails, oil (liquids) may leak into pipelines, leading to adverse effects, and these failures are caused by high temperatures and thermal aging. As a result, polymers used for gasket and sealing in aircraft engines must be stable and heat-resistant even at elevated temperatures.

Some special polymers used in aircraft gaskets and seals include nitrile butadiene rubber, fluoro rubber, fluorocarbon rubber, high-strength graphite gasket, polytetrafluoroethylene, fluorosilicone rubber, some copolymers such as tetrafluoroethylene–hexafluoropropylene, and so on (Wu et al., 2025; Drobny, 2023). Sometimes, these special polymers for gaskets and sealants are mixed with fibers such as boron nitride to improve oil resistance and better high-temperature performance. Researchers like Joshi (2024) showed that the flexibility and temperature performance of gaskets increase by incorporating heat-resistant fibers into polymers like polyurethane or epoxy.

14.3.2 Structural components

14.3.2.1 Fuselage panels

The main body structures of an aircraft are the fuselage panels, which house the passengers, cargo, and vital systems. High-temperature polymers are particularly useful in fuselage panels, which require materials that can survive harsh conditions such as high temperatures, aerodynamic pressures, and environmental deterioration (Mat Yazik and Ismail, 2025). The use of high-temperature polymers in fuselage panels provides various advantages over traditional metallic materials. High-temperature polymers and composites greatly reduce aircraft weight, resulting in increased fuel efficiency and cheaper operating costs (Kesarwani, 2017). These polymers can endure high temperatures, making them ideal for areas subjected to aerodynamic heating (Ma et al., no date). Polymers, unlike metals, are resistant to oxidation and chemical breakdown in severe settings. Polymers' moldability enables complicated designs and the incorporation of multifunctional components. Numerous high-temperature polymers have great fire resistance, lowering the risk of onboard fires (Uddin, 2016).

Researchers have extensively studied the use of several high-temperature polymers in aeronautical applications. PIs are widely employed in composite fuselage panels due to their excellent thermal and mechanical qualities (Yang and Ji, 2018). PEEK, a semicrystalline polymer with great strength and heat resistance, is frequently reinforced with carbon fibers to improve performance (Mat Yazik and Ismail, 2024). Phenolic resins are widely employed in fire-resistant applications. Phenolic-based composites offer exceptional thermal and mechanical stability (Qiao et al., 2025). Bismaleimide (BMI) resins have high temperature tolerance, making them suitable for structural aerospace components (Prasanaa Iyer and Arunkumar, 2020). Polymers based on silicone are utilized for coatings and thermal insulation. These materials protect panels of fuselages from high temperatures (Dharini and Krishnamoorthy, 2023).

Recent research and technology breakthroughs have improved the performance of high-temperature polymers in fuselage panels. The use of nanoparticles, such as graphene or CNT, has greatly increased the mechanical strength and heat resistance of polymer composites (Kumar et al., 2021). Advanced techniques of manufacturing allow for the creation of highly customized polymer-based fuselage components with increased performance (Goh et al., 2024). Integrating polymers with lightweight metal alloys has resulted in hybrid constructions that balance strength and weight (Li et al., 2025). Recent advances in self-repairing polymer solutions improve the durability and endurance of fuselage panels (Rb et al., 2024).

14.3.2.2 Wing component

Structures of aircraft, especially wing components, are exposed to harsh climatic conditions such as high-temperature conditions, mechanical strains, and aerodynamic loads (Mat Yazik and Ismail, 2024). Conventional materials, such as aluminum and titanium alloys, have been employed widely, but their high density and vulnerability to corrosion pose constraints (Sukiman, 2012). High-temperature polymers, such as PEEK, PIs, and thermosetting resins, have received interest as prospective alternatives.

PIs are utilized in cutting-edge parts and structural reinforcements due to their superior resistance to heat (Xu, 2023). PEEK is used in fasteners, brackets, and composite reinforcements due to its remarkable strength and chemical stability. Owing to its exceptional tolerance to chemicals and heat, polybenzimidazole (PBI) is used in high-stress environments (Wang et al., 2022a). Carbon fibers are frequently used to strengthen thermosetting resins (such as epoxy and BMIs) to produce sophisticated composite materials with increased durability (Chen et al., 2023a; Rajak et al., 2021).

High-temperature polymer composites are employed in advanced components to resist aerodynamic heating during flight at high speeds (Solntsev, 2011). Carbon fiber-reinforced polymer composites improve both the stiffness and strength of wing spars and ribs (Soutis, 2005; Ayaz et al., 2024). PI-based coatings are used to offer thermal insulation in high-temperature environments (Hicyilmaz and Celik Bedeloglu, 2021). PEEK-based fasteners and bushings provide excellent resistance to wear and prevent corrosion from galvanic attack at composite-metal interfaces (Mat Yazik and Ismail, 2024; Chen et al., 2023b).

14.3.2.3 Interior components

Aircraft interior parts must comply with strict safety, durability, and weight standards. Conventional materials like aluminum and steel have been extensively employed, but their high density and vulnerability to corrosion pose restrictions (Sukiman, 2012; Dursun and Soutis, 2014). Because of their flame resistance, lightweight characteristics, and mechanical strength, high-temperature polymers such as PIs, PEEK, and thermosetting resins have gained popularity as prospective alternatives for interior applications (Wu and Koo, 2022; Mat Yazik and Ismail, 2024; Parin and Demirci, 2024).

Due to their high-temperature and flame-retardant qualities (Liu et al., 2022b), PIs are utilized in paneling, insulation, and seat constructions. Seat frames and brackets are made of PEEK, which is well-known for its high strength-to-weight ratio, durability, and resistance to chemicals (Yilmaz, 2024). Epoxy, phenolics, and BMIs are examples of thermosetting resins, frequently reinforced with carbon fibers to produce interior constructions that are both robust and lightweight (Dufton, 2000).

Composites based on PI and PPS give wall panels structural stability while preserving their fire resistance (Parin and Demirci, 2024; Hughes, 2019). To ensure strength and longevity, PEEK-reinforced composites are utilized for tray tables, armrests, and seat frames

(Peters, 2013). The fire safety and durability of cargo liners and floors are enhanced by high-temperature polymer-based laminates. High-performance polymers are used in instrument panels and control housings for their lightweight and heat-resistant qualities (Agarwal et al., 2017). PI and PPS materials are utilized in air ducts due to their low smoke emissions and high tolerance to heat (Strong, 2008).

14.4 Advantages of high-temperature polymers in aviation

The aerospace sector, which is always pushing the limits of technology to reach new heights, is a monument to human ingenuity and advancement. The constant search for strong, lightweight materials that can endure the demands of flight while maintaining efficiency and safety is at the heart of this development. To meet these expectations, PMCs have become a ground-breaking technology that is transforming the way aircraft components are built. The aerospace industry's ongoing progress is highlighted by the dynamic interaction between thermoset and thermoplastic polymers (Mat Yazik and Ismail, 2024).

Polymer materials, due to their numerous engineering designable advantages—such as superior fatigue and fracture resistance compared to metals and ceramics, the ability to meet strict dimensional stability requirements, the ability to meet fast process cycles, and specific strength properties with weight savings of 20%–40%, are widely used for many aerospace applications (Ghori and Siakeng, 2018; Njuguna and Pielichowski, 2003; Toldy et al., 2011).

14.4.1 Light-weight and improved fuel efficiency

Composite materials have been shown to be highly effective in the aircraft sector in reducing component weight and improving fuel efficiency. The sandwich concept structure has grown in popularity as a result of the creation of synthetic cellular materials. A sandwich structure is made up of two thin, stiff, and strong skins (faces, facings, or covers); a thick, light core that serves to separate the skins and transfer loads between them; and an adhesive attachment that can transfer axial and shear loads to and from the core. The primary requirements of the aerospace sector are for sandwich structures to be both rigid and extremely lightweight (Kesarwani, 2017; Nayak, 2014).

The aviation sector is experiencing a surge in demand for lightweight materials due to increasing aviation fuel prices. Fuel consumption accounts for about half of operating costs in the aviation industry (Njuguna, 2016). The expense of a heavier system increases because it requires more fuel to take off. When it comes to various metallic materials utilized in aircraft components, polymer materials provide significant benefits over others (Mouritz, 2012). One of the main characteristics is its low weight and specific strength, which together reduce total weight by about 20%–40%. Additional characteristics include the capability for quick process cycles, the capacity to adhere to strict dimensional stability requirements, reduced thermal expansion characteristics, and superior fatigue and fracture resistance (Cantor and Assender, 2015). Through the use of polymer composite materials, military fighter aircraft of today have achieved a weight reduction of 30%.

14.4.2 Cost-effectiveness, enhanced durability, and reliability

High-temperature composite materials have a longer service life and are more corrosion-resistant than metals. They can reduce maintenance costs by up to 30% and operational expenses by up to 10% because fewer structural checks are necessary (Zolkin et al., 2021). Present-day aerospace applications have focused on using fiber polymer composites to replace the secondary structure, with carbon, glass, Kevlar, or combinations of these serving as the reinforcement media (Williams et al., 2007). High performance and the ability to customize material properties to satisfy particular needs have made polymeric materials the preferred option for numerous aerospace applications. These materials can be designed to be both strong and lightweight, resistant to corrosion from most chemicals, and long-lasting in the harshest environmental conditions. PEEK is a noteworthy illustration of a high-temperature polymer.

The semicrystalline thermoplastic PEEK is renowned for its superior mechanical qualities, resistance to chemicals, and thermal stability. It is strong and stiff even at temperatures higher than 200°C (392°F). Consequently, PEEK-based composites are utilized in parts like thrust reversers, heat shields, and engine parts that are subjected to high temperatures for extended periods of time. PEEK-based materials' remarkable thermal resilience helps reduce maintenance needs and increase engine efficiency. They guarantee the dependability of aircraft systems even under the most taxing flying conditions because of their capacity to retain structural integrity in the face of intense heat. Additionally, the lightweight nature of polymer composites enables improved airplane performance and fuel efficiency (Ching et al., 2018).

14.5 Future directions and innovations

Future developments in multifunctional nanocomposite design have the potential to significantly impact various industries, particularly in the construction, automotive, and aerospace sectors (Simões, 2024; Monteiro and Simões, 2024). Researchers and engineers are focusing on innovative ways to enhance the performance, sustainability, and usability of these composites as the demand for multifunctional materials that can accomplish multiple tasks simultaneously increases. Adding sophisticated nanomaterials to composite matrices is one intriguing approach. Adding metal-organic frameworks, graphene, and CNTs can greatly improve the mechanical, thermal, and electrical characteristics of composites (Lin et al., 2024; Latif et al., 2023). Recently, for example, graphene-based natural fiber composites have been shown to have superior mechanical and thermal properties, which makes them appropriate for a variety of multipurpose uses, including EMI shielding and structural health monitoring (Osman et al., 2022). Through the use of nanomaterial reinforcement, these composites' characteristics can be customized, creating new opportunities for high-performance applications (Rashid et al., 2024).

Furthermore, as environmental concerns gain prominence, the development of sustainable and biobased composites is gaining momentum. High-performing and ecofriendly composites can be created by combining natural fibers like hemp and jute with biobased resins (Mohammed et al., 2024; Skosana et al., 2025; Prasad et al., 2024). In line with the increased focus on sustainability in material design, these materials can offer superior mechanical qualities while lowering dependency on fossil fuels. It is anticipated that research into multifunctional composites, including biobased elements, would increase, especially in industries where environmental sustainability is a top concern (Nagaraja et al., 2024; Al-Gethami et al., 2024).

Developing smart composites that react to outside stimuli is another important trend (Li et al., 2023). More and more research is being done on materials that have shape memory effects, self-healing qualities, or the ability to alter their characteristics in reaction to environmental changes. For instance, creating multipurpose composites that use phase change materials for temperature control might improve buildings' and automobiles' energy efficiency (Amberkar and Mahanwar, 2023; Wang et al., 2022b; Malazi et al., 2025; Reddy et al., 2024).

By absorbing, storing, and releasing thermal energy, these materials can enhance performance in a range of applications. Another fascinating trend is the incorporation of multipurpose properties into textiles. Smart textiles that track environmental and health factors or even have the ability to collect energy can be created by integrating sensors and electronic components into textiles (Shi et al., 2020). Fiber-integrated materials have the ability to perform a variety of tasks, including sensing and structural support, which is a major development in multifunctional composites.

Additionally, creating and improving multifunctional composites with the use of artificial intelligence (AI) and machine learning is becoming a game-changing strategy (Alarifi, 2024; Hassan et al., 2024; Kaur et al., 2024; Nguyen et al., 2024). Finding the best material combinations and processing parameters can be made easier with AI, which will improve performance and save manufacturing costs. Advanced multifunctional composites should be developed more quickly if digital technologies are incorporated into material design procedures.

Engineering polymers face both obstacles and chances for further development on their path to more extensive uses in a range of sectors. Finding a balance between cost-effectiveness and performance is a major challenge. Despite having better qualities than conventional materials, high-performance polymers are frequently more expensive, which makes widespread adoption difficult, especially in industries where costs are a concern. Furthermore, the sustainability of polymers is still a major worry. The need for polymers made from renewable resources and that are recyclable or biodegradable has increased as the world shifts to a more environmentally conscious production paradigm (Ncube et al., 2023; Harun-Ur-Rashid and Imran, 2024). Continuous advancements in polymer chemistry and processing technologies are necessary to meet this demand. Furthermore, regulatory environments are changing to place more emphasis on compliance, safety, and environmental effects, which is stretching the limits of polymer research and development.

Looking ahead, advances in material science that address these issues will be crucial to the development of engineering polymers. This covers the creation of economical synthesis techniques, the identification of novel sustainable polymer materials, and improvements in recycling technology (Ncube et al., 2023; Radu et al., 2025).

Additionally, combining polymers with cutting-edge technologies like additive manufacturing and nanotechnology creates new possibilities for applications that were previously unthinkable (Harun-Ur-Rashid and Imran, 2024; Khaleel and Alsharif, 2025; Bahadur, 2024; Mohammadi, 2024). The potential for engineering polymers to revolutionize sectors ranging from electronics and healthcare to automotive and aerospace is increasing as researchers and engineers work through these obstacles, pointing to a future in which these materials will be essential for promoting sustainability and creativity.

To sum it up, combining cutting-edge materials, environmentally friendly procedures, intelligent features, and digital technology will define multifunctional composite design in the future (Sabet, 2024; Kumar et al., 2024b; Rane et al., 2023). As more academics investigate these possibilities, it becomes increasingly clear how multifunctional composites could transform various industries. Composites will perform better due to continuous advancements in material science, which will also help create a more efficient and sustainable future (Bhong et al., 2023).

14.5.1 Emerging high-temperature polymer technologies

14.5.1.1 Polyimides

The exceptional mechanical strength, low outgassing, and heat resistance of PIs make them one of the most popular high-temperature polymers in aerospace (Hicyilmaz and Celik Bedeloglu, 2021; Prakash et al., 2025). New PI grades are being created to function in even harsher environments; some of them can withstand temperatures above 400°C. There are uses for these polymers in:

- Composites with advanced properties for high-temperature structures
- Products for thermal insulation
- Laminates, foams, and films for insulation

14.5.1.2 Polyether–ether ketone

PEEK is renowned for its remarkable blend of mechanical properties, chemical resistance, and thermal stability (Luo et al., 2023; Ma et al., 2021). Due to its strength and ability to withstand temperatures as high as 260°C, it is an ideal fit for aerospace applications. Recent advancements in PEEK composites consist of the following:

- PEEK composites bonded with carbon fiber for structural elements
- Better PEEK coatings to withstand high temperatures
- Lightweight clips and fasteners

14.5.1.3 Polybenzimidazole

One of the best thermoplastic polymers, PBI, is renowned for its exceptional thermal stability and can function at constant temperatures of up to 400°C (Wang et al., 2022a). PBI is finding new uses in aviation, such as the following:

- Materials for thermal and acoustic insulation
- Materials for engine component sealing and bearings
- Clothing that protects technicians performing airplane maintenance

14.5.1.4 Liquid crystal polymers

Liquid crystal polymers (LCPs) have low moisture absorption, flame retardancy, and good temperature stability (Liu et al., 2023; Liu et al., 2022a). Due to their dimensional stability and dielectric properties, they are becoming a crucial component of airplane electronics. Uses for LCPs include the following:

- High-performance cables and connectors
- Packaging of microelectronics
- Thin-wall components for pumps and fuel systems

14.5.1.5 Silicone-based polymers

Silicones are frequently employed in high-temperature applications due to their electrical insulating qualities, flexibility, and temperature range (Dharini and Krishnamoorthy, 2023; Wu et al., 2024). More recent silicone formulations can operate at temperatures higher than 300°C due to their improved thermal resistance. Among the applications are the following:

- Insulation and thermal barriers
- Flexible hoses and tubing
- Gaskets and seals for high-temperature settings

14.5.1.6 Thermoplastic polyimides

Thermoplastic PIs (TPIs) combine the advantages of thermoplastic production with the thermal characteristics of traditional PIs (Valente et al., 2023). Compared to conventional thermosetting materials, TPIs require less processing, and new grades are being created for the following:

- High-temperature injection-molded parts
- Lightweight, strong structures for the airframe and powerplant
- Components of hybrid metal-polymers

14.5.2 Potential applications in next-generation aircraft

Future developments in material science and engineering, especially in the field of polymers, appear to be bright, signaling the start of a new era of technological breakthroughs and applications (Sharif Bakhsh et al., 2025; Hughes et al., 2025). Continued research into new polymer structures, the development of materials that respond to environmental cues, and the enhancement of sustainable polymers derived from renewable resources are all anticipated to yield potential breakthroughs. These developments have the potential to completely transform the energy and healthcare sectors by providing more effective, ecofriendly solutions that can change with the times.

These developments are largely the result of interdisciplinary studies that connect chemistry, physics, biology, and engineering to find new uses and insights for polymers (Al-Amiery et al., 2024). By combining expertise from these several domains, materials with previously unheard-of qualities and capabilities can be developed, creating new avenues for solving global issues. The influence of these advances is further enhanced by international cooperation, which makes it easier to share resources, ideas, and knowledge across national boundaries. Partnerships between government, business, and academia can scale up research and development initiatives, speeding up invention and its conversion into useful applications. Future advances in polymer technology and material science will be largely driven by interdisciplinary research and international collaboration, as these cooperative efforts are crucial for addressing the intricate, multidimensional issues that society faces today, from healthcare to climate change (Sabet, 2025b; Sarangi, 2025).

To overcome the technological, financial, and environmental challenges associated with using these materials, a concerted effort is necessary. The requirement for a closed-loop lifecycle, cost-effectiveness, and scalability continues to be the major obstacle. Moreover, a multidisciplinary strategy and strong international cooperation are necessary to close the gap between lab research and industry application, guaranteeing that developments in polymer technology can be converted into real societal advantages. The analysis emphasizes how crucial it is to continue innovating and adapting in the field of engineering polymers. For these materials to reach their full potential, stakeholders from all walks of life, from researchers and business leaders to legislators, must work together consistently. By accomplishing this, the engineering polymer industry can keep advancing a society that is more efficient, technologically sophisticated, and sustainable by addressing some of the most pressing problems of our day through the use of material science (Paraye and Sarviya, 2024; Abedsoltan, 2024; Sabet, 2025a). The revolutionary potential of engineering polymers is a testament to human creativity and a beacon of hope for both environmental stewardship and future technology.

High-temperature polymers in aviation are expected to concentrate on the following in the future (Parin and Demirci, 2024):

- Enhancing the mechanical and thermal stability of current polymers
- Creating polymers that can mend themselves
- Improving environmental sustainability and recyclability
- Increasing the production of high-temperature polymer products with additive manufacturing

14.5.3 High-temperature polymer valves in a cryogenic environment: a case study

Liquid oxygen (a cryogenic propellant) used in the launching of vehicles in space creates a challenging environment. Valves within the rocket's fluid control mechanism are subjected to high-temperature conditions as they regulate the flow of propellants. Hence, they are critical in ensuring a successful and safe mission. Considering the critical requirements for such valves, including loads, speeds, high pressure, and installation procedures, the valves must be designed to withstand the overall stress. This is to prevent the catastrophic consequences of valve failure.

As the propulsion of the system becomes more intense, the temperature of each element increases from −207°C to 398°C with pressure up to 200 bar. The operation of the valve is required to ensure low friction while regulating the fluid and thus ensuring smooth functionality of the launch vehicle and engine. To achieve this, a suitable material is required to withstand such a wide range of temperatures while ensuring low friction during system operation.

The Omniseal PFAS-Free* 103A spring-energized polymer seal is a custom-designed sealing solution that is engineered to handle the natural shrinkage and dilatation of polymers used in seal jackets at cryogenic to very high temperatures. This is a new PI-based material, the G45, with temperature resistance above the required temperature of 320°C. To address the growing concerns of the aviation industry, this high-temperature polymer valve, which is designed to accommodate different thermal behavior, is manufactured using PFAS-Free* material.

14.6 Conclusion

The application of high-temperature polymers in the aviation industry has revolutionized aircraft performance and design. Some of the properties offered by these polymers are light weight, mechanical strength, and, of course, exceptional thermal stability. These properties make them exceptional for use in various aircraft components, from TPS to engine parts. The progressing development and research in this field are focused on fabricating more thermally sustainable, ecofriendly, and efficient materials for aviation applications. In the future, high-temperature polymers are expected to play an important role in advancing aircraft technology and improving overall aircraft performance. Innovations in high-temperature polymer composites will promote further development in aviation, leading to durable, cost-effective, and safer aircraft designs.

References

Abedsoltan, H., 2024. Applications of plastics in the automotive industry: current trends and future perspectives. Polymer Engineering & Science 64 (3), 929–950. Available at https://doi.org/10.1002/pen.26604.

Agarwal, B.D., Broutman, L.J., Chandrashekhara, K., 2017. Analysis and Performance of Fiber Composites. John Wiley & Sons.

Aina, A.N., Rizal, M.A.M., Rased, M.F.A., Hassan, S.A., Ng, L.F., Rajeshkumar, L., et al., 2024. Fiber-reinforced thermoplastic composites for future use in aircraft radomes: biomimetic design approaches and its performances. Fibers and Polymers 25 (12), 4503–4527. Available at https://doi.org/10.1007/s12221-024-00776-1. https://www.springer.com/journal/12221.

Al-Amiery, A.A., Fayad, M.A., Abdul Wahhab, H.A., Al-Azzawi, W.K., Mohammed, J.K., Majdi, H.S., 2024. Interfacial engineering for advanced functional materials: surfaces, interfaces, and applications. Results in Engineering 22, 102125. Available at https://doi.org/10.1016/j.rineng.2024.102125.

Alarifi, I.M., 2024. Revolutionising fabrication advances and applications of 3D printing with composite materials: a review. Virtual and Physical Prototyping 19 (1). Available at https://doi.org/10.1080/17452759.2024.2390504, http://www.tandf.co.uk/journals/titles/17452759.asp.

Al-Gethami, W., Qamar, M.A., Shariq, M., Alaghaz, A.-N.M.A., Farhan, A., Areshi, A.A., et al., 2024. Emerging environmentally friendly bio-based nanocomposites for the efficient removal of dyes and micropollutants from wastewater by adsorption: a comprehensive review. RSC Advances 14 (4), 2804–2834. Available at https://doi.org/10.1039/d3ra06501d.

Amberkar, T., Mahanwar, P., 2023. Thermal energy management in buildings and constructions with phase change material-epoxy composites: a review. Energy Sources, Part A: Recovery, Utilization and Environmental Effects 45 (1), 727–761. Available at https://doi.org/10.1080/15567036.2023.2171514, http://www.tandf.co.uk/journals/titles/15567036.asp.

Arun, M., Gupta, A.M., Lodhe, M., 2020. Processing and properties of SiC/SiC composites. Advances in Intelligent Systems and Computing 1053, 815–823. Available at https://doi.org/10.1007/978-981-15-0751-9_74, http://www.springer.com/series/11156.

Ayaz, A., Rasool Memon, I., Ali, F., Malik, S., Shah, U.H., Ghafoor, U., et al., 2024. Aeroelastic tailoring framework of pazy wing with variable stiffness of composites spar. IEEE Access 12, 106042–106055. Available at https://doi.org/10.1109/access.2024.3435961.

Bahadur, P.S. 2024. Nano Technology: The Future of Innovation.

Bello, S.A., 2021. Carbon-fiber composites: development, structure, properties, and applications. Handbook of Nanomaterials and Nanocomposites for Energy and Environmental Applications. Springer International Publishing, pp. 63–84. Available at https://link.springer.com/book/10.1007/978-3-030-36268-3. https://doi.org/10.1007/978-3-030-36268-3_86.

Bhong, M., Khan, T.K.H., Devade, K., Vijay Krishna, B., Sura, S., Eftikhaar, H.K., et al., 2023. Review of composite materials and applications. Materials Today: Proceedings. Available at https://doi.org/10.1016/j.matpr.2023.10.026.

Bystry-King, F.A., King, J.J., 2020. Polyimides. Informa UK Limited, pp. 315–333. Available at https://doi.org/10.1201/9781003066156-13.

Cantor, B., Assender, H., Grant, P., 2015. Aerospace Materials Aerospace Materials. CRC Press, pp. 1–310. Available at http://www.tandfebooks.com/doi/book/10.1201/9781420034721. https://doi.org/10.1201/9781420034721.

Chen, F., Zhang, H., Li, S., Chen, Y., Liang, M., Heng, Z., et al., 2023a. Design of high-performance resin by tuning cross-linked network topology to improve CF/bismaleimide composite compressive properties. Composites Science and Technology 242, 110170. Available at https://doi.org/10.1016/j.compscitech.2023.110170.

Chen, Y., Zhang, J., Li, Z., Zhang, H., Chen, J., Yang, W., et al., 2023b. Manufacturing technology of lightweight fiber-reinforced composite structures in aerospace: current situation and toward intellectualization. Aerospace 10 (3), 206. Available at https://doi.org/10.3390/aerospace10030206.

Ching, Y.C., Gunathilake, T.M.S.U., Ching, K.Y., Chuah, C.H., Sandu, V., Singh, R., et al., 2018. Effects of High Temperature and Ultraviolet Radiation On Polymer Composites Durability and Life Prediction in Biocomposites, Fibre-Reinforced Composites and Hybrid Composites. Elsevier, pp. 407–426. Available at http://www.sciencedirect.com/science/book/9780081022900. https://doi.org/10.1016/B978-0-08-102290-0.00018-0.

Das, A., Chatham, C.A., Fallon, J.J., Zawaski, C.E., Gilmer, E.L., Williams, C.B., et al., 2020. Current understanding and challenges in high temperature additive manufacturing of engineering thermoplastic polymers. Additive Manufacturing 34, 101218. Available at https://doi.org/10.1016/j.addma.2020.101218.

Deng, Y., Zhang, F., Liu, Y., Leng, J., 2022. Design and synthesis of shape memory phenol-formaldehyde with good irradiation resistance, thermal, and mechanical properties. ACS Applied Polymer Materials 4 (8), 5789–5799. Available at https://doi.org/10.1021/acsapm.2c00719.

Dharini, T., Krishnamoorthy, A., Kuppusami, P., 2023. Springer Science and Business Media LLC, pp. 385–401. Available at https://doi.org/10.1007/978-3-031-45534-6_14, Silicone-Based Coatings for High-Temperature Applications 1843.

Drobny, J.G., 2023. Fluorocarbon Elastomers. Informa UK Limited, pp. 133–186. https://doi.org/10.1201/9781003204275-12.

Dufton, P.W., 2000. Lightweight Thermoset Composites: Materials in Use, Their Processing and Applications. iSmithers Rapra Publishing.

Dursun, T., Soutis, C., 2014. Recent developments in advanced aircraft aluminium alloys. Materials & Design (1980-2015) 56, 862–871. Available at https://doi.org/10.1016/j.matdes.2013.12.002.

Fahy, W.P., Chang, A., Wu, H., Koo, J.H. 2021. Recent developments of ablative thermal protection systems for atmospheric entry. AIAA Scitech 2021 Forum, pp. 1–19.

Garg, P., Jamwal, A., Kumar, D., Kumar Sadasivuni, K., Hussain, C.M., Gupta, P., 2019. Advance research progresses in aluminium matrix composites: manufacturing & applications. Journal of Materials Research and Technology 8 (5), 4924–4939. Available at https://doi.org/10.1016/j.jmrt.2019.06.028.

Ghori, S.W., Siakeng, R., Rasheed, M., Saba, N., Jawaid, M., 2018. The Role Of Advanced Polymer Materials in Aerospace Sustainable Composites for Aerospace Applications. Elsevier, pp. 19–34. Available at http://www.sciencedirect.com/science/book/9780081021316. https://doi.org/10.1016/B978-0-08-102131-6.00002-5.

Goh, G.D., Wong, K.K., Tan, N., Seet, H.L., Nai, M.L.S., 2024. Large-format additive manufacturing of polymers: a review of fabrication processes, materials, and design. Virtual and Physical Prototyping 19 (1). Available at https://doi.org/10.1080/17452759.2024.2336160, http://www.tandf.co.uk/journals/titles/17452759.asp.

Grilli, M.L., Valerini, D., Slobozeanu, A.E., Postolnyi, B.O., Balos, S., Rizzo, A., et al., 2021. Critical raw materials saving by protective coatings under extreme conditions: a review of last trends in alloys and coatings for aerospace engine applications. Materials 14 (7), 1656. Available at https://doi.org/10.3390/ma14071656.

Guo, M.F., Yu, K.X., Yang, J.P., Zhang, P., Zhang, Y.C., Zhu, D.X., 2024a. B2O3-reinforced ablative materials with superior and comprehensive ablation resistance used in aerospace propulsion thermal protection systems. Polymer Degradation and Stability 223, 110740. Available at https://doi.org/10.1016/j.polymdegradstab.2024.110740.

Guo, M.F., Du, J.F., Zhang, Y.C., 2024b. ZrO2-reinforced polymer-matrix composites used for thermal protection systems of ultra-high temperature aerospace propulsion. Aerospace Science and Technology 145, 108906. Available at https://doi.org/10.1016/j.ast.2024.108906.

Harun-Ur-Rashid, M., Imran, A.B., 2024. Emerging trends in engineering polymers: a paradigm shift in material engineering. Recent Progress in Materials 06 (03), 1–37. Available at https://doi.org/10.21926/rpm.2403024.

Hassan, M.S., Billah, K.M.M., Hall, S.E., Sepulveda, S., Regis, J.E., Marquez, C., et al., 2022. Selective laser sintering of high-temperature thermoset polymer. Journal of Composites Science 6 (2), 41. Available at https://doi.org/10.3390/jcs6020041.

Hassan, K., Thakur, A.K., Singh, G., Singh, J., Gupta, L.R., Singh, R., 2024. Application of artificial intelligence in aerospace engineering and its future directions: a systematic quantitative literature review. Archives of Computational Methods in Engineering 31 (7), 4031–4086. Available at https://doi.org/10.1007/s11831-024-10105-7, https://www.springer.com/journal/11831.

Hernandez-Carrillo, I., Wood, C.J., Liu, H., 2017. Advanced materials for the impeller in an ORC radial microturbine. Energy Procedia 129, 1047–1054. Available at https://doi.org/10.1016/j.egypro.2017.09.241, http://www.sciencedirect.com/science/journal/18766102.

Hernandez-Carrillo, I., Wood, C., Liu, H., 2019. Development of a 1000 W organic Rankine cycle micro-turbine-generator using polymeric structural materials and its performance test with compressed air. Energy Conversion and Management 190, 105–120. Available at https://doi.org/10.1016/j.enconman.2019.03.092, https://www.journals.elsevier.com/energy-conversion-and-management.

Hicyilmaz, Celik Bedeloglu, A., 2021. Applications of polyimide coatings: a review. SN Applied Sciences 3 (3).

Hughes, J. 2019. Mechanical response of carbon fiber/polyphenylene sulfide commingled fiber & polymethacrylimide foam core sandwich composites.

Hughes, K.J., Ganesan, M., Tenchov, R., Iyer, K.A., Ralhan, K., Diaz, L.L., et al., 2025. Nanoscience in action: unveiling emerging trends in materials and applications. ACS Omega 10 (8), 7530–7548. Available at https://doi.org/10.1021/acsomega.4c10929, pubs.acs.org/journal/acsodf.

Hussain, I., Immanuel, R.J., 2021. Composite Materials and Its Advancements for a Cleaner Engine of Future. Springer Science and Business Media LLC, pp. 169–191. Available at https://doi.org/10.1007/978-981-16-8337-4_9.

Joshi, A.M., 2024. Rubbers and Elastomers in Specialty Applications. Informa UK Limited, pp. 293–314. Available at https://doi.org/10.1201/9781003561361-15.

Kalaiyarasan, A., Ramesh, P., 2018. Paramasivam, study of advanced composite materials in aerospace application. International Journal of Scientific Research in Mechanical and Materials Engineering 2 (1), 25–34.

Kaur, S., Kumar, R., Singh, K., Huang, Y., 2024. Leveraging artificial intelligence for enhanced sustainable energy management. Journal of Sustainability for Energy 3 (1), 1–20. Available at https://doi.org/10.56578/jse030101.

Kesarwani, S., 2017. Polymer composites in aviation sector. International Journal of Engineering Research 6, 518–525.

Khaleel, M.M., Alsharif, A., 2025. Nanotechnology in materials engineering innovations in construction and manufacturing. The Open. European Journal of Applied Sciences (OEJAS) 51–64.

Koo, J.H., 2016. Fundamentals, Properties, and Applications of Polymer Nanocomposites. Cambridge University Press, pp. 1–697. Available at http://www.cambridge.org/us/academic/subjects/engineering/materials-science/fundamentals-properties-and-applications-polymer-nanocomposites, https://doi.org/10.1017/CBO9781139342766.

Kumar, A., Sharma, K., Dixit, A.R., 2021. A review on the mechanical properties of polymer composites reinforced by carbon nanotubes and graphene. Carbon Letters 31 (2), 149–165. Available at https://doi.org/10.1007/s42823-020-00161-x, https://www.springer.com/journal/42823.

Kumar, S., Ranjan, A., Manocha, L.M., Namburi, E.P., 2024a. High Temperature Composites for Aerospace Applications and Their (Select) Production Technologies. Springer Science and Business Media LLC, pp. 253–285. Available at https://doi.org/10.1007/978-981-99-9795-4_9.

Kumar, V., Parvin, N., Park, S.-S., Joo, S.W., Kumar Mandal, T., 2024b. Review on cutting-edge innovations in carbon nanomaterials reinforced silicone rubber composites for flexible electronics and healthcare devices. ACS Applied Polymer Materials 6 (23), 14235–14259. Available at https://doi.org/10.1021/acsapm.4c02511.

Latif, Z., Ali, M., Lee, E.-J., Zubair, Z., Lee, K.H., 2023. Thermal and mechanical properties of nano-carbon-reinforced polymeric nanocomposites: a review. Journal of Composites Science 7 (10), 441. Available at https://doi.org/10.3390/jcs7100441.

Lau, K.S.Y., 2014. High-Performance Polyimides and High Temperature Resistant Polymers Handbook of Thermoset Plastics. Elsevier Inc., pp. 297–424. Available at http://www.sciencedirect.com/science/book/9781455731077. https://doi.org/10.1016/B978-1-4557-3107-7.00010-5.

Laub B. Venkatapathy E. 2004 Thermal protection system technology and facility needs for demanding future planetary missions European Space Agency, (Special Publication) ESA SP 03796566 544 239-247.

Leicht, B.T., Bohan, B.T., Schauer, F., Kemnitz, R., Rueschhoff, L.M., Lam, B., et al., 2023. Manufacturing a ceramic turbine rotor for a compact jet engine. Journal of Turbomachinery 145 (8). Available at https://doi.org/10.1115/1.4062124, https://asmedigitalcollection.asme.org/turbomachinery.

Li, J., Ding, Q., Wang, H., Wu, Z., Gui, X., Li, C., et al., 2023. Engineering smart composite hydrogels for wearable disease monitoring. Nano-Micro Letters 15 (1). Available at https://doi.org/10.1007/s40820-023-01079-5.

Li, Q., Du, B., Liu, Y., Jian, Q., Feng, T., Zhang, X., et al., 2025. On structural behavior of lightweight steel-reinforced polymer/composite hybrid structures: a review of analytical, experimental, and numerical methods. Mechanics of Advanced Materials and Structures 1–30. Available at https://doi.org/10.1080/15376494.2025.2479207.

Lin, Y., Zhong, B., Chen, J., Zhang, B., Yu, Y., 2024. Reinforcement of microwave absorption and thermal conductivity of polydimethylsiloxane achieved by metal-organic framework-derived carbon nanotube/boron nitride flake heterogeneous structure. Carbon 227, 119242. Available at https://doi.org/10.1016/j.carbon.2024.119242.

Liu, B.W., Zhao, H.B., Wang, Y.Z., 2022a. Advanced flame-retardant methods for polymeric materials. Advanced Materials 34 (46). Available at https://doi.org/10.1002/adma.202107905, http://onlinelibrary.wiley.com/journal/10.1002/(ISSN)1521-4095.

Liu, Y.Y., Wang, Y.K., Wu, D.Y., 2022b. Synthetic strategies for highly transparent and colorless polyimide film. Journal of Applied Polymer Science 139 (28). Available at https://doi.org/10.1002/app.52604, http://onlinelibrary.wiley.com/journal/10.1002/(ISSN)1097-4628.

Liu, Z., Fan, X., Zhang, J., Chen, L., Tang, Y., Kong, J., et al., 2023. PBO fibers/fluorine-containing liquid crystal compound modified cyanate ester wave-transparent laminated composites with excellent mechanical and flame retardance properties. Journal of Materials Science & Technology 152, 16–29. Available at https://doi.org/10.1016/j.jmst.2023.01.007.

Lohawala, N., Wen, Z.P. 2024. Navigating sustainable skies: Challenges and strategies for greener aviation. Resources for the Future (RFF).

Luo, C., Liu, Y., Peng, B., Chen, M., Liu, Z., Li, Z., et al., 2023. PEEK for oral applications: recent advances in mechanical and adhesive properties. Polymers 15 (2), 386. Available at https://doi.org/10.3390/polym15020386.

Ma, Y., Li, Y., Zhu, S. New development and application on microwave-absorbing sandwich structure based on high-temperature resistant composite materials. First Aerospace Frontiers Conference.

Ma, H., Suonan, A., Zhou, J., Yuan, Q., Liu, L., Zhao, X., et al., 2021. PEEK (Polyether-ether-ketone) and its composite materials in orthopedic implantation. Arabian Journal of Chemistry 14 (3), 102977. Available at https://doi.org/10.1016/j.arabjc.2020.102977.

Ma, X., Ji, T., Zhang, J., Shen, S., Wang, S., Wang, J., et al., 2023. A double-decker silsesquioxane of norbornene and performance of crosslinking reactive modified EPDM ablation resistance composites. Composites Part A: Applied Science and Manufacturing 165, 107370. Available at https://doi.org/10.1016/j.compositesa.2022.107370.

Malazi, M.T., Afshari, F., Kaya, K., 2025. Phase change materials (pcms) for buildings and automotive applications: a review study. Heat Transfer Research 56 (1), 39–52. Available at https://doi.org/10.1615/heattransres.2024053911.

Mat Yazik, M.H., Ismail, I., 2024. Aerospace Structures and Engines from Polymer Composites Aerospace Materials: Novel Technologies and Practical Applications. Elsevier, pp. 349–383. Available at https://www.sciencedirect.com/book/9780443221187. https://doi.org/10.1016/B978-0-443-22118-7.00015-4.

Mat Yazik, M.H., Ismail, I., 2025. Aerospace Structures and Engines from Polymer Composites. Elsevier BV, pp. 349–383. Available at https://doi.org/10.1016/b978-0-443-22118-7.00015-4.

Mittal, V., 2011a. High Performance Polymers: An Overview High Performance Polymers and Engineering Plastics. Wiley, pp. 1–20. Available at https://doi.org/10.1002/9781118171950.ch1.

Mittal, V., 2011b. Synthesis and Characterization of Novel Polyimides High Performance Polymers and Engineering Plastics. Wiley, pp. 205–242. Available at https://doi.org/10.1002/9781118171950.ch7.

Mohammadi, A.T. 2024. Pioneering the Future: Exploring Cutting-edge Science Innovations and Emerging Technologies.

Mohammed, M., Oleiwi, J.K., Mohammed, A.M., Jawad, A.J.M., Osman, A.F., Adam, T., et al., 2024. A review on the advancement of renewable natural fiber hybrid composites: prospects, challenges, and industrial applications. Journal of Renewable Materials 12 (7), 1237–1290. Available at https://doi.org/10.32604/jrm.2024.051201, https://cdn.techscience.cn/files/jrm/2024/TSP_JRM-12-7/TSP_JRM_51201/TSP_JRM_51201.pdf.

Monteiro, B., Simões, S., 2024. Recent advances in hybrid nanocomposites for aerospace applications. Metals 14 (11), 1283. Available at https://doi.org/10.3390/met14111283.

Mouritz, A.P., 2012. Introduction to Aerospace Materials. Woodhead Publishing Limited. Available from, www.woodheadpublishing.com/9781855739468. https://doi.org/10.1533/9780857095152.

Nagaraj Goud, B., Sura, S., Aravind, P., Jawahar Lal, B., Sanskruti, K., Pavan, C., 2022. An experimental study on mechanical properties of Kevlar composite for aircraft structural applications. Materials Today: Proceedings 64, 909–916. Available at https://doi.org/10.1016/j.matpr.2022.06.053.

Nagaraja, S., Anand, P.B., Kumar, K.M., Ammarullah, M.I., 2024. Synergistic advances in natural fibre composites: a comprehensive review of the eco-friendly bio-composite development, its characterization and diverse applications. RSC Advances 14 (25), 17594–17611. Available at https://doi.org/10.1039/d4ra00149d, http://pubs.rsc.org/en/journals/journal/ra.

Natali, M., Kenny, J.M., Torre, L., 2016. Science and technology of polymeric ablative materials for thermal protection systems and propulsion devices: a review. Progress in Materials Science 84, 192–275. Available at https://doi.org/10.1016/j.pmatsci.2016.08.003.

Nayak, N.V., 2014. Composite materials in aerospace applications. International Journal of Scientific and Research Publications 4 (9), 1–10.

Ncube, A., Mtetwa, S., Bukhari, M., Fiorentino, G., Passaro, R., 2023. Circular economy and green chemistry: the need for radical innovative approaches in the design for new products. Energies 16 (4), 1752. Available at https://doi.org/10.3390/en16041752.

Nguyen, V.G., Sharma, P., Ağbulut, Ü., Le, H.S., Cao, D.N., Dzida, M., et al., 2024. Improving the prediction of biochar production from various biomass sources through the implementation of eXplainable machine learning approaches. International Journal of Green Energy 21 (12), 2771–2798. Available at https://doi.org/10.1080/15435075.2024.2326076, http://www.tandf.co.uk/journals/titles/15435075.asp.

Njuguna, J., 2016. Lightweight Composite Structures in Transport: Design, Manufacturing, Analysis and Performance Lightweight Composite Structures in Transport: Design, Manufacturing, Analysis and Performance. Elsevier Inc., pp. 1–453. Available at http://www.sciencedirect.com/science/book/9781782423256. https://doi.org/10.1016/C2014-0-02646-9.

Njuguna, J., Pielichowski, K., 2003. Polymer nanocomposites for aerospace applications: properties. Advanced Engineering Materials 5 (11), 769–778. Available at https://doi.org/10.1002/adem.200310101.

Osman, A., Elhakeem, A., Kaytbay, S., Ahmed, A., 2022. A comprehensive review on the thermal, electrical, and mechanical properties of graphene-based multi-functional epoxy composites. Advanced Composites and Hybrid Materials 5 (2), 547–605. Available at https://doi.org/10.1007/s42114-022-00423-4.

Paraye, P., Sarviya, R.M., 2024. Advances in polymer composites, manufacturing, recycling, and sustainable practices. Polymer-Plastics Technology and Materials 63 (11), 1474–1497. Available at https://doi.org/10.1080/25740881.2024.2339295.

Parin, F.N., Demirci, F., 2024. Durability of Polymer Composite Materials for High-temperature Applications Aging and Durability of FRP Composites and Nanocomposites. Elsevier, Turkey, pp. 135–170. Available at https://www.sciencedirect.com/book/9780443155451, https://doi.org/10.1016/B978-0-443-15545-1.00017-2.

Parker, D. 2000. Polymers, high-temperature. In: Ullmann's Encyclopedia of Industrial Chemistry.

Parveez, B., Kittur, M.I., Badruddin, I.A., Kamangar, S., Hussien, M., Umarfarooq, M.A., 2022. Scientific advancements in composite materials for aircraft applications: a review. Polymers 14 (22), 5007. Available at https://doi.org/10.3390/polym14225007.

Peters, S.T., 2013. Handbook of Composites. Springer Science & Business Media.

Pierce, R.S., Falzon, B.G., 2017. Simulating resin infusion through textile reinforcement materials for the manufacture of complex composite structures. Engineering 3 (5), 596–607. Available at https://doi.org/10.1016/J.ENG.2017.04.006, http://www.journals.elsevier.com/engineering/.

Prakash, P., Weerasinghe, J., Levchenko, I., Prasad, K., Alexander, K., 2025. Polyimide nanocomposites for next generation spacesuits. Materials Horizons. Available at https://doi.org/10.1039/d4mh01816h.

Prasad, V., Alliyankal Vijayakumar, A., Jose, T., George, S.C., 2024. A comprehensive review of sustainability in natural-fiber-reinforced polymers. Sustainability 16 (3), 1223. Available at https://doi.org/10.3390/su16031223.

Prasanaa Iyer, N., Arunkumar, N., 2020. Review on fiber reinforced/modified bismaleimide resin composites for aircraft structure application. IOP Conference Series: Materials Science and Engineering 923 (1), 012051. Available at https://doi.org/10.1088/1757-899x/923/1/012051.

Qiao, Y., Bian, X., Ke, H., Zhao, Z., Zhang, X., Zhang, X., et al., 2025. Phenolic-based porous composite with embedded short carbon fiber/hollow spheres for mechanical properties and thermostability. Composites Communications 55, 102305. Available at https://doi.org/10.1016/j.coco.2025.102305.

Radu, I.-C., Vadureanu, A.-M., Cozorici, D.-E., Blanzeanu, E., Zaharia, C., 2025. Advancing sustainability in modern polymer processing: strategies for waste resource recovery and circular economy integration. Polymers 17 (4), 522. Available at https://doi.org/10.3390/polym17040522.

Radyjowski, P.P. 2021. Additively manufactured Si-SiC combustion devices: the role of heat recirculation in counter-flowing reactors.

Rajak, D., Pagar, D., Menezes, P., Linul, E., 2019. Fiber-reinforced polymer composites: manufacturing, properties, and applications. Polymers 11 (10), 1667. Available at https://doi.org/10.3390/polym11101667.

Rajak, D.K., Wagh, P.H., Linul, E., 2021. Manufacturing technologies of carbon/glass fiber-reinforced polymer composites and their properties: a review. Polymers 13 (21). Available at https://doi.org/10.3390/polym13213721, https://www.mdpi.com/2073-4360/13/21/3721/pdf.

Rallini, M., Natali, M., Torre, L., 2019. An Introduction to Ablative Materials and High-Temperature Testing Protocols Nanomaterials in Rocket Propulsion Systems. Elsevier, pp. 529–549. Available at https://www.sciencedirect.com/book/9780128139080. https://doi.org/10.1016/B978-0-12-813908-0.00014-9.

Ramgobin, A., Fontaine, G., Bourbigot, S., 2019. Thermal degradation and fire behavior of high performance polymers. Polymer Reviews 59 (1), 55–123. Available at https://doi.org/10.1080/15583724.2018.1546736, http://www.tandf.co.uk/journals/titles/15583724.asp.

Rane, N., Choudhary, S., Rane, J., 2023. Leading-edge technologies for architectural design: a comprehensive review. SSRN Electronic Journal. Available at https://doi.org/10.2139/ssrn.4637891.

Rashid, A.B., Haque, M., Islam, S.M.M., Uddin Labib, K.M.R., 2024. Nanotechnology-enhanced fiber-reinforced polymer composites: recent advancements on processing techniques and applications. Heliyon 10 (2). Available at https://doi.org/10.1016/j.heliyon.2024.e24692, https://www.sciencedirect.com/science/journal/24058440.

Rb, J.R., M, J.A., D, R., A, S., Kennedy, S.M., 2024. A complete overview of self-healing composites including its models in aeronautical systems. Polymer-Plastics Technology and Materials 63 (9), 1143–1174. Available at https://doi.org/10.1080/25740881.2024.2325427, https://www.tandfonline.com/loi/lpte21.

Reddy, V.J., Ghazali, M.F., Kumarasamy, S., 2024. Innovations in phase change materials for diverse industrial applications: a comprehensive review. Results in Chemistry.

Rinaldi, M., Ferrara, M., Pigliaru, L., Allegranza, C., Nanni, F., 2023. Additive manufacturing of polyether ether ketone-based composites for space application: a mini-review. CEAS Space Journal 15 (1), 77–87. Available at https://doi.org/10.1007/s12567-021-00401-4.

Sabet, M., 2024. Unveiling advanced self-healing mechanisms in graphene polymer composites for next-generation applications in aerospace, automotive, and electronics. Polymer-Plastics Technology and Materials 63 (15), 2032–2059. Available at https://doi.org/10.1080/25740881.2024.2365282.

Sabet, M., 2025a. Advanced functionalization strategies for carbon nanotube polymer composites: achieving superior dispersion and compatibility. Polymer-Plastics Technology and Materials 64 (4), 465–494. Available at https://doi.org/10.1080/25740881.2024.2409312.

Sabet, M., 2025b. From self-assembly to sustainability: advanced polymerization techniques for energy, healthcare, and robotics. Polymer-Plastics Technology and Materials 64 (6), 763–793. Available at https://doi.org/10.1080/25740881.2024.2438044.

Sarangi, P.K., 2025. Machine learning in polymer science: emerging trends and future directions. Macromolecular Symposia. Wiley Online Library.

Shabbir, S., Garvey, S., Dakka, S.M. 2019. Aerospace sealing technology for maintenance, repair and overhaul of engines: a review. 24th International Symposium on Air Breathing Engines.

Shanmugam, V., Karthikbabu, N.B., Rajendran, S., Das, O., 2021. Applications of Various Types of Polymer-Based Composites. Informa UK Limited, pp. 139–149. Available at https://doi.org/10.1201/9781003126300-10.

Sharif Bakhsh, E., Tavakoli Dare, M., Jafari, A., Shahi, F., 2025. Light-activated nanofibers: advances in photo-responsive electrospun polymer technologies. Polymer-Plastics Technology and Materials 64 (4), 397–438. Available at https://doi.org/10.1080/25740881.2024.2408346.

Shi, J., Liu, S., Zhang, L., Yang, B., Shu, L., Yang, Y., et al., 2020. Smart textile-integrated microelectronic systems for wearable applications. Advanced Materials 32 (5). Available at https://doi.org/10.1002/adma.201901958, http://onlinelibrary.wiley.com/journal/10.1002/(ISSN)1521-4095.

Siengchin, S., 2023. A review on lightweight materials for defence applications: present and future developments. Defence Technology 24, 1–17. Available at https://doi.org/10.1016/j.dt.2023.02.025.

Simões, S., 2024. High-performance advanced composites in multifunctional material design: state of the art, challenges, and future directions. Materials 17 (23), 5997. Available at https://doi.org/10.3390/ma17235997.

Skosana, S.J., Khoathane, C., Malwela, T., 2025. Driving towards sustainability: a review of natural fiber reinforced polymer composites for eco-friendly automotive light-weighting. Journal of Thermoplastic Composite Materials 38 (2), 754–780. Available at https://doi.org/10.1177/08927057241254324, https://journals.sagepub.com/home/JTC.

Solntsev, S.S., 2011. High-temperature composite materials and coatings on the basis of glass and ceramics for aerospace technics. Russian Journal of General Chemistry 81 (5), 992–1000. Available at https://doi.org/10.1134/S1070363211050306.

Soutis, C., 2005. Carbon fiber reinforced plastics in aircraft construction. Materials Science and Engineering: A 412 (1-2), 171–176. Available at https://doi.org/10.1016/j.msea.2005.08.064.

Strong, A.B. 2008. Fundamentals of composites manufacturing: materials, methods and applications. Society of Manufacturing Engineers.

Sukiman, N., 2012. Durability and corrosion of aluminium and its alloys: overview, property space, techniques and developments. Aluminium Alloys-New Trends in Fabrication and Applications 5, 47–97.

Sutton, G.P., Biblarz, O., 2011. Rocket Propulsion Elements. John Wiley & Sons.

A. Tayade, Impact of Ceramic Matrix Composites on the Thermal Performance of Gas Turbine Engines. (2024).

Tayade, A., Lynch, S., 2025. Effect of ceramic matrix composites on the thermal efficiency of a power generation turbine. Journal of Turbomachinery 147 (5). Available at https://doi.org/10.1115/1.4067026.

Toldy, A., Szolnoki, B., Marosi, G., 2011. Flame retardancy of fibre-reinforced epoxy resin composites for aerospace applications. Polymer Degradation and Stability 96 (3), 371–376. Available at https://doi.org/10.1016/j.polymdegradstab.2010.03.021.

Uddin, F., 2016. Flame-retardant fibrous materials in an aircraft. Journal of Industrial Textiles 45 (5), 1128–1169. Available at https://doi.org/10.1177/1528083714540700.

Upadhyay, R.K., Mishra, A.K., Kumar, A., 2020. Mechanical degradation of 3D printed PLA in simulated marine environment. Surfaces and Interfaces 21. Available at https://doi.org/10.1016/j.surfin.2020.100778, http://www.journals.elsevier.com/surfaces-and-interfaces.

Valente, M., Rossitti, I., Sambucci, M., 2023. Different production processes for thermoplastic composite materials: sustainability versus mechanical properties and processes parameter. Polymers 15 (1), 242. Available at https://doi.org/10.3390/polym15010242.

Wang, K.Y., Weber, M., Chung, T.S., 2022a. Polybenzimidazoles (PBIs) and state-of-the-art PBI hollow fiber membranes for water, organic solvent and gas separations: a review. Journal of Materials Chemistry A 10 (16), 8687–8718. Available at https://doi.org/10.1039/d2ta00422d, http://pubs.rsc.org/en/journals/journal/ta.

Wang, X., Li, W., Luo, Z., Wang, K., Shah, S.P., 2022b. A critical review on phase change materials (PCM) for sustainable and energy efficient building: design, characteristic, performance and application. Energy and Buildings 260, 111923. Available at https://doi.org/10.1016/j.enbuild.2022.111923.

Williams, G., Trask, R., Bond, I., 2007. A self-healing carbon fibre reinforced polymer for aerospace applications. Composites Part A: Applied Science and Manufacturing 38 (6), 1525–1532. Available at https://doi.org/10.1016/j.compositesa.2007.01.013.

Wu, H., Koo, J.H., 2022. Characterization of High-temperature Polymers for Extreme Environments Analysis of Flame Retardancy in Polymer Science. Elsevier, pp. 299–331. Available at https://www.sciencedirect.com/book/9780128240458. https://doi.org/10.1016/B978-0-12-824045-8.00008-3.

Wu, Y.-Y., Wu, Z.-H., Chen, Z.-Y., Peng, L.-D., Guan, Z.-Q., Li, Y., et al., 2024. Large-scale and facile fabrication of phenyl-containing silicone foam materials with lightweight, wide-temperature flexibility and tunable pore structure for exceptional thermal insulation. Chemical Engineering Journal 492, 152183. Available at https://doi.org/10.1016/j.cej.2024.152183.

Wu, et al., 2025. Experimental study on thermal decomposition temperature and thermal expansion coefficient of typical nonmetallic materials in aeroengine components. Materials (Basel, Switzerland) 18 (6).

Xu, Y. 2023. Integration of Polyimide Flexible PCB Wings in Northeastern's Aerobat. *arXiv*. Available from: https://arxiv.org. https://doi.org/10.48550/arXiv.2309.14346.

Xu, H.-B., Fan, K.-Y., Yang, J.-X., Lian, X.-R., He, F.-M., Li, Z.-Y., 2022. Design and evaluation of variable porosity charring composite for thermal protection system of reentry vehicles. Case Studies in Thermal Engineering 37, 102305. Available at https://doi.org/10.1016/j.csite.2022.102305.

Yang, S.Y., Ji, M., 2018. Polyimide Matrices for Carbon fiber Composites Advanced Polyimide Materials: Synthesis, Characterization, and Applications. Elsevier, pp. 93–136. Available at http://www.sciencedirect.com/science/book/9780128126400. https://doi.org/10.1016/B978-0-12-812640-0.00003-2.

Yang, C., Shen, S.-L., Hou, D.-W., Liao, S.-M., Yuan, D.-J., 2018. Material properties of the seal gasket for shield tunnels: a review. Construction and Building Materials 191, 877–890. Available at https://doi.org/10.1016/j.conbuildmat.2018.10.021.

Yilmaz, 2024. High-performance thermoplastic nanocomposites for aerospace applications: a review of synthesis, production, and analysis. Journal of Reinforced Plastics and Composites.

Zirak, N., Shirinbayan, M., Deligant, M., Tcharkhtchi, A., 2022. Toward polymeric and polymer composites impeller fabrication. Polymers 14 (1), 97. Available at https://doi.org/10.3390/polym14010097.

Zolkin, A.L., Galanskiy, S.A., Kuzmin, A.M., 2021. Perspectives for use of composite and polymer materials in aircraft construction. IOP Conference Series: Materials Science and Engineering 1047 (1), 012023. Available at https://doi.org/10.1088/1757-899x/1047/1/012023.

15

Application of high-temperature polymers in aerospace applications

Mantsopa Koena Selatile Zamisa[1], Rita Okpuwhara[2], Makungu M. Madirisha[3], Tumelo Seadira[1]

[1]*DEPARTMENT OF CHEMICAL AND MATERIALS ENGINEERING, UNIVERSITY OF SOUTH AFRICA, JOHANNESBURG, SOUTH AFRICA* [2]*DEPARTMENT OF CHEMICAL ENGINEERING, UNIVERSITY OF JOHANNESBURG, JOHANNESBURG, SOUTH AFRICA* [3]*DEPARTMENT OF CHEMISTRY, UNIVERSITY OF DAR ES SALAAM, DAR ES SALAAM, TANZANIA*

15.1 Introduction

Aerospace applications encompass a broad range of technologies and materials essential for aviation and space exploration, aiming to enhance performance and safety, reduce costs, and increase sustainability. Aerospace materials serve as the structural components that bear the loads on an aircraft's airframe during flight. They are used in primary and secondary structural parts, including rocket motor casings, antenna dishes, engine nacelles, stabilizers, wing boxes, wings, pressure bulkheads, landing gear, doors, engine cowls, beams, cones, flap track panels, and other essential components. Certain materials, such as those used in wings, fuselage, and landing gear, are critical for safety. Aerospace materials are also utilized in jet engine components like turbine blades and other vital parts. In a nutshell, materials play a pivotal role in aerospace engineering, significantly impacting the design, performance, and safety of both aircraft and spacecraft.

The selection of appropriate materials is crucial due to the demanding operating conditions and performance requirements unique to aerospace components. Various advanced materials are employed in aerospace engineering, each with specific properties, advantages, and applications. While metals were traditionally the primary materials used in aerospace structures, advancements in materials science, particularly in composite materials, have led to the development of new materials for aerospace engineering.

Composites are hybrid materials formed by combining two or more substances to leverage the beneficial properties of each component. Composite materials are extensively used in aerospace applications due to their high strength-to-weight ratios, excellent fatigue resistance, and properties tailored to meet stringent industry requirements. In most common composites, the reinforcing phase and the matrix (continuous form) work together to enhance properties. Reinforcements improve the material's strength and stiffness and are generally anisotropic, offering different properties in various directions. The matrix, often a synthetic

High-Temperature Polymeric Materials. DOI: https://doi.org/10.1016/B978-0-443-32792-6.00012-9

polymer, binds the reinforcements and distributes loads. Composite materials can be classified according to their constituents, that is, base material and reinforcement material. The base material, also termed as a matrix or a binder material, binds or holds the reinforcement materials in composite structures. Depending on the matrix materials, composites may be classified as organic, metal, and ceramic matrix composites. Organic matrix composites could further be classified as carbon and polymer matrix composites. The reinforcing materials may be present in the form of fibers, particles, platelets, flakes, or whiskers of natural or synthetic materials. Based on these reinforcement materials, composites can be classified broadly as fiber-reinforced composites, particle-reinforced composites, and laminar or sheet-molded composites.

These materials, such as fiber-reinforced polymer (FRP) composites (Fig. 15-1), offer benefits like low weight, high stiffness, and corrosion resistance. These characteristics make composites ideal for structural components in aerospace engineering, ensuring optimal performance and durability under demanding conditions (Hu et al., 2023; John et al., 1999; Mizutani and Nishi, 2003).

Natural FRP composites are increasingly utilized in the aerospace industry due to their lightweight nature and high strength (Khan et al., 2018). Carbon fiber-reinforced polymer (CFRP) composites have gained significant traction in aerospace and automotive industries due to their superior properties, including high specific stiffness, specific strength, and

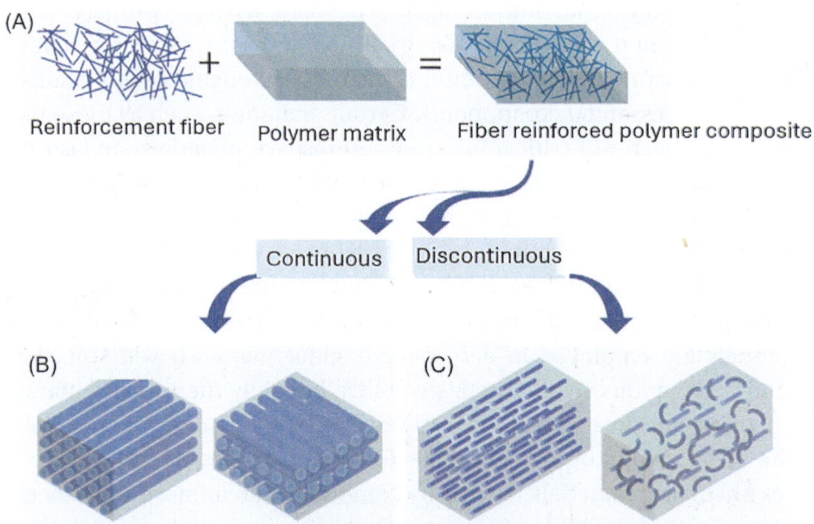

FIGURE 15-1 The general structural assembly of fiber-reinforced polymer (FRP) composites and their classification: (A) basic constituents of FRP composites, (B) continuous FRP, and (C) discontinuous FRP composites. *Reproduced under the terms of the CC-BY license. Wiley-VCH GmbH. The figure illustrates the fundamental structural constituents of fiber-reinforced polymer (FRP) composites, including reinforcing fibers, polymer matrices, and their combined architectures. It further classifies FRP systems into continuous and discontinuous fiber composites, highlighting typical fiber arrangements and structural configurations used in engineering applications. Islam, M.H., et al., 2022. Graphene and CNT-based smart fiber-reinforced composites: a review. Advanced Functional Materials 32 (40), 2205723.*

corrosion resistance (Jia et al., 2019; Fan et al., 2017). The use of CFRP composites in aerospace structures is further supported by their excellent fatigue resistance and high strength-to-weight ratios (Li et al., 2017). CFRP composites have been widely applied in the aerospace industry for various components, including aircraft structural elements, owing to their exceptional mechanical and physical properties (Yang et al., 2019; Jin and Park, 2015).

The consumption of composite materials in the aerospace industry, as depicted in Figs. 15-2 and 15-3 has increased by over 50%, driven by their lightweight, high strength, good fracture resistance, and high damage tolerance. In aviation, fuel consumption accounts for

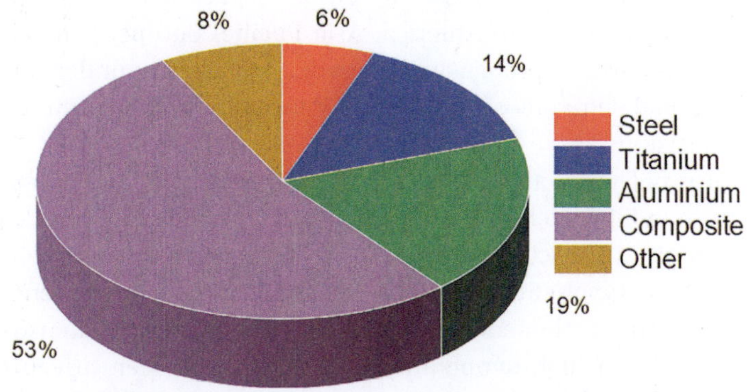

FIGURE 15–2 Materials used by weight percentage of Airbus A350-XWB. The figure presents the material composition of the Airbus A350-XWB by weight, showing the dominant contribution of composite materials relative to conventional metallic alloys such as aluminum, titanium, and steel. The high composite content reflects modern aircraft design strategies focused on weight reduction, fuel efficiency, and enhanced structural performance. *Data from MarketsandMarkets, Global Aerospace Composites Market Size & Forecast (https://www.marketsandmarkets.com/Market-Reports/aerospace-composites-market-246663558.html).*

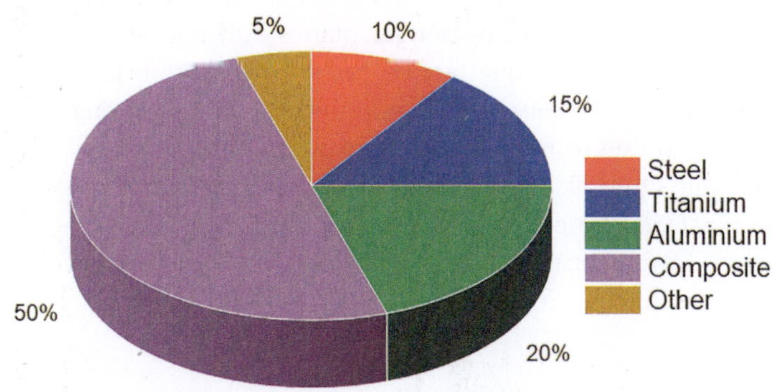

FIGURE 15–3 Materials used by weight percentage of Boeing 787/Dreamliner. *Reproduced under the terms of the CC-BY license.* International Journal of Engineering Research & Technology *(IJERT). The figure illustrates the material composition of the Boeing 787 Dreamliner, highlighting the dominant use of composite materials compared with conventional metallic alloys such as aluminum, titanium, and steel. The high composite fraction reflects design strategies aimed at weight reduction, improved fuel efficiency, and enhanced structural performance. Kesarwani, S. (2017). Polymer composites in aviation sector. International Journal of Engineering Research, 6(10), 590–592.*

around 50% of operational expenses. Heavier systems demand more fuel to achieve lift, thus increasing overall costs. Aerospace composites in the market are categorized based on several criteria, including fiber type, matrix type, manufacturing process, aircraft type, application, and region. In terms of fiber type, the market is segmented into glass fiber, carbon fiber, ceramic fiber, and other fiber types. The matrix type includes polymer matrix composites, metal matrix composites, and ceramic matrix composites. Manufacturing processes cover a range of techniques such as automated fiber placement/automated tape laying, lay-up, resin transfer molding, filament winding, and other processes. For aircraft types, the market is divided into commercial, business, and general aviation, civil helicopters, military aircraft, and other types. Applications vary across different aerospace needs, and the market is also analyzed by region, including North America, Asia Pacific, Europe, Latin America, and the Middle East and Africa. This comprehensive categorization helps in understanding the diverse applications and regional demands of aerospace composites, guiding industry stakeholders in making informed decisions.

High-temperature polymers are essential in the aerospace industry due to the extreme conditions they must endure, such as those experienced during space travel and atmospheric reentry. These materials are crucial for maintaining the structural integrity and performance of various aerospace components, including thermal protection systems (TPS), engine components, and structural elements of spacecraft and aircraft. Extensive research and development in the field of high-temperature polymers have been driven by the need for materials that can retain their properties and structural stability under such extreme conditions (Lu et al., 2023). A significant focus in aerospace materials research is on developing polymer-based composites that can effectively operate at cryogenic temperatures. These composites have garnered attention in the aerospace sector for their potential to offer lightweight and high-strength solutions for space exploration and satellite applications (Chen et al., 2021).

Overall, composites are highly versatile and used in all aircraft and spacecraft components, from hot air gondolas and gliders to passenger planes and fighter jets. For example, CFRP composites are known for their unique fatigue behavior but are brittle, as demonstrated by the Rolls-Royce incident in the 1960s, when an innovative RB211 jet engine with carbon fiber compressor blades failed due to bird strikes. In experimental programs, Boeing successfully used 150 composite parts to replace metal components in a helicopter. The use of composite-based components instead of metal in maintenance cycles is growing rapidly in commercial and leisure aviation. Overall, carbon fiber remains the most extensively used composite material in aerospace applications.

15.2 Polymer composite in aerospace engineering

The modern aerospace industry is notably advanced, and polymer composite materials play a crucial and transformative role in driving its progress and innovation. Polymer composites offer significant advantages over traditional metallic materials used in various aerospace components. A key benefit is their ability to maintain a lightweight while delivering specific strength, reducing overall weight by approximately 20%–40%. Additional benefits include rapid processing cycles, stringent dimensional stability, low thermal expansion properties, and superior fatigue and fracture resistance.

Polymer composites are among the most critical applications of polymers, whether natural or synthetic. These materials serve as multiphase solids where the phases interact in one, two, or three dimensions within different polymer matrices. They are ideal for high-performance aerospace applications, especially when the reinforcement's properties are significantly different from or better than the matrix. The most advanced polymer composites are those reinforced with polymers or thermoplastics, either naturally or synthetically. These materials can be fashioned into various shapes and sizes, offering great strength, stiffness, and corrosion resistance. They outperform traditional materials, providing up to 40% weight reduction while maintaining specific strength. Their additional properties, such as rapid processing cycles, stringent dimensional stability, low thermal expansion, and excellent fatigue and fracture resistance, make them highly desirable.

Current aerospace applications often use FRP composites for secondary structures, utilizing carbon, glass, Kevlar, or hybrids as reinforcement media. The need for materials that meet specific performance criteria has solidified polymer composites as a top choice for many aerospace applications. These materials can be engineered to offer high strength, low weight, chemical resistance, and long-term durability under harsh environmental conditions.

Modern military fighter jets have reduced their weight by 30% using polymer composites. These composites make up 80% of the structures of modern launch vehicles intended for satellites, encompassing essential satellite components like honeycomb structures, equipment panels, cylinder supports, and solar array substrates. For instance, the rocket motor casings of the space shuttle's solid booster comprise 30 t of graphite-reinforced epoxy composites. Another significant application of polymers is as adhesives for joining aircraft parts. Polymers can produce high-strength, durable joints without the need for traditional fasteners like rivets and screws. A thin layer of adhesive bonds aluminum and fiber polymer composite sheets, forming materials like GLARE (glass-reinforced aluminum laminate) used in the Airbus 380 fuselage. While elastomers are typically used for nonstructural parts requiring high flexibility and elasticity, such as seals and gaskets, polymers with metal or fiber reinforcements may face issues such as high viscosity, processing temperatures, and pressure, poor creep resistance, and others related to weight, agglomeration, improper stress distribution, and adhesion.

15.2.1 Advantages and disadvantages of polymer composite materials

Polymer composite materials offer numerous advantages in various applications, particularly in the aerospace industry. One significant benefit is weight reduction, with savings typically ranging from 20% to 50%. This reduction in weight contributes to improved fuel efficiency and overall performance of aerospace vehicles. Additionally, the mechanical properties of these composites can be tailored through specific design techniques, such as "lay-up" processes that adjust the thickness and orientation of reinforcing cloth and fibers to meet particular needs. Another advantage is high impact resistance. Materials like Kevlar (aramid) are used in armor shields and aerospace applications, providing protection and reducing damage from accidental impacts, such as those to engine pylons that house critical engine controls and fuel lines. Polymer composites also exhibit high damage tolerance, which enhances the survivability of aircraft in the event of accidents.

Polymer composites can prevent galvanic corrosion, an electrical corrosion problem that occurs when two dissimilar metals are in contact, especially in humid marine environments. The nonconductive nature of fiberglass in these composites helps avoid such issues, further extending the durability and lifespan of aerospace components. However, there are some disadvantages associated with polymer composite materials. These include higher recurring and nonrecurring costs, as well as higher material costs compared to traditional materials. Another drawback is the difficulty in detecting nonvisible impact damage, which can pose safety risks if not properly monitored. Additionally, repairs to polymer composites differ significantly from those to metal structures, requiring specialized techniques and skills. Lastly, isolation is needed to prevent adjacent aluminum parts from suffering galvanic corrosion when in contact with polymer composites, adding another layer of complexity to the maintenance and repair processes in aerospace applications. Despite these challenges, the benefits of polymer composite materials often outweigh the disadvantages, making them a valuable choice for high-performance and safety-critical applications in the aerospace industry.

15.3 Structural components

In the aerospace industry, the use of high-temperature polymers in both primary and secondary structures is crucial for ensuring the safety and efficiency of airframe parts. CFRP composites have gained significant attention due to their exceptional properties. These composites offer a high strength-to-weight ratio, making them ideal for applications where weight reduction is critical, such as in aircraft fuselages (Feito et al., 2016). The use of CFRP composites in fuselages is well-established, with these materials being preferred for their excellent mechanical properties and durability.

One of the key considerations in the selection of materials for airframe parts is their ability to withstand elevated temperatures. For instance, in the aerospace industry, aluminum alloys like 7085-T7452 are used for structural airframe parts that are exposed to slightly elevated temperatures due to aerodynamic heating or engine-generated heat. However, CFRPs have emerged as preferred materials for airframe components due to their high-temperature resistance, mechanical characteristics, and lightweight nature (Al Zahmi et al., 2022). These materials offer superior performance in terms of specific strength, fatigue resistance, and corrosion resistance, making them highly desirable for aerospace applications. Additionally, the thermal properties of hybrid composites, such as kenaf/carbon fiber–reinforced epoxy panels, have been extensively studied using techniques like thermogravimetric analysis and differential scanning calorimetry to ensure their suitability for high-temperature applications.

Carbon fiber-reinforced epoxy has become prevalent in wings, fuselages, and empennage components due to its superior mechanical properties and lightweight characteristics.

Furthermore, the use of advanced sustainable woven natural fiber polymer composites has also been explored. These composites, which incorporate thermoset polymers, exhibit properties that make them suitable for high-temperature applications, such as those encountered in aircraft components (Aisyah et al., 2019). Moreover, the incorporation of carbon nanotubes (CNTs) in polymer composites has shown promise for enhancing the mechanical properties of these materials, making them suitable for specific applications in the aircraft industry. The reinforcement of polymers with micro- and nanocellulose has also been investigated, leading to the production of parts with precise dimensions and low cycle times, which are essential considerations in airframe manufacturing. The design and manufacturing of airframe parts involve intricate processes to ensure the structural integrity and performance of the final components. For example, the drilling of CFRPs requires special tool geometries to maintain the integrity of the material and achieve precise results. Additionally, the effects of microstructure on the machining of unidirectional CFRP composites have been studied to optimize the manufacturing processes and enhance the functional performance of these materials in aerospace structures.

Overall, the utilization of high-temperature polymers, particularly CFRP composites, in primary and secondary airframe structures is a critical aspect of aerospace engineering. These materials offer exceptional mechanical properties, thermal resistance, and lightweight characteristics that are essential for ensuring the safety, efficiency, and performance of airframe parts. Through advanced manufacturing techniques, material enhancements, and rigorous testing methods, the aerospace industry continues to push the boundaries of innovation to develop airframe components that meet the stringent requirements of modern aircraft design and operation.

15.3.1 Wing and tail structures, fuselage, and cabin components

High-temperature polymers are essential in aerospace applications, particularly in manufacturing structural components like wing and tail structures, fuselage, and cabin components. These polymers possess unique properties that make them suitable for demanding aerospace environments. Research has indicated that the phase behavior of liquid crystalline polymers is influenced by factors such as the length of side chains and the presence of specific substituents. For example, polymers with butyl substituents can form columnar nematic phases at high temperatures, which disappear upon cooling to ambient temperature. This behavior is crucial for ensuring the stability and performance of polymer-based components in aerospace applications. Furthermore, the reorganization of thermoresponsive block copolymers into various 3D structures upon temperature changes showcases their potential for aerospace structural components (Kessel et al., 2011). The adaptability of these polymers to temperature variations can enhance the resilience and durability of aerospace materials. Additionally, conjugated polymers have demonstrated versatility in applications ranging from transistors to medical devices, indicating their potential for aerospace structural components.

Their unique properties make them promising candidates for improving the functionality and performance of aerospace materials.

Moreover, the synthesis of hydrophilic copolymers with twin-tail tadpole shapes and their responsive behavior to temperature changes offers insights into designing polymers for aerospace applications (Li et al., 2009). These polymers exhibit specific temperature-dependent characteristics that can be customized to meet the requirements of high-temperature aerospace environments. Additionally, the modulation of hydrophobic interactions in polymer solutions through the addition of beta-cyclodextrin and temperature variations demonstrates the potential for controlling polymer behavior in aerospace applications. Such control over interactions is vital for ensuring the stability and reliability of polymer-based aerospace components. The impact of the main chain on the phase behaviors of liquid-crystalline polymers with specific mesogens further emphasizes the significance of molecular design in aerospace polymer applications. By adjusting the main chain structure, researchers can manipulate the phase transitions of polymers, making them suitable for aerospace structural components.

Additionally, the synthesis and characterization of novel liquid crystalline polycarbonates with varying tail lengths provide valuable insights into the mesophase behavior of polymers at different temperatures (Jansen et al., 1995). Understanding these behaviors is crucial for developing high-temperature polymers capable of withstanding the extreme conditions experienced in aerospace operations. Additionally, the reversible clustering of pH- and temperature-responsive nanoparticles highlights the potential for creating complex polymer structures with tailored properties for aerospace applications. These nanoparticles can self-assemble into clusters at specific temperatures, offering opportunities for designing advanced aerospace materials. The temperature-directed morphology transformation method for tailoring the self-assembly of block copolymers into asymmetric nanoparticles further demonstrates the versatility of polymer structures in aerospace engineering. By controlling the self-assembly process, researchers can create polymer nanostructures with diverse geometries suitable for aerospace structural components.

Furthermore, the synthesis of polyacetylenes with specific mesogens as pendants and their phase behaviors at elevated temperatures provides valuable insights into designing polymers for aerospace applications. Understanding how polymer structures respond to temperature changes is crucial for ensuring the reliability and performance of aerospace materials. The temperature-directed formation of anisotropic nanostructures without the need for a swelling-induced solvent offers a novel approach to creating polymer structures for aerospace applications. This method allows for precise control of polymer morphology, enabling the design of materials tailored to aerospace requirements. In general, research on high-temperature polymers and their behaviors in response to temperature variations provides valuable knowledge for developing advanced aerospace structural components. By leveraging the unique properties and phase behaviors of polymers, researchers can design materials that meet the stringent demands of aerospace applications. The ability to control polymer structures at the molecular level and manipulate their responses to temperature changes opens up new possibilities for enhancing the performance and durability of aerospace materials.

15.4 Thermal protection systems

High-temperature polymers are essential in aerospace applications, serving as heat shields, insulation materials, and ablative materials for reentry vehicles. Composite materials made from high-temperature-resistant polymers are being extensively researched for their potential use as protective shells for space vehicles during reentry phases (Levinskak et al., 2011). The development of high-efficiency charring composites is crucial for designing effective heat shields in the TPS of reentry vehicles, ensuring their ability to withstand the intense heat generated during reentry. Charring ablative materials are a key focus in aerospace research for TPS of reentry vehicles experiencing high aerodynamic heat loads. These materials are vital for safeguarding spacecraft and probes during atmospheric entry by effectively dissipating heat. Additionally, polymers are being explored for their potential in providing insulation to solid rocket motors at high temperatures, highlighting their versatility in aerospace applications (George et al., 2018). Polymeric ablative materials are crucial for passively cooling rocket combustion chambers and offering insulation to solid rocket motors, demonstrating their importance in aerospace engineering.

In the realm of TPS, the development of ablative composites is vital for safeguarding both the payload and the vehicle itself during reentry into the atmosphere. Understanding the thermal conductivity of pyrolysis layers within charring materials is critical for optimizing the design of TPS in reentry vehicles, emphasizing the importance of material properties in ensuring effective heat dissipation. Moreover, the lightweight and robust nature of CNT/polyimide (PI) foams makes them ideal for efficient and heat-resistant electromagnetic interference (EMI) shielding, a crucial requirement in aerospace applications. The pursuit of high-performance EMI shielding materials in aerospace has led to the exploration of various polymer composites with enhanced shielding capabilities. By combining materials like cellulose nanofibers, graphene nanoplatelets, and MXene, composite films with excellent EMI shielding performance and high thermal conductivity can be achieved, meeting the stringent requirements of aerospace applications (Chu et al., 2022). The development of conductive polymer composites, which offer flexibility, lightweight properties, and adjustable conductivity, is essential for effective EMI shielding and Joule heating in aerospace systems. The integration of graphene into polymer composites has shown promise in enhancing both EMI shielding and thermal management in aerospace applications. These dual-functional graphene composites can effectively spread heat and protect against electromagnetic waves, addressing key challenges in aerospace thermal and electromagnetic shielding.

Furthermore, the utilization of high-temperature polymers in the form of lightweight and thermally conductive composites is crucial for achieving efficient EMI shielding and thermal management in aerospace systems. The ability of these materials to control or mitigate electromagnetic radiation pollution underscores their significance in aerospace technology. In general, high-temperature polymers are indispensable in aerospace applications, serving as heat shields, insulation materials, and ablative materials for reentry vehicles. Ongoing research and development in composite materials, charring ablative materials, and polymer composites with enhanced EMI shielding capabilities are driving innovation in aerospace engineering. By harnessing the unique properties of polymers and advanced composites, the aerospace industry can continue to push the boundaries of space exploration while ensuring the safety and efficiency of spacecraft during reentry and beyond.

15.5 Electrical and insulation materials

The choice of polymer for any application depends on the performance requirement and operating environment. Polymer materials used in the aerospace industry incorporate those for engineering designs, aircraft, and spacecraft (Ghori et al., 2018). The materials used for electrical insulation and assemblies in the aerospace industry are made of polymers that are majorly used for wiring and cable insulation, circuit boards, radar, and communication systems. Polymers are often exposed to high temperatures and harsh thermal conditions during operations. Hence the polymer material needs to be high-temperature polymers for electrical and insulation purposes.

Consequently, high-temperature polymers that can withstand temperatures exceeding 150°C were explored by the aviation sectors around the middle and later part of the 20th century (Lopez, 2021). Around the 1960s and 1980s eras, some high-temperature polymers were discovered to transmit signals and electric power (Ebnesajjad, 2016). This includes polytetrafluoroethylene (PTFE, trade name is Teflon) insulated cable, ethylene tetrafluoroethylene (ETFE, trade name is Tefzel), perfluoroalkoxy (PFA), and fluorinated ethylene propylene (FEP) (Cardoso et al., 2018). PI (Kapton) was commonly used as cable insulation in military aircraft, civil aircraft, and space shuttles. Unfortunately, Kapton has been banned for aircraft applications in the United States since 1988 due to suspected accidents involving cables and connectors manufactured from it. In the early 1900s, to prevent the usual Kapton from moisture and decrease arc-flash hazards, it was upgraded to KT, that is, (Kapton/Teflon). The upgraded KT was made of PI that has been protected with a thin layer of FEP combined with PTFE to limit the exposure of Kapton to moisture and decrease arc-flash hazards (Lopez, 2021). Recently, polymers such as polyfluoroalkyl (PFA) materials are suggested as high-temperature polymers for aeronautic cables. However, clarification of the European Chemicals Agency REACH (Registration, Evaluation, Authorization, and Restriction of Chemicals) is ongoing for approval based on its polyfluoroalkyl content to avoid environmental pollution (Kemari and Belijar, 2023).

Over the past year, some high-temperature polymers have been synthesized, and they comprise PIs, ETFE, FEP, perfluoroalkoxy, cross-linked ethylene vinyl acetate, PTFE, polyether–ether ketone (PEEK), liquid crystal polymers, polyphenylene sulfide, polyetherimides (PEIs), and silicone rubber.

High-temperature polymers are used for electrical insulation, particularly for wiring and cable insulation. The cables are used to transmit electric power between various components and systems, especially those exposed to extreme and harsh temperature conditions during operation. Transmission of signals for avionics communication, control systems, motor and actuator wiring, power transfer between battery cells, modules, and electrical heating elements for antiicing systems. In the aerospace system, high-temperature polymer cables are mostly used for engine compartments and power distribution cables for high electric current. For instance, the more recent aircraft engines have ultrahigh bypass ratio, which require higher-temperature polymers as insulation materials for cables to maintain performance integrity and contrast with the conventional turbofan engines subsequently used in the aerospace industry (Kemari and Belijar, 2023).

The use of high-temperature polymers for the insulation of critical parts, such as circuit boards, communication cables, and gadgets, ensures that the aerospace systems can withstand high electrical current, signals during transmission, and high electrical stresses for overall safety, reliability, and proficiency of the aerospace vehicles. The thermal stability of the materials used is required to maintain their integrity and performance when exposed to critical environments such as engine heat and friction.

Some characteristics of high-temperature polymers are that they should have high thermal stability to withstand high temperatures exceeding about 150°C to avoid degradation during operation (Giesa and Schmidt, 2001). They should proffer electrical insulation properties with high dielectric strength, low dielectric constant, and high electrical resistivity for superior performance in insulation and maintain signal integrity to avoid short circuits. Other factors are lightweight, which contributes to the overall weight reduction of the aerospace industry, leading to lower emissions, consumption, and efficiency; a high strength-to-weight ratio; good mechanical properties; and durability to endure mechanical stress, abrasion, and vibration. In addition, chemical resistance is required to prevent deterioration when exposed to chemicals such as fuel, hydraulic fluids, and lubricants (Yuan et al., 2019). It is important for the polymers used for wiring cable insulation to have excellent dielectric properties. Several properties that qualify these polymers for electrical insulation can be categorized as displayed in Table 15-1.

In the near future, novel high-temperature advanced cable materials will be required in the aerospace industry as electrification and propulsion systems in aerospace continue to develop. The polymers should provide high thermal stability and electrical performance. Therefore it tolerates exposure to very high temperatures and retains electrical integrity without degrading in performance. One high-temperature polymer of interest to look out for in the future in a progressive aerospace industry is PEEK. Its high chemical and mechanical stability for electrical and insulation purposes and its good anticorrosion properties also make it a replacement for metal materials for the construction of sensitive parts. The more recent aircraft engines, in contrast to the conventional engines used in the aerospace industry, necessitate higher-temperature polymers. Therefore high-temperature insulation materials are required for wires, cables, circuit boards, and communication systems.

15.6 Satellites and interior applications

The harsh complex space environment conditions (either in orbit around Earth or during transmission between the planets) have led to stringent material selection strategies to avoid

Table 15–1 Properties of high-temperature polymers used as electrical and insulation materials.

Type of polymer	Thermal stability range	Mechanical properties	Chemical and other properties	Electrical properties	Application in aerospace	Refs.
PI (Kapton)	400°C	High mechanical strength	Chemical and radiation resistance	Good electrical insulation	Insulating films, tapes for wiring, and cables for high-temperature conditions. Flexible PCBs. Dielectric layers in microelectronic	Lopez (2021)
PTFE (Teflon)	260°C	High flexibility and flexural strength. Good mechanical properties, extreme low and high temperatures	Excellent chemical resistance. Weak wear and abrasion resistance. Good flame resistance, cryogenic properties	Good electrical and insulating properties. Low dielectric constant, high dielectric strength, and low friction properties	High-frequency and high-voltage wire and insulation signal cables, coatings for electronic components, and power cables for high-temperature areas. Substrates for microwave circuits, insulating bushing, and spacer	Deli et al. (1989), Ebnesajjad (2020)
ETFE (Tefzel)	150°C	Enhanced toughness, impact strength, and stress cracking	Good chemical and abrasion resistance	High electric insulation	Engine wiring for aircraft, high-voltage power cable, signal and data cables, lightweight wiring harnesses	Ebnesajjad (2017)
FEP	200°C	Flexible and durable. Good elongation at break, tensile modulus, and strength, but the mechanical properties are poor at high temperatures	Good chemical resistance	Good thermal stability, high electric insulation	Data transmission cables, high voltage power cables, signal cables, and general aircraft wiring	Ebnesajjad (2020)

PFA	260°C	Good flexibility and impact strength	Good chemical Inertness. Prone to radiation attack and low coefficient of friction. Low smoke and nonflammable. Nontoxic and biocompatible	Good thermal and insulation stability. Low dissipation factor and low dielectric constant	Insulating of high-temperature cables and wires, connectors, terminals, and tubing and protecting cables and wires. Insulation for PCB in avionics. Provides insulation, coating, and encapsulation of various electronic components and connectors	Ebnesajjad (2017)
XL-EVA	125°C	High elasticity toughness, resilience, impact, and abrasion resistance	Good resistance to most solvents, oils, and chemicals except fluoropolymers. Flame retardant, low water absorptivity	Excellent electric insulation properties, good dielectric strength	Power distribution cables, general purpose wiring, battery cables, insulated control cables, connectors and terminals, seals and gaskets	
PEEK	260°C, short operation (300°C)	High modulus tensile strength, and impact strength	Excellent chemical resistance, good fatigue, creep, and radiation resistance, low coefficient of friction and wear resistance, biocompatible, low moisture absorption, flame retardancy	Good electric insulation, high dielectric strength	High-temperature areas near the engine. High-performance electrical connectors, plugs, and cable insulation, PCB, insulation for downhole, wafer holder dielectric films, transformer insulation component, insulation for high-temperature areas near the engine	Ling et al. (2020)
LCP	280°C	High modulus and strength. Dimensional stability	Low moisture absorption	Good electrical properties, low flammability,	High frequency connectors, high-speed connectors, and high-density interconnect applications. Surface mount technology insulation, flexible circuits	Lyu et al. (2020)

(Continued)

Table 15–1 (Continued)

Type of polymer	Thermal stability range	Mechanical properties	Chemical and other properties	Electrical properties	Application in aerospace	Refs.
PPS	260°C	Excellent toughness, impact resistance, high modulus, mechanical strength, and dimensional stability	Chemical resistance	High thermal stability and flame retardant	Electrical insulators and connectors, Insulating components in avionics, high-temperature environment, enclosure for electronic components	Chen et al. (2021), Rahate et al. (2013)
PEI	200°C, short operations (340°C)	Exclusive blend of high flexibility, rigidity, and specific strength, outstanding impact resistance, tensile and flexural strength, even at high temperatures	Good hydrolytic stability and flame resistance	Excellent insulating properties, low dissipation factor, low conductivity, high dielectric strength, low dielectric loss factor, flame retardancy	Insulators and connectors for electronic components, high-performance electrical connectors	Hasan (2020)
SiR	200°C	High flexibility	Hydrophobicity, radiation resistance	Good electrical insulation, flame retardant	Insulated wires for moving parts, for high voltage ignition wires	Abed et al. (2021), Thangabalan et al. (2020)

ETFE, Ethylene tetrafluoroethylene; *FEP*, fluorinated ethylene propylene; *LCP*, liquid crystal polymer; *PCB*, printed circuit board; *PEI*, polyetherimide; *PEEK*, polyether–ether ketone; *PI*, polyimide; *PPS*, polyphenylene sulfide; *PTFE*, polytetrafluoroethylene; *SiR*, silicone rubber; *XL-EVA*, cross-linked ethylene vinyl acetate.

adverse effects on the spacecraft material mechanics and physical and chemical properties. The key shortcomings of using polymers in space emanate from their poor resistance to the space environment. For example, thermal cycling and the high-temperature gradients experienced under vacuum can range between −100°C and 120°C, which can thus lead to the spacecraft material mechanical failure and cause deformation, cracking, and detachment from the spacecraft surface (Chen et al., 2016).

Only a few polymers and composite materials, including fiber metal laminates, polymer-metal composites, polymer, and ceramic composites, have been the effective leading materials in the aerospace industry, including components such as satellites, launch vehicles, and space centers, due to their reduced weight, dimensional stability, high specific strength, thermal resistance, and tribological properties (Abiri et al., 2022).

The exceptional properties of polymer materials, including their high toughness and elasticity, good insulation, low density, high or low melting point, and ease of molding, have led to their interest in aerospace engineering and technology (Katnam et al., 2013). The key features of aerospace materials required for aerospace include lightweight, high strength, high resistance to fracture and damage, and high-temperature tolerance (Joshi and Chatterjee, 2016).

Polymer composites, owing to their low density and low cost, compared to the various metallic materials, are becoming more dominant in the aircraft industry since lighter weight is the primary factor in aircraft design. In addition, owing to their excellent properties such as a high strength-to-weight ratio, and resistance against chemical and corrosion attacks (Muhammad et al., 2021), much-concerted effort has been devoted to research and their application in one of the most critical aerospace applications such as material and coating technologies for applications, including turbine engine hot section components, rocket engine combustion chamber liners, high-temperature components of advanced space power systems, and atmospheric reentry vehicle surfaces (Trzepieciński et al., 2021). Polymer composite materials (such as FRP composites that replaced the metal parts), and later, with the advances in materials science, polymer nanocomposites (PNCs) developed as advanced materials (Joshi and Chatterjee, 2016) and used extensively to construct most of the commercial aircraft's/spacecraft's exterior structural body parts (e.g., jet's epoxy wings, fuselage, flaperons, and rudder). Due to the delamination, fatigue, and cracking drawbacks, PNCs replaced FRPs, and since then, PNCs have been highly beneficial toward the overall reduction in the weight of spacecraft by more than 30%, thus improving fuel consumption and, eventually, reducing operation costs (Joshi and Chatterjee, 2016). This is due to their resistance to fatigue and corrosion/oxidation, and their mechanical, tribological, and structural properties that lead to weight reduction (Karalis et al., 2021; Tzounis et al., 2018). PNCs are thus considered high-performance materials due to their multifunctional properties, such as thermal stability, fire retardancy, electronic properties, field emission, optical properties, improved material durability, high impact resistance, energy absorption, and magnetic features, which are highly beneficial as construction materials for aerospace structures (Santos et al., 2020). PNCs are therefore molded into internal fittings and furniture for passenger aircraft or used in other forms such as sandwiched structures, foams, laminates, radar, satellites, and sensors (Rathod et al., 2017).

High costs associated with launching satellites to lower orbits are one of the challenges facing aerospace technology. The introduction of low Earth orbit (LEO) and very LEO satellites used for observation and communication (Weerasinghe et al., 2023). However, these are faced with challenges such as exposure to space environment, such as damage from space debris, meteorites, thermal variations, atomic oxygen (AO), space radiation, high vacuum, microgravity, plasma, and thermal management in vacuum (Phillips and Diston, 2011; Raja et al., 2024; Zoei et al., 2021).

Solar UV radiation: Total energy from the sun, with a solar constant of 1366 W m^{-2}, while solar UV radiation is ~8% of the solar radiation (Zhao et al., 2005). The UV photon energy is sufficiently high to cause degradation in the polymer materials of spacecraft. Polymer degradation by UV occurs when the amount of energy absorbed exceeds the polymer bond energy. The bond energy of various polymer functional groups (e.g., 420–560, 300–720, and 500–700 KJ mol^{-1} for the C–H, C–C, and C=O bonds, respectively) is lower than the UV radiation experienced in LEO (Wypych, 2020). The absorbed energy that surpasses the bond energy of the polymer molecules leads to bond breakage, thus causing polymer degradation by scission (Chen et al., 2016). Crosslinking is induced by UV radiation through photochemical reactions, leading to new bonds forming. The cross-linked polymer material will become hard and brittle. Ionization radiation, on the other hand, is confined radiation and solar event particles, which include protons, high-energy electrons, and photons, and results in chain scissions and crosslinking (Smith et al., 2021; Zhao et al., 2005).

Thermal cycling is the recurring temperature changes on the material surface when exposed to the space atmosphere of large temperature fluctuations that occur during the day–night transitions in orbit (e.g., −120/120°C) (Tan et al., 2019). The material degradation occurs due to the weakening of material bond strength (Tan et al., 2019; Zhao et al., 2005). Thermal cycling results in cracks and delamination in materials such as fiber/matrix composites and coated surfaces due to the mismatch in the material thermal expansions, thus deteriorating the mechanical properties (Earp et al., 2021).

High-temperature polymers, therefore, can withstand extreme temperatures and electrical field conditions in space environments. These polymers are designed to have high-temperature stability and resistance to heat, allowing them to maintain their structural integrity and mechanical properties even at elevated temperatures of 300°C minimum. They are commonly applied as TPS, structural components, and insulation materials (Kubota et al., 2019). Examples of these polymers include polybenzimidazole (PBI), PI, PEI (Nair et al., 2019), Polyamide-imide, PTFE, which have been applied as high-performance polymers used in aviation and aerospace (Huang et al., 2014).

PBI is a high-performance polymer due to its ability to withstand degradation while maintaining its properties under harsh conditions. The rigid structure with strong intermolecular forces emanates from the monomers containing highly stable benzimidazole rings. Additionally, as a smart polymer material (e.g., shape memory polymer [SMP]), high-temperature SMPs are characterized by high glass-transition temperatures (T_g) of above 100°C (Boudjellal et al., 2022). High-performance thermoplastic polymers with a high T_g have good mechanical properties and can be used for elevated temperature purposes. SMPs are tailored to cater for materials undergoing UV or ionizing radiation and mechanical stresses resulting from exposure to temperatures exceeding 150°C on the solar side of the low-earth orbit region while dropping to −150°C on the shadow side (Boudjellal et al., 2022).

PI's are a class of high-temperature thermoset and thermoplastic with outstanding thermal and mechanical properties and resistance to radiation. The applications include flexible solar arrays, coatings, and high-temperature adhesives (Gouzman et al., 2019). The aromatic PI's are preferred for aerospace applications. Kapton, a PI film, when used as a multilayer insulator, functions as a thermal blanket to protect satellites from solar radiation (Kimoto et al., 2016). An attack from AO and UV in the LEO environment on surfaces such as the exterior of satellites is one of the major concerns regarding polymeric materials (Dodd et al., 2009). Protective coatings such as SiO_2, TiO_2, and Al_2O_3 have been applied on the polymer surfaces for protection against AO erosion, and the degree of attack increases with an increase in temperature. However, the susceptibility of the ceramic coatings under the oxidative LEO conditions led to the development of advanced hybrid material blends (organic–inorganic hybrid copolymers and blends) by the inclusion of polyhedral oligomeric silsesquioxane (POSS) into the polymeric chains to impart self-passivating properties (Gouzman et al., 2019). The SiO_2 passivation layer develops upon interaction with the AO, thus preventing an AO attack on the polymer. Hybrid coatings such as polysiloxane/POSS hybrid coatings on Kapton (Wu et al., 2020; Duo et al., 2013), and the incorporation of hyperbranched polysiloxane into the PI polymer chains through condensation reactions also offered self-healing through the degradation of the PI layer to form a protective SiO_2 layer (Lei et al., 2015).

15.6.1 Structural parts for satellites

Satellites are devices that are launched into the aerospace environment and orbit the Earth and other planets and are usually launched into the harsh LEO environment (i.e., an altitude of 200–1000 km) (Barth, 2004). These harsh conditions include a combination of vacuum, UV/proton/electron radiations, thermal cycling, AO, high-speed μm-sized particles, and charging (Smith et al., 2021). Satellites are mainly used for aerospace navigation, communication, and astronomy, and use subsystems (depending on the satellite mission, these include an antenna, cameras, scientific sensors, and a power source such as a solar panel or battery). The subsystems and their location within the satellites are shown in Fig. 15–4. As these delicate components need to be protected from the harsh space atmosphere, they are therefore installed or located inside a protective covering (commonly known as satellite primary structures). This covering offers a mechanical structure to the delicate satellite subsystems and therefore requires durable material to withstand the harsh conditions experienced during the launch in orbit. Similar to aircraft construction materials, satellites also undergo material advancements in terms of lightweight and reducing the costs of manufacturing and launching (Aborehab et al., 2020; Reda et al., 2023). Due to the extreme space environment conditions such as the high pressures, and impacts from space debris (e.g., smaller asteroids), in addition to the thermal fluctuations, they also experience extreme loading conditions, namely, static lading (longitudinal and lateral accelerations

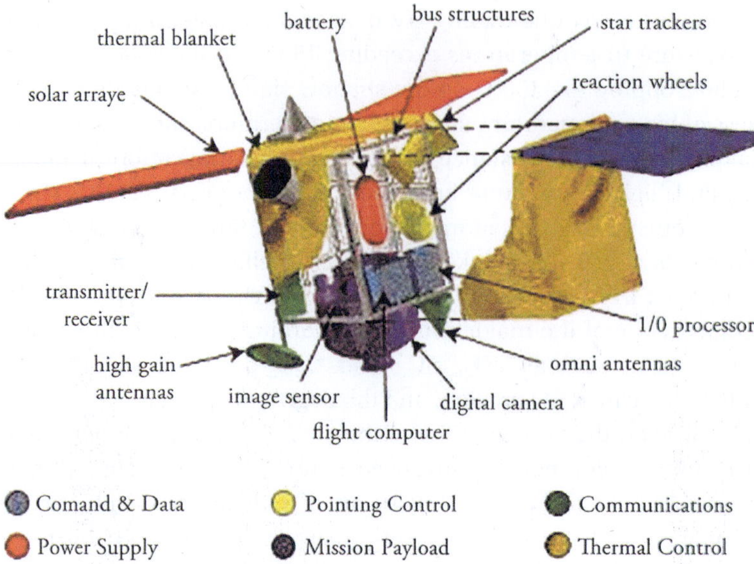

FIGURE 15–4 The different arrangements of the main satellite subsystems. *Reproduced under the Creative Commons Attribution–NonCommercial–NoDerivatives 4.0 License. Sciendo. The figure illustrates the typical configuration and spatial arrangement of key satellite subsystems, including the payload, power system, thermal control, attitude determination and control, communication units, and structural components, highlighting their functional integration within a spacecraft platform. Reda, R., Ahmed, Y., Magdy, I., Nabil, H., Khamis, M., Refaey, A., et al., 2023. Basic principles and mechanical considerations of satellites: a short review. Transactions on Aerospace Research 2023 (3), 40–54. Reproduced under the Creative Commons Attribution–NonCommercial–NoDerivatives 4.0 License.*

during take-off), dynamic loading (due to the vibrations caused by the noise of launching vehicle engines, mixing of exhaust and atmospheric gases, and the forces acting upon the satellite inside its cage), and mechanical loading (due to the separation stresses as the satellites become released). During the launching process, the satellite polymer material (e.g., resin) is subjected to vibrations, thermal expansion, and accelerations, thus leading to material deterioration and failure (Stevens, 2002). The resin matrix of the composite material can degrade under the hash LEO conditions; therefore the material used to build the satellite must consequently be capable of resisting space radiation and protecting its internal components from damage (Madry et al., 2018). Moreover, as the first step of material selection, the materials need to be space-grade and thus listed on the National Aeronautics and Space Administration (NASA) list of satellite materials. Such materials usually possess properties such as good thermal conductivity, insulation, and expansion, while avoiding material deformation characteristics such as thermal stresses as a result of thermal expansion (Reda et al., 2023).

Due to their high specific modulus, fiber composites have a stiffness of up to five times higher than aluminum. The transfer of flexural shear loads from the polymer matrix to the axial loads of the high-strength fibers, stiffness-to-weight ratios (capable of damping unwanted vibrations), which are superior to those of metallic materials, thermal expansions with a negative axial coefficient that prevents deformation structures when exposed to extreme conditions, and higher thermal conductivity than metals such as copper enables lightweight thermal handling and heat sinks, are all the properties of polymer composites have to offer (Reda et al., 2023).

Polymer films have been extensively used in the construction of satellites and applied as solar array blankets, thermal insulation, and inflatable or deployable structures as thin films (Weerasinghe et al., 2023). Recently, for improved strength, polymer composite materials have become more popular, especially for solar panels, deployable structures, trusses, frames, brackets, support structures, etc. (Weerasinghe et al., 2023; Barra et al., 2023). A few polymers, for example, carbon epoxy and bismaleimide for thermosets and PEEK for thermoplastics, have been used (as fiber-reinforced composites) due to their known behavior to withstand harsh space environmental conditions (Rinaldi et al., 2021). However, under extreme conditions like a high vacuum and thermal cycle, other factors such as outgassing have always been a concern for organic materials, causing alterations in material composition, dimensions, and eventually performance of materials (Grossman and Gouzman, 2003). Under these detrimental environmental conditions, polymer materials produce harmful gaseous species under high vacuum conditions. These gases, as outgassing products, can negatively impact the sensitive components and performance of a spacecraft due to material properties degradation and molecular contamination. Depending on the type and extent of outgassing, polymers can undergo severe changes in their properties. At the same time, contamination can damage spacecraft surfaces such as the thermal control surfaces (and also affect the performance of optical instruments, as condensed outgassing products may obscure surfaces and cause pollution and corrosion), instruments (readings can be affected by local clouds), and solar cells (Pastore et al., 2020).

Polymer composites, however, are ideal as materials for secondary structures, rather than the delicate primary structures. A vast amount of advanced polymer composites has been applied not only for commercial or military aircraft and spacecraft but also for satellite components (Ghori et al., 2018). These components include antennas, honeycomb structures, equipment panes, cylinder support structures, and solar array substrates (Baturkin, 2005). Polybenzoxazine is a thermoset polymer with high aromatic content, good thermal and mechanical properties, minimum water uptake, and low shrinkage (near-zero) (Fig. 15–5). It has been employed as a space-grade epoxy resin upon the incorporation of special additives into the resin system; these include cashew nut-based monobenzoxazine, 3,3′-thiodipropionic acid, and POSS, to improve thermal and mechanical properties and the AO resistance (He, 2020; Gohardani et al., 2014).

Nanomaterials such as CNTs are electrically conductive additives with the ability to develop a conducting network within a polymer matrix and give uniform conductivity throughout the molded aircraft components due to their extremely isotropic nature. CNTs possess good field emission and optical properties and are used as field emitters. They are thus highly suited to be utilized as field emission electric propulsion, colloids, microion thrusters, electrodynamic tethers, etc., as high-voltage, low-current electrical power applications (Lee et al., 2006, 2016). Field emitters based on CNTs are well-suited for structures such

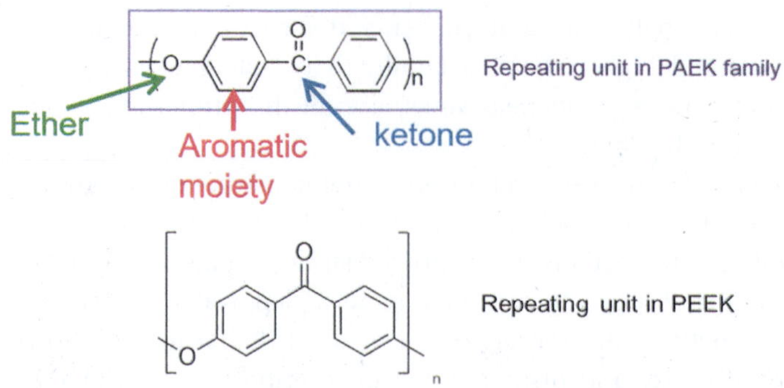

FIGURE 15–5 Structure of benzoxazine monomer. *Hamerton, I., et al., 2013. Examining the initiation of the polymerization mechanism and network development in aromatic polybenzoxazines. Macromolecules 46 (13), 5117–5132. https://doi.org/10.1021/ma401014h.*

as satellites, particularly microsatellites and nanosatellites (NanoSats), due to their reduced mass, volume, and power requirements.

NanoSats are composed of similar subsystems of larger spacecraft nanosats and are a class of small-sized, low-weight satellites (typically between 1 and 10 kg, while CubeSats generally are 1 kg or less), yet high-performing structures and typically used for educational and research project purposes (Rinaldi et al., 2021; Kopacz et al., 2020). They are produced by 3D printing and layer-by-layer procedures (additive manufacturing [AM] fused deposition modeling), and their construction requires the use of advanced polymers that can tolerate the space environment and comply with the appropriate standards, that is, the European Cooperation for Space Standardization (ECSS) (Rinaldi et al., 2021). Polymers, such as PEEK and PEI, are well known as technopolymers and meet such requirements owing to their excellent mechanical properties while meeting the outgassing requirements and withstanding the harsh conditions associated with outer space environment, such as vacuum (Rinaldi et al., 2021). PEEK is a semicrystalline thermoplastic polymer with high-temperature and chemical stability and superior mechanical properties (elastic modulus of 3.6 GPa and ultimate strength of 100 MPa) (Rinaldi et al., 2017). It originates from the high-performance poly (aryl ether ketone)s family, the low-density thermoplastics with good mechanical properties with low outgassing, properties highly suited for high-operation temperature, chemical, and radiation resistance requirements (Rinaldi et al., 2023). The in-chain ether and ketone functional groups that link the aromatic molecules play a role in the rigidity of the repeat units, while the linkage ratio of the ether and ketone affects the T_g and crystallization properties (Wang et al., 2016).

PEEK (Victrex) is one of the low-outgassing polymers listed in NASA's spacecraft materials selection (Rinaldi et al., 2023). However, 3D printing via AM (especially in the *z*-direction, as compared to the *x*–*y* axis, similar to injection molded material) comes with complexities that require proper assessment of the influence of the process parameters (e.g., heating and cooling rates, and nozzle size) on the resulting material properties, such as mechanical properties, and surface roughness. This is due to factors such as interlayer contact, spreadability, and polymer diffusion across the interface between the polymer being deposited and the substrate. In addition, 3D printing is limited in terms of size and, therefore, requires the joining of smaller printed materials to form larger objects or NanoSats (Rinaldi et al., 2023). Micro- and nanofillers such as carbon fibers, graphene, and CNTs have been utilized to improve the mechanical and thermal properties of 3D-printed PEEK (Rinaldi et al., 2021).

15.7 Fire resistance

The thermal stability, thermal conductivity, and fire retardancy/resistivity properties of the materials of the aircraft interiors (cabin) are continuously undergoing improvements, as these are essential for the reduction in costs, as well as the market-driven or state-of-the-art performance for the aerospace environment (George et al., 2019). The interior of the air cabin is mostly made of materials such as glass-reinforced phenolic composites or phenolic FRP (accounts for about 80% of fittings such as ceiling panels, wall panels, partitions, galley erections, cabinet walls, structural flooring, and overhead storage compartments), as they generate less smoke and flames during combustion. However, due to unsatisfactory fire-resistant properties, such as a low ignition time of 146 seconds, there is still a need for further fire retardancy improvement. Approaches such as coating with high-performance polymers or fire-resistant polymers like PBI, an ultrahigh-performance polymer with very high thermal resistance and thermo-oxidative decomposition due to the existence of heterocyclic and aromatic rings in its backbone configuration (Iqbal et al., 2017).

15.8 Conclusion

This chapter underscores the transformative role of advanced materials in addressing the unique challenges of aerospace engineering. The advent of high-temperature polymers and composites has redefined the performance benchmarks of structural, thermal, and electrical components in aerospace applications. Their lightweight nature, superior mechanical properties, and resilience under extreme environmental conditions have positioned them as indispensable in both traditional and cutting-edge aerospace technologies. Key innovations, such as CFRPs and PBI composites, highlight the ongoing evolution of materials tailored to withstand harsh environments, from atmospheric reentry to low Earth orbit. Moreover, the integration of nanomaterials, such as CNTs, has opened avenues for multifunctional applications, enhancing durability, thermal conductivity, and electromagnetic shielding.

Despite challenges like outgassing and complex manufacturing processes, these materials have demonstrated remarkable potential, particularly in satellite systems and aircraft interiors. As the aerospace sector continues to push boundaries, future advancements in high-performance polymers will play a pivotal role in achieving greater sustainability, efficiency, and innovation.

References

Abed, M.E.A., Hadi, H., Slama, M.E.A., Belarbi, A.W., Zekri, N., 2021. Experimental investigation on dielectric/electric properties of aged polymeric insulator: correlation of permittivity and dielectric loss tangent with surface hydrophobicity. Electrical Engineering 1–14.

Abiri, B., Arya, M., Bohn, F., Fikes, A., Gal-Katziri, M., Gdoutos, E., et al. 2022. A lightweight space-based solar power generation and transmission satellite. *arXiv*. Available from: https://arxiv.org. https://doi.org/10.48550/arXiv.2206.08373.

Aborehab, A., Kassem, M., Nemnem, A.F., Kamel, M., Kamel, H., 2020. Configuration design and modeling of an efficient small satellite structure. Engineering Solid Mechanics 8 (1), 7–20. Available at https://doi.org/10.5267/j.esm.2019.9.002.

Aisyah, H.A., Paridah, M.T., Sapuan, S.M., Khalina, A., Berkalp, O.B., Lee, S.H., et al., 2019. Thermal properties of woven kenaf/carbon fibre-reinforced epoxy hybrid composite panels. International Journal of Polymer Science 2019, 1–8. Available at https://doi.org/10.1155/2019/5258621.

Al Zahmi, S., Alhammadi, S., ElHassan, A., Ahmed, W., 2022. Carbon fiber/PLA recycled composite. Polymers 14 (11), 2194. Available at https://doi.org/10.3390/polym14112194.

Barra, G., Guadagno, L., Raimondo, M., Santonicola, M.G., Toto, E., Vecchio Ciprioti, S., 2023. A comprehensive review on the thermal stability assessment of polymers and composites for aeronautics and space applications. Polymers 15 (18) Available from. Available at https://doi.org/10.3390/polym15183786, http://www.mdpi.com/journal/polymers.

Barth, J.L., 2004. Space and atmospheric environments: from low earth orbits to deep space. In: Jacob I., Kleiman, Zelina, Iskanderova (Eds.), Protection of Materials and Structures from Space Environment: ICPMSE-6. Springer.

Baturkin, V., 2005. Micro-satellites thermal control—concepts and components. Acta Astronautica 56 (1-2), 161–170. Available at https://doi.org/10.1016/j.actaastro.2004.09.003.

Boudjellal, A., Trache, D., Khimeche, K., Hafsaoui, S.L., Bougamra, A., Tcharkhtchi, A., et al., 2022. Stimulation and reinforcement of shape-memory polymers and their composites: a review. Journal of Thermoplastic Composite Materials 35 (11), 2227–2260. Available from. Available at https://doi.org/10.1177/0892705720930775, https://journals.sagepub.com/home/JTC.

Cardoso, V.F., Correia, D.M., Ribeiro, C., Fernandes, M.M., Lanceros-Méndez, S., 2018. Fluorinated polymers as smart materials for advanced biomedical applications. Polymers 10 (2) Available from. Available at https://doi.org/10.3390/polym10020161, http://www.mdpi.com/2073-4360/10/2/161/pdf.

Chen, J., Ding, N., Li, Z., Wang, W., 2016. Organic polymer materials in the space environment. Progress in Aerospace Sciences 83, 37–56. Available at https://doi.org/10.1016/j.paerosci.2016.02.002.

Chen, D., Li, J., Yuan, Y., Gao, C., Cui, Y., Li, S., et al., 2021. A review of the polymer for cryogenic application: methods, mechanisms and perspectives. Polymers 13 (3), 320. Available at https://doi.org/10.3390/polym13030320.

Chu, Q., Lin, H., Ma, M., Chen, S., Shi, Y., He, H., et al., 2022. Cellulose nanofiber/graphene nanoplatelet/mxene nanocomposites for enhanced electromagnetic shielding and high in-plane thermal conductivity. ACS Applied Nano Materials 5 (5), 7217–7227. Available at https://doi.org/10.1021/acsanm.2c01126.

Deli, G., Qunji, X., Hongli, W., 1989. Study of the wear of filled polytetrafluoroethylene. Wear 134, 283–295. Available at https://doi.org/10.1016/0043-1648(89)90131-2.

Dodd, J.A., Baker, P.M., Hwang, E.S., Sporleder, D., Stearns, J.A., Chambreau, S.D., et al., 2009. Hyperthermal atomic oxygen source for near-space simulation experiments. Review of Scientific Instruments 80 (9). Available at https://doi.org/10.1063/1.3212676.

Duo, S., Chang, Y.C., Liu, T., Zhang, H., 2013. Atomic oxygen erosion resistance of polysiloxane/poss hybrid coatings on kapton. Physics Procedia 50, 337–342. Available at https://doi.org/10.1016/j.phpro.2013.11.053.

Earp, B., Hubbard, J., Tracy, A., Sakoda, D., Luhrs, C., 2021. Electrical behavior of CNT epoxy composites under in-situ simulated space environments. Composites Part B: Engineering 219, 108874. Available at https://doi.org/10.1016/j.compositesb.2021.108874.

Ebnesajjad, S., 2016. Expanded PTFE Applications Handbook: Technology, Manufacturing and Applications. Elsevier Inc, pp. 1–286. Available at http://www.sciencedirect.com/science/book/9781437778557.

Ebnesajjad, S., 2017. Introduction to fluoropolymers. In: Myer, Kutz (Ed.), Applied Plastics Engineering Handbook. Elsevier.

Ebnesajjad, S., 2020. Introduction to Fluoropolymers: Materials, Technology, and Applications. William Andrew.

Fan, W., Li, J.L., Fan, S.H., Liu, X., Sun, R.J., Meng, J.G., et al., 2017. Random process model of mechanical property degradation in carbon fiber-reinforced plastics under thermo-oxidative aging. Journal of Composite Materials 51 (9), 1253–1264. Available from. Available at https://doi.org/10.1177/0021998316672089, http://jcm.sagepub.com/content/by/year.

Feito, N., Diaz-Álvarez, A., Cantero, J.L., Rodríguez-Millán, M., Miguélez, H., 2016. Experimental analysis of special tool geometries when drilling woven and multidirectional CFRPs. Journal of Reinforced Plastics and Composites 35 (1), 33–55. Available from. Available at https://doi.org/10.1177/0731684415612931, http://jrp.sagepub.com/content/by/year.

George, K., Panda, B.P., Mohanty, S., Nayak, S.K., 2018. Recent developments in elastomeric heat shielding materials for solid rocket motor casing application for future perspective. Polymers for Advanced Technologies 29 (1), 8–21. Available from. Available at https://doi.org/10.1002/pat.4101, http://onlinelibrary.wiley.com/journal/10.1002/(ISSN)1099-1581.

George, P., Bhowmik, S., Abraham, M., Sriram, P.K., Pitchan, M.K., Ajeesh, G., 2019. High-performance fire-resistant polymeric nanocomposite for aerospace applications. Proceedings of the Institution of Mechanical Engineers, Part L: Journal of Materials: Design and Applications 233 (2), 97–108. Available from. Available at https://doi.org/10.1177/1464420716660874, http://pil.sagepub.com/content/by/year.

Ghori, S.W., Siakeng, R., Rasheed, Saba, N., 2018. 2 - The Role Of Advanced Polymer Materials in Aerospace Sustainable Composites for Aerospace Applications. Woodhead Publishing.

Giesa, R., Schmidt, H.-W., 2001. High-Temperature Stable Polymers. Elsevier BV, pp. 3803–3806. Available at http://doi.org/10.1016/b0-08-043152-6/00677-x.

Gohardani, O., Elola, M.C., Elizetxea, C., 2014. Potential and prospective implementation of carbon nanotubes on next generation aircraft and space vehicles: a review of current and expected applications in aerospace sciences. Progress in Aerospace Sciences 70, 42–68. Available at https://doi.org/10.1016/j.paerosci.2014.05.002.

Gouzman, I., Grossman, E., Verker, R., Atar, N., Bolker, A., Eliaz, N., 2019. Advances in polyimide-based materials for space applications. Advanced Materials 31 (18) Available from. Available at https://doi.org/10.1002/adma.201807738, http://onlinelibrary.wiley.com/journal/10.1002/(ISSN)1521-4095.

Grossman, E., Gouzman, I., 2003. Space environment effects on polymers in low earth orbit. Nuclear Instruments and Methods in Physics Research Section B: Beam Interactions with Materials and Atoms 208 (1-4), 48–57. Available at https://doi.org/10.1016/S0168-583X(03)00640-2.

Hasan, Z., 2020. Tooling for Composite Aerospace Structures: Manufacturing and Applications. Butterworth-Heinemann. Butterworth-Heinemann.

He, Y., 2020. Improving atomic oxygen resistance of composite materials for flexible deployable. Mercury 254.

Hu, Y., Huang, Z., Fan, K., Zhang, Y., 2023. Isogeometric pre-buckling analysis of laminated shallow shells. Journal of Physics: Conference Series 2551 (1), 012016. Available at https://doi.org/10.1088/1742-6596/2551/1/012016.

Huang, X.L., Martinez-Vega, J., Malec, D., 2014. Morphological evolution of polytetrafluoroethylene in extreme temperature conditions for aerospace applications. Journal of Applied Polymer Science 131 (3). Available at https://doi.org/10.1002/app.39841.

Iqbal, H.M.S., Bhowmik, S., Benedictus, R., 2017. Performance evaluation of polybenzimidazole coating for aerospace application. Progress in Organic Coatings 105, 190–199. Available at https://doi.org/10.1016/j.porgcoat.2017.01.005.

Jansen, J.C., Addink, R., Mijs, W.J., 1995. Synthesis and characterization of novel side-chain liquid crystalline polycarbonates. Molecular Crystals and Liquid Crystals Science and Technology. Section A. Molecular Crystals and Liquid Crystals. 261 (1), 415–426. Available at https://doi.org/10.1080/10587259508033486.

Jia, Z., Bai, Y., Wang, F., Ma, J., Cheng, D., Zhang, Z., 2019. Effect of drill flute direction on delamination at the exit in drilling carbon fiber reinforced plastic. Polymer Composites 40 (2) E1434, Available from. Available at https://doi.org/10.1002/pc.25040, http://onlinelibrary.wiley.com/journal/10.1002/(ISSN)1548-0569.

Jin, F.L., Park, S.J., 2015. Preparation and characterization of carbon fiber-reinforced thermosetting composites: a review. Carbon Letters 16 (2), 67–77. Available from. Available at https://doi.org/10.5714/CL.2015.16.2.067, http://carbonlett.org/Upload/files/CARBONLETT/%5B067-077%5D-01.pdf.

John, R., Zawada, L.P., Kroupa, J.L., 1999. Stresses due to temperature gradients in ceramic-matrix-composite aerospace components. Journal of the American Ceramic Society 82 (1), 161–168. Available from. Available at https://doi.org/10.1111/j.1151-2916.1999.tb01736.x, http://onlinelibrary.wiley.com/journal/10.1111/(ISSN)1551-2916.

Joshi, M., Chatterjee, U., 2016. Polymer nanocomposite: an advanced material for aerospace applications. In: Sohel, Rana, Raul, Fangueiro (Eds.), Advanced Composite Materials for Aerospace Engineering. Elsevier.

Karalis, G., Tzounis, L., Dimos, E., Mytafides, C.K., Liebscher, M., Karydis-Messinis, A., et al., 2021. Printed single-wall carbon nanotube-based joule heating devices integrated as functional laminae in advanced composites. ACS Applied Materials and Interfaces 13 (33), 39880–39893. Available from. Available at https://doi.org/10.1021/acsami.1c10001, http://pubs.acs.org/journal/aamick.

Katnam, K.B., Da Silva, L.F.M., Young, T.M., 2013. Bonded repair of composite aircraft structures: a review of scientific challenges and opportunities. Progress in Aerospace Sciences 61, 26–42. Available at https://doi.org/10.1016/j.paerosci.2013.03.003.

Kemari, Y., Belijar, G., Valdez-Nava, Z., Forget, F., Diaham, S., 2023. Review on high-temperature polymers for cable insulation: State-of-the-art and future developments. In: Jun-Wei, Zha, Zhi-Min, Dang (Eds.), High Temperature Polymer Dielectrics: Fundamentals and Applications in Power Equipment. Wiley, pp. 103–148. Available at http://doi.org/10.1002/9783527841059.ch4, https://www.wiley.com/en-in/High+Temperature+Polymer+Dielectrics%3A+Fundamentals+and+Applications+in+Power+Equipment-p-9783527351824.

Kessel, S., Urbani, C.N., Monteiro, M.J., 2011. Mechanically driven reorganization of thermoresponsive diblock copolymer assemblies in water. Angewandte Chemie - International Edition 50 (35), 8082–8085. Available at https://doi.org/10.1002/anie.201102651.

Kesarwani, S., 2017. Polymer Composites in Aviation Sector. International Journal of Engineering Research 6 (10). Available at https://doi.org/10.17577/ijertv6is060291.

Khan, T., Hameed Sultan, M.T.B., Ariffin, A.H., 2018. The challenges of natural fiber in manufacturing, material selection, and technology application: a review. Journal of Reinforced Plastics and Composites 37 (11), 770–779. Available from. Available at https://doi.org/10.1177/0731684418756762, http://jrp.sagepub.com/content/by/year.

Kimoto, Y., Fujita, T., Furuta, N., Kitamura, A., Suzuki, H., 2016. Development of space-qualified photocurable-silsesquioxane-coated polyimide films. Journal of Spacecraft and Rockets 53 (6), 1028–1034. Available at https://doi.org/10.2514/1.a33488.

Kopacz, J.R., Herschitz, R., Roney, J., 2020. Small satellites an overview and assessment. Acta Astronautica 170, 93–105. Available from. Available at https://doi.org/10.1016/j.actaastro.2020.01.034, http://www.journals.elsevier.com/acta-astronautica.

Kubota, Y., Miyamoto, O., Aoki, T., Ishida, Y., Ogasawara, T., Umezu, S., 2019. New thermal protection system using high-temperature carbon fibre-reinforced plastic sandwich panel. Acta Astronautica 160, 519–526. Available at https://doi.org/10.1016/j.actaastro.2019.02.004.

Lee, S.E., Kang, J.H., Kim, C.G., 2006. Fabrication and design of multi-layered radar absorbing structures of MWNT-filled glass/epoxy plain-weave composites. Composite Structures 76 (4), 397–405. Available at https://doi.org/10.1016/j.compstruct.2005.11.036.

Lee, S.E., Lee, W.J., Oh, K.S., Kim, C.G., 2016. Broadband all fiber-reinforced composite radar absorbing structure integrated by inductive frequency selective carbon fiber fabric and carbon-nanotube-loaded glass fabrics. Carbon 107, 564–572. Available from. Available at https://doi.org/10.1016/j.carbon.2016.06.005, http://www.journals.elsevier.com/carbon/.

Lei, X., Qiao, M., Tian, L., Chen, Y., Zhang, Q., 2015. Evolution of surface chemistry and morphology of hyperbranched polysiloxane polyimides in simulated atomic oxygen environment. Corrosion Science 98, 560–572. Available at https://doi.org/10.1016/j.corsci.2015.05.060.

Levinskak, R., Kezelis, R., Brinkiene, K., Grigaitiene, V., Valincius, V., Lukosiute, I., et al., 2011. High temperature ablation of composite material under plasma jet impact. Materials Science 17 (4). Available at https://doi.org/10.5755/j01.ms.17.4.781.

Li, L.Y., He, W.D., Li, J., Han, S.C., Sun, X.L., Zhang, B.Y., 2009. Synthesis of twin-tail tadpole-shaped hydrophilic copolymers and their thermo-responsive behavior. Journal of Polymer Science, Part A: Polymer Chemistry 47 (24), 7066–7077. Available from. Available at https://doi.org/10.1002/pola.23746, http://www3.interscience.wiley.com/cgi-bin/fulltext/122679506/PDFSTART.

Li, Z., Zhang, D., Qin, W., Geng, D., 2017. Feasibility study on the rotary ultrasonic elliptical machining for countersinking of carbon fiber–reinforced plastics. Proceedings of the Institution of Mechanical Engineers, Part B: Journal of Engineering Manufacture 231 (13), 2347–2358. Available at https://doi.org/10.1177/0954405415626086.

Ling, X., Jing, X., Zhang, C., Chen, S., 2020. Polyether ether ketone (PEEK) properties and its application status. IOP Conference Series: Earth and Environmental Science 453 (1), 012080. Available at https://doi.org/10.1088/1755-1315/453/1/012080.

Lopez, G., 2021. High-performance polymers for aeronautic wires insulation: current uses and future prospects. Recent progress in Materials 3, 1–15.

Lu, Y., Sun, P., Yang, X., Guo, X., Li, X., Ming, W., 2023. Review of research progress in nontraditional machining of ultrahigh-temperature ceramic matrix composites. Coatings 13 (1), 187. Available at https://doi.org/10.3390/coatings13010187.

Lyu, X., Xiao, A., Shi, D., Li, Y., Shen, Z., Chen, E.Q., et al., 2020. Liquid crystalline polymers: discovery, development, and the future. Polymer 202 Available from. Available at https://doi.org/10.1016/j.polymer.2020.122740, http://www.journals.elsevier.com/polymer/.

Madry, S., Martinez, P., Laufer, R., 2018. Innovative Design, Manufacturing and Testing of Small Satellites. Springer.

Mizutani, A., Nishi, Y., 2003. Improved strength in carbon fiber reinforced plastics due after electron beam irradiation. Materials Transactions 44 (9), 1857–1860. Available at https://doi.org/10.2320/matertrans.44.1857.

Muhammad, A., Rahman, M.R., Baini, R., Muhammad Bin Bakri, K., 2021. Applications of Sustainable Polymer Composites in Automobile and Aerospace Industry. Elsevier BV, pp. 185–207. Available at http://doi.org/10.1016/b978-0-12-820338-5.00008-4.

Nair, S., Pitchan, M.K., Bhowmik, S., Epaarachchi, J., 2019. Development of high temperature electrical conductive polymeric nanocomposite films for aerospace applications. Materials Research Express 6 (2), 026422. Available at https://doi.org/10.1088/2053-1591/aaefa8.

Pastore, R., Delfini, A., Albano, M., Vricella, A., Marchetti, M., Santoni, F., et al., 2020. Outgassing effect in polymeric composites exposed to space environment thermal-vacuum conditions. Acta Astronautica 170, 466–471. Available at https://doi.org/10.1016/j.actaastro.2020.02.019.

Phillips, P., Diston, D., 2011. A knowledge driven approach to aerospace condition monitoring. Knowledge-Based Systems 24 (6), 915–927. Available at https://doi.org/10.1016/j.knosys.2011.04.008.

Raja, S., Al-Tmimi, H.M., Ghadir, G.K., Mustafa, M.A., Alani, Z.K., Rusho, M.A., et al., 2024. An analysis of polymer material selection and design optimization to improve structural integrity in 3D printed aerospace components. Applied Chemical Engineering 7 (2) Available from. Available at https://doi.org/10.59429/ace.v7i2.1875, https://ace.as-pub.com/index.php/ACE/article/view/3879.

Rahate, A.S., Nemade, K.R., Waghuley, S.A., 2013. Polyphenylene sulfide (PPS): state of the art and applications. Reviews in Chemical Engineering 29. Available at https://doi.org/10.1515/revce-2012-0021.

Rathod, V.T., Kumar, J.S., Jain, A., 2017. Polymer and ceramic nanocomposites for aerospace applications. Applied Nanoscience 7 (8), 519–548. Available at https://doi.org/10.1007/s13204-017-0592-9.

Reda, R., Ahmed, Y., Magdy, I., Nabil, H., Khamis, M., Refaey, A., et al., 2023. Basic principles and mechanical considerations of satellites: a short review. Transactions on Aerospace Research 2023 (3), 40–54. Available at https://doi.org/10.2478/tar-2023-0016.

Rinaldi, M., Puglia, D., Dominici, F., Cherubini, V., Torre, L., Nanni, F., 2017. Melt processing and mechanical property characterization of high-performance poly(ether ether ketone)-carbon nanotube composite. Polymer International 66 (12), 1731–1736. Available at https://doi.org/10.1002/pi.5451.

Rinaldi, M., Cecchini, F., Pigliaru, L., Ghidini, T., Lumaca, F., Nanni, F., 2021. Additive manufacturing of polyether ether ketone (PEEK) for space applications: a nanosat polymeric structure. Polymers 13 (1), 11. Available at https://doi.org/10.3390/polym13010011.

Rinaldi, M., Ferrara, M., Pigliaru, L., Allegranza, C., Nanni, F., 2023. Additive manufacturing of polyether ether ketone-based composites for space application: a mini-review. CEAS Space Journal 15 (1), 77–87. Available at https://doi.org/10.1007/s12567-021-00401-4.

Santos, L.F.P., Ribeiro, B., Hein, L.R.O., Alderliesten, R., Zarouchas, D., Botelho, E.C., et al., 2020. The effect of temperature on fatigue strength of poly(ether-imide)/multiwalled carbon nanotube/carbon fibers composites for aeronautical application. Journal of Applied Polymer Science 137 (39) Available from. Available at https://doi.org/10.1002/app.49160, http://onlinelibrary.wiley.com/journal/10.1002/(ISSN)1097-4628.

Smith, C.T.G., Delkowki, M., Anguita, J.V., Cox, D.C., Haas, C., Silva, S.R.P., 2021. Complete atomic oxygen and UV protection for polymer and composite materials in a low earth orbit. ACS Applied Materials and Interfaces 13 (5), 6670–6677. Available from. Available at https://doi.org/10.1021/acsami.0c21552, http://pubs.acs.org/journal/aamick.

Stevens, C.L., 2002. Design, Analysis, Fabrication, and Testing Of A Nanosatellite Structure. Virginia Tech.

Tan, Q., Li, F., Liu, L., Liu, Y., Yan, X., Leng, J., 2019. Study of low earth orbit ultraviolet radiation and vacuum thermal cycling environment effects on epoxy-based shape memory polymer. Journal of Intelligent Material Systems and Structures 30 (18-19), 2688–2696. Available at https://doi.org/10.1177/1045389x19873398.

Thangabalan, B., Sarathi, R., Harid, N., Griffiths, H., 2020. Analysis of space charge and charge trap characteristics of gamma irradiated silicone rubber nanocomposites. IET Nanodielectrics 3 (2), 44–52. Available at https://doi.org/10.1049/iet-nde.2019.0041.

Trzepieciński, T., Najm, S.M., Sbayti, M., Belhadjsalah, H., Szpunar, M., Lemu, H.G., 2021. New advances and future possibilities in forming technology of hybrid metal–polymer composites used in aerospace applications. Journal of Composites Science 5 (8) Available from. Available at https://doi.org/10.3390/jcs5080217, https://www.mdpi.com/2504-477X/5/8/217/pdf.

Tzounis, L., Hegde, M., Liebscher, M., Dingemans, T., Pötschke, P., Paipetis, A.S., et al., 2018. All-aromatic SWCNT-Polyetherimide nanocomposites for thermal energy harvesting applications. Composites Science and Technology 156, 158–165. Available from. Available at https://doi.org/10.1016/j.compscitech.2017.12.030, http://www.journals.elsevier.com/composites-science-and-technology/.

Wang, Y., Beard, J.D., Evans, K.E., Ghita, O., 2016. Unusual crystalline morphology of Poly Aryl Ether Ketones (PAEKs). RSC Advances 6 (4), 3198–3209. Available from. Available at https://doi.org/10.1039/c5ra17110e, http://pubs.rsc.org/en/journals/journalissues.

Weerasinghe, J., Prasad, K., Mathew, J., Trifoni, E., Baranov, O., Levchenko, I., et al., 2023. Carbon nanocomposites in aerospace technology: a way to protect low-orbit satellites. Nanomaterials 13 (11), 1763. Available at https://doi.org/10.3390/nano13111763.

Wu, H., Zhang, Y., Guo, Y.D., Qi, H.R., An, Y.C., Jia, Y.J., et al., 2020. Preparation and properties of intrinsically atomic-oxygen resistant polyimide films containing polyhedral oligomeric silsesquioxane (Poss) in the side chains. Polymers 12 (12), 1–15. Available from. Available at https://doi.org/10.3390/polym12122865, https://www.mdpi.com/2073-4360/12/12/2865/pdf.

Wypych, G., 2020. Handbook of UV Degradation and Stabilization Handbook of UV Degradation and Stabilization. Elsevier, pp. 1–506. Available at http://doi.org/10.1016/B978-1-927885-57-4.50001-2, https://www.sciencedirect.com/book/9781927885574.

Yang, X., Zhan, L., Jiang, C., Zhao, X., Guan, C., Chang, T., 2019. Evaluating random vibration assisted vacuum processing of carbon/epoxy composites in terms of interlaminar shear strength and porosity. Journal of Composite Materials 53 (17), 2367–2376. Available at https://doi.org/10.1177/0021998319829531.

Yuan, C., Xie, C., Li, L., Xu, X., Gubanski, S.M., Zhou, Y., et al., 2019. Space charge behavior in silicone rubber from in-service aged HVDC composite insulators. IEEE Transactions on Dielectrics and Electrical Insulation 26 (3), 843–850. Available from. Available at https://doi.org/10.1109/TDEI.2019.8726032, https://ieeexplore.ieee.org/servlet/opac?punumber=94.

Zhao, X.H., Shen, Z.G., Xing, Y.S., Ma, S.L., 2005. An experimental study of low earth orbit atomic oxygen and ultraviolet radiation effects on a spacecraft material - Polytetrafluoroethylene. Polymer Degradation and Stability 88 (2), 275–285. Available at https://doi.org/10.1016/j.polymdegradstab.2004.11.002.

Zoei, M.S., Gorabi, H., Khorasani, M.A., Asghari, S., Mirahmadi, S.J., 2021. Investigation of atomic oxygen corrosion of satellite components coating under equivalent LEO orbit conditions. Technology and Applications 1, 135.

16

Application of high-temperature polymers in building and construction

Evral Ntsa[1], Vinny N. Katambwe[2,3], Rosisky M. Kalombe[1], Babatunde Oladipo[2], Leslie F. Petrik[4], Tunde Victor Ojumu[2]

[1]DEPARTMENT OF CIVIL AND MECHANICAL ENGINEERING, UNIVERSITY OF MISSOURI-KANSAS CITY, KANSAS CITY, MISSOURI, UNITED STATES [2]DEPARTMENT OF CHEMICAL ENGINEERING, CAPE PENINSULA UNIVERSITY OF TECHNOLOGY, CAPE TOWN, WESTERN CAPE, SOUTH AFRICA [3]DEPARTMENT OF ENVIRONMENTAL STUDIES FOR ADVANCED SOCIETY, GRADUATE SCHOOL OF ENVIRONMENTAL STUDIES, TOHOKU UNIVERSITY, SENDAI, JAPAN [4]DEPARTMENT OF CHEMISTRY, UNIVERSITY OF THE WESTERN CAPE, CAPE TOWN, WESTERN CAPE, SOUTH AFRICA

16.1 Introduction

As human society develops, the requirements for building materials increase, and the building industry has played an important role in human history. Using polymer materials in architecture has greatly enhanced and broadened building material functions (Shen et al., 2020). In recent decades, extensive efforts have been made to develop and design these polymers for use in various valuable applications in challenging areas such as aerospace, defense, energy, electronics, and automotive, as well as in the building and construction industries. For instance, polyimides, polytetrafluoroethylene, and polyether–ether ketone are well-known for their ability to perform under extreme thermal conditions (Shen et al., 2020). Unlike conventional materials such as steel and concrete, high-temperature polymers can withstand severe conditions without significantly degrading. In the building and construction industry, polymers are becoming increasingly useful due to their unique characteristics, such as high thermal stability, outstanding mechanical strength, high corrosion resistance to fluids and chemicals, and lightweight properties (Halliwell, 2002). The development of polymer-based construction materials for civil engineering applications has been studied over the years. There are three main types of polymer-based materials available today: (1) synthetic fiber-reinforced (cement) composites (FRCs), (2) fiber-reinforced polymer (FRP), and (3) polymer concrete (PC) (Ramgobin et al., 2019; Yoo et al., 2019). One emerging high-temperature polymer in the construction sector is geopolymer. Geopolymers are covalently bonded networks of inorganic polymers, typically made of aluminosilicates as precursors. These aluminosilicate materials are primarily derived from materials such as metakaolin (MK), ground granulated blast furnace slag, coal fly ash (FA), which is a byproduct of coal

High-Temperature Polymeric Materials. DOI: https://doi.org/10.1016/B978-0-443-32792-6.00009-9

combustion in power plants, etc. Coal FA (CFA)-based geopolymers are highly thermally stable, making them ideal for applications such as thermal insulation and fire resistance (Davidovits, 1989).

High-temperature polymers are advanced materials engineered to maintain their properties and performance in extreme thermal environments (Ramgobin et al., 2019). One of the primary advantages of a high-temperature polymer, such as geopolymer, is its ability to retain structural performance or provide fire protection at elevated temperatures in low and high technologies, often exceeding 200°C up to 1000°C, thus significantly outperforming traditional construction materials. FA-based geopolymer has been integrated into the field of industries, such as automobile and aerospace, nonferrous foundries and metallurgy, civil engineering, and plastic industries, based on their Si:Al ratio. Additionally, as an inorganic material, geopolymers also contribute significantly to fireproofing due to their inherent noncombustibility and ability to form thermally stable ceramics (Davidovits, 2017). Materials exhibiting such thermal performance may be well-suited for applications necessitating fireproof components subjected to high temperatures to adhere to safety regulations, ensuring stability, integrity, and insulation properties. FA-based geopolymers also offer a sustainable and ecofriendly alternative with comparable advantages. Moreover, the mechanical and durability properties are another significant benefit of FA-based geopolymers. Compared to heavier, currently being used construction materials such as OPC concrete, FA-based geopolymers tend to exhibit higher strength-to-weight ratios. Based on the results of previous studies (Ntsa, 2023), its usage in composite materials may allow structural elements to be lighter and stronger at the same time. This not only reduces the overall weight of the building but also enhances ease of handling and installation. Consequently, this minimizes the overall weight of buildings while streamlining handling and installation processes.

As technology advances, the use of high-temperature polymers may be expected to expand, leading to further innovation and efficiency as well as providing safety in the building and construction industry. Throughout this chapter, we will explore the commonly utilized high-temperature polymers, their applications, and benefits over traditional materials. Additionally, we will put forward their future prospects in the field of building and construction.

16.2 Commonly used high-temperature polymers in building and construction

Numerous high-temperature polymers are commonly used in building and construction due to their unique thermal properties. Compared to concrete and steel, FRP strengthening materials exhibit exceptional performance under high-temperature exposure (as in fire situations) (Kodur et al., 2019). On the other hand, other polymers such as FRC and PC have also been widely used. In recent years, FRC has gained much attention because of its superior mechanical properties. The performance of the material under high temperatures is affected by a number of factors, such as the use of replacement materials and the reinforcement of fibers. Additionally, the introduction of replacement materials, such as silica fume (SF), FA, grounded granulated blast furnace slag, MK, and slag powder, has significantly improved the performance of FRC at high temperatures (Kalifa et al., 2001; Xu et al., 2001; Han and Hwang, 2005; Kizilkanat et al., 2015; Serrano et al., 2016; Wu et al., 2020). PC has been used to improve the performance of a concrete structure, like in the case of lightweight PC material manufacturing. The exposure to fire of concrete-based PC samples at temperatures of 23°C, 50°C, and 400°C, in 50°C intervals, reveals that the mechanical properties of the material, such as compressive strength, flexural strength, and residual elastic modulus, are significantly impacted by temperatures above 200°C (Ataabadi et al., 2021). FA-based geopolymer concrete (GP) is an innovative polymer that enhances sustainability in construction by utilizing FA, a waste byproduct. It provides a cost-effective, fire-resistant material with excellent stability, integrity, and insulation. When tested under ISO 834 standards, it withstands temperatures up to 850°C (Ntsa, 2023).

Polymers play a crucial role in construction, offering diverse applications to enhance the performance and durability of structures. FRP strengthens concrete structures, while PC improves durability and reduces permeability. Polymeric coatings and sealants protect against corrosion and water damage, and geopolymers provide sustainable alternatives for concrete binding. Polymeric waterproofing membranes prevent water ingress, safeguarding building integrity (Yoo et al., 2019). These polymer-based solutions contribute significantly to the construction advancement, ensuring safer, more resilient structures.

16.3 Applications in building and construction

Composite materials are made from the combination of two or more materials, which, when used individually, may not possess the necessary properties for construction but, when combined and maintained at an identifiable interface surface, may produce a new material that combines the best properties of both original materials (Correia, 2015). The utilization of polymer-based composite materials in the building and construction sector has gained momentum in recent years due to their multibeneficial properties such as lightweight, wear resistance, high strength, heat insulation, anticorrosion, sound insulation, improved ductility, high durability, reduced maintenance requirements, low cure shrinkage, good adhesive properties, low permeability, ecofriendly characteristics, vibration-dampening properties, excellent weathering properties, low density, high design flexibility, thermal stability, and resistance to thawing and freezing, corrosion, abrasion, and chemicals. The attractiveness of replacing conventional materials (such as aluminum, steel, and concrete) with polymer composite materials resides in the properties mentioned (Nguyen et al., 2016; Oladele et al., 2020). Polymers in building and construction can be for structural (beams, columns, trusses, decks, etc.) or nonstructural (façade, cladding, etc.) purposes. This chapter will focus on the structural purposes.

16.3.1 Structural components

Different materials are under investigation to partially or fully replace cement in the manufacturing of construction materials. The environmental issues associated with cement production are well known. It has been reported that the cement production process is energy intensive, next only to steel and aluminum (Hardjito and Rangan, 2005). Also, releasing CO_2 in the atmosphere is one of the significant issues associated with cement production. Ntsa (2023) reported that for every ton of cement produced, an equivalent amount of CO_2 is emitted into the atmosphere. For ages, cement has been used as the primary binder in the manufacturing of concrete. The need for new materials that are more durable and sustainable is of paramount importance.

16.3.2 Composite materials

16.3.2.1 Polymer concrete

As an alternative to cement concrete, PC offers a number of benefits, as mentioned earlier. These characteristics make PC an excellent choice for various civil engineering applications (such as underground construction, floor drains, bridge decks, building cladding, sewer pipes, utility boxes, drinking water filtration slabs, manholes, industrialized flooring, hydraulic structures, precast products, trenches, geothermal energy, acid tanks, airport runways, etc.) (Oladele et al., 2020). It is worth noting that epoxy resin is the predominant binder used in PC development, while a variety of fillers (such as palm oil fuel ash, PET fiber, SF, bagasse ash, FA, marble waste, and calcium carbonate) are used as fillers (Sarde and Patil, 2019; Oladele et al., 2020).

16.3.2.2 Geopolymer concrete

Davidovits described geopolymer in 1978 as a semicrystalline gel with a three-dimensional framework (Davidovits, 1991). Geopolymers are linked to organic polymers. Chemically, they are similar to natural zeolite; however, their microstructure is amorphous rather than crystalline (Hardjito and Rangan, 2005). Geopolymers are inorganic alumino-silicate compounds synthesized from materials of geological origin (clay, MK) or byproduct materials (CFA, volcanic rock powders) rich in silicon and aluminum. Thus, the source materials (e.g., CFA) and alkaline solution (e.g., NaOH solution) are the two most crucial components (Davidovits, 2008).

The polycondensation or polymerization process here is a three-dimensional polymeric chain and ring structure made up of Si–O–Al–O bonds that occurs in a very rapid chemical reaction on Si–Al minerals under highly alkaline conditions (Davidovits, 1999, 2008):

$$Mn\ [-(SiO_2)\ z\text{-}AlO_2]\ n.\ w\ H_2O \tag{1}$$

where M designates a basic cation such as Ca, Na, or K; the symbol (–) indicates the presence of a bond; n represents the degree of polymerization or polycondensation; and z is 1, 2, 3, or higher, up to 32.

Table 16–1 shows various applications of geopolymer materials based on the Si/Al molar ratio proposed by Davidovits (1999).

Cement and clay are the two commonly used materials for building and construction. However, using FA-based geopolymer has shown the potential to offer environmental benefits, such as reduction of natural resource consumption and net CO_2 emissions

Table 16–1 Application of geopolymer materials (Davidovits, 1999).

Si/Al	Application
1	Ceramics, bricks, fire protection
2	Concrete, low-CO_2 cement, radioactive and toxic waste encapsulation
3	Heat-resistant composites, fiber-glass composites, foundry equipment
>3	Sealants for industry
20<Si / Al<35	Fire- and heat-resistant fiber composites

Source: From Davidovits, J., 1999. Chemistry of geopolymeric systems, terminology. Geopolymer 99(292), 9–39.

compared to cement. Several studies have been conducted on the synthesis of geopolymers (Hardjito and Rangan, 2005; Nyale et al., 2013; Nyale, 2014). Barbosa and Mackenzie (2003) reported using clay-based geopolymers as fireproof building components and heat insulators due to their potential to resist high temperatures. The feasibility of preparing high-compressive-strength geopolymers (35–44 MPa) has been demonstrated (Katambwe, 2022; Ntsa, 2023). This was achieved by incorporating a mixture of CFA, bottom ash, and fluidized bed combustion ash into a highly alkaline solution containing NaOH and Na_2SiO_3. The resulting composite was cured at 65°C for 48 h, yielding the desired mechanical properties. These geopolymers can sustain high temperatures (up to 1300°C–1400°C) and are freeze–thaw resistant. Izquierdo et al. (2009) stated that these high compressive strength CFA-based geopolymers exhibit long-term durability, good physical performance, and, in some instances, acid resistance and could be utilized to stabilize some metallic and radioactive wastes as well as industrial effluent (Katambwe, 2022). GP has found application in building pavements, retaining walls, water tanks, and precast concrete products such as bridge decks, railway sleepers, electric power poles, and parking tiles (Katambwe, 2022). It has been reported that when CFA-based GP was exposed to temperatures as high as 1200°C, only a few cracks were observed, indicating limited fire-induced damage to the specimen (Katambwe, 2022). In contrast, traditional concrete exhibited severe cracking when the temperature exceeded 800°C. This demonstrate that GP is a suitable material for applications where fire resistance and structural performance are critical. Ntsa (2023) synthesized different geopolymer products, such as reinforced concrete (RC) precast wall panel geopolymer (RCPWPG) on which a fire test behavior was performed, and paving bricks (Fig. 16–1). A fire test by Ntsa (2023) on the fire resistance of RCPWPG is shown in Fig. 16–2.

SANS 10177-2 standard guidelines were followed to perform the fire resistance ratio on the RCPWPG (Fig. 16–2). The test was based on four criteria, namely integrity, stability, insulation, and pollution observation. The author reported that prior to the testing, the panel showed some thermal cracking due to improper covering of the product using a plastic film during curing at ambient temperature. In Fig. 16–2, "U" denotes the unexposed surface, and "E" represents the fire-exposed surface.

The author observed that during the first 10 min of the test, there was water egress on the unexposed surface of the panel. This was due to the sealant applied on the panel to cover the thermal curing cracks. The visual observations and findings from the test are tabulated in Table 16–2.

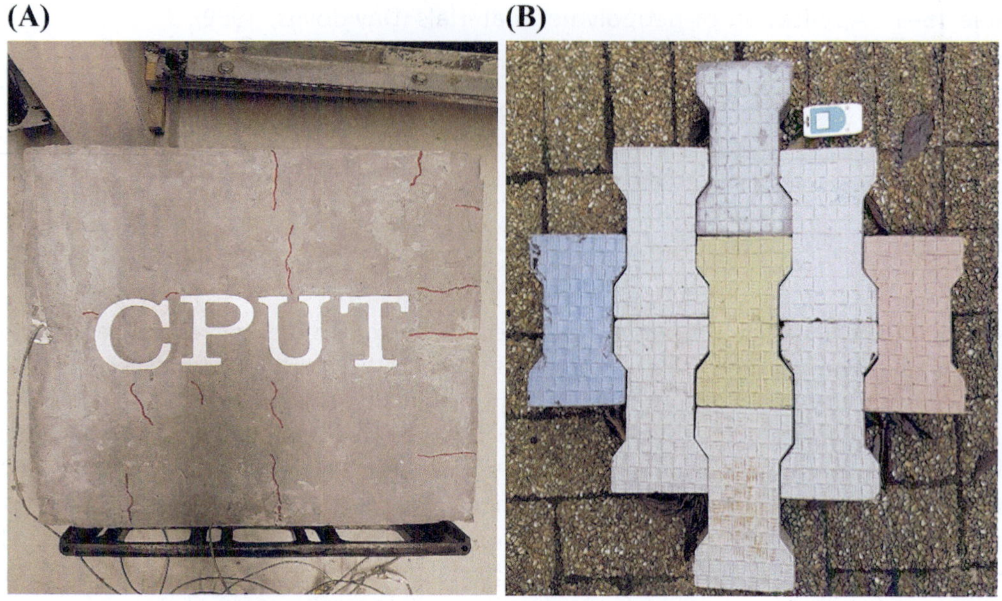

FIGURE 16–1 (A) Geopolymer concrete panel and (B) paving bricks (Ntsa, 2023).

The test lasted for about 65 min. The observations made are shown in Table 16–3, which is the interpretation of the results illustrated in Fig. 16–2. At the end of the test, the panel was destroyed to check how the steel behaved, and no sign of degradation was observed. The steel reached a temperature of 378.8°C, therefore surpassing the 375°C required for stability failure. It was concluded that the panel passed the fire rating of 1 h with respect to the criteria such as integrity, stability, insulation, and pollution gas according to the SANS 10177-2 fire testing standard.

16.3.3 Reinforcement materials

16.3.3.1 Fiber-reinforced polymers

FRP materials consist of two phases: The first one is the fiber reinforcement, which contributes to the strength and stiffness of the material, and the other phase is the polymeric matrix (such as polyester, epoxy, and vinyl ester resins), which serves as the glue of the composite material, transferring loads between fibers as well as between the applied loads and the composite (Correia, 2015). Besides the resin, polymeric matrixes of FRP materials contain fillers (such as calcium carbonate, aluminum silicate, alumina trihydrate, aluminum oxide, and calcium sulfates) and additives (such as fire retardants, colorants, and catalysts) that may lower the cost of production and enhance the manufacturing process as well as the final product characteristics (Bernard and Potyrała 2011; Correia, 2015; Kossakowski and Wciślik, 2022).

Glass, carbon, and aramid are the three types of fibers commonly used in structural engineering (ACI Committee 440, 2006). FRP composites made with these fibers are, namely,

FIGURE 16–2 Fire behavior test of reinforced concrete precast wall panel geopolymer (Ntsa, 2023).

Table 16–2 Reinforced concrete precast wall panel geopolymer (RCPWPG) fire failure evaluation criteria (Ntsa, 2023).

Criteria	Failure description	Time of failure (min)
Stability	Reinforced steel reached above 378.8°C	64
Integrity	No cracks of 6 mm with a 150-mm length were observed	65+
Insulation	No 140°C or 180°C plus the ambient temperature was reached	65+
Pollution observation	No pollutant gases were observed	65+

Table 16–3 Visual observation during fire testing (Ntsa, 2023).

Time (min)	Description	Observation face or side	Fig. 16–2
1	Test started	U	A
10	Evaporation of water via the initial thermal curing cracks (water egress)	U	B
25	Observation of sealant lines on the panel	U	C
42	0.84 mm center deflection measured	U	D
50	Reinforced steel reaches an average temperature of 232.6°C, and a 2.80-mm center deflection IS measured	U	–
53	The disappearance of sealant lines	U	–
59	About 130 mm crack long measured when reinforced steel reached 341.95°C	U	F
65	Steel reinforcement was assumed to reach a failing temperature of 375°C, as 378.8°C was obtained at 65 min	E	–
65	Data logger stopped	E	G

glass FRP (GFRP), carbon FRP (CFRP), and aramid FRP (Correia, 2015). However, for the construction industry, glass fibers happen to be the most often used fiber reinforcement due to the relatively low cost compared to carbon and aramid fibers (Correia, 2015; Kossakowski and Wciślik, 2022). FRP materials are utilized in structural profiles, reinforcing bars, and strengthening applications (Qureshi, 2022). FRP materials have found application in four primary fields in the building and construction industry, including FRP rebars for internal reinforcement, repair and strengthening (mostly external), new hybrid structures, and all new structures (Keller, 2002a; Correia, 2015). FRP materials have mainly been applied to repair and construct pedestrian and vehicular bridges. FRP has several benefits, including a high strength-to-weight ratio, being lightweight, superior resistance to corrosion, and simplicity of installation. Because FRP composites are lightweight, they add very little to a dead load of structure, need no formwork, and require little to no scaffolding for installation. They are also less expensive to transport. Since FRP composites are so strong, repairing beams and columns requires only a few thin layers that barely change the original dimensions. The main benefit is corrosion resistance since steel reinforcement corrosion has long been acknowledged as a severe and expensive maintenance issue (Bakis et al., 2002; Wu and Eamon, 2017).

16.3.3.1.1 Rebars for internal reinforcement

Concrete structure components are reinforced or prestressed using FRP-reinforcing bars, grids, and prestressing tendons. Although FRP-reinforcing bars had been utilized in construction since the 1970s, their use had become more widespread by the late 1980s (Bank, 2006). FRP reinforcements in concrete are used for a variety of reasons. The key factor is the increased longevity of FRP rebars; further factors include the high strength of the GFRPs, lightweight, and electromagnetic neutrality. Because FRP is noncorrosive, it is likely to find structural uses in chemical and industrial plants, as well as in or near maritime environments (e.g., piers) (Correia, 2015; Qureshi, 2022). FRP rebars are most commonly used internally in areas where traditional steel reinforcement corrosion poses a financial and safety risk (Rossini et al., 2019). As per Ortiz et al. (2023), FRP bars possess significantly higher strength (GFRP bars have the potential to have an ultimate tensile strength ranging from 550 to 1380 MPa) (Emparanza and Kampmann, 2017). Their density is approximately one-fourth that of steel, which means that less scaffolding and heavy equipment are needed for the structure, and they can also achieve a longer service life (Goyal et al., 2018). Moreover, FRP bars have been utilized in retention walls and tunneling projects (such as shaft walls) in underground constructions. According to Correia (2015), the primary benefits of FRP bars over steel bars in these applications are their increased resistance to corrosion; their lightweight nature, which makes it easier to lift and handle reinforcement cages; and their ease of cutting, which makes tunnel drilling operations easier.

16.3.3.1.2 Repair and strengthening of existing structures

Traditional methods of reinforcing and restoring steel bridges include replacing damaged structural elements, fixing corroded beam ends, adding stiffeners, and applying protective coatings (Wipf and Fanous, 2003; Wu and Eamon, 2017). To repair or increase the load-carrying capacity of structural components, waterproofing, cathodic protection against reinforcement corrosion, cathodic sealing, spot-patching damaged sections, and epoxy injection sealing are some of the rehabilitation techniques used on concrete bridges. Still, there are intrinsic drawbacks to older methods. Furthermore, these methods, especially for bridges, are seen to be costly and disruptive to traffic (Wu and Eamon, 2017).

According to Wu and Eamon (2017), using FRP composite materials as a substitute technique for repairing and strengthening bridge structural components has consequently gained traction. FRP has been used in a number of industries, such as aerospace, military, maritime, and automotive, for strengthening applications. To strengthen RC components subjected to bending, shearing, or axial loading, FRP composites are widely applied as externally bonded reinforcement in the form of plates, sheets, or strips (Triantafillou, 1998; Kotynia et al., 2008; Yang, 2021). Meier first proposed CFRP strip bonding for structural repair and strengthening applications in Switzerland in 1984 (Wu and Eamon, 2017). Later, in the early 1990s, the California Department of Transportation, or Caltrans, upgraded California bridge columns seismically by using GFRP fabrics (Wu and Eamon, 2017). The application of GFRP marked the beginning of the use of FRP composite materials in the United States.

Structure engineers can find FRP in a range of shapes and sizes, including bars, grids, sheets, and prestress tendons. FRP material applications for strengthening have grown significantly in the last several years (Günaslan et al., 2016). According to Günaslan et al. (2016), they are a reference material in many strengthening applications because of their lightweight, high strength, resistance to corrosion, speed and ease of application, and ability to be shaped into various forms on-site. These applications include using corrugated GFRP laminates to increase the flexural strength of RC beams (Aravind et al., 2015); using FRP wraps to reduce corrosion activity and concrete cracking in cylinders of chloride-contaminated concrete (El Maaddawy et al., 2006); using CFRP laminates as external bonds to prolong the service life of corroded RC beams (El Maaddawy, Soudki, 2005); and using FRP jacketing systems to repair fire-damaged RC columns (Zhou and Wang, 2019).

16.3.3.1.3 New hybrid structures

These types of FRP materials have either been combined with conventional materials or have taken their place in certain new structure components, such as external cables, girders, or even slabs, particularly in bridge decks. The full potential of FRP materials is not always utilized in most applications of this type, especially when the design is primarily focused on using traditional materials (Keller, 2002b). In addition to the numerous pedestrian and vehicular bridges that have already been constructed using FRP decks, a number of bridges have been constructed using external FRP cables (Correia, 2015). A hybrid bridge is most commonly described as a structure with an FRP bridge deck attached to steel or concrete girders. FRP bridge decks were first offered as a solution because they were lightweight, simple to install, and might be resistant to chemical and environmental deterioration (Tuakta, 2005). Bernard and Potyrała (2011) have given an exhaustive list of hybrid FRP bridge structures, with detailed explanation given on the following: footbridge over road no. 11 in Gądki (Poland); King Stormwater Channel Bridge on California State Route 86 near Salton Sea (United States); and Friedberg bridge over B3 highway in Friedberg near Frankfurt (Germany).

16.3.3.1.4 All new fiber-reinforced polymer structure

Structures constructed for this kind of application are made entirely of FRP. Some of these innovative building designs make better use of the possibilities of FRP materials by incorporating concepts tailored to the specific properties of the materials. The first all-composite bridge was constructed in Miyun, China, in 1982 with a span of 20.7 m and constituting six hand-laminated glass fiber/polyester sandwich girders (Keller, 2003). These structural uses are fundamentally related to decks in bridge construction because traditional materials are still used to build piers and abutments. GFRP profiles were used in the construction of two lateral trusses and a deck of the Pontresina Bridge in Switzerland in 1997. This bridge is a prime example of such an application. FRP materials have been employed in secondary structures and nonstructural components of buildings, in addition to the core structural applications just described. A variety of novel all-FRP structures and bridges have been constructed using pultruded FRP profiles (Bernard and Potyrała, 2011; Correia, 2015).

Details and specifications of all new FRP bridge structures, such as West Mill Bridge, ApATeCh arched footbridge, Aberfeldy footbridge, and Lleida footbridge, are given by Bernard and Potyrała (2011). Various types of structures made of all FRP composites are available in literature (Bernard and Potyrała 2011).

16.3.3.2 Fire behavior of fiber-reinforced polymer composites

The fire reaction behavior of construction materials used in buildings must be sufficient to prevent flame escalation, fire deflagration, and excessive release and propagation of smoke. Correia (2015) mentioned that structural components must possess adequate fire resistance to ensure a structure remains intact long enough to allow for the safe evacuation of occupants during a fire. The application of FRP materials in building construction has been impeded by the fire behavior of these materials in structural applications. When these composite materials (FRP) are exposed to high temperatures (300°C–500°C), there is a decomposition in their matrix that occurs, therefore releasing smoke, soot, and some toxic volatiles. These materials soften, creep, and distort when heated to moderate temperatures (>100°C–200°C), therefore affecting the mechanical properties, which suffer considerably (Correia, 2015). Although these properties have limited the use of FRP materials in buildings; however these materials exhibit some excellent attributes in case of fire. In comparison to steel, FRP composites are very good heat insulators and present excellent burn-through resistance. Due to its slower rate of heat conduction, this results in slowing the spread of heat from room to room (Correia, 2015). Therefore, providing an effective protective barrier against flame, heat, smoke, and toxic fumes. To improve the fire performance of FRP composite materials, different measures, such as active and passive, can be incorporated (Correia, 2015).

16.4 Benefits over traditional materials

High-temperature polymers exhibit excellent mechanical properties, including high strength-to-weight ratios, which allow for the construction of lightweight yet durable structural components. For instance, FRPs are extensively utilized in construction due to their high mechanical properties and lightweight characteristics, making them suitable for construction applications. Furthermore, these polymers demonstrate outstanding chemical resistance, which significantly extends their lifespan and reduces maintenance costs compared to traditional materials like steel and conventional concrete. Moreover, integrating high-temperature polymers into composite materials drives advancements in structural components by providing lightweight and robust alternatives to conventional materials. FRCs, FRPs, and PCs improve fire safety in building envelopes while offering enhanced structural integrity and resilience (Ramgobin et al., 2019; Yoo et al., 2019). Depending on the application, their usage in composite materials may be customized to meet a specific performance requirement, considering the cost and environmental impact. FA-based geopolymers offer an environment-friendly alternative to traditional Portland cement (OPC) concrete, utilizing industrial byproducts to create sustainable construction materials. These geopolymers possess high compressive strength, excellent thermal stability, and remarkable durability, making them suitable for a wide range of construction applications. They also contribute to reducing carbon emissions associated with cement production, thereby addressing environmental concerns and promoting sustainable construction practices (Davidovits, 1989). As advancements in material science progress, the adoption of high-temperature polymers is expected to grow, driving innovation and further enhancing the performance and sustainability of modern building practices.

16.5 Future for high-temperature polymers in building and construction

The future outlook for high-temperature polymers in building and construction is promising, given the advancements in materials science and growing awareness of their benefits. As previously mentioned, integrating high-temperature polymers like FRC and FRP into composite materials is anticipated to drive advancements in structural components. These polymers not only enhance fire safety within building envelopes but also provide a robust alternative to traditional OPC concrete. Additionally, a polymer such as PC provides better lightweight performance to OPC concrete, and this may reduce the cost associated with product transportation. As a result, FA-based geopolymer as a construction material has garnered significant attention from researchers. This interest stems not only from its ability to provide strong and fire-resistant materials but also from its promotion of waste material usage compared to other polymers. FA-based geopolymer offers a promising future by significantly reducing the large quantities of global waste generated by power production and lowering CO_2 emissions associated with cement production. This will contribute to the development of sustainable and resilient infrastructure while ensuring environmental safety and cleanliness. Furthermore, it may minimize the fire devastation concerns, which are a major issue in current building design.

High-temperature polymers are likely crucial in addressing emerging challenges, such as climate change mitigation and infrastructure resilience. Based on their current research attention, they may become more versatile and cost-effective, enabling innovative construction techniques and designs. Based on current global waste management issues, it has been reviewed that the utilization of various solid wastes as aggregates in concrete production results in green concrete by eliminating the need to mine and process natural aggregates (Sivakrishna et al., 2020). From this, incorporating FA-based geopolymer with cardboard as aggregates to produce a lightweight geopolymeric insulation material may offer a sustainable construction solution, reducing environmental impact while maintaining performance standards. A study by Nyabanga (2019) has resulted in materials with densities ranging from 929.9 to 760.5 kg m^{-3} by adding 10%–40% cardboard in the mix with corresponding compressive strengths of 5.82–0.98 MPa, respectively. The cost of the material produced was found to be lower than that of similar gypsum board products, although the fire performance of the product remains an outstanding concern.

16.6 Conclusion

The applications of high-temperature polymers in building and construction represent a significant advancement toward more resilient, efficient, and sustainable structures. These polymers, including FPRs, FRCs, PCs, and FA-based geopolymers, are innovative polymers. They offer superior performance characteristics such as thermal stability, mechanical strength, and chemical resistance compared to conventional building and construction materials. These polymers play crucial roles across various applications, from insulation materials providing fire protection and thermal insulation to structural components enhancing durability and load-bearing capacity. Their benefits over traditional materials are substantial, including reduced weight, improved durability, and enhanced environmental sustainability through a reduced carbon footprint. As the field continues to evolve with advancements in material science and engineering, the future outlook for high-temperature polymers in construction may be promising. Continued innovation is expected to further optimize their properties and expand their applications, driving efficiency and sustainability in modern building practices. The integration of high-temperature polymers in building and construction applications represents a pivotal shift toward more resilient and environment-friendly construction materials. Their diverse applications and superior performance characteristics make them indispensable in shaping the future of building and construction, where efficiency, durability, and sustainability are paramount considerations. The potential combination of FA-based geopolymer with paper is intriguing; however, further testing is necessary. Key areas for investigation include fire resistance, insulation properties, mechanical strength, durability, and cost-effectiveness. These tests are essential to ensure safety, evaluate longevity, and assess the economic feasibility of a building or construction project requiring their usage.

References

ACI Committee 440, 2006. Guide for the design and construction of structural concrete reinforced with FRP bars. ACI 440.1R-06. American Concrete Institute, Farmington Hills, MI.

Aravind, N., Samanta, A.K., Roy, D.K.S., Thanikal, J.V., 2015. Flexural strengthening of reinforced concrete (RC) beams retrofitted with corrugated glass fiber reinforced polymer (GFRP) laminates. Curved and Layered Structures 2 (1).

Ataabadi, H.S., Zare, A., Rahmani, H., Sedaghatdoost, A., Mirzaei, E., 2021. Lightweight dense polymer concrete exposed to chemical condition and various temperatures: an experimental investigation. Journal of Building Engineering 34, 101878.

Bakis, C.E., Bank, L.C., Brown, V., Cosenza, E., Davalos, J.F., Lesko, J.J., et al., 2002. Fiber-reinforced polymer composites for construction—state-of-the-art review. Journal of composites for construction 6 (2), 73–87.

Bank, L.C., 2006. Composites for Construction: Structural Design with FRP Materials. John Wiley & Sons.

Barbosa, V.F., Mackenzie, K.J., 2003. Synthesis and thermal behaviour of potassium sialate geopolymers. Materials Letters 57 (9-10), 1477–1482.

Bernard Potyrała, P., 2011. Use of fibre reinforced polymer composites in bridge construction. State of the art in hybrid and all-composite structures (Master's thesis, Universitat Politècnica de Catalunya).

Correia, J.R., 2015. Fibre-reinforced polymer (FRP) composites. Materials for Construction and Civil Engineering: Science, Processing, and Design 501–556.

Davidovits, J., 1989. Geopolymers and geopolymeric materials. Journal of Thermal Analysis 35, 429–441.

Davidovits, J., 1991. Geopolymers. Journal of Thermal Analysis 37 (8), 1633–1656.

Davidovits, J., 1999. Chemistry of geopolymeric systems, terminology. In Geopolymer 99 (292), 9–39 sn.

Davidovits, J., 2008. Geopolymer chemistry and applications. Geopolymer Institute, France.

Davidovits, J., 2017. State of Geopolymer. France.

Emparanza, A.R., Kampmann, R., De Caso y Basalo, F., 2017. State-of-the-practice of global manufacturing of FRP rebar and specifications. American Concrete Institute Fall Convention 327 45-1.

Goyal, A., Pouya, H.S., Ganjian, E., Claisse, P., 2018. A review of corrosion and protection of steel in concrete. Arabian Journal for Science and Engineering 43, 5035–5055.

Günaslan, S.E., Uğurlu, M.A., Karaşin, A. 2016. Strengthening concrete sections by FRP materials. International Journal of Theoretical and Applied Mechanics.

Halliwell, S., 2002. iSmithers Rapra Publishing Polymers in building and constructionVol. 154.

Han, C.G., Hwang, Y.S., Yang, S.H., Gowripalan, N., 2005. Performance of spalling resistance of high performance concrete with polypropylene fiber contents and lateral confinement. Cement and Concrete Research 35 (9), 1747–1753.

Hardjito and Rangan, 2005. Development and Properties of Low-Calcium Fly Ash-Based Geopolymer Concrete. Curtin University of Technology.

Izquierdo, M., Querol, X., Davidovits, J., Antenucci, D., Nugteren, H., Fernández-Pereira, C., 2009. Coal fly ash-slag-based geopolymers: microstructure and metal leaching. Journal of Hazardous Materials 166 (1), 561–566.

Kalifa, P., Chene, G., Galle, C., 2001. High-temperature behaviour of HPC with polypropylene fibres: From spalling to microstructure. Cement and Concrete Research 31 (10), 1487–1499.

Katambwe, V.N., 2022. Performance and optimisation of a jet loop reactor's pilot plant for the treatment of acid mine water using South African coal fly ash (Masters dissertation, Cape Peninsula University of Technology).

Keller, T., 2002a. Fibre reinforced polymer materials in bridge construction. International Association for Bridge and Structural Engineering. In IABSE Symposium Report 86(7): 119-126).

Keller, T., 2002b. Overview of fibre-reinforced polymers in bridge construction. Structural Engineering International 12 (2), 66–70.

Keller, T., 2003. Use of fibre reinforced polymers in bridge construction. IABSE Structural Engineering Documents No 7.

Kizilkanat, A.B., Kabay, N., Akyüncü, V., Chowdhury, S., Akça, A.H., 2015. Mechanical properties and fracture behavior of basalt and glass fiber reinforced concrete: an experimental study. Construction and Building Materials 100, 218–224.

Kodur, V.K.R., Bhatt, P.P., Naser, M.Z., 2019. High temperature properties of fiber reinforced polymers and fire insulation for fire resistance modeling of strengthened concrete structures. Composites Part B: Engineering 175, 107104.

Kossakowski, P.G., Wciślik, W., 2022. Fiber-reinforced polymer composites in the construction of bridges: opportunities, problems and challenges. Fibers 10 (4), 37.

Kotynia, R., Abdel Baky, H., Neale, K.W., Ebead, U.A., 2008. Flexural strengthening of RC beams with externally bonded CFRP systems: test results and 3D nonlinear FE analysis. Journal of Composites for Construction 12 (2), 190–201.

El Maaddawy, T., Soudki, K., 2005. Carbon-fiber-reinforced polymer repair to extend service life of corroded reinforced concrete beams. Journal of Composites for Construction 9 (2), 187–194.

El Maaddawy, T., Chahrour, A., Soudki, K., 2006. Effect of fiber-reinforced polymer wraps on corrosion activity and concrete cracking in chloride-contaminated concrete cylinders. Journal of Composites for Construction 10 (2), 139–147.

Nguyen, Q.T., Ngo, T., Tran, P., Mendis, P., Zobec, M., Aye, L., 2016. Fire performance of prefabricated modular units using organoclay/glass fibre reinforced polymer composite. Construction and Building Materials 129, 204–215.

Ntsa, E., 2023. Synthesis of geopolymer from South Africa CFA for construction application (Masters dissertation, Cape Peninsula University of Technology).

Nyabanga, N.M.S., 2019. Development of lightweight geopolymeric materials using coal fly ash and Wastepaper (Masters dissertation, Cape Peninsula University of Technology).

Nyale, S.M., Babajide, O.O., Birch, G.D., Böke, N., Petrik, L.F., 2013. Synthesis and characterization of coal fly ash-based foamed geopolymer. Procedia Environmental Sciences 18, 722–730.

Nyale, S.M., 2014. Geopolymers from South African Fly Ash: Synthesis and Characteristics. Doctoral dissertation, Chemistry Department, University of the Western Cape.

Oladele, I.O., Omotosho, T.F., Adediran, A.A., 2020. Polymer-based composites: an indispensable material for present and future applications. International Journal of Polymer Science 2020 (1), 8834518.

Ortiz, J.D., Khedmatgozar Dolati, S.S., Malla, P., Nanni, A., Mehrabi, A., 2023. FRP-reinforced/strengthened concrete: state-of-the-art review on durability and mechanical effects. Materials 2023 (16), 1990.

Qureshi, J., 2022. Fibre-reinforced polymer (FRP) in civil engineering. IntechOpen.

Ramgobin, A., Fontaine, G., Bourbigot, S., 2019. Thermal degradation and fire behavior of high performance polymers. Polymer Reviews 59 (1), 55–123.

Rossini, M., Saqan, E., Nanni, A., 2019. Prediction of the creep rupture strength of GFRP bars. Construction and Building Materials 227, 116620.

Sarde, B., Patil, Y.D., 2019. Recent research status on polymer composite used in concrete-an overview. Materials Today: Proceedings 18, 3780–3790.

Serrano, R., Cobo, A., Prieto, M.I., de las Nieves González, M., 2016. Analysis of fire resistance of concrete with polypropylene or steel fibers. Construction and Building Materials 122, 302–309.

Shen, J., Liang, J., Lin, X., Lin, H., Yu, J., Yang, Z., 2020. Recent progress in polymer-based building materials. International Journal of Polymer Science 2020 (1), 8838160.

Sivakrishna, A., Adesina, A., Awoyera, P.O., Kumar, K.R., 2020. Green concrete: a review of recent developments. Materials Today: Proceedings 27, 54–58.

Triantafillou, T.C., 1998. Shear strengthening of reinforced concrete beams using epoxy-bonded FRP composites. ACI Structural Journal 95, 107–115.

Tuakta, C., 2005. Use of fiber reinforced polymer composite in bridge structures (MSc dissertation, Massachusetts Institute of Technology).

Wipf, T.J., Fanous, F.S., Klaiber, F.W., Eapen, A.S., 2003. Evaluation of appropriate maintenance, repair and rehabilitation methods for Iowa Bridges (No. TR-429), Department. of Transportation, Iowa.

Wu, H.C., Eamon, C.D., 2017. Strengthening of concrete structures using fiber reinforced polymers (FRP): design, construction and practical applications.

Wu, H., Lin, X., Zhou, A., 2020. A review of mechanical properties of fibre reinforced concrete at elevated temperatures. Cement and Concrete Research 135, 106117.

Xu, Y., Wong, Y.L., Poon, C.S., Anson, M., 2001. Impact of high temperature on PFA concrete. Cement and Concrete Research 31 (7), 1065–1073.

Yang, J., 2021. Strengthening reinforced concrete structures with FRP composites. (Doctorate thesis, Chalmers University of Technology).

Yoo, D.Y., Banthia, N., Gupta, R., Kim, Y.H., Bhutta, A., 2019. Polymer-based construction materials for civil engineering. International Journal of Polymer Science 1–2.

Zhou, J., Wang, L., 2019. Repair of fire-damaged reinforced concrete members with axial load: a review. Sustainability 11 (4), 963.

17

Application of high temperature biopolymers in building and construction

Chewe Kambole

DEPARTMENT OF CIVIL ENGINEERING AND CONSTRUCTION, SCHOOL OF ENGINEERING, THE COPPERBELT UNIVERSITY, KITWE, ZAMBIA

17.1 Introduction

One of the current challenges in the building and construction industries is the pressing need for alternative materials that can be used to negate the adverse environmental impacts brought about by conventionally used fossil fuel–based polymers. This requires research and development of highly stable, low toxic, and eco-friendly biopolymers, which have the potential to reduce the carbon footprint (Shasiya et al., 2023). Some biopolymers like chitosan, gelatin, and collagen are known to be biodegradable and biocompatible. Biodegradation relates to a process of changes in the chemical composition of a substance triggered by microorganisms, leading to the formation of natural end metabolites. Biocompatibility refers to nontoxicity and the inability of a substance to cause any known allergic reactions (Safin et al., 2021).

Biodegradability and biocompatibility render biopolymers appropriate for use in various applications that include emulsions, food industry packaging materials, medical, and pharmaceutical industries (Baranwal et al., 2022). Biopolymers are projected to replace fossil fuel–based polymers by 30%–90% in the near future (Aaliya et al., 2021). Niaounakis (2015) summarizes biopolymers used in the construction and building industries as listed in Table 17–1

Biobased products derived from biomass and agricultural feedstock can be successfully used in the building and construction industry, just like the commonly used petroleum-based products. Accordingly, biopolymer composites (bioadmixtures) are a sustainable alternative to petroleum-based reinforced composites (Tusnim and Hoque, 2020).

To qualify for use in building and construction, biopolymers need to have a combination of various, often opposing, characteristics, such as high strength and stiffness, low density, low cost, environmental durability, and biodegradability. In the United States, biopolymers must conform to the requirements for structural and flammability performance, as described in the International Building Code (IBC), to be used as building materials (Niaounakis, 2015). It must be noted, however, that the IBC does not explicitly stipulate guidelines for the use of biopolymers in construction. Nonetheless, IBC requirements such as the need for exterior walls incorporating

High-Temperature Polymeric Materials. DOI: https://doi.org/10.1016/B978-0-443-32792-6.00004-X

Table 17–1 Biopolymers used in the construction and building industries.

Biopolymer	Types and examples
Biobased polymers	Polymers derived from plants (natural biopolymers): • Polysaccharides (e.g., starch- and cellulose-polymers), vegetable resins (e.g., rosin), and natural rubbers or gums: Polymers derived from animals: • Collagen/gelatin and casein Polymers derived from microorganisms: • Polyhydroxyalkanoates
Bioderived polymers	Poly(lactic acid) or poly(trimethylene terephthalate)
Biodegradable polymers derived from fossil fuel resources	Poly(glycolic acid), poly(ε-caprolactone), poly(butylene succinate), and poly(vinyl alcohol)

Source: Adapted from Niaounakis, M., 2015. Chapter 10—Building and construction applications. In: Niaounakis, M. (Ed.), Biopolymers: Applications and Trends. William Andrew Publishing, Oxford. https://doi.org/10.1016/B978-0-323-35399-1.00010-7.

fiber-reinforced polymer to have a flame spread index (FSI) of 25 or less (INTERNATIONAL CODE COUNCIL, 2020) would be applicable to biopolymers used as construction materials. An FSI of 25 (or less) is the highest fire-resistance rating of building materials against the spread of fire flames. Nazrun et al. (2024) asserts that recent studies have established that some biopolymer-based materials do indeed have high tensile strength, modulus, and impact resistance. With these characteristics, biopolymers have great potential to replace conventional materials like steel, concrete, or wood in constructing civil engineering structures.

High-temperature biopolymers can be described as unconventional materials capable of remaining structurally and functionally stable at high temperatures. These are materials that can withstand high heat, even exceeding 200°C. They are strong, tough, durable, and biocompatible and can withstand several cycles and doses of all types of radiation. They have been successfully used in medical devices and coatings (Sastri, 2014).

This chapter particularly focuses on high-temperature biodegradable and biocompatible polymers. These biopolymers exhibit unique characteristics such as renewability, low toxicity, and eco-friendliness. This focus is in tandem with the world's push for the use of eco-friendly materials in construction in a bid to address pressing environmental challenges such as climate change, deforestation, pollution, and natural resources depletion. Nonbiodegradable (biobased and fossil fuel–based polymers) are not discussed in the chapter.

17.2 Types of high-temperature biopolymers for building and construction applications

17.2.1 Cellulose-based biopolymers

Cellulose is the supporting material of plant cell walls and the most abundant biopolymer in nature. It is considered an ideal candidate material to replace petroleum-based plastics (Teng et al., 2023; Wang et al., 2020, 2021). Cellulose consists of a homopolymer of D-glucose linked

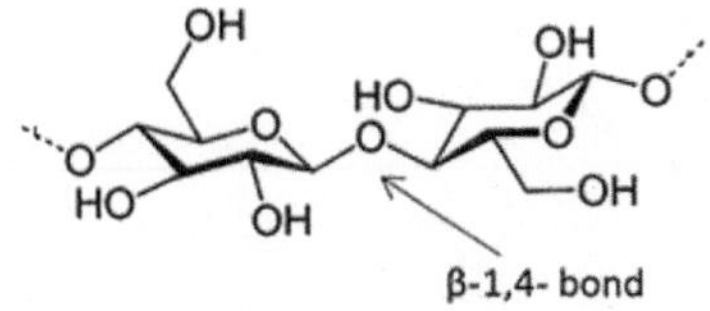

FIGURE 17–1 Chemical structure of cellulose showing two glucose rings in chain confromations connected by a beta-1,4- glycosidic bond with dashed lines indicating polymer extension (You and Zhang, 2017). *From You, C., Zhang, Y.H.P., 2017. Ex Vivo Enzymatic Conversion of Non-Food Cellulose Biomass to Starch. Springer Nature, pp. 129–142. https://doi.org/10.1007/978-3-662–53704-6_6.*

through β-1,4 glycoside bonds. Sources of cellulose include wood, agricultural residues, and perennial plants (Moon et al., 2011; You and Zhang, 2017). Fig. 17–1 presents the chemical structure of cellulose, primarily composed of β-D-glucose units linked by β(1→4) glycoside bonds, forming long linear chains. It is characterized by broad intra- and intermolecular hydrogen bonding, which significantly influences its crystallinity, crystal size, and overall stability.

Cellulose can be processed into various forms, like powders, films, and gels. It can also be chemically modified to produce cellulose-derivative biopolymers, like cellulose acetate and cellulose nitrate, with desirable properties for several applications (Yu et al., 2020). Unlike plastic or synthetic polymers, cellulose cannot be reshaped due to the strong connection between its molecules and hydrogen bonds (Galiano et al., 2018).

Cellulose-based polymers present extraordinary biodegradability rates in the environment. They also have remarkable thermal stability. They exhibit glass transition temperatures of between 220°C and 250°C, melting temperatures of between 260°C and 350°C, and crystallinity percentages in the range of 40–80. They can withstand moderate temperatures without undergoing significant degradation, a quality that makes them suitable where heat resistance is a vital requirement (Nazrun et al., 2024). Additionally, the presence of carboxylic acid and hydroxyl functional groups along the cellulose chain allows further functional group modifications (Teng et al., 2023). This affords the possibility of producing cellulose-based biopolymers that can specifically be applicable in building and construction works.

Pintiaux et al. (2019) investigated the production of renewable and sustainable all-cellulose materials. The study aimed at preparing all-cellulose materials by uniaxial high-pressure (above 50 MPa) thermocompression (high-temperature cellulose consolidation) of cellulose. Cellulose was assessed at low moisture content (MC) of between 0% and 8% MC (in 2% increment stages) and high temperature of between 175°C and 250°C (in 25°C increment stages). The best mechanical properties were obtained at 2% MC and 250°C. These included modulus and maximum stress in bending of 70.1 MPa and 8.1 GPa, respectively, and modulus and maximum tensile stress of 31.1 MPa and 1.9 GPa, respectively. The produced materials were orthotropic, showing better properties in bending than in tension.

Sampling points on test specimens for X-ray microtomography and SEM microscopy are shown in Fig. 17–2. SEM microscopy (Fig. 17–3) as well as X-ray microtomography revealed a

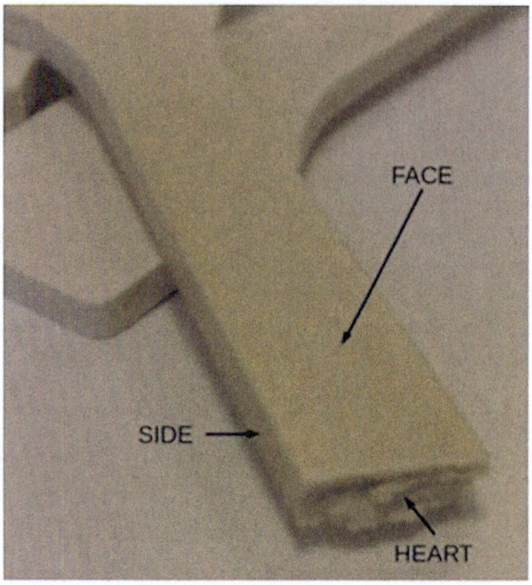

FIGURE 17–2 Test specimen showing sampling points for X-ray microtomography and SEM microscopy with arrows indicating the face, side and heart sampling areas (Pintiaux et al., 2019). *From Pintiaux, T., Heuls, M., Vandenbossche, V., Murphy, T., Wuhrer, R., Castignolles, P., Gaborieau, M., Rouilly, A., 2019. Cellulose consolidation under high-pressure and high-temperature uniaxial compression. Cellulose 26 (5), 2941–2954. https://doi.org/10.1007/s10570-019-02273-8.*

very specific morphology with a more porous heart, a dense but very thin skin on the faces (orthogonal to the compression axis), and thick and extremely dense sides where no more fibers were discernible (at 2% MC and 250°C).

This study showed that thermocompression of cellulose at high temperature with limited moisture could produce strong, renewable, and sustainable all-cellulose materials for application in biobased plastic substitutes such as binderless boards, with potential applications in building and construction projects.

Sen et al. (2022) devised a simple procedure for synthesizing cellulose nanofibers from biomass-derived wood pulp and iron oxide nanoparticles to make a nanohybrid. This was done in an effort to develop ecofriendly and cost-effective thermally insulating materials that can improve the energy efficiency of buildings. Functional characteristics like mechanical strength and flammability in these nanohybrids were greatly improved, indicating that the limitations of the commercially available thermal insulating materials in terms of their stability and durability were addressed. Most notably, these nanohybrids exhibited very low thermal conductivity (as low as 0.024 $W\ m^{-1}\ K^{-1}$), implying that they possess better insulating potential compared to other conventional insulating materials like polystyrene and mineral wool. These results point to the potential and sustainable use of this renewable thermal insulating bionano polymer material for applications in smart buildings. Furthermore, with approximately 10% of the global total energy needed to maintain a pleasant atmosphere inside buildings, thermal insulation with this bionano polymer can play a crucial role in reducing energy consumption in buildings.

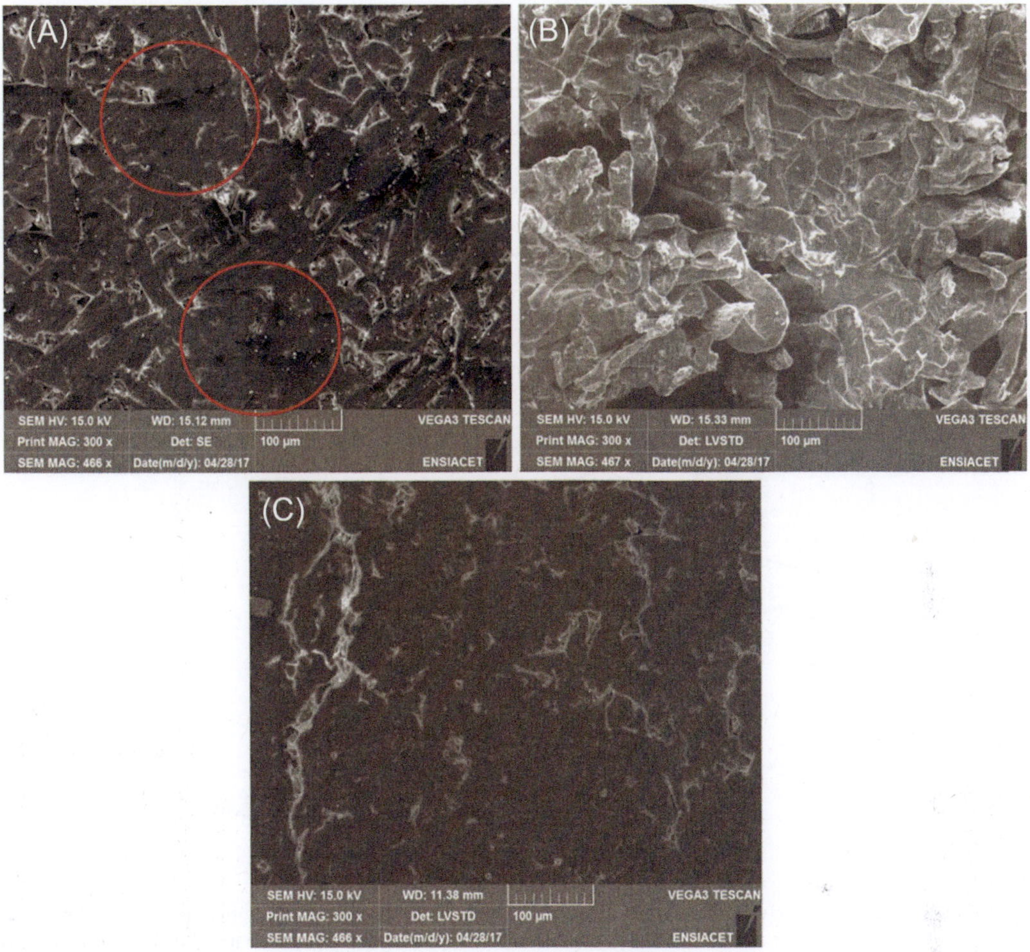

FIGURE 17–3 SEM micrographs of face (A), heart (B), and sides (C) of 2%—250°C sample (Pintiaux et al., 2019). *From Pintiaux, T., Heuls, M., Vandenbossche, V., Murphy, T., Wuhrer, R., Castignolles, P., Gaborieau, M., Rouilly, A., 2019. Cellulose consolidation under high-pressure and high-temperature uniaxial compression. Cellulose 26 (5), 2941–2954. https://doi.org/10.1007/s10570-019-02273-8.*

17.2.2 Starch-based biopolymers

Starch is a carbohydrate and the main constituent of the biopolymer structure that belongs to the polysaccharides group. It is considered the second most crucial renewable resource after cellulose (Diyana et al., 2021). It comprises two polysaccharide macromolecules, that is, soluble amylose and insoluble amylopectin. Amylose is made up of linear α1–4 linked polymer chains of glycosidic bonds, whereas amylopectin is made of larger, branched, and complex polymer chains of primarily D-glucose units with α1–4 and α1–6 bonds (Figs. 17–4 and 17–5).

Starch-based biopolymers are largely derived from natural renewable sources, like corn, potatoes, cassava, wheat, and rice, and are principally biodegradable or compostable

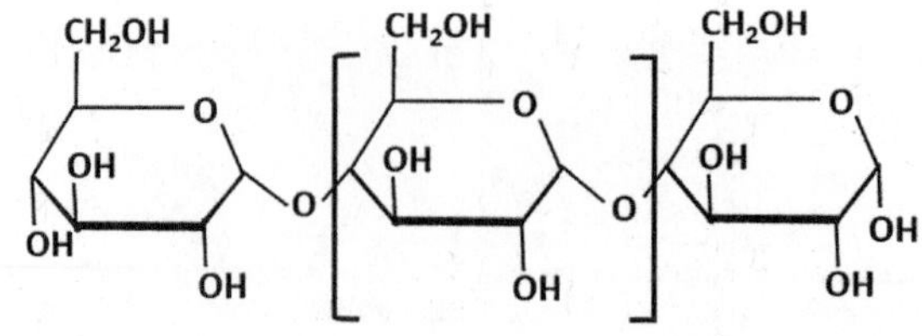

FIGURE 17–4 Chemical structure of amylose polymer chain showing three glucose units connected by alpha-1,4-glycosidic bonds (Tusnim and Hoque, 2020). *From Tusnim, J., Hoque, M.E., Biswas, M. C., 2020. Chapter 12—Biopolymers in building materials. In: Advanced Processing, Properties, and Applications of Starch and Other Bio-Based Polymers. Elsevier, Amsterdam.*

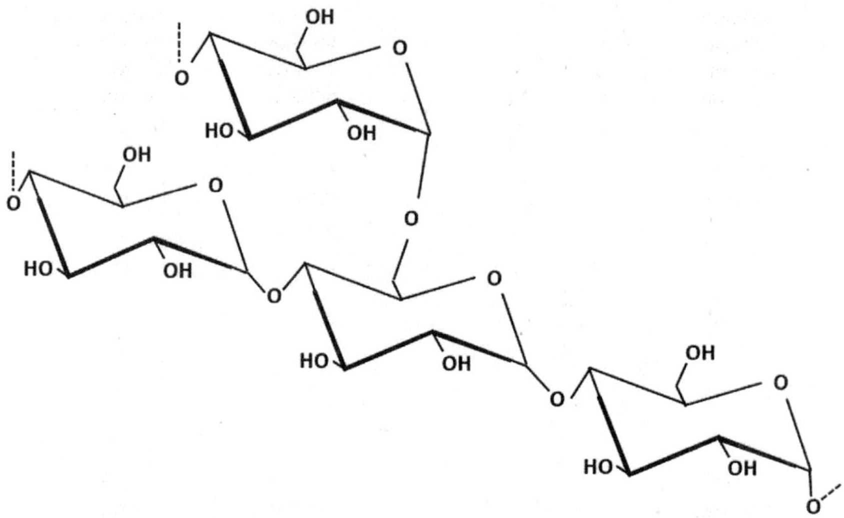

FIGURE 17–5 Chemical structure of amylopectin segment showing glucose rings in chair conformations connected by alpha-1,4-glycosidic bonds and an alpha-1,6-glycosidic branch point (Tusnim and Hoque, 2020). *From Tusnim, J., Hoque, M.E., Biswas, M. C., 2020. Chapter 12—Biopolymers in building materials. In: Advanced Processing, Properties, and Applications of Starch and Other Bio-Based Polymers. Elsevier, Amsterdam.*

(Datta, 2011; Nanda et al., 2022; Nazrun et al., 2024; Singh et al., 2024). Considerable attention has been given to starch as a source of sustainable materials due to its total biodegradability—leaving no toxic residues, renewable nature, easy availability, and low cost. Disadvantages associated with the starch-based biopolymers include their low mechanical properties compared with most conventional synthetic polymers. This significantly limits their application (Bumanis et al., 2020; Singh et al., 2024; Tusnim and Hoque, 2020).

In construction, starch-based biopolymers are largely used in the pregelatinized form in starch admixture fabrication. Pregelatinized starch, under high temperature, is converted to carboxymethyl starch, containing a negatively charged functional group (CH_2COO^-). The resulting starch structure is disturbed by the incorporation of carboxymethyl starch, producing starch with decreased gelatinization temperature, increased solubility, and improved storage stability. This product has temperature stability of up to 120°C, which is better than pregelatinized starch. Furthermore, as an alkaline material, it can also be used in oil well drilling as an additive in place of carboxymethyl cellulose. Starch-based, as well as cellulose-based biopolymers, have become attractive for the production of building products in advanced technologies like machine plastering, in place of conventional construction materials (Tusnim and Hoque, 2020).

The application of starch and its derivatives (often reinforced with natural fibers like hemp, sisal, jute, and cotton) in the building and construction industry is very promising. These biocomposite polymer materials have been widely used as binding material to improve mechanical, rheological, degradation, thermal insulation, and other physical properties of conventionally used materials like cement, concrete, and asphalt (Mansour et al., 2020a; Zoumaki et al., 2019).

Starch is furthermore widely used as a geopolymer binder for biocomposites. Compared to biocomposites produced using lime and cement binders, starch biocomposites are ecological building materials that exhibit better thermal insulation properties and relatively high compressive strength. This is attributed to their low density and high porosity (approximately 89%) as well as their thermal conductivity of between 0.0634 and 0.0738 $W\,m^{-1}\,K^{-1}$. To enhance fire resistance and biodegradation, appropriate additives need to be included in the production of the biocomposite materials (Bumanis et al., 2020).

Mansour et al. (2020b) studied the feasibility of making starch-sandstone biopolymer composite building materials using gelatinized starch as a binder and sand as filler. The making of this biopolymeric material was based on the physical modification of starch and the formation of a gelatinized starch matrix that bound sand grains, forming a rapid hardening starch-sand composite structure. The resulting biopolymetric material exhibited physical properties similar to conventional cement mortars and some common natural rocks. Thermogravimetric analysis (TGA) of the material indicated that the initial mass change occurred at around 100°C and was associated with the loss of absorbed water. Further heating did not lead to further removal of moisture. This is attributed to the trapping of the water molecules within the gelatinized starch matrix. Greater mass change above 200°C was thus attributed to the loss of starch content and is related to the thermal degradation of the starch matrix. The maximum mass loss rate occurred at around 290°C (the burning temperature of native corn starch), and degradation was recorded at 450°C.

Worth noting further is that the biodegradability, renewability, melting temperature between 130°C and 180°C, and low carbon footprint of starch-based biopolymers make these materials an appealing choice for external wall cladding in buildings (Nazrun et al., 2024).

17.2.3 Gellan gum

Gellan gum, a microorganism-based biopolymer, is a thermogelatin anionic polysaccharide produced through bacterial fermentation of *Sphingomonas elodea* (Das et al., 2023; Jansson et al., 1983). Gellan gum is made up of a tetrasaccharide unit of two residues of β-D-glucose, one of β-D-glucuronate, and another of α-L-rhamnose, corresponding to about 60% of glucose, 20% of rhamnose, and 20% of glucoronate (Giavasis et al., 2000; Gomes et al., 2023).

Two main types of gellan gum are native gellan gum, also called high acyl gellan gum, and deacetylated gellan gum, known as low acyl gellan gum. Their chemical structures are shown in Figs. 17–6 and 17–7. The major difference relates to the two acyl groups, glycerate and acetate, that native gellan gum has bound to the same β-D-glucose residue. Glycerate is found at every repeated tetrasaccharide unit, while acetate is found at two repeated tetrasaccharide units. The two acyl groups can be removed from native gellan gum through chemical or enzymatic processes, such as treatment with an esterase capable of deacylating the gellan gum, producing deacetylated gellan gum.

The removal of acyl groups has a significant impact on the rheological properties of the resulting gels (Bemiller and Gellans, 2019). An example is a change from soft, elastic, deformable gels made with native gellan gum to harder, brittle gels exhibiting greater thermal stability and optimal gelling conditions (Fialho et al., 2008).

Commercial applications of gellan gum include its use as a deacylated polymer, which is soluble in water at temperatures higher than 90°C, and then forms a gel when the temperature is reduced (Chang et al., 2015a). The resulting gels can be of very high-quality and also stable at high temperatures and low pH conditions (Chang et al., 2015b). These characteristics make gellan gum a suitable biopolymer additive for soil improvement (Fatehi et al., 2021). Furthermore, the thermal stability of gellan gum makes it suitable for potential use in thermal insulating plasters in building and construction works.

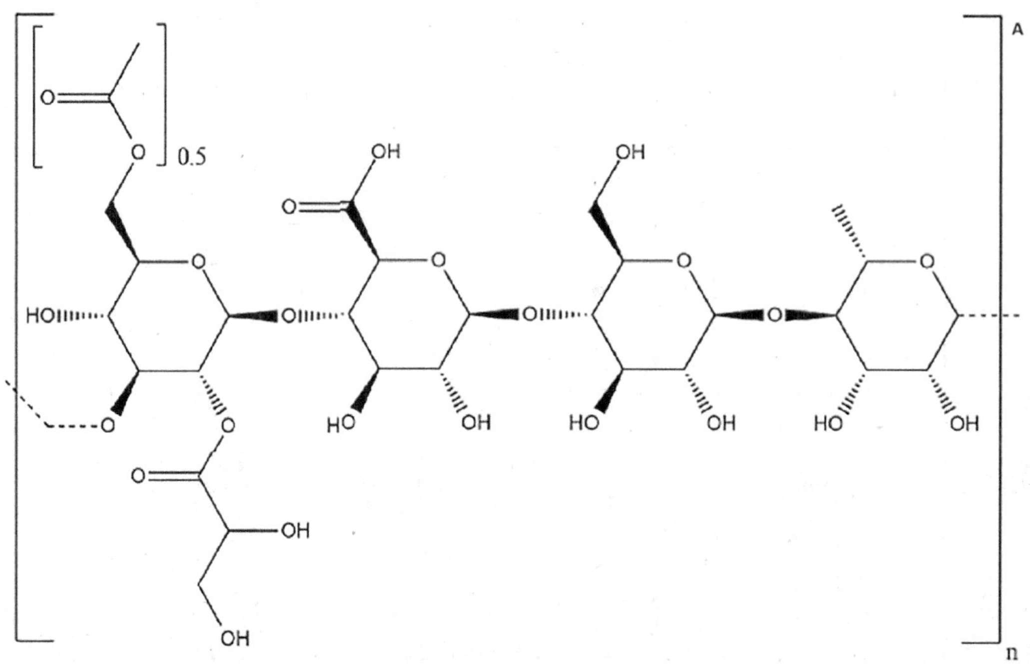

FIGURE 17–6 Chemical structure of native gellan gum showing a tetrasaccharide repeating unit with O-acetyl and L-glycerate substitutes (Gomes et al., 2023). *From Gomes, D., Batista-Silva, J.P., Sousa, A., Passarinha, L.A., 2023. Progress and opportunities in gellan gum-based materials: a review of preparation, characterization and emerging applications. Carbohydrate Polymers 311, 120782. https://doi.org/10.1016/j.carbpol.2023.120782.*

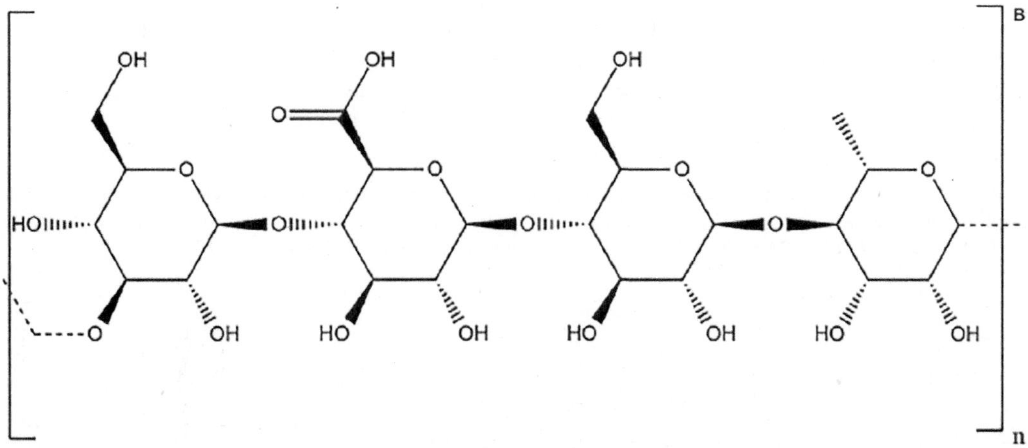

FIGURE 17–7 Chemical structure of deacetylated gellan gum showing a linear tetrasaccharide repeating unit of glucose, glucuronic acid, and rhamnose (Gomes et al., 2023). *From Gomes, D., Batista-Silva, J.P., Sousa, A., Passarinha, L.A., 2023. Progress and opportunities in gellan gum-based materials: a review of preparation, characterization and emerging applications. Carbohydrate Polymers 311, 120782. https://doi.org/10.1016/j.carbpol.2023.120782.*

Gao (2016) investigated the potential use of high acyl gellan gum as an additive for drilling mud. The properties investigated included the impacts of temperature on mud characteristics. Bentonite clay, used as base mud, was mixed with gellan gum and then heated to 70°C, cooled down to room temperature (25°C), and heated up again. Another part of the investigation involved heating mud samples containing gellan gum and some salts (NaCl, KCl, and $CaCl_2$) to 80°C for 6 hours, and then evaluating various mud properties. The study results revealed that mud properties remained very stable at 80°C. These results mean that gellan gum boosts drilling mud viscosity and makes the gellan gum-mud biopolymer composite stable at moderately high temperatures.

17.2.4 Polybutylene succinate

Polybutylene succinate (PBS) is a semicrystalline polymer produced through the synthesis of succinic acid and 1,4-butanediol (Alias and Marsilla, 2020; Gigli et al., 2016). Compared with traditional petrochemical plastics, PBS is an ecofriendly material that degrades slowly in the environment, leaving no toxic materials behind (Barrino et al., 2023; Nazrun et al., 2024). Fig. 17–8 shows the chemical structures of succinic acid, 1,4-butanediol, and PBS. The low glass transition temperature (T_g) of −32°C and appreciably high melting temperature (T_m) of 113°C–115°C of PBS contribute to its high thermal stability (Nazrun et al., 2024; Su et al., 2019).

In building and construction applications, PBS can be used to make weather-resistant cladding, siding, and insulation panels. Its high heat resistance and favorable mechanical properties make it an appropriate construction material. Its flame retardant and other properties can further be improved by blending it with polylactic acid (PLA), to make ecofriendly exterior cladding panels, interior wall panels, and roofing materials (Nazrun et al., 2024; Yang et al., 2022).

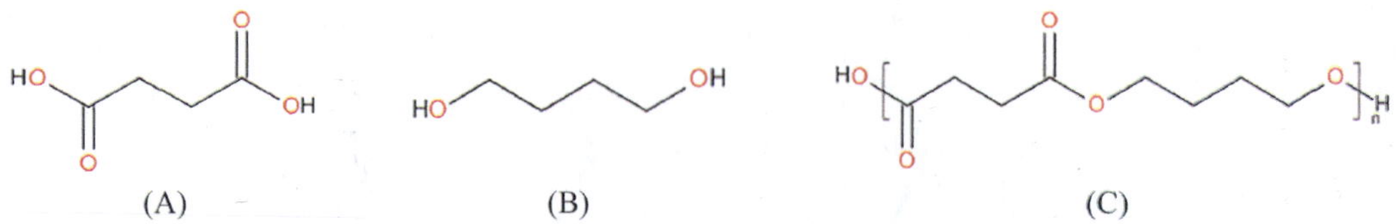

FIGURE 17–8 Chemical structure of (A) succinic acid, (B) 1,4-butanediol, and (C) polybutylene succinate (Su et al., 2019). *From Su, S., Kopitzky, R., Tolga, S., Kabasci, S., 2019. Polylactide (PLA) and its blends with poly(butylene succinate) (PBS): a brief review. Polymers 11 (7), 1193. https://doi.org/10.3390/polym11071193.*

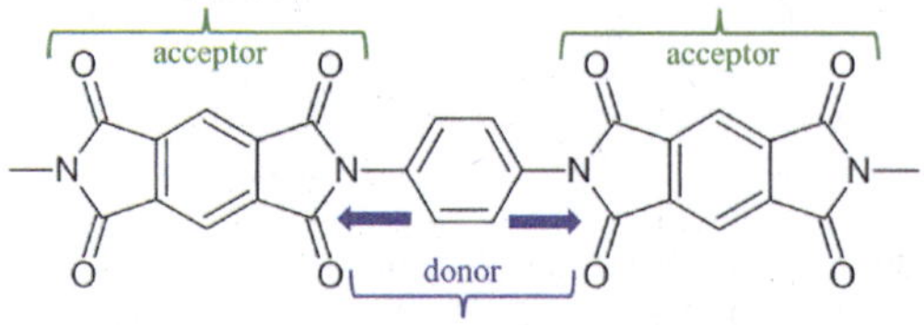

FIGURE 17–9 Polyimide intrachain interactions: Donor–acceptor arrangements that occur within polyimide chains, creating charge transfer complexes (CTCs) (Bigham, 2019). *From Bigham, K.J., 2019. Focus on Polyimides. Resinate. Zeus Industrial Products, Inc. https://www.zeusinc.com/wp-content/uploads/2019/07/RESINATE-No7-Focus-On-Polyimides.pdf.*

17.2.5 Polyimides and polyimide fibers

Polyimides (PIs) fibers are polymeric materials comprising imide rings in macromolecular chains. They have a highly conjugated structure, which gives the material remarkable mechanical property, solvent corrosion resistance, good UV resistance, and reasonably high thermal stability (Bigham, 2019; Zhang and Dong, 2018; Zhang et al., 2022).

A crucial characteristic of PIs is their unique chain interactions resulting from their repeated imide moieties. Particularly, aromatic heterocyclic PIs benefit from the inter- and intrachain interactions. With the nitrogen atoms of the imide moieties possessing larger electron density than the carbonyl (C=O) groups, the electron withdrawing power of the carbonyl oxygen pulls the electron density away from the nitrogen. This electron density shift further impacts neighboring atoms beyond the nitrogen, to the resulting "donor" and "acceptor" arrangements (Fig. 17–9), referred to as charge transfer complexes (CTCs) (Jung et al., 2013; Yang et al., 2015b). The CTCs phenomenon within the polymer chain are largely responsible for the stiffness of PIs. The donor–acceptor interactions also occur between adjacent polymer chains (Fig. 17–10). Moreover, stacking interactions of C6 aromatic (benzene-type) rings arising from positive and negative faced quadrupole moments further attract neighboring chains. The PI chains are held very tightly together by these interactions and arrangements, thereby leading to negligible chain movement or slippage. Conjointly, CTCs and stacking interactions produce a potent cumulative effect on aromatic heterocyclic PIs, resulting in an extremely strong and stiff polymer with very high glass transition (T_g) and melting temperature (T_m). Their T_m can exceed beyond 300°C (572°F), to as high as 400°C (752°F) (Bigham, 2019).

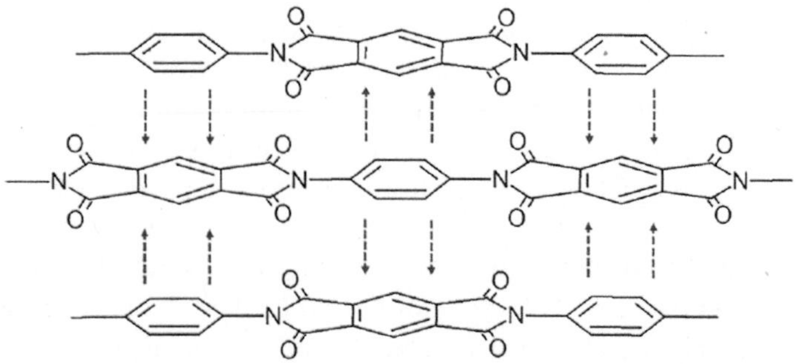

FIGURE 17–10 Three stacked polyimide interchain interactions with vertical dashed arrows indicating interchain charge transfers (Bigham, 2019). *From Bigham, K.J., 2019. Focus on Polyimides. Resinate. Zeus Industrial Products, Inc. https://www.zeusinc.com/wp-content/uploads/2019/07/RESINATE-No7-Focus-On-Polyimides.pdf.*

PIs have been used in many high-temperature applications due to their outstanding heat resistance (thermogravimetry of more than 300°C) and low linear thermal expansion coefficient ($2\sim4\times10^{-5}$/°C). An example is the replacement of metal as a liner material in aircraft engines with DuPont's SP series of PI composites in the early 1980s. Also, PI fibers made from P84 series by Evonik GMBH, capable of withstanding temperatures of more than 300°C, have been used to make fire-protective clothing (Ma et al., 2019; Yan et al., 2024). The P84 PI fiber has, furthermore, been widely used in high-temperature filtration applications (Zhang and Dong, 2018).

Other qualities of PIs include low thermal conductivity (can be as low as 0.03 $W\,m^{-1}\,K^{-1}$), good flame retardancy, as well as resistance and heat oxidation resistance. PIs are self-extinguishing materials with a loss on ignition (LOI) of about 35–50, and do not burn or melt. Furthermore, due to their high thermal stability and acid resistance over other heat-resistant organic fibers, PI fibers can be used for a long time under extremely harsh and high temperature or high corrosive gas environmental conditions. Moreover, the cost-effectiveness and good dust removal characteristics of PI fibers have enabled their successful application in boiler dust removal in ferroalloy and coal-fired industries. They have also been successfully used for dust removal in the cement production industry (Zhang and Dong, 2018; Zhang et al., 2022).

He et al. (2023) investigated high-temperature mechanical behavior of commercial PI under different strain rates. Uniaxial compression experiments under different temperatures and strain rates were done on PI samples. The study results revealed that PI is able to carry some load at temperatures as high as 350°C, indicating PI's superiority compared to other polymer materials.

The low thermal conductivity and self-extinguishing properties of PIs are particularly of importance to building and construction. These properties entail that PIs can make very effective insulators for application in climate-resilient structures.

17.2.6 Polyether–ether ketone

Polyether–ether ketone (PEEK) is a semicrystalline thermoplastic high-performance biopolymer. It exhibits exceptional chemical characteristics and mechanical strength as well as high thermal stability. Its tensile strength ranges between 90 and 100 MPa. The biopolymer has a rigid Young's modulus of 3.6 GPa. PEEK is highly resistant to thermal degradation as well as degradation from harsh environments (Gunatilake and Adhikari, 2016; McKeen, 2012).

Due to its relatively high glass transition temperature (~145°C) and high melting point (~334°C), PEEK is extensively used in high-temperature engineering applications. The material's outstanding thermal stability is attributed to its relatively stiff backbone (Rangari et al., 2019; Surendran et al., 2023).

A typical chemical structure of PEEK is given in Fig. 17–11—which typically has aromatic rings (C_6H_4) connected by ether (O) and ketone (CO) linkages. This structure gives PEEK its exceptional thermal stability, chemical resistance, and mechanical properties. Considering that PEEK exhibits these exceptional characteristics even at temperatures up to 240°C, it is one of the polymers considered for metal replacement in high-temperature applications. Accordingly, PEEK has been successfully used in automotive, aerospace, oil and gas, and space applications (Arif et al., 2018).

A study by Park et al. (2021) investigated the practical utilization of PEEK via three-dimensional (3D) printing and postprocessing. A high-temperature 3D printing system equipped with a thermally efficient extruder and hot chamber, developed after appropriate analytical and experimental considerations, was used. With controlled temperature conditions of the extruder and chamber, structurally stable and mechanically reinforced constructs were produced. The as-printed specimens showed that the tensile strength of the 3D printed specimen measured up to 89% compared to the original material property of PEEK. An observation that this could be increased up to 93% by a postheat treatment around the crystallization temperature was made. The study confirmed the high-level crystallization of the high-temperature 3D printed and postprocessed structures. The 3D printed and tested prototypes of wafer jig, a lathe chuck part, and plastic injection molds yielded excellent PEEK performance relating to chemical resistance, high strength, and heat resistance. This demonstrated the potential application of the PEEK 3D printing technique in manufacturing and in a wide range of other industries.

Siraj et al. (2022) assert that due to their outstanding mechanical strength, thermal stability, and biocompatibility, PEEK and PEEK-based composites have great potential for future applications in structural components in building and construction.

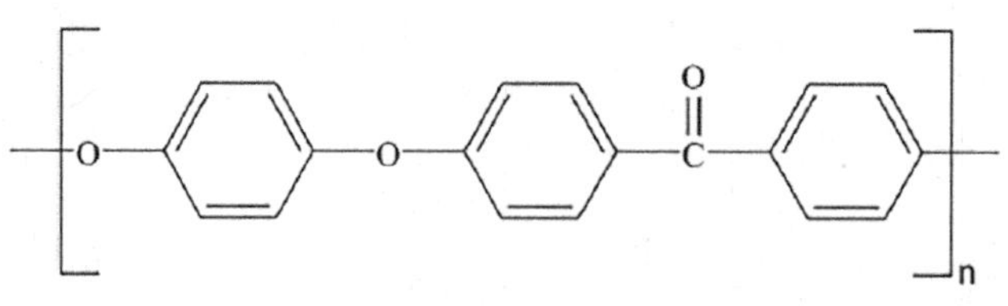

FIGURE 17–11 Structure of polyether–ether ketone indicating the repeating units of PEEK, containing three rings linked by two ether oxygens and one carbonyl group (McKeen, 2012). *From McKeen, L.W., 2012. Chapter 11—High-temperature and high-performance polymers. In: McKeen, L.W. (Ed.), Permeability Properties of Plastics and Elastomers, 3rd ed. William Andrew Publishing, pp. 233–250. https://doi.org/10.1016/B978-1-4377-3469-0.10011-6.*

17.2.7 Lignosulfonate

Lignosulfonate was introduced in the 1920s as a concrete biopolymer additive for concrete plastification (Plank, 2005). Lignosulfonates are obtained from sulfite pulping processes in the wood and pulp industry, involving the extraction of cellulose. This process comprises mixing sulfur dioxide (SO_2) with an aqueous base solution to produce the raw liquor for cooking the wood. In water, the SO_2 forms sulfurous acid (H_2SO_3), which degrades and eventually sulfonates the lignin by replacing a hydroxyl group with a sulfonate group, allowing it to be solubilized and separated from the cellulose in nonprecipitated form (Niaounakis, 2015; Anukam et al., 2021). The spent sulfite liquor contains lignosulfonate and sugars, primarily monosaccharides that can be removed or destroyed to enable effective use of the lignosulfonate as a water-reducing concrete additive (Niaounakis, 2015). Fig. 17–12 presents a typical lignosulfonate structure showing sulfonic groups attached to benzylic carbon atoms—a highly aromatic structure, reflecting lignosulfonate's origin from lignin, a major component of plant cell walls.

Lignosulfonate can be used as an alternative to conventional additives like lime, cement, and fly ash in soil stabilization techniques, leading to improved strength and swell characteristics of stabilized soils (Sharmila et al., 2021). A study by Huang et al. (2018) has shown that

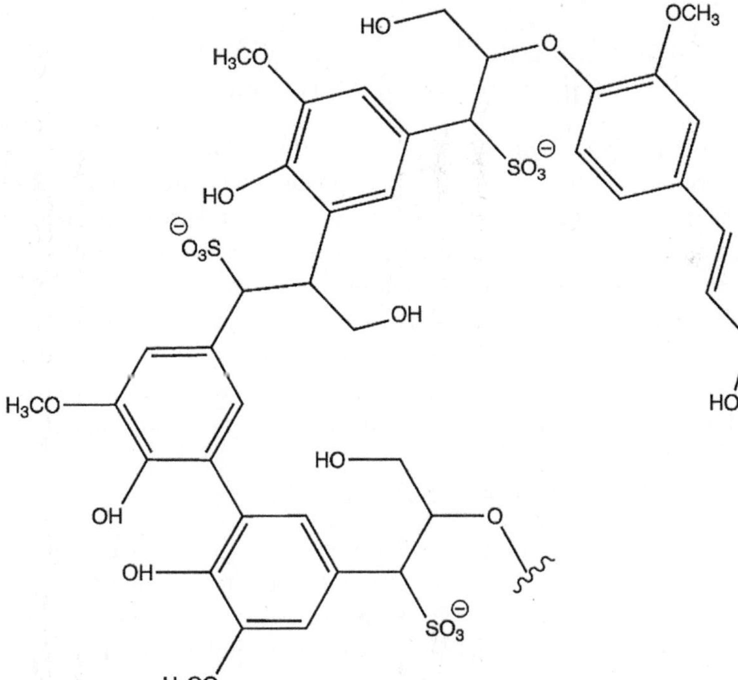

FIGURE 17–12 Typical lignosulfonate structure showing sulfonic groups attached to benzylic carbon atoms (Anukam et al., 2021). *From Anukam, A., Berghel, J., Henrikson, G., Frodeson, S., Ståhl, M., 2021. A review of the mechanism of bonding in densified biomass pellets. Renewable and Sustainable Energy Reviews 148, 111249. https://doi.org/10.1016/j.rser.2021.111249.*

lignosulfonate obtained from sulfomethylation of biorefinery lignin can be used as a water reducer for concrete, demonstrating its potential to enhance the fluidity of cement paste and increase compressive strength of concrete mixes. These findings indicate that lignosulfonate has high potential for use as a sustainable and effective additive in many construction applications that involve concrete production and soil stabilization. Niaounakis (2015) affirms that lignosulfonate is also used as a dispersing agent in the production of brick or tile.

In oil well construction, lignosulfonates have commonly been used as retarders for oil well cementing. Retarding oil well cementing is necessary for avoiding premature cement set induced by high temperature in the well-bore. Lignosulfonates function effectively as set retarders when used with low-C_3A cements to about 125°C bottom-hole circulating temperature (Plank, 2005).

Lignosulfonates can also be synthesized with other polymeric materials to form other high-temperature performance materials. Song et al. (2022), for example, modified lignosulfonate to produce lignin phenol formaldehyde foams. These foams exhibited high thermal stability, a maximum carbon residue rate of 43.76% at 800°C, good thermal conductivity (maximum thermal value of 0.0344 $W\ m^{-1}\ K^{-1}$), and a maximum oxygen index of 39.1%, qualifying them for thermal insulation applications in buildings.

17.3 Some high-temperature biopolymer composites for application in building and construction

High-temperature biopolymer composites applicable in building and construction are blends of biopolymers with other materials. They are meant to create durable, sustainable, and ecofriendly alternative materials in place of conventional construction materials for various building/structural engineering applications. These composites should withstand high temperatures and harsh environmental conditions.

Díez-Pascual (2023) acknowledges that some biopolymers present poor mechanical characteristics and limited processing capabilities, resulting in their limited applications. Consequently, to address these drawbacks, biopolymers can be blended or reinforced with, for example, fillers or nanofillers and thus create innovative materials for various applications.

17.3.1 Polyhydroxyalkanoates based bioplastics

Polyhydroxyalkanoates (PHAs) are a family of polyesters that are naturally synthesized by microorganisms. The resulting polymers enable the production of other materials possessing hard to extremely elastic characteristics. Some of these PHA-based materials have the potential to exceed the performance achieved by synthetic polymers. PHAs are biodegradable and biocompatible. Decomposition of PHAs by microorganisms happens in an active microbiological environment (i.e., in soil, fresh water, seawater, compost, etc.), even under anaerobic conditions. The PHAs biocompatibility implies that they are nontoxic and cannot cause any known allergic reactions (Safin et al., 2021).

Production of PHAs from industrial by-products and waste, including agricultural waste, used vegetable oils, or wastewater has been a focus of various researchers. Worth mentioning here is that production of PHAs from inexpensive waste carbon sources can reduce their cost of production and increase their sustainability (Safin et al., 2021; Mimini et al., 2019).

Safin et al. (2021) studied the production of PHA-based biocomposite samples with a lignin filler for potential use in the production of building materials. The study involved hydrolysis of wood pine sawdust in 4% sulfurous acid at 170°C, followed by polymer matrix production from liquid hydrolyzates. Evaluation of mechanical properties of the composite material from a dried solid residue and a polymer matrix produced by microorganisms indicated that not more than 40% of the filler content was needed to produce useful building materials. The produced biocomposites can potentially be used as finishing materials and as biodegradable formwork in building and construction projects.

17.3.2 Polylactic acid—lignin blends

The use of biodegradable materials or bioplastics derived from renewable biological resources or from biomass is a sustainable alternative solution to the use of environmentally nuisance permanent plastics, which are largely produced from petroleum sources (Lora and Glasser, 2002; Mimini et al., 2019).

Polyactic acid (PLA) is commercially available bioplastic that is biodegradable and biocompatible. It is produced via bacterial fermentation of carbohydrates like starch extracted from corn, or sugars extracted from sugar beet, whey, and molasses. The chemical structure of PLA is presented in Fig. 17-13. It typically consists of repeating units of lactic acid, which can be derived from the fermentation of sugars and starches. PLAs can also be chemically synthesized through ring opening polymerization of cyclic lactide, resulting in the structure shown in Fig. 17-14. Applications of PLAs are limited by their various unsatisfactory properties, such as modest strength and modulus, brittleness, low heat distortion temperature, low glass transition temperature, and moisture sensitivity. They are also relatively expensive and can only be minimally used as a single polymer in hot packaging (Li and Susan Sun, 2011; Murariu and Dubois, 2016). To enhance some of PLA's properties, such as modulus and biodegradability, and to reduce its cost, PLA has been blended with different low-cost natural biopolymers such as lignin, soy flour, soy protein, sugarcane, and starch (Fang et al., 2009; Li and Susan Sun, 2011; Li et al., 2010; Silva et al., 2019).

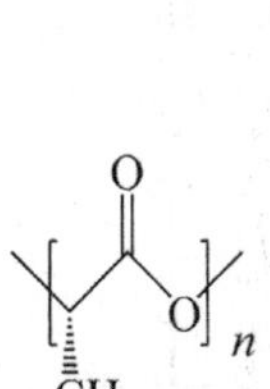

FIGURE 17–13 Structure of the repeating unit of polylactic acid (PLA) (Shi et al., 2023). *From Shi, K., Liu, G., Sun, H., Weng, Y., 2023. Polylactic acid/lignin composites: a review. Polymers 15 (13), 2807. https://doi.org/10.3390/polym15132807.*

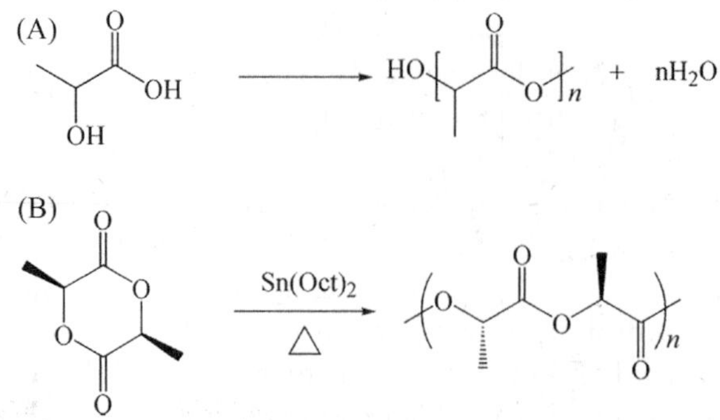

FIGURE 17–14 Structures of polylactic acid synthesized through polycondensation of lactic acid (A) and ring opening polymerization of cyclic lactide (B) (Shi et al., 2023). *From Shi, K., Liu, G., Sun, H., Weng, Y., 2023. Polylactic acid/lignin composites: a review. Polymers 15 (13), 2807. https://doi.org/10.3390/polym15132807.*

Lignin is the second largest renewable bioresource after cellulose in the plant world. It has an amorphous three-dimensional network structure comprising *p*-hydroxyphenyl alcohol (H), guaiacyl (G), and syringyl (S) phenylpropane units (Fig. 17–15).

The composition of lignin depends on its source. In hardwood, the proportion of lignin ranges between 20% and 25%, comprising mainly G and S units. In coniferous wood, the composition ranges between 25% and 35%, comprising largely of G units and a small portion of H units. In grass, the lignin proportion is between 15% and 25%, and all (H, G, and S) units are present (Beaucamp et al., 2022; Sugiarto et al., 2022).

In addition to being abundantly available, lignin also has the advantages that it is a low cost and appreciably biodegradable material with good antioxidant activity. These qualities give lignin a wide application potential (Ma et al., 2020; Shi et al., 2023).

Blending PLA and lignin renders the possibility of creating ecofriendly PLA-based biopolymer composites with reduced cost and enhanced properties. The high carbon content, low density, and hardness of lignin are the characteristics that are key to improving the mechanical properties of PLA. However, when lignin is used in its unmodified form in the PLA-lignin blend, it tends to self-aggregate, leading to poor compatibility and ultimately poor tensile strength and elongation at break (Anwer et al., 2015; Hambardzumyan et al., 2012; Shi et al., 2023; Spiridon and Tanase, 2018). Lignin, especially modified via graft copolymerization, effectively improves the compatibility of lignin and PLA, resulting in new lignin properties that subsequently improve PLA-lignin biopolymer composites (Shi et al., 2023).

Yang et al. (2015a) studied the effects of lignin nanoparticles (LNPs) on the mechanical, thermal, and crystallization properties of the PLA-LNP biopolymer composites. The study results indicate that at 1 wt.% LNP dosing, LNPs were uniformly dispersed in the PLA matrix. The LNPs' heterogeneous nucleation sites triggered the nucleation of PLA, resulting in increased crystallinity of the composite from 15.0% to 22.5%. An increase in tensile strength from 44.3 MPa of neat PLA to 48.7 MPa was observed. This was attributed to the uniform distribution of LNP in PLA, thus enabling effective stress transfer in PLA. An increase in elongation at break from 16.8% to 26.7% was also recorded, contrary to behavior of many composites containing rigid nanoparticles, which may be linked to the special chemical

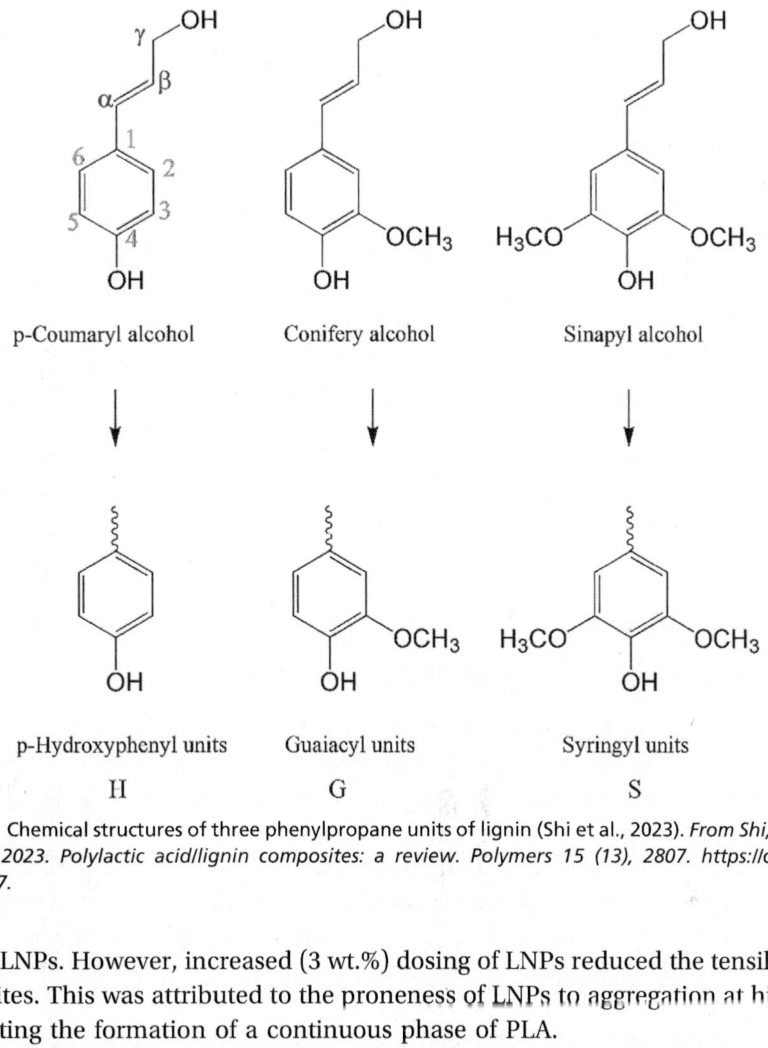

FIGURE 17–15 Chemical structures of three phenylpropane units of lignin (Shi et al., 2023). *From Shi, K., Liu, G., Sun, H., Weng, Y., 2023. Polylactic acid/lignin composites: a review. Polymers 15 (13), 2807. https://doi.org/10.3390/polym15132807.*

structure of LNPs. However, increased (3 wt.%) dosing of LNPs reduced the tensile strength of the composites. This was attributed to the proneness of LNPs to aggregation at high contents, thus preventing the formation of a continuous phase of PLA.

A study by Ouyang et al. (2012) evaluated PLA and cellulolytic enzyme lignin (CEL) mixes prepared by twin-screw extrusion blending at temperatures of between 160°C and 180°C. The morphology of the blends, as well as their mechanical and thermal properties, was examined. The study results showed significantly high Young's modulus of the PLA-CEL blends compared to that of the neat PLA. There was also evidence of improved shore hardness. However, a slight decrease in the tensile strength, elongation at break, and the impact strength was observed. TGA showed insignificant change in the thermal stability of the PLA as a result of blending it with CEL, even with 40 wt.% CEL.

The foregoing shows that PLA-lignin blends can result in biopolymer composites with enhanced mechanical, thermal, and other properties compared to PLA-only and lignin-only biopolymers. These enhanced material attributes indicate potential application of PLA-lignin biopolymer composites in building and construction works.

17.3.3 Encapsulated *Lysinibacillus sphaericus* bacteria–cellulose biopolymer

A study conducted by Igbokwe et al. (2022) whose findings were further documented by Igbokwe et al. (2024) investigated the healing of millimeter scale cracks in ordinary cement concrete through encapsulating *Lysinibacillus sphaericus* bacteria (LSB), known for its strong biocalcification abilities, in cellulose fibers. Using a 0.45% by volume of mortar, the optimal concentration of cellulose fiber, approximately 238 million bacteria were immobilized in each gram of cellulose fibers. The resulting cellulose fiber-encapsulated *LSB* enabled up to 25% increase in concrete compressive and split tensile strengths. The self-healing concrete was observed to heal cracks up to 2.5 mm wide in fully wet environments and up to 1.5 mm wide in wet-dry conditions. Furthermore, after crack healing, the self-healing concrete still had higher mechanical strength than the control concrete. The cellulose fibers were considered to act as a protective environment for LSB, enabling LSB to withstand harsh environmental conditions, including temperatures of up to 160°C. These results point to greater potential use of cellulose fiber-blended biopolymer composites in building and construction high-temperature applications.

17.4 Future trends in the application of high-temperature biopolymers in building and construction

There is replete literature that addresses the application of high-temperature biopolymers in biomedical, pharmaceutical, packaging, electronics, automotive, aviation, aerospace, and military industries. However, only an appreciable amount of literature pertaining to the application of biopolymers (in general) in building and construction exists. Some of the notable literature addressing biopolymers (in general) in building construction between the years 2005 and 2023 include Plank (2005), Niaounakis (2015), Shasiya et al. (2023), Cho and Chang (2021), Datta (2011), Fatehi et al. (2021), Gowda et al. (2023), Mansour et al. (2020a), Shen et al. (2020), and Tusnim and Hoque (2020). On the other hand, the subject dealing with application of high-temperature biopolymers has only been sparsely documented and, in many instances, only indirectly addressed and discussed in literature. Consequently, this subject has largely been inferred in this chapter.

It is, nonetheless, noteworthy that the use of high-temperature biopolymers in the building and construction industry has great potential to address the industry's need and push for durable, ecofriendly, and sustainable materials. It also addresses environmental nuisance brought about by nonrenewable petroleum-based polymeric construction materials. The biodegradability, ability to reduce carbon footprint, and potential for recycling characteristics of these biopolymers fit into this trend.

Literature further evinces that researchers continue to explore ways of optimizing and enhancing the properties of biopolymers. This assures future development of different high-temperature biopolymers that can be successfully applied in harsh and high-temperature environments within the construction sector, offering sustainable solutions to the environmental challenges brought about by climate change. These high-temperature biopolymer materials are thus poised to play a substantial role in shaping the future of environmentally responsive buildings.

Buildings account for a significant amount of global energy consumption. High-temperature biopolymers are poised to lead to the production of durable, lightweight insulating materials that reduce heating and cooling needs in buildings, thus contributing to efficiency in energy consumption. They are also likely to be integrated into energy-efficient building systems to ensure appropriate responses of buildings to environmental conditions.

Other possible future innovations include integration of high-temperature biopolymers into other emerging technologies in construction, like 3D printing and smart materials. High-temperature biopolymers could, for example, be used as feedstock for 3D-printed building components or as matrices for embedded sensors in building materials.

Low-carbon 3D printing concrete is an emerging technology and an innovative technique for sustainable construction. It addresses carbon emissions reduction and enhanced construction efficiency. The use of high-temperature biopolymers and composites in this emerging concrete technology is likely to be vigorously explored by researchers to ensure that the technology makes a significant contribution to the reduction of the carbon footprint.

Additionally, high-temperature biopolymers are poised to be engineered to meet specific performance requirements, offering versatility in building and construction applications. They can be reinforced with natural fibers, nanoparticles, or other additives to achieve desired material properties. They are also likely to be used to manufacture structural components such as lightweight yet strong beams, columns, and panels, providing architects and engineers with flexible design options.

17.5 Conclusion

There is replete literature that addresses the application of high temperature and high performance biopolymers in biomedical, packaging, electronics, automotive, aviation, aerospace, and military industries. On the other hand, while an appreciable amount of literature is available about the use of biopolymers (in general) in construction, the application of high-temperature biopolymers in building and construction is sparse and, in many cases, only indirectly addressed in literature.

Relevant literature surveyed indicates that high-temperature biopolymers and biopolymer composites can be subjected to harsh environmental conditions, including temperatures higher than 160°C, and still remain structurally and functionally stable. They can thus be successfully used in various ways in building and construction. For example, they can be used in situations where heat resistance is a vital requirement, as effective insulators for application in climate resilient structures, and in thermal insulation, with the end desire of ensuring efficient energy consumption in buildings.

Furthermore, high-temperature biopolymers can be used in the manufacturing of fire-resistant materials for erecting interior wall panels/partitions, external wall cladding in buildings subjected to high temperatures, machine plastering, and self-healing concrete. In oil well construction, some biopolymers have been used as retarders for oil well cementing and for boosting drilling mud viscosity.

High-temperature biopolymers are poised to foster sustainable innovation in building and construction. The combination of their successful performance under harsh and elevated heat conditions and the environmental benefits associated with their use makes them a future viable and sustainable choice for future use as ecofriendly alternative construction materials. They are poised to play a very significant role in the manufacture of structural and nonstructural units in the building industry, providing architects and engineers with flexible design options.

AI disclosure

During the preparation of this work the author(s) used Scispace to help identify various literature sources about the subject matter. After using this tool/service, the author(s) reviewed and extracted the needed content and take(s) full responsibility for the content of the publication.

References

Aaliya, B., Sunooj, K.V., Lackner, M., 2021. Biopolymer composites: a review. International Journal of Biobased Plastics 3 (1), 40–84. Available at https://doi.org/10.1080/24759651.2021.1881214.

Alias, N.F., Marsilla, K.I.K., 2020. Chapter 6 - Processes and Characterization for Biobased Polymers from Polybutylene Succinate. Elsevier, Amsterdam, Netherlands.

Anukam, A., Berghel, J., Henrikson, G., Frodeson, S., Ståhl, M., 2021. A review of the mechanism of bonding in densified biomass pellets. Renewable and Sustainable Energy Reviews 148, 111249. Available at https://doi.org/10.1016/j.rser.2021.111249.

Anwer, M.A.S., Naguib, H.E., Celzard, A., Fierro, V., 2015. Comparison of the thermal, dynamic mechanical and morphological properties of PLA-Lignin & PLA-Tannin particulate green composites. Composites Part B: Engineering 82, 92–99. Available at https://doi.org/10.1016/j.compositesb.2015.08.028.

Arif, M.F., Kumar, S., Varadarajan, K.M., Cantwell, W.J., 2018. Performance of biocompatible PEEK processed by fused deposition additive manufacturing. Materials and Design 146, 249–259. Available at https://doi.org/10.1016/j.matdes.2018.03.015.

Baranwal, J., Barse, B., Fais, A., Delogu, G.L., Kumar, A., 2022. Biopolymer: a sustainable material for food and medical applications. Polymers 14 (5). Available at https://doi.org/10.3390/polym14050983, https://www.mdpi.com/2073-4360/14/5/983/pdf.

Barrino, F., De La Rosa-Ramírez, H., Schiraldi, C., López-Martínez, J., Samper, M.D., 2023. Preparation and characterization of new bioplastics based on polybutylene succinate (PBS). Polymers 15 (5). Available at https://doi.org/10.3390/polym15051212, http://www.mdpi.com/journal/polymers.

Beaucamp, A., Muddasar, M., Amiinu, I.S., Moraes Leite, M., Culebras, M., Latha, K., et al., 2022. Lignin for energy applications - state of the art, life cycle, technoeconomic analysis and future trends. Green Chemistry 24 (21), 8193–8226. Available at https://doi.org/10.1039/d2gc02724k, http://pubs.rsc.org/en/journals/journal/gc.

Bemiller, J.N., Gellans, Curdlan, 2019. Levans dextrans, pullulan. Carbohydrate Chemistry for Food Scientists. AACC International Press, St. Paul, MN.

Bigham K.J. 2019 Zeus Industrial Products, Inc Focus on Polyimides. Resinate https://www.zeusinc.com/wp-content/uploads/2019/07/RESINATE-No7-Focus-On-Polyimides.pdf.

Bumanis, G., Vitola, L., Pundiene, I., Sinka, M., Bajare, D., 2020. Gypsum, geopolymers, and starch-alternative binders for bio-based building materials: a review and life-cycle assessment. Sustainability (Switzerland) 12 (14). Available at https://doi.org/10.3390/su12145666, https://res.mdpi.com/d_attachment/sustainability/sustainability-12-05666/article_deploy/sustainability-12-05666-v2.pdf.

Chang, I., Prasidhi, A.K., Im, J., Cho, G.C., 2015a. Soil strengthening using thermo-gelation biopolymers. Construction and Building Materials 77, 430–438. Available at https://doi.org/10.1016/j.conbuildmat.2014.12.116.

Chang, I., Jeon, M., Cho, G.C., 2015b. Application of microbial biopolymers as an alternative construction binder for earth buildings in underdeveloped countries. International Journal of Polymer Science 2015. Available at https://doi.org/10.1155/2015/326745, http://www.hindawi.com/journals/ijps/.

Cho, G.C., Chang, I., Im, J., 2021. Microbial biopolymers as an alternative construction binder. Microbial Polymers: Applications and Ecological Perspectives. Springer Singapore, South Korea, pp. 581–617. Available at https://doi.org/10.1007/978-981-16-0045-6_23, https://link.springer.com/book/10.1007/978-981-16-0045-6.

Das, A., Ringu, T., Ghosh, S., Pramanik, N., 2023. A comprehensive review on recent advances in preparation, physicochemical characterization, and bioengineering applications of biopolymers. Polymer Bulletin 80 (7), 7247–7312. Available at https://doi.org/10.1007/s00289-022-04443-4, https://www.springer.com/journal/289.

Datta, C., 2011. Starch as a biopolymer in construction and civil engineering. Handbook of Bioplastics and Biocomposites Engineering Applications. John Wiley and Sons, India, pp. 317–344. Available at https://doi.org/10.1002/9781118203699.ch11, http://onlinelibrary.wiley.com/book/10.1002/9781118203699.

Díez-Pascual, A.M., 2023. Biopolymer composites 2022. International Journal of Molecular Sciences 24 (7). Available at https://doi.org/10.3390/ijms24076430, http://www.mdpi.com/journal/ijms.

Diyana, Z.N., Jumaidin, R., Selamat, M.Z., Ghazali, I., Julmohammad, N., Huda, N., et al., 2021. Physical properties of thermoplastic starch derived from natural resources and its blends: a review. Polymers 13 (9), 5–20. Available at https://doi.org/10.3390/polym13091396, https://www.mdpi.com/2073-4360/13/9/1396/pdf.

Fang, K., Wang, B., Sheng, K., Sun, X.S., 2009. Properties and morphology of poly(lactic acid)/soy protein isolate blends. Journal of Applied Polymer Science 114 (2), 754–759. Available at https://doi.org/10.1002/app.30446China, http://www3.interscience.wiley.com/cgi-bin/fulltext/122414347/PDFSTART.

Fatehi, H., Ong, D.E.L., Yu, J., Chang, I., 2021. Biopolymers as green binders for soil improvement in geotechnical applications: a review. Geosciences 11 (7), 291. Available at https://doi.org/10.3390/geosciences11070291.

Fialho, A.M., Moreira, L.M., Granja, A.T., Popescu, A.O., Hoffmann, K., Sá-Correia, I., 2008. Occurrence, production, and applications of gellan: current state and perspectives. Applied Microbiology and Biotechnology 79 (6), 889–900. Available at https://doi.org/10.1007/s00253-008-1496-0.

Galiano, F., Briceño, K., Marino, T., Molino, A., Christensen, K.V., Figoli, A., 2018. Advances in biopolymer-based membrane preparation and applications. Journal of Membrane Science 564, 562–586. Available at https://doi.org/10.1016/j.memsci.2018.07.059, http://www.elsevier.com/locate/memsci.

Gao, C.H., 2016. Preliminary evaluation of gellan gum as mud thickener. Journal of Petroleum Exploration and Production Technology 6 (4), 857–861. Available at https://doi.org/10.1007/s13202-015-0223-5, http://www.springerlink.com/content/2190-0566/.

Giavasis, I., Harvey, L.M., McNeil, B., 2000. Gellan gum. Critical Reviews in Biotechnology 20 (3), 177–211. Available at https://doi.org/10.1080/07388550008984169.

Gigli, M., Fabbri, M., Lotti, N., Gamberini, R., Rimini, B., Munari, A., 2016. Poly(butylene succinate)-based polyesters for biomedical applications: a review. European Polymer Journal 75, 431–460. Available at https://doi.org/10.1016/j.eurpolymj.2016.01.016.

Gomes, D., Batista-Silva, J.P., Sousa, A., Passarinha, L.A., 2023. Progress and opportunities in Gellan gum-based materials: a review of preparation, characterization and emerging applications. Carbohydrate Polymers 311, 120782. Available at https://doi.org/10.1016/j.carbpol.2023.120782.

Gowda, Y., G., T., Sharath, Puttegowda, M., Verma, A., Rangappa, S.M., et al., 2023. Biopolymer-based composites: an eco-friendly alternative from agricultural waste biomass. Journal of Composites Science 7 (6). Available at https://doi.org/10.3390/jcs7060242, https://www.mdpi.com/2504-477X/7/6/242.

Gunatilake, P.A., Adhikari, R., 2016. Chapter 2 - Nondegradable synthetic polymers for medical devices and implants. In: Poole-Warren, L., Martens, P., Green, R. (Eds.), Biosynthetic Polymers for Medical Applications. Woodhead Publishing, Cambridge, UK. Available at https://doi.org/10.1016/B978-1-78242-105-4.00002-X.

Hambardzumyan, A., Foulon, L., Chabbert, B., Aguié-Béghin, V., 2012. Natural organic UV-absorbent coatings based on cellulose and lignin: designed effects on spectroscopic properties. Biomacromolecules 13 (12), 4081–4088. Available at https://doi.org/10.1021/bm301373b.

He, H., Yin, J., Song, L., Deng, Q., 2023. High-temperature mechanical responses of the polyimide at a high strain rate. 2nd International Conference on Applied Mechanics and Advanced Materials. Available at https://doi.org/10.1088/1742-6596/2535/1/012016/meta, https://iopscience.iop.org/article/10.1088/1742-6596/2535/1/012016/meta.

Huang, C., Ma, J., Zhang, W., Huang, G., Yong, Q., 2018. Preparation of lignosulfonates from biorefinery lignins by sulfomethylation and their application as a water reducer for concrete. Polymers 10 (8), 841. Available at https://doi.org/10.3390/polym10080841.

Igbokwe, E., Ibekwe, S., Mensah, P., Agu, O., Abedin, R. 2022. Efficacy of cellulose fiber as bacteria carrier for self-healing concrete and 3D polymer microstructure. In: Proceedings of the American Society for Composites – 37th Technical Conference (ASC 2022). DEStech Publications Inc., United States. ISBN 9781605956909.

Igbokwe, E., Ibekwe, S., Mensah, P., Agu, O., Li, G., 2024. Self-healing of macroscopic cracks in concrete by cellulose fiber carried microbes. Journal of Building Engineering 90, 109383. Available at https://doi.org/10.1016/j.jobe.2024.109383.

INTERNATIONAL CODE COUNCIL, 2020. 2021 International Building Code. International Code Council. Country Club Hills, Illinois, United States. Available at https://codes.iccsafe.org/content/IBC2021P2/copyright.

Jansson, P.E., Lindberg, B., Sandford, P.A., 1983. Structural studies of gellan gum, an extracellular polysaccharide elaborated by Pseudomonas elodea. Carbohydrate Research 124 (1), 135–139. Available at https://doi.org/10.1016/0008-6215(83)88361-X.

Jung, Y., Yang, Y., Kim, S., Kim, H.S., Park, T.G., Yoo, B.W., 2013. Structural and compositional effects on thermal expansion behavior in polyimide substrates of varying thicknesses. European Polymer Journal 49 (11), 3642–3650. Available at https://doi.org/10.1016/j.eurpolymj.2013.07.034.

Li, Y., Susan Sun, X., 2011. Mechanical and thermal properties of biocomposites from poly(lactic acid) and DDGS. Journal of Applied Polymer Science 121 (1), 589–597. Available at https://doi.org/10.1002/app.33681.

Li, Y., Venkateshan, K., Sun, X.S., 2010. Mechanical and thermal properties, morphology and relaxation characteristics of poly(lactic acid) and soy flour/wood flour blends. Polymer International 59 (8), 1099–1109. Available at https://doi.org/10.1002/pi.2834, http://onlinelibrary.wiley.com/doi/10.1002/pi.2834/pdf.

Lora, J.H., Glasser, W.G., 2002. Recent industrial applications of lignin: a sustainable alternative to nonrenewable materials. Journal of Polymers and the Environment 10 (1-2), 39–48. Available at https://doi.org/10.1023/A:1021070006895.

Ma, P., Dai, C., Wang, H., Li, Z., Liu, H., Li, W., et al., 2019. A review on high temperature resistant polyimide films: heterocyclic structures and nanocomposites. Composites Communications 16, 84–93. Available at https://doi.org/10.1016/j.coco.2019.08.011, http://www.journals.elsevier.com/composites-communications.

Ma, M., Dai, L., Xu, J., Liu, Z., Ni, Y., 2020. A simple and effective approach to fabricate lignin nanoparticles with tunable sizes based on lignin fractionation. Green Chemistry 22 (6), 2011–2017. Available at https://doi.org/10.1039/d0gc00377h, http://pubs.rsc.org/en/journals/journal/gc.

Mansour, G., Zoumaki, M., Tsongas, K., Tzetzis, D., 2020a. Starch-sandstone materials in the construction industry. Results in Engineering 8, 100182. Available at https://doi.org/10.1016/j.rineng.2020.100182.

Mansour, G., Zoumaki, M., Tzetzis, D., Mansour, G., Tzetzis, D., Kyratsis, P., 2020b. Starch sandstones in building bio-materials. MATEC Web of Conferences 318, 01046. Available at https://doi.org/10.1051/matecconf/202031801046.

McKeen, L.W., 2012. Chapter 11 - high-temperature and high-performance polymers. In: McKeen, L.W. (Ed.), Permeability Properties of Plastics and Elastomers. William Andrew Publishing, Oxford, UK. Available at https://doi.org/10.1016/B978-1-4377-3469-0.10011-6.

Mimini, V., Sykacek, E., Hashim, S.N.A., Holzweber, J., Hettegger, H., Fackler, K., et al., 2019. Compatibility of kraft lignin, organosolv lignin and lignosulfonate with PLA in 3D printing. Journal of Wood Chemistry and Technology 39 (1), 14–30. Available at https://doi.org/10.1080/02773813.2018.1488875, http://www.tandfonline.com/loi/lwct20.

Moon, R.J., Martini, A., Nairn, J., Simonsen, J., Youngblood, J., 2011. Cellulose nanomaterials review: structure, properties and nanocomposites. Chemical Society Reviews 40 (7), 3941–3994. Available at https://doi.org/10.1039/c0cs00108b.

Murariu, M., Dubois, P., 2016. PLA composites: from production to properties. Advanced Drug Delivery Reviews 107, 17–46. Available at https://doi.org/10.1016/j.addr.2016.04.003, http://www.elsevier.com/locate/drugdeliv.

Nanda, S., Patra, B.R., Patel, R., Bakos, J., Dalai, A.K., 2022. Innovations in applications and prospects of bioplastics and biopolymers: a review. Environmental Chemistry Letters 20 (1), 379–395. Available at https://doi.org/10.1007/s10311-021-01334-4, http://springerlink.metapress.com/app/home/journal.asp?wasp=d86tgdwvtg0yvw9gvkwp&referrer=parent&backto=browsepublicationsresults,140,541.

Nazrun, T., Hassan, M.K., Hossain, M.D., Ahmed, B., Hasnat, M.R., Saha, S., 2024. Application of biopolymers as sustainable cladding materials: a review. Sustainability (Switzerland) 16 (1). Available at https://doi.org/10.3390/su16010027, http://www.mdpi.com/journal/sustainability/.

Niaounakis, M., 2015. Chapter 10 - Building and Construction Applications. Biopolymers: Applications and Trends. William Andrew Publishing, Oxford. Available at https://doi.org/10.1016/B978-0-323-35399-1.00010-7.

Ouyang, W., Huang, Y., Luo, H., Wang, D., 2012. Poly(Lactic Acid) blended with cellulolytic enzyme lignin: mechanical and thermal properties and morphology evaluation. Journal of Polymers and the Environment 20 (1), 1–9. Available at https://doi.org/10.1007/s10924-011-0359-4.

Park, S.J., Lee, J.E., Park, J., Lee, N.K., Son, Y., Park, S.H., 2021. High-temperature 3D printing of polyetheretherketone products: perspective on industrial manufacturing applications of super engineering plastics. Materials and Design 211. Available at https://doi.org/10.1016/j.matdes.2021.110163, https://www.journals.elsevier.com/materials-and-design.

Pintiaux, T., Heuls, M., Vandenbossche, V., Murphy, T., Wuhrer, R., Castignolles, P., et al., 2019. Cellulose consolidation under high-pressure and high-temperature uniaxial compression. Cellulose 26 (5), 2941–2954. Available at https://doi.org/10.1007/s10570-019-02273-8.

Plank, J., 2005. Applications of biopolymers in construction engineering. In: Plank, J. (Ed.), Biopolymers Online. Wiley. Available at https://doi.org/10.1002/3527600035.bpola002.

Rangari, V.K., Biswas, M.C., Tiimob, B.J., Umerah, C., 2019. Biodegradable polymer blends for food packaging applications. In: da Cruz, R.M.S. (Ed.), Food Packaging. CRC Press. Available at https://doi.org/10.1201/9780429023101-7.

Safin, R., Galyavetdinov, N., Salimgaraeva, R., Ilalova, G., Saerova, K. 2021. Use of filled bioplastics in construction. In: E3S Web of Conferences – 2nd International Scientific Conference on Socio-Technical Construction and Civil Engineering. https://doi.org/10.1051/e3sconf/202127404013

Sastri, V.R., 2014. Chapter 8 - High-temperature engineering thermoplastics: Polysulfones, polyimides, polysulfides, polyketones, liquid crystalline polymers, and fluoropolymers. Plastics in Medical Devices. William Andrew Publishing, Oxford. Available at https://doi.org/10.1016/B978-1-4557-3201-2.00008-2.

Sen, S., Singh, A., Kailasam, K., Bera, C., Roy, S., 2022. Biomass-derived cellulose nanofibers and iron oxide-based nanohybrids for thermal insulation application. Nanoscale Advances 4 (16), 3381–3390. Available at https://doi.org/10.1039/d2na00010e.

Sharmila, B., Bhuvaneshwari, S., Landlin, G., 2021. Application of lignosulphonate—a sustainable approach towards strength improvement and swell management of expansive soils. Bulletin of Engineering Geology and the Environment 80, 6395–6413. Available at https://doi.org/10.1007/s10064-021-02323-1, https://doi.org/10.1007/s10064-021-02323-1.

Shasiya, P.S., Simi Pushpan, K., Nair, A.B., 2023. Biopolymers as engineering materials. In: Thomas, S., AR, A., Chirayil, C.J., Thomas, B. (Eds.), Handbook of Biopolymers. Springer, Singapore. Available at https://doi.org/10.1007/978-981-19-0710-4_23.

Shen, J., Liang, J., Lin, X., Lin, H., Yu, J., Yang, Z., 2020. Recent progress in polymer-based building materials. International Journal of Polymer Science 2020. Available at https://doi.org/10.1155/2020/8838160, http://www.hindawi.com/journals/ijps/.

Shi, K., Liu, G., Sun, H., Weng, Y., 2023. Polylactic acid/lignin composites: a review. Polymers 15 (13), 2807. Available at https://doi.org/10.3390/polym15132807.

Silva, T.Fd, Menezes, F., Montagna, L.S., Lemes, A.P., Passador, F.R., 2019. Effect of lignin as accelerator of the biodegradation process of poly(lactic acid)/lignin composites. Materials Science and Engineering: B 251. Available at https://doi.org/10.1016/j.mseb.2019.114441, https://www.journals.elsevier.com/materials-science-and-engineering-b.

Singh, P., Pandey, V.K., Singh, R., Singh, K., Dash, K.K., Malik, S., 2024. Unveiling the potential of starch-blended biodegradable polymers for substantializing the eco-friendly innovations. Journal of Agriculture and Food Research 15. Available at https://doi.org/10.1016/j.jafr.2024.101065, https://www.sciencedirect.com/science/journal/26661543.

Siraj, N., Hashmi, S.A.R., Verma, S., 2022. State-of-the-art review on the high-performance poly (ether ether ketone) composites for mechanical, tribological and bioactive characteristics. Polymers for Advanced Technologies 33 (10), 3049–3077. Available at https://doi.org/10.1002/pat.5795, http://onlinelibrary.wiley.com/journal/10.1002/(ISSN)1099-1581.

Song, S., Xu, Z., Zhen, X., Wang, Z., Ge, T., 2022. Preparation of lignosulfonate-based phenol formaldehyde foam with excellent thermal performance. Macromolecular Chemistry and Physics 223 (20). Available at https://doi.org/10.1002/macp.202200159, http://onlinelibrary.wiley.com/journal/10.1002/(ISSN)1521-3935.

Spiridon, I., Tanase, C.E., 2018. Design, characterization and preliminary biological evaluation of new lignin-PLA biocomposites. International Journal of Biological Macromolecules 114, 855–863. Available at https://doi.org/10.1016/j.ijbiomac.2018.03.140, http://www.elsevier.com/locate/ijbiomac.

Su, S., Kopitzky, R., Tolga, S., Kabasci, S., 2019. Polylactide (PLA) and its blends with poly(butylene succinate) (PBS): a brief review. Polymers 11 (7), 1193. Available at https://doi.org/10.3390/polym11071193.

Sugiarto, S., Leow, Y., Tan, C.L., Wang, G., Kai, D., 2022. How far is Lignin from being a biomedical material? Bioactive Materials 8, 71–94. Available at https://doi.org/10.1016/j.bioactmat.2021.06.023, http://www.keaipublishing.com/en/journals/bioactive-materials/.

Surendran, R., Pavithran, S.S., Balachandran, A., Vijayan, S., Madanan, K., Brabazon, D., 2023. A review of the current state of the art of polyether ether ketone (PEEK) composite based 3D-printed biomedical scaffolds. Designs 7 (6). Available at https://doi.org/10.3390/designs7060128, http://www.mdpi.com/journal/designs.

Teng, C.P., Tan, M.Y., Toh, J.P.W., Lim, Q.F., Wang, X., Ponsford, D., et al., 2023. Advances in cellulose-based composites for energy applications. Materials 16 (10). Available at https://doi.org/10.3390/ma16103856, http://www.mdpi.com/journal/materials.

Tusnim, J., Hoque, M.E., Biswas, M.C., 2020. Biopolymers in building materials. Advanced Processing, Properties, and Applications of Starch and Other Bio-Based Polymers. Elsevier, Amsterdam, The Netherlands.

Wang, X., Pang, Z., Chen, C., Xia, Q., Zhou, Y., Jing, S., et al., 2020. All-natural, degradable, rolled-up straws based on cellulose micro- and nano-hybrid fibers. Advanced Functional Materials 30 (22). Available at https://doi.org/10.1002/adfm.201910417, http://onlinelibrary.wiley.com/journal/10.1002/(ISSN)1616-3028.

Wang, X., Xia, Q., Jing, S., Li, C., Chen, Q., Chen, B., et al., 2021. Strong, hydrostable, and degradable straws based on cellulose-lignin reinforced composites. Small 17 (18). Available at https://doi.org/10.1002/smll.202008011, http://onlinelibrary.wiley.com/journal/10.1002/(ISSN)1613-6829.

Yan, Y., Zhang, B., Wang, J., Cao, C., Yan, F., 2024. MWCNT–polyimide fiber-reinforced composite for high-temperature tribological applications. Coatings 14 (2). Available at https://doi.org/10.3390/coatings14020181, https://www.mdpi.com/2079-6412/14/2/181.

Yang, W., Fortunati, E., Dominici, F., Kenny, J.M., Puglia, D., 2015a. Effect of processing conditions and lignin content on thermal, mechanical and degradative behavior of lignin nanoparticles/polylactic (acid) bionanocomposites prepared by melt extrusion and solvent casting. European Polymer Journal 71, 126–139. Available at https://doi.org/10.1016/j.eurpolymj.2015.07.051.

Yang, Y., Jung, Y., Cho, M.D., Lee, S.G., Kwon, S., 2015b. Transient color changes in oxidative-stable fluorinated polyimide film for flexible display substrates. RSC Advances 5 (71), 57339–57345. Available at https://doi.org/10.1039/c5ra06066d, http://pubs.rsc.org/en/journals/journal/ra.

Yang, Y., Haurie, L., Wang, D.Y., 2022. Bio-based materials for fire-retardant application in construction products: a review. Journal of Thermal Analysis and Calorimetry 147 (12), 6563–6582. Available at https://doi.org/10.1007/s10973-021-11009-5, http://www.springer.com/sgw/cda/frontpage/0,11855,1-40109-70-35752391-0,00.html.

You, C., Zhang, Y.H.P., 2017. Ex Vivo Enzymatic Conversion of Non-food Cellulose Biomass to Starch. Springer Nature, pp. 129–142. Available at https://doi.org/10.1007/978-3-662-53704-6_6.

Yu, Z., Ji, Y., Bourg, V., Bilgen, M., Meredith, J.C., 2020. Chitin- and cellulose-based sustainable barrier materials: a review. Emergent Materials 3, 919–936. Available at 10.1007/s42247-020-00147-5.

Zhang, Q.-H., Dong, J., Wu, D.-Z., 2018. Advanced polyimide fibers. In: Yang, S.-Y. (Ed.), Advanced Polyimide Materials. Elsevier, Amsterdam, The Netherlands. Available at https://doi.org/10.1016/B978-0-12-812640-0.00002-0.

Zhang, Q., Li, X., Dong, J., Zhao, X., 2022. Chapter 6: high-performance polyimide fibers. Advanced Industrial and Engineering Polymer Research 5 (2), 107–116. Available at https://doi.org/10.1016/j.aiepr.2022.03.004, http://www.keaipublishing.com/en/journals/advanced-industrial-and-engineering-polymer-research/.

Zoumaki, M., Tzetzis, D., Mansour, G. 2019. Development and characterization of starch-based nanocomposite materials. IOP Conference Series: Materials Science and Engineering, 564(1). Institute of Physics Publishing. Available at https://doi.org/10.1088/1757-899X/564/1/012037.

Index

Note: Page numbers followed by "*f*" and "*t*" refer to figures and tables, respectively.

I

9780443327926